Lecture Notes in Physics

For information about Vols. 1–67, please contact your bookseller or Springer-Verlag.

Vol. 68: Y. V. Venkatesh, Energy Methods in Time-Varying System Stability and Instability Analyses. XII, 256 pages. 1977.

Vol. 69: K. Rohlfs, Lectures on Density Wave Theory. VI, 184 pages. 1977.

Vol. 70: Wave Propagation and Underwater Acoustics. Edited by J. Keller and J. Papadakis. VIII. 287 pages. 1977.

Vol. 71: Problems of Stellar Convection. Proceedings 1976. Edited by E. A. Spiegel and J. P. Zahn. VIII, 363 pages. 1977.

Vol. 72: Les instabilités hydrodynamiques en convection libre forcée et mixte. Edité par J. C. Legros et J. K. Platten. X, 202 pages. 1978.

Vol. 73: Invariant Wave Equations. Proceedings 1977. Edited by G. Velo and A. S. Wightman. VI, 416 pages. 1978.

Vol. 74: P. Collet and J.-P. Eckmann, A Renormalization Group Analysis of the Hierarchical Model in Statistical Mechanics. IV, 199 pages. 1978.

Vol. 75: Structure and Mechanisms of Turbulence I. Proceedings 1977. Edited by H. Fiedler. XX, 295 pages. 1978.

Vol. 76: Structure and Mechanisms of Turbulence II. Proceedings 1977. Edited by H. Fiedler. XX, 406 pages. 1978.

Vol. 77: Topics in Quantum Field Theory and Gauge Theories. Proceedings, Salamanca 1977. Edited by J. A. de Azcárraga. X, 378 pages 1978.

Vol. 78: Böhm, The Rigged Hilbert Space and Quantum Mechanics. IX, 70 pages. 1978.

Vol. 79: Group Theoretical Methods in Physics. Proceedings, 1977. Edited by P. Kramer and A. Rieckers. XVIII, 546 pages. 1978.

Vol. 80: Mathematical Problems in Theoretical Physics. Proceedings, 1977. Edited by G. Dell'Antonio, S. Doplicher and G. Jona-Lasinio. VI, 438 pages. 1978.

Vol. 81: MacGregor, The Nature of the Elementary Particle. XXII, 482 pages. 1978.

Vol. 82: Few Body Systems and Nuclear Forces I. Proceedings, 1978. Edited by H. Zingl, M. Haftel and H. Zankel. XIX, 442 pages. 1978.

Vol. 83: Experimental Methods in Heavy Ion Physics. Edited by K. Bethge. V, 251 pages. 1978.

Vol. 84: Stochastic Processes in Nonequilibrium Systems, Proceedings, 1978. Edited by L. Garrido, P. Seglar and P. J. Shepherd. XI, 355 pages. 1978

Vol. 85: Applied Inverse Problems. Edited by P. C. Sabatier. V, 425 pages. 1978.

Vol. 86: Few Body Systems and Electromagnetic Interaction. Proceedings 1978. Edited by C. Ciofi degli Atti and E. De Sanctis. VI, 352 pages. 1978.

Vol. 87: Few Body Systems and Nuclear Forces II, Proceedings, 1978. Edited by H. Zingl, M. Haftel, and H. Zankel. X, 545 pages. 1978.

Vol. 88: K. Hutter and A. A. F. van de Ven, Field Matter Interactions in Thermoelastic Solids. VIII, 231 pages. 1978.

Vol. 89: Microscopic Optical Potentials, Proceedings, 1978. Edited by H. V. von Geramb. XI, 481 pages. 1979.

Vol. 90: Sixth International Conference on Numerical Methods in Fluid Dynamics. Proceedings, 1978. Edited by H. Cabannes, M. Holt and V. Rusanov. VIII, 620 pages. 1979.

Vol. 91: Computing Methods in Applied Sciences and Engineering, 1977, II. Proceedings, 1977. Edited by R. Glowinski and J. L. Lions. VI, 359 pages. 1979.

Vol. 92: Nuclear Interactions. Proceedings, 1978. Edited by B. A. Robson. XXIV, 507 pages. 1979.

Vol. 93: Stochastic Behavior in Classical and Quantum Hamiltonian Systems. Proceedings, 1977. Edited by G. Casati and J. Ford. VI, 375 pages. 1979.

Vol. 94: Group Theoretical Methods in Physics. Proceedings, 1978. Edited by W. Beiglböck, A. Böhm and E. Takasugi. XIII, 540 pages. 1979.

Vol. 95: Quasi One-Dimensional Conductors I. Proceedings, 1978. Edited by S. Barišić, A. Bjeliš, J. R. Cooper and B. Leontič. X, 371 pages. 1979.

Vol. 96: Quasi One-Dimensional Conductors II. Proceedings 1978. Edited by S. Barišić, A. Bjeliš, J. R. Cooper and B. Leontič. XII, 461 pages. 1979.

Vol. 97: Hughston, Twistors and Particles. VIII, 153 pages. 1979.

Vol. 98: Nonlinear Problems in Theoretical Physics. Proceedings, 1978. Edited by A. F. Rañada. X, 216 pages. 1979.

Vol. 99: M. Drieschner, Voraussage – Wahrscheinlichkeit – Objekt. XI, 308 Seiten. 1979.

Vol. 100: Einstein Symposion Berlin. Proceedings 1979. Edited by H. Nelkowski et al. VIII, 550 pages. 1979.

Vol. 101: A. Martin-Löf, Statistical Mechanics and the Foundations of Thermodynamics. V, 120 pages. 1979.

Vol. 102: H. Hora, Nonlinear Plasma Dynamics at Laser Irradiation. VIII, 242 pages. 1979.

Vol. 103: P. A. Martin, Modèles en Mécanique Statistique des Processus Irréversibles. IV, 134 pages. 1979.

Vol. 104: Dynamical Critical Phenomena and Related Topics. Proceedings, 1979. Edited by Ch. P. Enz. XII, 390 pages. 1979.

Vol. 105: Dynamics and Instability of Fluid Interfaces. Proceedings, 1978. Edited by T. S. Sørensen. V, 315 pages. 1979.

Vol. 106: Feynman Path Integrals, Proceedings, 1978. Edited by S. Albeverio et al. XI, 451 pages. 1979.

Vol. 107: J. Kijowski, W. M. Tulczyjew, A Symplectic Framework for Field Theories. IV, 257 pages. 1979.

Vol. 108: Nuclear Physics with Electromagnetic Interactions. Proceedings, 1979. Edited by H. Arenhövel and D. Drechsel. IX, 509 pages. 1979.

Vol. 109: Physics of the Expanding Universe. Proceedings, 1978. Edited by M. Demiański. V, 210 pages. 1979.

Vol. 110: D. A. Park, Classical Dynamics and Its Quantum Analogues. VIII, 339 pages. 1979.

Vol. 111: H.-J. Schmidt, Axiomatic Characterization of Physical Geometry. V, 163 pages. 1979.

Vol. 112: Imaging Processes and Coherence in Physics. Proceedings, 1979. Edited by M. Schlenker et al. XIX, 577 pages. 1980.

Vol. 113: Recent Advances in the Quantum Theory of Polymers. Proceedings 1979. Edited by J.-M. André et al. V, 306 pages. 1980.

Lecture Notes in Physics

Edited by J. Ehlers, München, K. Hepp, Zürich
R. Kippenhahn, München, H. A. Weidenmüller, Heidelberg
and J. Zittartz, Köln
Managing Editor: W. Beiglböck, Heidelberg

142

Recent Progress in Many-Body Theories

Proceedings of the
Second International Conference
Held at Oaxtepec, Mexico,
January 12–17, 1981

Edited by J. G. Zabolitzky, M. de Llano,
M. Fortes and J. W. Clark

Springer-Verlag
Berlin Heidelberg GmbH 1981

Editors

J. G. Zabolitzky
Inst. f. Theoret. Physik II, Ruhr-Universität Bochum
D-4630 Bochum, West Germany

M. de Llano
M. Fortes
Instituto de Fisica, UNAM
Apdo. Postal 20-364, Mexico 20, D.F., Mexico

J. W. Clark
Dept. of Physics, Washington University
St. Louis, MO 63130, USA

ISBN 978-3-540-10710-1 ISBN 978-3-540-38677-3 (eBook)
DOI 10.1007/978-3-540-38677-3

Originally published by Springer-Verlag Berlin Heidelberg New York in 1981

2153/3140-543210

Preface

The present volume contains the invited talks delivered at the Second International Conference on Recent Progress in Many-Body Theories held at Oaxtepec, Morelos, Mexico, January 12 to 17, 1981. It succeeded the first such international conference held in Trieste, Italy, in 1978, which in turn succeeded a number of more-or-less regular meetings on microscopic many-body theories, the first of which took place at the Stevens Institute of Technology, Hoboken, New Jersey, in 1957.

It is the aim of this series of conferences to foster the exchange of ideas and techniques among physicists from various fields of application of many-body theories, ranging over nucleon-nucleon interactions, nuclear physics, astrophysics, atomic physics, quantum fluids and solid-state physics. That this goal has been achieved is demonstrated by the diversity of talks collected in this volume - as well as the many fruitful discussions not reproduced herein.

These proceedings were processed by the photo-offset technique. To facilitate rapid publication, editing of the papers submitted by the speakers was kept to a minimum.

We would like to express our gratitude to the Instituto de Fisica, Universidad Nacional Autõnoma de Mexico and CONACyT, Mexico, for financial support, and to the many local scientists whose efforts contributed to the warm, friendly and stimulating atmosphere of the conference.

Bochum/Mexico City/St. Louis, February 1981

The Organizing Committee

J.W. Clark

M. Fortes

M. de Llano

J.G. Zabolitzky

Table of Contents

MEAN-FIELD APPROXIMATIONS TO THE MANY-BODY S-MATRIX*

S.E. Koonin,† Y. Alhassid,†† and K.R. Sandhya-Devi
W.K. Kellogg Radiation Laboratory
California Institute of Technology
Pasadena, California 91125/USA

Abstract: We discuss non-perturbative approximations to the excitation of a many-body system by a time-dependent one-body perturbation. For both exclusive and inclusive processes, a stationary-phase approximation to a functional integral representation of the interaction-picture evolution operator yields mean-field equations similar to TDHF. The character of the approximations is illustrated by application to the forced Lipkin model and atomic p + He collisions.

Introduction

The many outstanding contributions to this conference emphasize the considerable progress that has been made in understanding the equilibrium or near-equilibrium properties of many-body systems on a microscopic basis. Most of the experimental data on these systems have been obtained in reaction situations where some type of beam is scattered. For weakly interacting probes, the many-body information desired can be cleanly separated from the reaction mechanism. However, strongly interacting probes can create highly non-equilibrium conditions and so require a formulation of many-body scattering commensurate with that used to describe stationary states.

In this contribution we discuss some recent progress in treating many-body reaction phenomena using the time-dependent mean-field method. Because the concept of the mean-field is central to many methods for treating stationary states, it is a natural starting point in any attempt to describe scattering. The simplest implementation of the independent-particle picture for reactions is Time-Dependent Hartree-Fock (TDHF), which has had some success as an inclusive description of heavy-ion collisions.[1] However, cross sections to specific channels (i.e., elements of the S-matrix) cannot be calculated, even in principle, because of the inherent non-linear nature of the method.

Our work is aimed at formulating tractable mean-field approximations to the many-body S-matrix. As a first step, we consider the transitions induced in a many-body system by a time-dependent external one-body field. Although this is only a crude prototype of a full scattering problem, it nevertheless illustrates several of the difficulties involved and their potential solutions. Moreover, it is not without physical interest, as certain atomic and nuclear collisions can be formulated in this context.

Detailed presentations of this work appear elsewhere[2-4] and only its broad outlines can be given here. We first discuss the functional integral representation of the many-body evolution operator central to our methods. The approximations to elements of the many-body S-matrix and to inclusive averages of few-body observables are then presented. Finally, two illustrative examples are considered: the two-level Lipkin model and the more realistic situation of atomic p + He elastic and charge-transfer collisions.

---

*Supported in part by the National Science Foundation [PHY77-21602 and PHY79-23638].
†Alfred P. Sloan Foundation Fellow.
††Chaim Weizmann Post-doctoral Fellow.

The Time-Evolution Operator

We consider a system with a local two-body interaction, v, perturbed by a time-dependent one-body field, V, which vanishes as $|t| \to \infty$. Thus, the Hamiltonian can be written as:

$$H(t) = K + \tfrac{1}{2}(\rho, v\rho) + (V(t),\rho) \equiv H_0 + V(t). \tag{1}$$

Here, we have adopted a simple inner-product notation for spatial integrations [i.e., $(\rho,v\rho) = \int dx dx' \; \rho(x)v(x-x')\;\rho(x')$ and $(V(t),\rho) = \int dx \; V(x,t)\;\rho(x)$], $\rho(x) = \psi^{\dagger}(x)\psi(x)$ is the one-body density, and $K = \int dx(\nabla\psi^{\dagger}\cdot\nabla\psi)/2m - v(0)\int dx\rho(x)/2$ is the kinetic energy corrected for self-interaction. Our goal is to calculate transitions between the eigenstates of H_0 induced by V.

The Hubbard-Stratonovich representation of the many-body evolution operator[5-7] from time -t to +t is

$$U(t,-t) = \int \mathcal{D}[\sigma]\; \exp\left[i/2\int_{-t}^{t} d\tau(\sigma(\tau),v\sigma(\tau))\right] U_{\sigma}(t,-t) \tag{2}$$

where the functional integral is over all c-number fields $\sigma(x,\tau)$ and

$$U_{\sigma}(t,-t) = T\;\exp\left[-i\int_{-t}^{t} d\tau\; H_{\sigma}(\tau)\right] \tag{3}$$

is the evolution operator for the time-dependent one-body Hamiltonian $H_{\sigma}(\tau) = K + (\sigma(\tau),v\rho) + (V(\tau),\rho)$. Thus, (2) expresses the many-body evolution as a super-position of infinitely many one-body evolutions.

The utility of (2) is apparent when matrix elements of U between many-body states $|\beta'\rangle$ and $|\beta\rangle$ are considered. The functional integral may then be evaluated in a stationary phase approximation, where the phase of the integrand, $\tfrac{1}{2}\int d\tau(\sigma,v\sigma) + \arg\langle\beta'|U_{\sigma}|\beta\rangle$, is made stationary as a functional of σ. This implies that σ satisfies the equation

$$\sigma(x,\tau) = \mathrm{Re}\left[\frac{\langle\beta'(\tau)|\rho(x)|\beta(\tau)\rangle}{\langle\beta'(\tau)|\beta(\tau)\rangle}\right] \tag{4}$$

where $|\beta(\tau)\rangle = U_{\sigma}(\tau,-t)|\beta\rangle$ and $\langle\beta'(\tau)| = \langle\beta|U_{\sigma}(t,\tau)$ are the states which evolve forward or backward under H_{σ}. Note that (4) is similar to the TDHF approximation ($\sigma(x,\tau) = (|\beta(\tau)\rangle)^{\dagger}\rho(x)|\beta(\tau)\rangle$) in that σ depends upon states whose evolution is determined by σ, but differs from it in that σ depends upon both β and β'. Once σ has been determined from (4), the many-body evolution is approximated by the integrand at the stationary field,

$$\langle\beta'|U(t,-t)|\beta\rangle \approx \exp\left[i/2\int_{-t}^{t} d\tau(\sigma(\tau),v\sigma(\tau))\right]\langle\beta'|U_{\sigma}(t,-t)|\beta\rangle\;. \tag{5}$$

Pre-exponential factors depending on the second-variation of the phase can also be computed.

Approximations to the S-Matrix

If the system is prepared in a state $|\beta\rangle$ far in the past, the amplitude for it to be found in the state $|\beta'\rangle$ long after V(t) ceases to act is given by the S-matrix element, $S_{\beta'\beta} = \lim_{t\to\infty}\langle\beta'|U(t,-t)|\beta\rangle$. The most straightforward approach is to solve (4) and hence determine $S_{\beta'\beta}$ from (5). Unfortunately, this is unacceptable, as the magnitude of $S_{\beta'\beta}$ so determined would oscillate unphysically in time, even long before or after V acts. This is so for two reasons. First, in the ideal situation, $|\beta'\rangle$ and $|\beta\rangle$ are eigenfunctions of the many-body Hamiltonian H_o, and, if U is treated exactly, S only has a trivial time-dependent phase depending on the energies of β and β'. However, since the precise specification of β and β' requires solving the many-body problems, approximations are necessary. Thus, β' and β are more often than not wave-packets in the space of exact eigenstates, so that S oscillates with t. Second, even if β and β' were exact eigenstates of H_o, the approximate, non-linear mean-field evolution would induce oscillations in S.

The above considerations require that we seek a time-independent approximate S-matrix element involving necessarily approximate 'channel' states. This can be obtained by considering the interaction-picture evolution operator[2]

$$S_{\beta'\beta} = \lim_{t\to\infty} \langle\beta'|U^{(o)}(o,t)U(t,-t)U^{(o)}(-t,o)|\beta\rangle \qquad (6)$$

where $U^{(o)}$ is the evolution operator for the unperturbed many-body Hamiltonian H_o. We now introduce the representation (2) for each of the three evolution operators, and upon calling the integration variables σ_i, σ, and σ_f, obtain

$$S_{\beta'\beta} = \lim_{t\to\infty}\iiint D[\sigma_i]\,D[\sigma]\,D[\sigma_f]\,\exp\left[\frac{i}{2}\oint(\sigma,v\sigma)\right]\langle\beta'|U_{\sigma_f}(o,t)U_\sigma(t,-t)U_{\sigma_i}(-t,o)|\beta\rangle, \qquad (7)$$

where

$$\oint d\tau(\sigma,v\sigma) = \int_o^{-t} d\tau(\sigma_i,v\sigma_i) + \int_{-t}^{t} d\tau(\sigma,v\sigma) + \int_t^{o} d\tau(\sigma_f,v\sigma_f). \qquad (8)$$

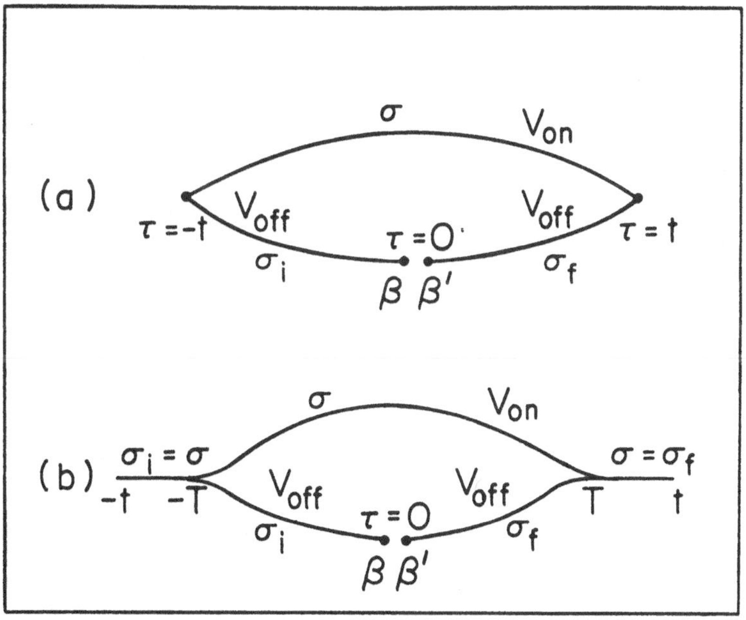

Fig. 1. (a) The loop. (b) The "collapsed" loop. $V(\tau) \neq 0$ for $|\tau| < T$, with $T < t$. (See text for details.)

The integrand of Eq. (7) can be visualized by the simple diagram shown in Fig. 1a. Evolution is represented by motion along a loop, which consists of 3 sections: (i) The "preparation" process: the system starts at $\tau = 0$ in the lower middle and

moves backward in time along the lower left-hand side of the loop (to time -t) with the mean field σ_i and with the interaction V turned off.
(ii) The interaction process: the system evolves forward from -t to t along the upper section of the loop with the field σ and the interaction V turned on.
(iii) The "analysis" process: the system moves backward from t to 0 along the lower right-hand portion of the loop, with a field σ_f and the interaction V turned off.

Upon applying the stationary phase approximation to (7), conditions similar to (4) can be found which must be satisfied by the σ's. These involve the states which evolve clockwise or counter-clockwise around the loop under a one-body Hamiltonian depending on the σ's. Most remarkably, the approximate S-matrix element which results (analogous to (5)) has the property of being independent of t, when t is larger than the interaction time. For the stationary fields, σ coincides with σ_i before the interaction starts and with σ_f after the interaction ceases, so that the evolution caused by U_σ is cancelled by U_{σ_i} for very early times and by U_{σ_f} for very late times. This can be illustrated by the "collapsed" ends of the loop as in Fig. 1b. It is also interesting to note that the above mean field approximation, unlike the usual TDHF, preserves any time reversal symmetry present in the exact problem. Specifically, both the exact and approximate S-matrix have $S_{\beta'\beta} = S_{\beta\beta'}$. For a T-reversal invariant situation, it can also be shown that for elastic propagation, $\beta' = \beta$, $\sigma_i(-\tau) = \sigma_f(\tau)$ and $\sigma(\tau) = \sigma(-\tau)$.

In the approximation described above, the optimal mean-field for a given S-matrix element depends upon the initial and final channels and hence must be calculated anew for every transition of interest. While this might be acceptable for exclusive (or nearly exclusive) measurements, it is evidently a considerable complication in describing any inclusive measurement which averages over a large number of exit channels. In these cases, the full S-matrix contains far more information than is needed to describe experiment. It is therefore appropriate to consider an alternative implementation of the mean-field approximation[3] which calculates directly inclusive observables, i.e., the final expectation values of self-adjoint few-body operators for a given initial channel.

The inclusive expectation value of any local one-body observable which is a constant of the unperturbed motion can be expressed in terms of the inclusive one-body density,

$$\langle \rho(x) \rangle = \lim_{t \to \infty} \langle \beta | U^\dagger(t,-t) \rho(x) U(t,-t) | \beta \rangle \; . \tag{9}$$

Similarly, the energy can be written as

$$\langle H \rangle = \lim_{t \to \infty} \frac{i}{2} \left(\frac{\partial}{\partial t_1} - \frac{\partial}{\partial t_2} \right) \langle \beta | U^\dagger(t_2,-t) U(t_1,-t) | \beta \rangle \Big|_{t_1 = t_2 = t} \; . \tag{10}$$

Here $|\beta\rangle$ is the initial state and (10) follows from the evolution equation for U, $i\, \partial U/\partial t = HU$.

Approximations to (9) and (10) can be generated by introducing the Hubbard-Stratonovich representation for each of the U's and invoking the stationary-phase approximation. The resulting consistency equations for the σ's are identical to TDHF, thus justifying this approximation for inclusive observables.

Several aspects of the application of mean-field methods to this time-dependent problem deserve comment. In both the exclusive and inclusive cases, the formulation has a loop structure (in the inclusive case, there is evolution from the far past to the far future and then back to the past). This is a very positive feature, since errors introduced by the mean-field approximation to each U are largely canceled by similar errors in the other U's. We have also tacitly assumed that there is only one stationary field configuration; if there are several, their contributions to S must be added coherently. Pre-exponential factors arising from gaussian integrals about each stationary field can also be derived; these involve time-dependent analogues of the usual RPA and are quite intractable in practical applications. However, because of the self-cancellation mentioned above, the "quadratic" corrections are usually unimportant; they can be significant when two or more stationary fields are similar.

Applications

In calculating mean-field approximations to S-matrix elements, the determination of the stationary σ-fields for any states β, β' presents a self-consistent time-dependent problem. One method of solution is to guess the form of the wave functions at each time around the loop in Fig. 1 (for example, the TDHF solutions), and use these to define the σ-fields through the stationary phase conditions similar to (4). New wave functions can then be found by evolving β and β' around the loop with the corresponding one-body H_σ. Repeated iterations of this process presumably converge to a solution. This method of solution has been applied to the realistic atomic problem discussed below. However, for a number of non-trivial model Hamiltonians, where the relevant operators form a finite Lie algebra, the mean-field equations can be reduced to a set of time local evolution equations for the group parameters. These equations must be solved self-consistently, but only with a small, finite number of iterative parameters. The ease of solution and transparency of these models makes them useful examples of the mean-field techniques.

One model Hamiltonian we have investigated[2] is the forced Lipkin model,[8] a non-trivial many-body system composed of N distinguishable fermions with pair-wise interactions of strength v. Each fermion can occupy single-particle levels with energy $\pm\,\varepsilon/2$. This system can be discussed in terms of a set of SU(2) quasi-spin operators, $\vec{J}$, the unperturbed Hamiltonian being

$$H_o = J_z + v\,(J_x^{\,2} - J_y^{\,2}). \tag{11}$$

For the perturbation, we take $V(t) = \vec{f}(t)\cdot\vec{J}$ and for β,β' choose either exact eigenstates of H_0 (obtained by numerical diagonalization), or the HF approximation to the ground state.

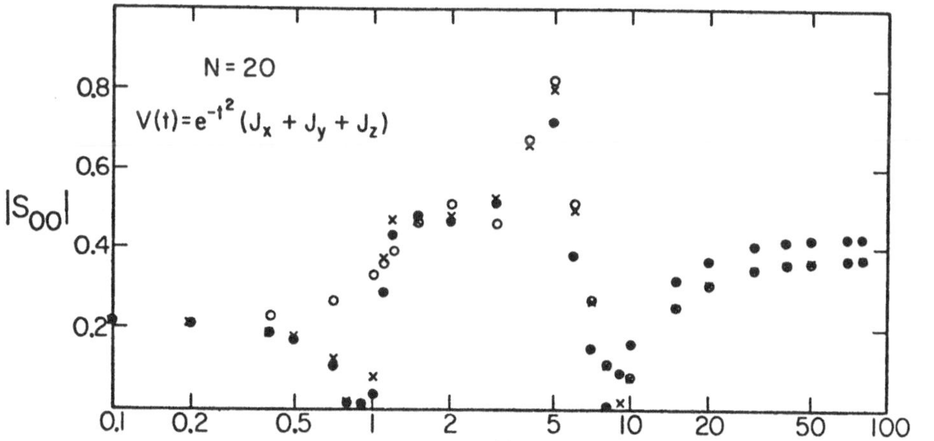

Fig. 2. Ground-state to ground-state S-matrix element in the forced Lipkin model. o - exact evolution of the exact ground state; x - exact evolution of the HF ground state; • - mean-field evolution of the HF ground state.

Figure 2 shows a typical result for the magnitude of the ground state to ground-state amplitude as a function of the coupling parameter, $\chi = Nv$. Results for the phase are of a similar quality. The complex structure of the exact results is reproduced by the mean-field calculations by the interference of several stationary solutions. Figure 3 shows the typical agreement obtained for off-diagonal S-matrix elements. The validity of the mean-field approximation improves with increasing N, and it is quite accurate for even relatively small systems. TDHF calculations of the inclusive excitation energy are also in good agreement with the exact results.[3]

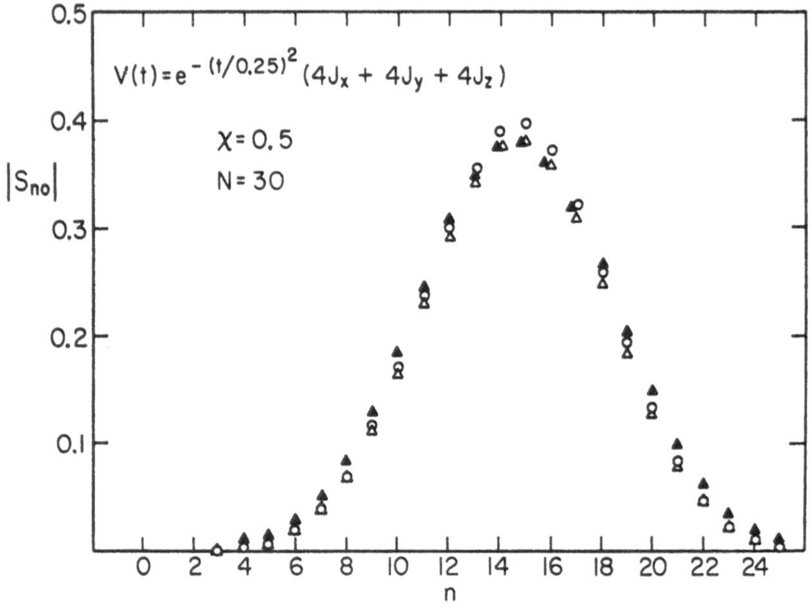

Fig. 3. Moduli of the amplitudes to excite the nth state in the forced Lipkin model. o - exact results; Δ - mean-field approximation; $\blacktriangle$ - mean field with $\sigma = \sigma_{TDHF}$.

Our mean-field methods have also been applied to the more realistic situation of atomic p + He scattering.[4] Here, the two electrons bound to the He nucleus are the "many-body" system under the time-dependent influence of the coulomb field of the proton, which is assumed to move on an unperturbed Rutherford trajectory. The mean-field equations are solved by the iterative scheme discussed above by using finite difference numerical techniques developed for TDHF calculations of nuclear collisions.[1] The problem is reduced to two spatial dimensions by constraining the electron wave functions to be axially symmetric about the line joining the two nuclei, and the initial wave function is taken to be the He Hartree state.

For most proton energies and impact parameters, the elastic S-matrix element converges after only a few iterations to a value only a few percent different from that at the first iteration. This is not too surprising, since the electron-electron interaction is relatively weak. An experimental observable which can be calculated is the inclusive single charge-transfer probability. Here, the final state is taken to be a hydrogen atom and He^+. A comparison of our calculations with the appropriate data at an input parameter of 0.1 Å is shown in Fig. 4. The agreement is good at lower energies, while the discrepancy at higher proton energies is most likely due to a breakdown of the assumption of axial symmetry. Calculations of inelastic and charge-transfer reactions to specific final channels are in progress.

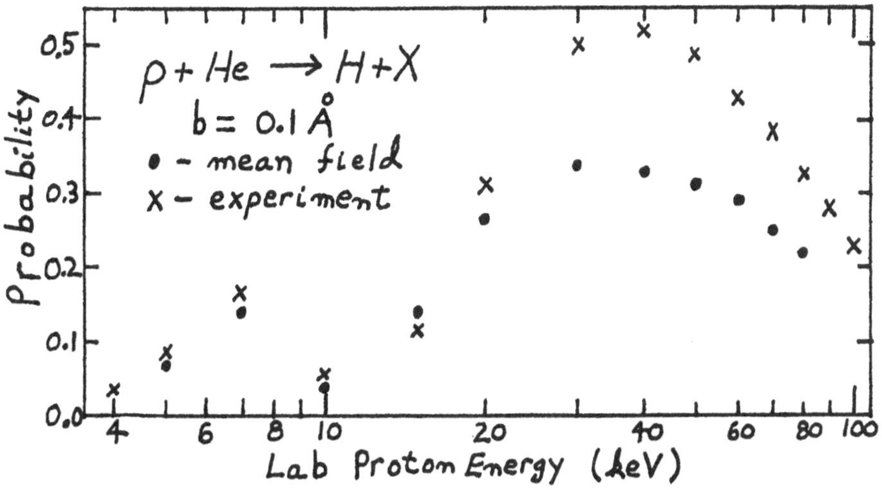

Fig. 4. Inclusive single charge transfer probability in p + He collisions.

References

1. See, for example, K.T.R. Davies, K.R. Sandhya-Devi, and M.R. Strayer, Phys. Rev. Lett. 44, 23 (1980) or S.E. Koonin in Progress in Nuclear and Particle Physics, Vol. 4, ed. Sir Denys Wilkinson, F.R.S. (Pergamon Press, Oxford, 1980) pp. 283-321.
2. Y. Alhassid and S.E. Koonin, Caltech Preprint MAP-14, Phys. Rev. C, in press.
3. Y. Alhassid, B. Müller, and S.E. Koonin, Caltech Preprint MAP-16, Phys. Rev. C, in press.
4. K.R. Sandhya-Devi and S.E. Koonin, to be published.
5. J. Hubbard, Phys. Rev. Lett. 3, 77 (1959).
6. R.L. Stratonovich, Dokl. Akad. Nauk SSSR 115, 1097 (1957), [Sov. Phys.—Dokl. 2, 416 (1958)].
7. S. Levit, Phys. Rev. C 21, 1594 (1980).
8. H.F. Lipkin, N. Meshkov, and A.J. Glick, Nucl. Phys. 62, 188 (1965); N. Meshkov, A.J. Glick, and H.F. Lipkin, Nucl. Phys. 62, 199 (1965); D. Agassi, H.J. Lipkin, and N. Meshkov, Nucl. Phys. 86, 321 (1966).

QUANTUM MEAN-FIELD THEORY OF COLLECTIVE DYNAMICS AND TUNNELING

J.W. Negele
Center for Theoretical Physics and Department of Physics
Massachusetts Institute of Technology
Cambridge, Massachusetts, 02139/USA

Introduction

A fundamental problem in quantum many-body theory is formulation of a microscopic theory of collective motion. For self-bound, saturating systems like finite nuclei described in the context of non-relativistic quantum mechanics with static interactions, the essential problem is how to formulate a systematic quantal theory in which the relevant collective variables and their dynamics arise directly and naturally from the Hamiltonian and the system under consideration. In collaboration with Shimon Levit and Zvi Paltiel, significant progress has been made recently in formulating the quantum many-body problem in terms of an expansion about solutions to time-dependent mean-field equations. The technical details of this approach are presented in detail in Refs. 1-3, and only the essential ideas, principal results, and illustrative examples will be summarized here.

The mean-field is an obvious candidate to communicate collective information. Possessing the infinite number of degrees of freedom of the one-body density matrix, it has access to all the shape and deformation degrees of freedom one intuitively believes to be relevant to nuclear dynamics. The static mean-field theory with appropriate effective interactions, commonly referred to as the Hartree Fock approximation, quantitatively reproduces the radial distributions and shapes of spherical and deformed nuclei throughout the periodic table. The time-dependent Hartree Fock (TDHF) approximation and its RPA limit for infinitesimal fluctuations similarly yields a reasonable description of transition densities to excited states, fusion cross sections in heavy ion reactions, and strongly damped collisions.

Whereas the mean field is thus a compelling foundation for a microscopic theory of collective motion, the TDHF initial value problem is an inappropriate starting point for a systematic quantum theory. Stimulated by developments in quantum field theory in which systematic expansions are developed about the solution to the corresponding classical field equations, we have developed a conceptually unambiguous quantum theory of collective motion. An exact expression for an observable of interest is written using a functional integral representation for the evolution operator, tractable time-dependent mean field equations are obtained by application of the stationary-phase approximation (SPA) to the functional integral, and corrections to the lowest-order theory may be systematically enumerated.

Outline of Approach

The essential steps in the method are as follows. First, one selects a few-body operator corresponding to a physical observable of interest and then one expresses its expectation value in terms of the evolution operator. For example, to calculate the bound state spectrum and the expectation value of any few-body operator $\mathcal{O}$ in any bound state, one may evaluate the poles and residues of the following expression:

$$-i\int dTe^{iET}\mathrm{tr}\mathcal{O}e^{-iHT} = \sum_n \frac{\langle n|\mathcal{O}|n\rangle}{E-E_n+i\varepsilon} \quad . \tag{1}$$

Next, one utilizes an appropriate functional integral representation for the many-body evolution operator. One particularly simple choice is the Hubbard-Stratonovich[4] transformation used in Ref. 5

$$Te^{-\frac{i}{2}[\hat{\rho}v\hat{\rho}]} = \int D[\sigma]e^{\frac{i}{2}[\sigma v\sigma]}Te^{-i[\sigma v\hat{\rho}]} \quad , \tag{2}$$

where the brackets denote the following integral

$$[\sigma v\hat{\rho}] \equiv \int dx_1 dx_2 dx_3 dx_4 dt\sigma(x_1,x_3;t)v(x_1x_2x_3x_4)\hat{\rho}(x_2,x_4;t) \quad , \tag{3}$$

$\hat{\rho}$ is the interaction representation operator

$$\rho(x,x';t) \equiv e^{iH_0t}\psi^+(x)\psi(x')e^{-iH_0t} \quad , \tag{4}$$

and T denotes a time ordered product. The evolution operator corresponding to a Hamiltonian containing two-body interactions is thus replaced by an integral over an infinite set of evolution operators containing only one-body operators. A second alternative is to break the evolution into very small time steps between each of which an overcomplete set of Slater determinants is inserted[6,7]

$$\langle\Psi_f|e^{-HT}|\Psi_i\rangle = \langle\Psi_f|\ldots e^{-iH\Delta T}\int du(z)|\Psi(z)\rangle\langle\Psi(z)|e^{-iH\Delta T}|\Psi_i\rangle \quad . \tag{5}$$

The theory is rendered manageable by virtue of a simple choice of the measure du(z) which efficiently handles the overcompleteness. A third alternative is to use Grassman variables as in field theory,[8] so that the trace of the exponential of the action becomes[9]

$$\mathrm{tr}e^{iS} = \int D[Z^*,Z]e^{i\left[\int Z^*\left(i\frac{\partial}{\partial t}-T\right)Z-\frac{1}{2}\int Z^*Z^*vZZ\right]} \quad . \tag{6}$$

Finally, for any of these functional integral representations when suitably generalized to include exchange, application of the SPA yields TDHF equations plus a systematic hierarchy of corrections.

The essence of the program is exemplified by applying it to the trivial problem of one-dimensional quantum mechanics in the potential shown in Fig. 1, for which case we may write[2]

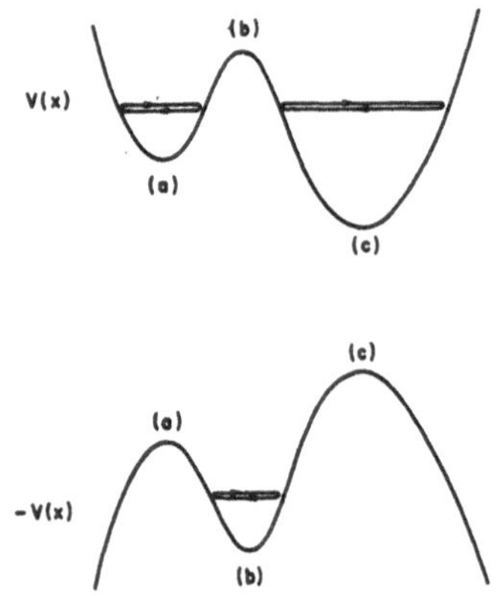

Fig. 1 Sketch of a double well with two classically allowed regions separated by one classically forbidden region.

$$\mathrm{Tr}\frac{1}{H-E} = i\int dT e^{iET}\int dq\langle q|e^{-iHT}|q\rangle$$

$$= i\int^{\infty} dT e^{-ET}\int dq\int D[q(t)]e^{iS[q(t)]}\Big|_{q(t)=q(0)=q} \quad , \tag{7}$$

where S[q(t)] in the Feynman path integral denotes the classical action. Application of the SPA to $\int D[q(t)]$ requires that q(t) must satisfy the classical equation of motion

$$m\frac{d^2}{dt^2}q = -\nabla V \tag{8}$$

and application of the SPA to $\int dq$ requires that the momentum at time T equal that at time 0. Thus, we obtain

$$\mathrm{Tr}\frac{1}{H-E} = i\int_0^{\infty} dt \sum_{q_{c\ell}} e^{i(ET+S(T))} \equiv i\int dT \sum_{q_{c\ell}} e^{iW(T)} \quad , \tag{9}$$

where S(T) is the action for a periodic solution to the classical equation of motion and the sum $\sum_{q_{c\ell}}$ includes all such periodic classical solutions.

Finally, the SPA is applied to the time integral in Eq. (9), giving rise to both real and complex stationary values of the period. Real periods simply correspond to multiples of the fundamental periods for classical oscillations around minima (a) and (c) in Fig. 1 such that the classical energy equals E. The period and contribution to the reduced action W(T) of Eq. (9) for periodic solutions in region a (and similarly for region c) are

$$T_a = 2\int dq\sqrt{\frac{m}{2(E-V(q))}} \quad , \tag{10}$$

and

$$W_a = \oint p\dot{q}dt = 2\int\sqrt{2m(E-V(q))}dq \quad . \tag{11}$$

The meaning of classical solutions for imaginary time is most evident if one simply replaces (it) by τ in the equation of motion. The two resulting factors of i in Eq. (8) are then equivalent to reversing the sign of V(q). As sketched in Fig. 1, this has the effect of interchanging classically allowed and forbidden regions, so one now has periodic solutions in region b with imaginary period and reduced action

$$iT_b = \bar{T}_b = 2\int dq\sqrt{\frac{m}{2(V(q)-E)}} \quad , \tag{12}$$

and

$$iW_b(E) \equiv \bar{W}_b(E) = 2\int\sqrt{2m(V(q)-E)dq} \quad , \tag{13}$$

Combining all integral numbers of periods in the three regions thus yields an infinite sequence of stationary points $T_{\ell mn} = \ell T_a + mT_c - in\bar{T}_b$ and the corresponding sum over classical periodic trajectories in Eq. (18) yields multiple geometric series which sum to

$$Tr\frac{1}{H-E} = \frac{e^{iW_a}+e^{-\bar{W}_b}+e^{iW_c}-2e^{i(W_a+W_c)}}{\left(1-e^{iW_a}\right)\left(1-e^{iW_c}\right)-e^{\bar{W}_b}} \quad . \tag{14}$$

For the case of a single well, in which case regions (b) and (c) don't exist, this yields poles at energies E_n such that

$$W_1(E_n) = \int pdq = 2n\pi \quad . \tag{15}$$

Eq. (15) differs from the usual Bohr-Sommerfeld quantization condition $(2n+1)\pi$ only because we have neglected phase factors arising from quadratic corrections to the SPA. In the case of spontaneous decay of a quasi-stationary state, region (c) is elongated to extend throughout an arbitrarily large normalization box, and one observes that W_c then yields a vanishing contribution to the smoothed level density[2]

$$P_\gamma \equiv \frac{1}{\pi}\, ImTr\frac{1}{H-E+i\gamma} \propto \left[\left(\frac{e^{-\bar{W}_b}}{2}\right)^2 + \left(\sin\frac{W_a}{2}\right)^2\right]^{-1} \tag{16}$$

The level density, Eq. (16), exhibits quasi-stationary states with energies given by Eq. (15) and widths

$$\Gamma_n = 2\left(\frac{\partial W_a}{\partial E}\right)e^{-\bar{W}_b(E_n)} = T_a e^{-\bar{W}_b(E_N)} \tag{17}$$

which agree with the familiar WKB result to within a factor 1/2 discussed in Ref. 10.

Application to Many-Body Problem

Straightforward application of the same program to the many-body problem results in application of the SPA to the T and σ integrals in an expression of the form

$$\int dTe^{iET}tre^{-iHT} \equiv \int dTe^{iET}\, D[\sigma]e^{iS[\sigma]} \quad , \tag{18}$$

where

$$S[\sigma] = \frac{1}{2}[\sigma v \sigma] - i\,\ell n\{\mathrm{tr}T\exp(-i[\sigma v \hat{\rho}])\} \quad ,$$

and yields three distinct classes of solutions.

Time-independent solutions to the SPA equations reproduce familiar HF theory. The quadratic corrections to SPA produce the RPA ground state correlations, and the systematic evaluation of higher corrections generates standard perturbation theory.[3] Aside from providing a terse and elegant derivation of perturbation theory, this functional integral approach has the additional advantage of dealing efficiently with constraints, such as those arising in gauge theories.[11]

A second class comprises time-dependent solutions with real period which correspond to eigenfunctions of large-amplitude collective motion. A set of N single-particle wave functions obey the following eigenvalue equation

$$\left[-i\frac{\partial}{\partial t}+K+\mathrm{tr}\sigma v\right]\phi_i(x,t) = \alpha_i \phi_i(x,t) \tag{19}$$

subject to the periodic boundary condition

$$\phi_i\left(x,\frac{T}{2}\right) = \phi_i\left(x,-\frac{T}{2}\right) \quad , \tag{20}$$

where the self-consistent mean field satisfies

$$\sigma(x,x',t) = \sum_i \phi_i^*(x',t)\phi_i(x,t) \quad , \tag{21}$$

K denotes the kinetic energy operator and the allowed values of the period are specified by the quantization condition

$$\int dx \int_{-T/2}^{T/2} dt\,\phi^*(x,t)\, i\frac{\partial}{\partial t}\phi(x,t) = n2\pi \quad . \tag{22}$$

Clearly the non-linear differential Eqs. (19-21) in four space-time dimensions have the same general structure as the static Hartree equations in three space dimensions, and they may be solved by the usual iterative procedure. Application of this method to the ground state multiplet of the spectrum of the Lipkin model yields the results shown in Fig. 2. Further discussion of large amplitude collective motion using this general approach may be found in Ref. 1.

The third class of solutions is made up of time-dependent solutions with imaginary period corresponding to tunneling phenomena in classically forbidden domains. In this case, the single-particle Equations (19) are replaced by

$$\left[\frac{\partial}{\partial \tau}+K+\mathrm{tr}\sigma v\right]\phi_i(x,\tau) = \alpha_i \phi_i(x,\tau) \quad , \tag{23}$$

with the same periodic boundary condition (20) and the self-consistent mean field

$$\sigma(x,x',\tau) = \sum_i \phi_i(x',-\tau)\phi_i(x,\tau) \quad . \tag{24}$$

Of particular physical interest are solutions which in the limit as $\pm T/2 \to \pm\infty$ approach the HF stationary local minimum for a fissioning nucleus and evolve near $T \sim 0$ toward the entrance to the classically allowed domain near the scission point for two fission fragments. Such solutions will be denoted "bounces", following Coleman,[10] and bear great formal similarity to the "pseudoparticles" and "instantons"[12,13] investigated extensively in field theory. Whereas the Euclidean solutions arising in field theory have trivial spatial dependence, being either constant or spherically symmetric in space-time, the non-trivial spatial dependence of the present "bounce" solutions is crucial to the physics and precludes analytic solution even for schematic models. Furthermore, for a nucleus possessing many decay channels such as symmetric fission, asymmetric fission, alpha, proton, or neutron decay, there will exist several distinct well-separated bounces, and the analog of the width Γ in Eq. (17) is the sum of partial widths:

$$\Gamma = \sum_m \Gamma^{(m)} , \qquad (25)$$

where each partial width is calculated from the action determined for the bounce solution for the appropriate channel

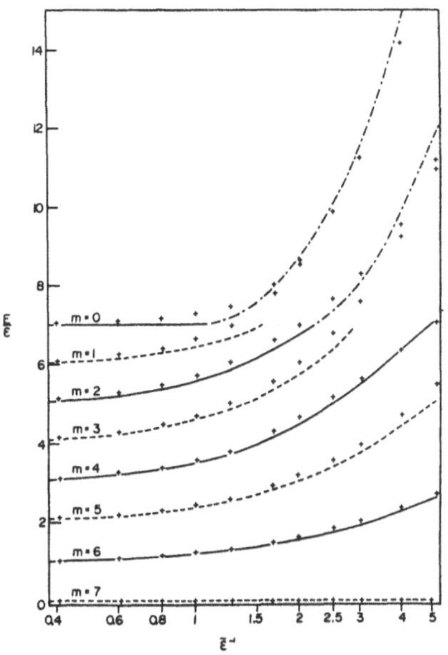

Fig. 2 Exact Lipkin spectrum (crosses) compared with the mean-field approximation as a function of $\tilde{\varepsilon}^{-1}=NV/\varepsilon$. The particle number N in this case is 14, v is the strength of the interaction coupling pairs of particles in the two levels, and ε is the energy separation of the two levels. The dot-dash curves denote doubly degenerate approximate solutions and the other curves are non-degenerate.

$$\Gamma^{(m)} = 2T_m e^{-\int_{-T/2}^{T/2} d\tau \phi(x,-\tau)\frac{\partial}{\partial\tau}\phi(x,\tau)} . \qquad (26)$$

To make these bounce solutions more concrete, it is useful to consider a saturating model system of nuclei in one spatial dimension interacting with an effective interaction of the Skyrme form.[2] The analog of the Coulomb force is adjusted such that a 16-particle system is unstable with respect to fission into two 8-particle daughters which are in turn stable with respect to further decay into 4-particle granddaughters. The constrained HF energy as a function of $\langle x^2\rangle$ for the 16-particle system is shown in Fig. 3, and displays the expected form of a fission barrier. The self-consistent single-particle solutions to Eqs. (23), assuming spin-isospin degeneracy 4, are shown in Fig. 4 at the two turning points, $\tau=\pm T/2$ and $\tau=0$. As expected, the determinant of these wave functions corresponds to the 16-particle HF static solution at $= T/2$ and closely approximates the product of two 8-particle determinants for nearly-separated fragments at $\tau=0$. The corresponding density, $\sigma(x,t)$ is shown in Fig. 5 for successive times between $\tau=-T/2$ and $\tau=0$.

Fig. 3 The constrained energy of a 16-particle model system as a function of $\langle x^2 \rangle$.

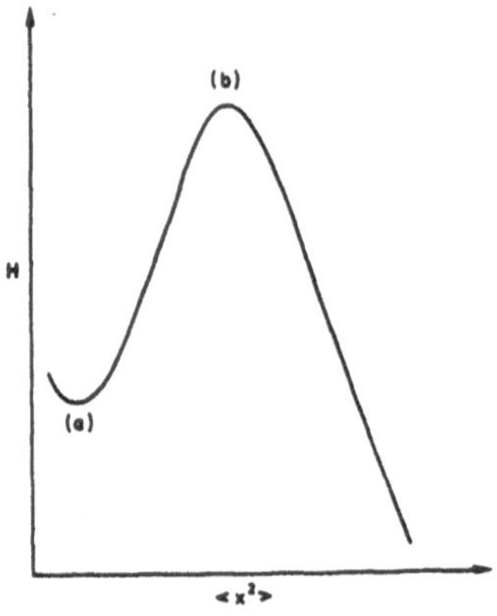

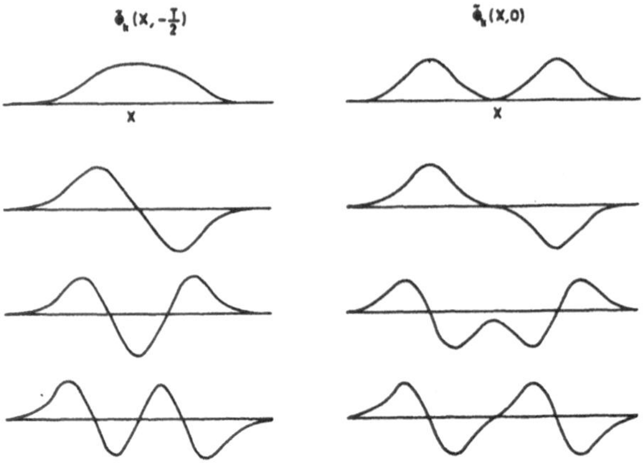

Fig. 4 Self consistent single-particle wave functions as a function of x at times $\tau=-T/2$ and $\tau=0$ for the bounce solution for spontaneous fission of a 16-particle model system.

Solution of Eqs. (22) in four space-time dimensions is obviously computationally more cumbersome, but has been accomplished for a range of nuclei up to A=32. In these calculations, the proton charge has been increased to obtain appropriate values of the fissility, and preliminary results for the fission of ^{8}Be are shown in Fig. 6. Although spurious cm motion problems prevent quantitative comparison of this particular calculation with experiment, this result does demonstrate the feasibility of obtaining bounce solutions with the appropriate properties and shows that all the relevant shape degrees of freedom are incorporated in this self-consistent theory.

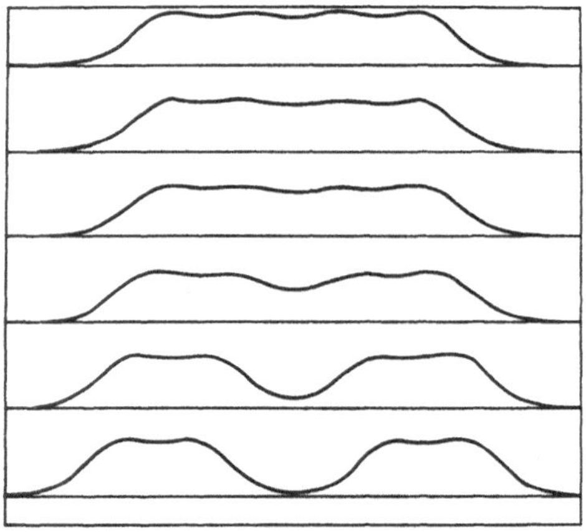

Fig. 5 The density $\sigma(x,\tau)$ for the same system as in Fig. 4 as a function of x at successive times from $\tau=-T/2$ to $\tau=0$.

Outlook

Clearly many other applications of quantum mean-field approximations arising from such functional integral expressions are possible. One should eventually be able to quantitatively understand the systematics of fission lifetimes in heavy nuclei, including shell effects and the competition between symmetric and asymmetric decay channels. Similarly, excited states of soft transitional nuclei involving very large amplitude collective vibrations should be well described by the present theory. Reaction theory poses many important and challenging problems. Although it is possible to write exact functional integral expressions for S-matrix elements,[14] the key to a meaningful reaction theory is finding an appropriate functional integral expression for relevant expectation values of few-body operators, such as mean fragment charge, mass, or excitation energy, which yields numerically tractable mean-field equations. In contrast to the TDHF initial value problem, which describes the most probable outcome, such functional integral expressions for specific observables can address specific components of interest, even those which are exponentially small relative to the most probable component. This, then, is a natural language to address such diverse and important questions as superheavy nucleus formation in heavy ion collisions, and tunneling phenomena in light-ion collisions associated with quasi-molecular states and the resonance behavior in such systems as ^{24}Mg. Generalization to finite temperature is straightforward and offers an ideal framework from which to consider the equation of state of hot matter at subnuclear density in neutron stars, as well as a variety of other finite temperature many-body systems.

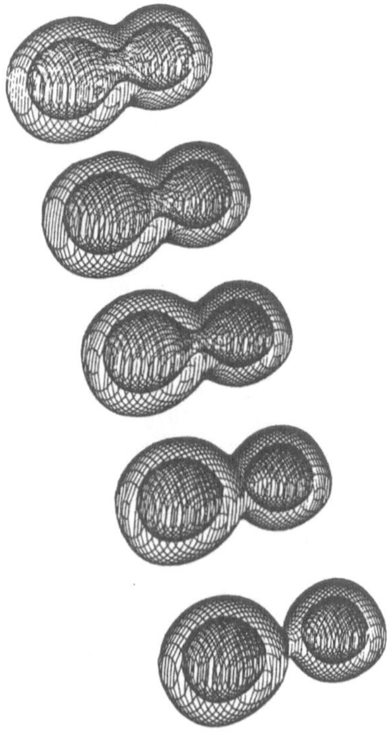

Fig. 6 Three dimensional perspective plots of surfaces of constant density for fission of ^{8}Be. The inner and outer surfaces correspond to densities of 1/3 and 2/3 nuclear matter density respectively and the sequence of shapes run from $\tau=-T/2$ to $\tau=0$.

In summary, the quantum mean-field theory presented here offers promise in a variety of applications in non-relativistic many-body theory. The principal unresolved challenges at present are understanding the validity and accuracy of the SPA and developing more powerful approximation techniques to deal with the resulting time-dependent mean field equations.

References

1. S. Levit, J.W. Negele, and Z. Paltiel, Phys. Rev. C21, 1603 (1980).
2. S. Levit, J.W. Negele, and Z. Paltiel, Phys. Rev. C22, 1979 (1980).
3. J.W. Negele, Proc. International School of Physics "Enrico Fermi" LXXVII, Varenna (1979).
4. J. Hubbard, Phys. Lett. 3, 77 (1959); R.L. Stratonovich, Sov. Phys. Doklady 2, 416 (1958).
5. S. Levit, Phys. Rev. C21, 1594 (1980).
6. J.P. Blaizot and H. Orland, J. Physique Lettres, 41, 53 (1980).
7. H. Kuratsuji and T. Suzuki, Phys. Lett. 92B, 19, (1980).
8. H. Reinhardt, J. Phys. G5 L91 (1979); Nucl. Phys. in press; Proc. International Summer School "Critical Phenomena in Heavy Ion Collisions", Aug. 25-Sept. 10, 1980, Poiana Brasov, Romania.
9. Y. Ohnuki and T. Kashiwa, Prog. Theor. Phys. 60, 548, (1978).
10. S. Coleman, lectures at International School of Subnuclear Physics, Ettore Majorana (1977).
11. L.D. Faddeev and V.N. Popov, Phys. Lett. 25B 29 (1967); L.D. Faddeev, Theor. Mat. Fiz. 1, 3 (1970), (Theor. Math. Phys. 1, 1 (1979)).
12. A.M. Polyakov, Nucl. Phys. B 121 429 (1977).
13. G. 't Hooft, Phys. Rev. Lett. 37, 8 (1976).
14. J. Alhassid and S.E. Koonin, Cal. Tech. Preprint MAP-14, (1980).

MODERN THEORIES OF IRREVERSIBLE PHENOMENA AND ENTROPY; SOME APPLICATIONS TO NUCLEAR PHYSICS.

E.S. Hernández (Departamento de Fisica, Facultad de Ciencias Exactas y Naturales, Universidad de Buenos Aires, 1428 Buenos Aires, Argentina).

The purpose of this contribution is to incorporporate some current concepts of irreversible dynamics into the study of many-nucleon systems. We believe this is an important issue, in view of the fact that we are presently attending to renewed interest in the study of equilibration and transport processes in the nuclear matter fluid, whose understanding might lead as well to a different comprehension of some aspects of the equilibrium distribution of the many-nucleon system. On the other hand, there have been presented some arguments towards relating the foundations of quantum mechanics to the distinguishability between reversible and irreversible processes [1], the latter being those through which we actually observe the elementary components of a many-body configuration.

The operator of irreversible dynamics.

In first place, we will present the so-called "operators of irreversible dynamics" and develop an example of their possible use in the Hilbert space representation of a many-fermion system. We will be interested in the microscopic Lyapounov operator [2], whose importance resides in the fact that its existence, for macroscopic systems whose density vectors evolve in time according to a Liouville equation, is a symptom of dissipative behaviour (relaxation towards thermodynamical equilibrium). Let us say that M is a Lyapounov operator in Hilbert space if

i) $M^+ = M$; (1)

ii) If $|\psi\rangle \in \mathcal{D}(H)$ ($\mathcal{D}(H)$ = domain of the operator H that generates the time evolution of our system) then $M|\psi\rangle \in \mathcal{D}(H)$.

iii) $M \geq 0$; (2)

iv) $i[H, M] = \dot{M} \leq 0$, $\dot{M}$ hermitian; (3)

v) $\langle\psi|M|\psi\rangle = 0$ if and only if $|\psi|^2$ is uniform.

In other words the functional $\Omega(|\psi\rangle)$,

$$\Omega = \langle \psi | M | \psi \rangle \tag{4}$$

is positive-definite and monotonically decreasing with time,

$$\dot{\Omega} = \langle \psi | \dot{M} | \psi \rangle \leqslant 0. \tag{5}$$

The existence of Ω entails a physical distinction between positive and negative times. In other words, we could expect that those systems for which a functional Ω exists, might possess some internal, intrinsic means of measuring time intervals. This property could be formalized in terms of a <u>time operator T</u>[3] that plays the role of a canonically conjugate partner to the generator of evolution, namely,

i) $T = T^{+}$; (6)

ii) If $|\psi\rangle \in \mathcal{D}(H)$, then $T|\psi\rangle \in \mathcal{D}(H)$

iii) $[T, H] = i$, or $i[H, T] = \dot{T} = I$. (7)

Integration of eq. (7) allows us to display the meaning of T as an internal time-measuring device, since

$$\langle \psi(t) | T | \psi(t) \rangle = t - t_0. \tag{8}$$

We see that if a time-operator exists, then a Lyapounov operator can be chosen as any monotonic, non-increasing function of T.

It has been shown [3] that it is not possible to find a Lyapounov operator consistently with the Hamiltonian formulation of quantum mechanics. (In the Liouville representation, the existence of such an operator can be guaranteed in a variety of situations). The conclusion in ref. [3] was attained under the hypothesis that the Hamiltonian is bounded from below. We will show below that if this condition is relaxed, operators of irreversible dynamics can be constructed that can be a valuable aid in the analysis of the time-evolution of the many body system [4].

Quasispin systems

We will refer to nuclear matter described by quasispin operators. Let J_z, $J_\pm$ be the usual generators of a SU(2) algebra and assume the Hamiltonian

$$H = N\,h \tag{9}$$

of a saturating fermion system to be a functional of these generators. In the standard Lipkin-Meshkov-Glick model [5] we have

$$H = \varepsilon J_z + \frac{1}{2} V\,(J_+^2 + J_-^2), \tag{1o}$$

where $\lim_{N\to\infty} VN = v$ in order to ensure convergence in the thermodynamic limit. The so-called ground-state phase transition takes place when $|v| = \varepsilon$ and indicates that particles in the low-lying level can be scattered across the gap of width ε by the self-consistent one-body field. In this case, the spherical ($J_z = -J = -N/2$) ground-state undergoes a transition towards a deformed shape.

For our purposes here, an important property of quasispin systems is a consequence of the process of group contraction [6]. Let $\{h_+, h_-, h_z, h_0\}$ be a set of operators related to the U(2) algebra generators $\{J_+, J_-, J_z, J_0\}$ by a non-singular transformation.

$$\begin{pmatrix} h_+ \\ h_- \\ h_z \\ h_0 \end{pmatrix} = \begin{pmatrix} N^{-1/2} & 0 & 0 & 0 \\ 0 & N^{-1/2} & 0 & 0 \\ 0 & 0 & 1 & N/2 \\ 0 & 0 & 0 & 1 \end{pmatrix} \begin{pmatrix} J_+ \\ J_- \\ J_z \\ J_0 \end{pmatrix} \tag{11}$$

Let us redefine the unperturbed ground-state energy as $\varepsilon/2$; we thus have,

$$H = \varepsilon\,(J_z + J + 1/2) + (1/2)\,V\,(J_+^2 + J_-^2) =$$

$$= \varepsilon(h_z + 1/2) + v/2\,(h_+^2 + h_-^2) \qquad (12)$$

When $N \to \infty$, the commutation relations between the h's are preserved. If we identify them as follows,

$$\lim_{N\to\infty} h_z = \lim_{N\to\infty} (J_z + J) = n = a^+ a; \qquad (13a)$$

$$\lim_{N\to\infty} h_+, h_- = a^+, a; \qquad (13b)$$

the a-operators commute like the generators of the H_4-algebra of the harmonic oscillator and we obtain,

$$H = \varepsilon\left[n + 1/2 + \chi\,(a^{+2} + a^2)\right] = \\ = 1/2\;\varepsilon\,(1 + \chi)\,x^2 + 1/2\;\varepsilon\,(1 - \chi)\,p^2 \qquad (14)$$

with $\chi = v/\varepsilon$ and $a, a^+ = 1/\sqrt{2}\,(x \pm i\,p)$. The Hamiltonian H can be diagonalized in the form,

$$H = \omega_0\,(A^+A + 1/2) \quad \text{with } \omega_0 = \varepsilon\sqrt{1 - \chi^2}\,; \qquad (15a)$$

with

$$A, A^+ = 1/\sqrt{2}\,(x/\sigma_0 \pm i\,\sigma_0\,p); \qquad (15b)$$

and

$$\sigma_0 = \left(\frac{1-\chi}{1+\chi}\right)^{1/4}. \qquad (15c)$$

The phase transition for $|\chi| = 1$ accounts for a singular point in the operators A, A^+ due to the fact that either the effective stiffness vanishes ($\chi < 0$) or the effective inertia becomes infinite ($\chi > 0$). Beyond the transition the frequency is an imaginary one and the Hamiltonian becomes an unbounded operator (particle in presence of a quadratic barrier for attractive fermion interaction). We can write,

$$H = \omega / 2 \quad (G_1G_2 + G_2G_1) \tag{16a}$$

with
$$G_1 = A/\sqrt{i}; \tag{16b}$$

$$\omega = \sqrt{\chi^2 - 1}\ ; \tag{16c}$$

and
$$\left[G_2, G_1\right] = i. \tag{16d}$$

Here G_1 and G_2 are related to x and p through a canonical transformation and satisfy,

$$\left[H, G_1\right] = - i \omega G_1; \tag{17a}$$

$$\left[H, G_2\right] = i \omega G_2. \tag{17b}$$

One sees that the operator $M = G_2^2$ verifies,

i) $M = M^+$; (18a)

ii) $M \geqslant 0$;

iii) $i\left[H, M\right] = -2 \omega M \leqslant 0$ (18c)

Then it could be regarded as a Lyapounov operator. In addition, if we choose,

$$T = -1/\omega \ \ln G_2, \tag{19}$$

it is easily seen that T behave like a time operator in the sence of eqs. (6) and (7).

Thus the infinite-fermion system lying above the ground-state phase transition can be expected to display some features of irreversible evolution. This can be examined in the diagonal representation of G_2; let

$$G_2 \,|\lambda\rangle = \lambda \,|\lambda\rangle \tag{20}$$

One finds,

$$\psi_\lambda(x) = \langle x | \lambda \rangle \quad \propto \quad \exp\left(i\, x^2/2\sigma^2 - i\sqrt{2}\,\lambda\, x/\sigma\right), \tag{21}$$

and

$$H(\lambda) = \omega/2 \left[\lambda^2 + i\left(\lambda \frac{\partial}{\partial\lambda} + \frac{\partial}{\partial\lambda}\lambda\right)\right] \tag{22}$$

The Schrödinger equation in λ-space can be straightforwardly integrated and gives

$$\psi(\lambda, t) = \psi(\lambda(t), 0) \; \exp\left(-i\, \lambda^2(t)/2\right) \; (\lambda/\lambda(t))^{1/2} \tag{23a}$$

where
$$\lambda(t) = \lambda \exp(-\omega t) \tag{23b}$$

and the moments,

$$\langle \lambda^n(t) \rangle = \exp(-\omega t) \langle \lambda^n \rangle . \tag{23c}$$

We realize that the time-evolution concentrates wave packets around the value $\lambda = 0$ (maximum "entropy"). Consequently, the asymptotic solutions of the equation of motion are, in the x-representation, the eigenfunctions of the Lyapounov operator for $\lambda = 0$ eigenvalue;

$$\psi(x, t) \xrightarrow[t\to\infty]{} \begin{cases} \psi_0(x) \propto \exp\left(i\, x^2/2\sigma^2\right) & \text{if } \psi(t=0) \text{ is even;} \\ x\, \psi_0(x) & \text{if } \psi(t=0) \text{ is odd.} \end{cases} \tag{24}$$

If expanded with respect to the harmonic oscillator basis $|n\rangle$, the corresponding coefficients of the wave functions (24) read, with $\varphi = \arg(\varepsilon - i\sigma^2)$,

$$(c_n/c_0)_{\text{even}} = \exp(i\, n\, \varphi) \sqrt{\frac{\Gamma\left(\frac{n+1}{2}\right)}{\Gamma\left(\frac{n}{2}+1\right)}} \underset{n\to\infty}{\propto} \exp(i n \varphi)\, n^{-1/4}$$

and

$$(c_n/c_1)_{Odd} = \exp(i\,n\,\varphi)\sqrt{\frac{\Gamma\left(\frac{n}{2}+1\right)}{\Gamma\left(\frac{n+1}{2}\right)}} \underset{n\to\infty}{\propto} \exp(in\varphi)\,n^{1/4} \quad (25)$$

In either case the asymptotic state spreads over all eigenstates of the harmonic oscillator, with amplitudes that tend to favour the population of the unperturbed ground-state ($n = 0$) in the even case, and the highest excited states ($n \to \infty$) in the odd one.

With this example we find that the existence of a regime of Hamiltonian dynamics characterized by the availability of irreversibility operators, is possible in view of the spectral properties of the structural operators of the many-fermion system. We must be aware that in this case we are not in presence of "true" kinetic irreversibility, indeed, all the equations of motion are time-reversal invariant. With this in mind, we can anyway set a thermodynamic analogy and consider that the final stage of the time evolution of the many-body system can be related to a maximum "entropy" λ^{-1}, as expressed by the spreading of the wave function over the microscopic configurations compatible with the constraints.

We must observe as well that the final wave function of the evolving system is not that of the true ground-state; we are considering here a closed system whose evolution is in the sense of maximizing entropy, rather than of minimizing energy. The latter evolution will be that performed by an open system coupled to a heat reservoir R at zero temperature. Some clues for this study could be given, in close analogy with the theories of generalized master equations [7-8], in what follows.

Dispersion relation for a quantal open system.

Let us assume that the total Hamiltonian for the composite system (object of interest S plus reservoir R) is

$$H = H_S + H_R + H_{RS}\,; \quad (26)$$

and that at any time t the total wave function can be decomposed as

$$\Psi = \Psi_0 + \Psi_c = \varphi_S \varphi_R + \Psi_c \qquad (27)$$

with $\langle \Psi_0 | \Psi_c \rangle = 0$. The "vacuum" or uncorrelated part Ψ_0 can be obtained out of a projection,

$$\Psi_0 = P \ \Psi \qquad = |\varphi_R\rangle\langle\varphi_R| \ \Psi . \qquad (28)$$

In this case φ_R represents an unspecified reference state that is assumed not to vary in time and selected so as to yield $\langle \varphi_R | \ H_{RS} | \varphi_R \rangle = 0$. If we let $Q = 1 - P$, $PQ = QP = 0$ and define $H_{00} = P H P$, $H_{0C} = P H Q$, etc, we can find the P and Q-projections of Schrödinger equation,

$$i \dot{\Psi}_0 = H_{00} \Psi_0 + H_{0C} \Psi_C \qquad (29a)$$

$$i \dot{\Psi}_C = H_{C0} \Psi_0 + H_{CC} \Psi_C \qquad (29b)$$

Eq. (29b) can be formally integrated and substituted into eq. (29a), giving

$$i \dot{\Psi}_0 = H_{00} \Psi_0 + H_{0C} \ \exp(-i H_{CC} \ t) \Psi_C(0) - i \int_0^t d\tau \, \Psi(\tau) \, \Psi_0(t-\tau) \qquad (30)$$

where the "collision" operator

$$\Psi(t) = H_{0C} \exp(-i H_{CC} \ t) \ H_{C0}, \qquad (31)$$

plays the role of the memory kernel that links the current $\Psi_0(t)$ to its past values. If we further assume the existence of a "kinetic" regime, i.e., of a regime that extends in time beyond the decay of the initial correlations (vanishing of the inhomogeneous term in (30)) we can write the equation of motion

$$i \dot{\Psi}_0(t) = \mathcal{H} \ \Psi_0 (t) \qquad (32)$$

where the effective generator of evolution satisfies the integral equation

$$\mathcal{H} = H_{00} - i \int_0^{\infty} dt\, \Psi(t) \exp(i \mathcal{H} t) \qquad (33)$$

It can be seen that this $\mathcal{H}$ is not hermitian if Ψ possesses a time-odd term, thus giving rise to the line-width or finite life-time of the system of interest, whose equation of motion can be found taking the scalar product of (32) with $\langle \varphi_R |$

$$i \dot{\psi}_S(t) = \mathcal{H}_S \psi_S(t) \qquad (34a)$$

$$\mathcal{H}_S = H_S + \langle \varphi_R | H_R | \varphi_R \rangle - i \int_0^{\infty} \langle \varphi_R | \Psi(t) | \varphi_R \rangle \exp(i \mathcal{H}_S t)\, dt \qquad (34b)$$

In this way we find that the spectral problem for $\mathcal{H}_S$ can be solved via the dispersion equation for $\mathcal{E}_S = \mathcal{E} + i\ \tau^{-1}$,

$$\mathcal{E}_S = \mathcal{E}_S^0 + H_R - \langle \varphi_S \varphi_R | H_{OC} \frac{1}{H_{CC} - \mathcal{E}_S} H_{CO} | \varphi_S \varphi_R \rangle . \qquad (35)$$

References.

1) C. George, F. Henin, F. Mayné and I. Prigogine, Hadronic J. 1 (1978) 52o.

2) I. Prigogine, C. George, F. Henin and L.Rosenfeld, Chem.Scr. 4 (1973) 5.

3) B. Misra, I. Prigogine and M. Courbage, Proc.Natl.Acad.Sci. USA 76 (1979) 4768.

4) E.S. Hernândez and H.G. Solari, (1980) to be published.

5) H.J. Lipkin, N. Meshkov and A.J. Glick, Nucl. Phys. 62 (1965) 188.

6) R. Gilmore, Ann. Phys. 74 (1972) 391.

7) F. Haake, Springer Tracts in Modern Physics, vol 66 (1973).

8) C. Dorso and E.S. Hernândez, (198o) to be published.

We appreciate the valuable collaboration of Drs. H.G. Solari and C. Dorso in the development of the subjects here presented.

NON TRIVIAL HARTREE-FOCK STUDIES IN THE THERMODYNAMIC LIMIT

G. Gutiérrez, A. Plastino
Physics Department, National University, C.C. 67,
La Plata (1900), Argentina

and

M. de Llano
Instituto de Fisica, Universidad Nacional Autónoma
de México, Apd. 364, México 20 D.F, México.

1- Introduction

The exact solution, analytically or numerically, of an N-fermion problem with realistic interparticle interactions is apparently far away.

Two general microscopic approaches to the problem are available: (1) variational[1], of the Jastrow-type, Fermi-hypernetted-chain approximation method, etc, and (2) perturbation theory[2], based mainly on diagrammatic methods of the "ladder", "ring", or other infinite partial summations.

Both of these general approaches begin with an assumed one-particle (or one quasi-particle) state, about which one then perturbs in one manner or another. This non-perturbed ground vacuum state can be chosen in several ways, i.e., a conjecture in this respect must be made. The usual such state is a single Slater determinant of plane-wave one-particle "orbitals" with occupied k-vectors spanning a spherical Fermi sea in k-space (the "Fermi sphere" of radius k_F, the Fermi vector, which is related to the particle density).

However, the "ideal" unperturbed, one-particle hamiltonian is the one that leads to the Hartree-Foch problem. This is non-linear (highly so, for N fermions) and so gives rise to not one, but several solutions which may display totally different qualitative properties. The plane-waves orbitals constitute just one of these solutions[3] (a trivial one, indeed). The investigation of the properties that alternative HF solutions may display may thus prove to be of interest.

The HF problem is only a first step, but a very important one in dealing with the full many-body problem, i.e., with correlations. The size, as well as the nature of these correlations will depend critically on the HF (unperturbed) state presupposed, not to speak of the convergence or not of the calculated perturbed state to the fully interacting ground state.

The purpose of these notes is to present a brief review of some work recently performed concerning non-plane-wave (called sometimes "non-trivial") HF states in the thermodynamic limit.

2- Heuristic Hartree-Fock solutions

2.1- One dimension

As is often the case, the one-dimensional case is quite useful in illustrating, in a simple fashion, how one is to deal with the non-trivial HF states we are interested in. Consider a system of $N\gg 1$ fermions (spin 1/2) in one dimension, placed inside a box of length L, and apply periodic boundary conditions. We shall concern ourselves with a set of single-particle, orthonormal orbitals $\phi_{k_i\sigma_i}(x,\sigma_{3_j})$ that satisfy the HF equations for orbitals occupied in the determinant

$$\Phi_0=(N!)^{-1/2}\ \det\{\phi_{k_i\sigma_i}(x_i,\sigma_{3_i})\} \qquad (1)$$

For the sake of a lighter notation, we shall use italic "k" to stand for k,σ and whatever additional quantum number may be relevant, and proceed in the same fashion with the coordinates.

Several different choices for the ϕ_k, that satisfy the (matrix) HF equations, have been recently shown[4] to yield more binding per particle than the PW orbitals in the case of a classical one-dimensional problem (the so-called attractive-delta gas one[5]).

We have a)

$$\begin{aligned} &\phi_k(x)= C\ e^{ikx}\ \{1+\alpha(\cos qx)^{n_1}\ e^{in_2qx}\}^{n_3},\quad \alpha \text{ complex} \\ &-k_F<k<k_F\ ,\quad q\equiv 2k_Fm\quad (m=\pm1,\ \pm2,\ \ldots) \qquad (2)\\ &n_1,n_2,n_3=0,1,2,\ldots \end{aligned}$$

where C is a normalization constant and m and α are two parameters (to be eventually varied). Subfamilies of special interest[4] are

$$(n_1,n_2,n_3) = \begin{cases} (0,0,0) & \text{Plane waves (PW)} \\ (0,1,n) & \text{"density wave" (DW-n)} \\ (1,0,n) & \text{"density standing wave" (DSW-n),} \end{cases} \qquad (3)$$

with $n=0,1,2,\ldots$

b) Overhauser orbitals[4,6]

$$\phi_k(x) = L^{-1/2}\,(u_k+v_k\,e^{-iqx})\,e^{ikx}$$

$$u_k^2+v_k^2=1\ ,\quad -k_F<k<k_F \qquad (4)$$

$$|q|>2k_F\ ,\quad \mathrm{sign}(k)=\mathrm{sign}(q).$$

c) Exponential orbitals[4]

$$\phi_k(x) = \{LI_0(2\alpha)\}^{-1/2}\,e^{ikx}\,e^{\alpha\cos qx}$$

$$-k_F<k<k_F\ ,\quad q>2k_F\ ,\quad \alpha>0\ , \qquad (5)$$

where I_0 is a modified Bessel function.

Except for the PW case, long-range order, i.e., perfect peridicity, appears in the single-particle local density

$$\rho(x) = \sum_{k(occ)} |\phi_k(x)|^2\ , \qquad (6)$$

which, for the PW case just gives $\rho(x)=\rho_0=2k_F/\pi$, where $2k_F$ is the length of the Fermi sea.

2.2- Three dimensions

In this case each of the non-trivial HF orbitals of the previous subsection is now a product of x,y and z orbitals. In the 3-dimensional attractive delta gas case, they all yield more binding per particle than the PW state (at least for some value of the order parameters α or q ;q yields, in all cases, a minimum when it equals $2k_F$). These orbitals have also been employed in nuclear matter calculations, using the Skyrme force as parametrized by Vautherin and Brink[7]. In this way, it has been shown that these orbitals, that give now a periodic structure of α particles, produce states that have lower energy than those corresponding to homogeneous nuclear matter (PW) at subnuclear densities[8] within the range 0.04 to 0.9 fm^{-3}. At such a density waves spontaneously appear, a feature which gives us some insight into the early stages of cluster formation in nuclei, and may be connected with the likelihood of finding α-particle clusters on the surface of the nuclei, a location where the density is expected to be low.

Very recently[9] semi-realistic nuclear potentials, like the so-called Bethe homework V_1 and V_2 potentials[10] have been used with non-plane-waves HF orbitals, similar to those of Ref.11, but sligtly generalized. The "corrugated-sheet-density-wave" (CSDW) as well as the "corrugated-sheet-spin-density-wave" (CSSDW) of Ref.11, but generalized, were used. The former has spin-up and spin-down populations overlapping spatially, while the latter displaces them by half a wavelength,i.e., the spin-up maximize where the spin-down minimize. The net density is not constant in space, i.e., there is still a density oscillation. Both neutron (2 species) as well as nuclear (4 species of particles) matter were investigated. The CSSDW produces a small but definite energy lowering (with respect to PW HF) at densities of about 0.025 fm^{-3} and above, i.e., including nuclear saturation density (≈0.17 fm^{-3}). No energy gain is obtained with the CSDW HF orbitals.

3- A systematic HF procedure in the thermodynamic limit

When one employs any of the HF orbitals of the previous section, one is making a definite, heuristic conjecture as to what the non-perturbed ground vacuum state is. Quite recently a systematic approach has been developed[12] in order to look for non-trivial HF states, which uses the HF equations themselves

$$\langle\phi_{k'}|t|\phi_k\rangle+\sum_{l(occ)}\langle\phi_{k'}\phi_l|v_{12}|\phi_k\phi_l-\phi_l\phi_k\rangle=\varepsilon_k\,\delta_{kk'} \qquad (7)$$

as a guide. In particular, and in view of the results commented upon in Sec. 2, we will look for periodic solutions that are able to give rise to a "lattice" structure with vectors $\vec{a}_1,\vec{a}_2$ and $\vec{a}_3$. This lattice should be invariant under translations $\vec{R}=n_1\vec{a}_1+n_2\vec{a}_2+n_3\vec{a}_3$, so that

$$T_{\vec{R}}|\phi_{\vec{k}}\rangle=e^{i\vec{k}\vec{R}}\,|\phi_{\vec{k}}\rangle\ , \qquad (8)$$

where $T_{\vec{R}}$ represents a unitary operator that commutes both with the total Hamiltonian H and with the Hartree-Fock hamiltonian h. Moreover, $T_{\vec{R}}$ should not affect the particular occupation that one may chose for a given Slater determinant (cf.Eq. (1)).

Those solutions of (7) that are also eigenstates of $T_{\vec{R}}$ will be Bloch functions

$$\phi_{\vec{k}}(\vec{x}) = e^{i\vec{k}.\vec{x}}\, u_{\vec{k}}(\vec{x}) \ , \tag{9}$$

with $u_{\vec{k}}(\vec{x}+\vec{R}) = u_{\vec{k}}(\vec{x})$, i.e., u has the same periodicity as the lattice. Expanding ϕ in plane waves

$$|\phi_{\vec{k},n}\rangle = \sum_{\vec{G}} C_{n,\vec{G}}(\vec{k})\ |\vec{k}+\vec{G}\rangle \ , \tag{10}$$

where $|\vec{k}+\vec{G}\rangle = e^{i(\vec{k}+\vec{G}).\vec{x}}$, $\vec{G}$ is a reciprocal lattice vector and $\vec{k}$ belongs to the first Brillouin zone, our problem reduces to that of employing the HF equations (7) in order to find the coefficients C in Eq.(10). The index n labels the rows of the unitary matrix corresponding to the transformation (10).

The approach can then be summarized as follows: i)one starts with a complete, orthonormal set, the PW, that satisfies the HF equations, ii) subject the set formed with a) $e^{i\vec{k}.\vec{x}}$ (with $\vec{k}$ in the first Brillouin zone) and b) (N-1) vectors of the reciprocal lattice $\vec{G}$ to a unitary transformation, given by the coefficients $C_{n,\vec{G}}(\vec{k})$ of Eq.(10), in order to obtain a new (complete, if plane waves are added beyond the N-th zone) orthonormal set $|\phi_{\vec{k},n}\rangle$. iii) Select the unitary trasformation in such a way that the $|\phi_{\vec{k},n}\rangle$ constitute, within the subspace reached by the trasformation, a HF set.

Introduction of the expansion (10) into the HF equations (7) leads to the eigenvalue problem

$$\sum_{\vec{G}} \langle\vec{k}+\vec{G}'|h|\vec{k}+\vec{G}\rangle\ C_{n,\vec{G}}(\vec{k}) = \varepsilon_n(\vec{k})\ C_{n,\vec{G}'}(\vec{k}) \ , \tag{11}$$

where

$$\langle\vec{k}+\vec{G}|h|\vec{k}+\vec{G}'\rangle = \langle\vec{k}+\vec{G}|t|\vec{k}+\vec{G}'\rangle + \sum_{lm(occ)} \langle\vec{k}+\vec{G},\phi_{lm}|v_{12}|\vec{k}+\vec{G}',\phi_{lm}\rangle \ , \tag{12}$$

where the matrix element of v_{12} is assumed to be a properly antisymmetrized one. The set (11) present us with a non-linear problem entirely similar to the one proposed in the usual Roothaan's scheme for the solution of the HF equations, when one deals with a finite number of fermions. The solution is obtained in the same fashion, namely, by an iterative diagonalization of the Hartree-Fock hamiltonian h. Enlarging the single-particle basis in the finite case has here its counterpart in taking a larger value N of vectors of the reciprocal lattice.

The method described in this section[12] has been applied in Ref. 12 to the one-dimensional delta gas. The results are displayed in the following figure

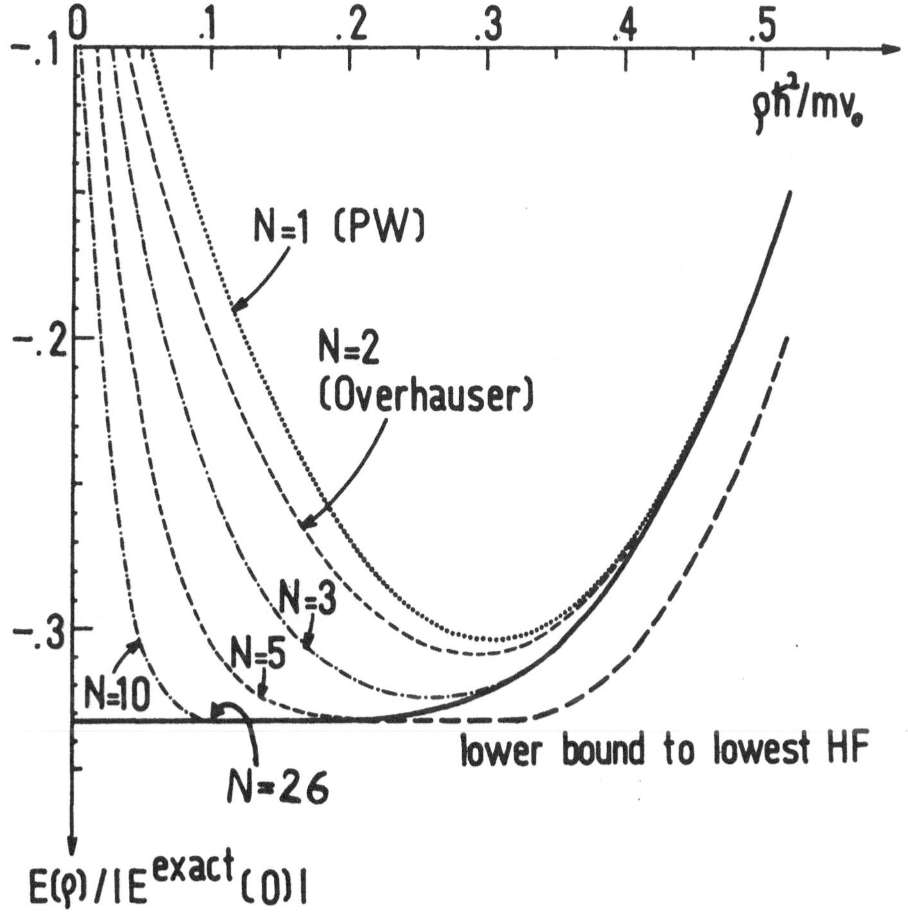

Fig. 1: Results for the one-dimensional attractive delta gas case. N denotes the number of vectors of the recipracal lattice considered in the HF calculation. The zero density limit can be obtained analytically whith the method of section 3.

Notice that by taking N large enough, a solution is found which satisfies the monotonicity of the ground state energy per particle as a function of the density[13]. To our knowledge, this is the only instance in which such requirement is satisfied in a microscopic calculation in the themodynamic limit. An application of the techniques described here to neutron matter in connection with the Bethe homework problem[14] is illustrated in Fig. 2. Notice that the approach of Ref. 12 is able, already for N=3, to bind neutron matter for the V_1 homework potential, while both the PW and Overhauser HF solutions yield rather large, positive energies.

4-Conclusions

The non-relativistic microscopic calculations of the ground state properties of a many-body system can be formulated via modern many-body diagramatic perturbation theory. However, it is to be emphasided that how the total hamiltonian is split into "unperturbed" and "perturbed" parts may be crucial for a meaningful description of a system subject to phase transitions in its ground states. The simplest unperturbed "vacuum state" is a PW determinant (for fermions) occupied in such a way that the kinetic energy expectation is minimized. Non-plane wave solutions to the HF equations exist, however, that may constitute a better vacuum in some circumstances. Some of these situations have been reviewed here. In addition, we have described a systematic approach (in contraposition to the usual heuristic one) that enables one to obtain non-trivial HF solutions in the thermodynamic limit. these solutions arise, as in Roothan's, by recourse to an iterative process in which a matrix is repeatedly diagonalized until selfconsistency is reached.

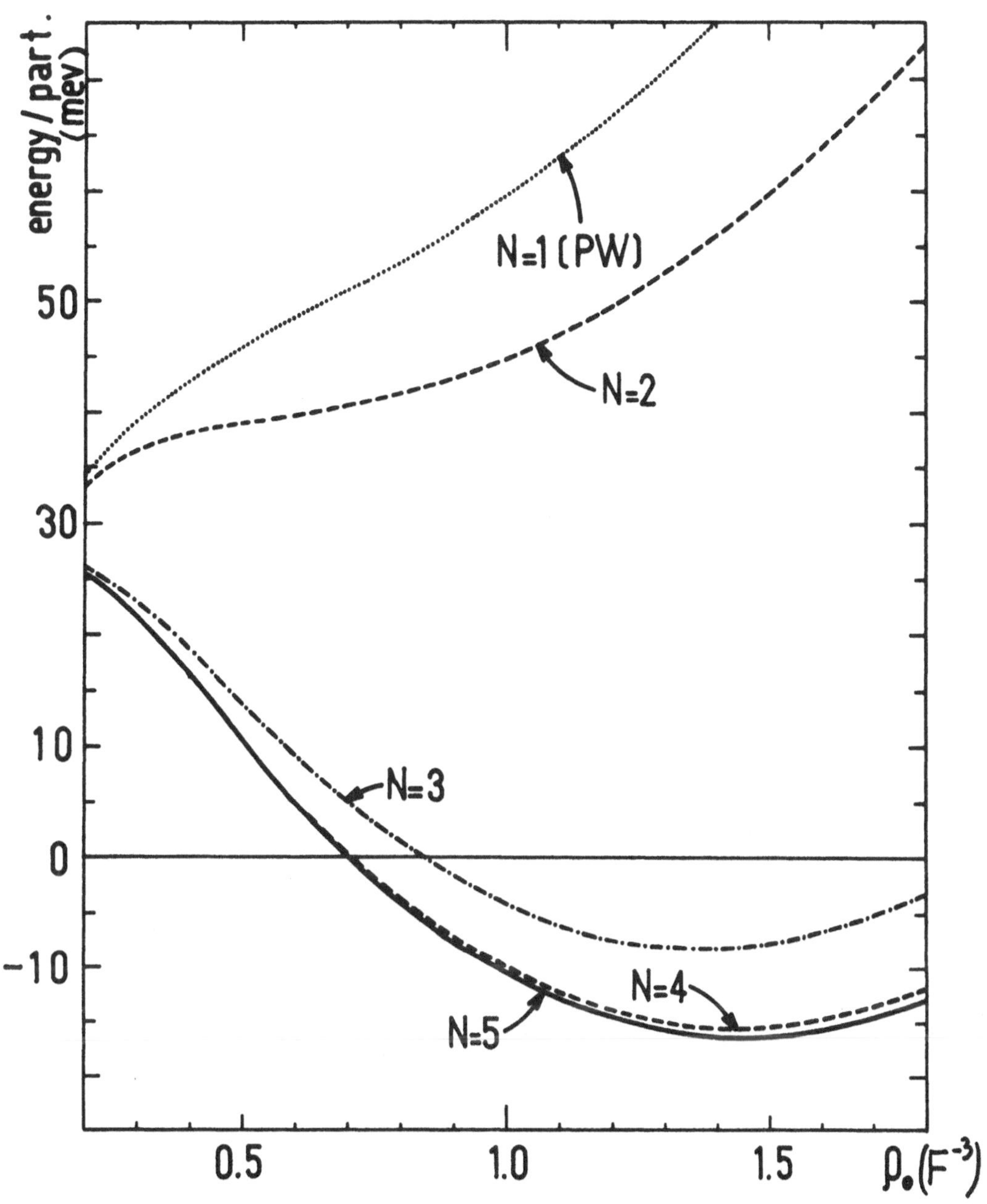

Fig. 2: HF energy per particle (Mev) in a neutron matter calculation with the Bethe homework V_1. N denotes the size of the transformation (10), i.e. the number of vectors of the reciprocal lattice considered.

References

1- E. Feenberg, Theory of Quantum Fluids (Academic, New York, 1969); J. W. Clark, Prog.Part.Nucl.Phys. (to be published)
2- B. Day, Rev.Mod.Phys. 39 (1967) 719.
3- A. L. Fetter, J. D. Walecka, Quantum Theory of Many-particle Systems (Mc. Graw-Hill, N.Y., 1971).
4- V. C. Aguilera, M. de Llano, S. Peltier, A. Plastino, Phys. Rev. A15 (1977) 1256; L. Dohnert, M. de Llano, A. Plastino, Phys.Rev. A17 (1978) 767.
5- J. B. Mc Guire, J.Math.Phys. 5 (1964) 622.
6- A. W. Overhauser, Phys.Rev.Lett. 4 (1960) 415.
7- D. Vautherin, D. M. Bring, Phys.Rev. C5 (1972) 626.
8- V. Aguilera et al, Phys.Rev. C16 (1977) 1642 and Phys.Rev. C16 (1977) 2081; L. Dohnert, M. de Llano,, A. Plastino, Nucl.Phys. A291 (1977) 45; O. Civitarese, A. Plastino, A. Faessler, Z.Phys. A291 (1979) 239.
9- G. Gutiérrez, A. Plastino, M. de Llano, Phys.Rev. C20 (1979) 2453.
10- B. Day, Rev.Mod.Phys. 50 (1978) 495.
11- M. de Llano, A. Plastino, Phys.Rev. A13 (1976) 1633.
12- G. Gutiérrez, A. Plastino, Ann. of Phys.(NY) (to be published).
13- E. Lieb, M. de Llano, J.Math.Phys. 19 (1978) 860.
14- G. Gutiérrez, A. Plastino (to be published).

NUCLEAR FORCES AND NUCLEAR MATTER INCLUDING PIONS AND ISOBARS

F. Coester
Argonne National Laboratory*
Argonne, IL 60439/USA

Conventional nuclear theory is based on the rather successful model which stipulates that nuclei consist of nucleons with two-body interactions. Extensive use of pion beams as probes of nuclear structure suggests a minimal extension of this model to include pion degrees of freedom explicitly in a manner which allows production and absorption of pions without dressing isolated nucleons with a pion cloud. Some properties of such models are the subject of this talk.

The parameters of the models are fitted to properties of the pion-nucleon and nucleon-nucleon systems. The nucleon is endowed with a $\frac{3}{2}\,\frac{3}{2}$ excited state, the Δ isobar, which can decay by pion emission. The pion-nucleon interaction consists of two-body potentials in addition to the Δ-decay interaction which provides the mechanism for pion production and absorption. There is no πNN vertex in the Hamiltonian and therefore no pion cloud attached to a single nucleon.

The two-body Hamiltonians are used to construct the many-body Hamiltonian which is of the general form

$$H = H_o + H_{int} \tag{1}$$

where

$$H_o = \sum_{\tau=1/2,3/2} \int d^3x\, c^\dagger(x,\tau)\left[m_\tau - m - \frac{\nabla^2}{2m_\tau}\right]c(x,\tau) + \int d^3k\, \alpha^\dagger(k)\sqrt{k^2+\mu^2}\,\alpha(k) \quad , \tag{2}$$

m is the nucleon mass, $m_{\frac{1}{2}} = m$ and

$$\begin{aligned} H_{int} = \frac{1}{4} & \sum_{\tau_1'\tau_2'} \sum_{\tau_1\tau_2} \int d^3x_1' \int d^3x_2' \int d^3x_1 \int d^3x_2 \\ & \times c^\dagger(x_1'\tau_1')c^\dagger(x_2'\tau_2')(x_2'\tau_2'x_1'\tau_1'|V|x_1\tau_1x_2\tau_2)c(x_2\tau_2)c(x_1\tau_1) \\ & + \int d^3x \int d^3y \int d^3k [\alpha^\dagger(k)c^\dagger(x,\tfrac{1}{2})(x,k|V_{\pi\Delta}|y)c(y,\tfrac{3}{2}) \\ & \quad + c^\dagger(y,\tfrac{3}{2})(y|V_{\pi\Delta}|k,x)c(x,\tfrac{1}{2})\alpha(k)] \\ & + \int d^3k' \int d^3x' \int d^3k \int d^3x\, \alpha^\dagger(k')c^\dagger(x',\tfrac{1}{2})(x',k'|V_{\pi N}|k,x)c(x,\tfrac{1}{2})\alpha(k) \quad . \end{aligned} \tag{3}$$

---

*This work was performed under the auspices of the U. S. Dept. of Energy.

The velocities of nucleons in nuclei are predominantly nonrelativistic, but the relevant velocities of the pions are not. Also, for nucleon-nucleon scattering above the pion threshold (which is used to fit the potentials) the relative velocities of the nucleons are not small compared to one. It is therefore best to construct relativistic models for the two-nucleon system[1] and expand in powers of the baryon velocities when appropriate.

In relativistic quantum mechanics Lorentz transformations as well as translations and rotations are unitary transformations of the Hilbert space of states. The generators of the infinitesimal translations $\vec{P}$, H, rotations $\vec{J}$ and Lorentz boosts $\vec{K}$ satisfy the commutation relations

$$[P_i,P_j] = [P_i,H] = [J_i,H] = 0 \quad , \tag{4}$$

$$[J_i,F_j] = i \sum_k \varepsilon_{ijk} F_k \quad , \tag{5}$$

for $\vec{F} = \vec{P}$, $\vec{J}$ or $\vec{K}$.

$$[K_j,H] = i P_j \quad , \tag{6}$$

$$[K_i,P_j] = i \delta_{ij} H \quad , \tag{7}$$

$$[K_i,K_j] = -i \sum_k \varepsilon_{ijk} K_k \quad . \tag{8}$$

If we start with a non-interacting system and add an interaction term to the Hamiltonian, then according to (6) and (7) either $\vec{K}$ or $\vec{P}$ or both must acquire interaction terms. In canonical field theories the Poincaré generators are realized by integrals over components of the energy-momentum tensor $T^{\mu\nu}(x)$,[2]

$$P^\mu = \int d^3x \, T^{o\mu}x \tag{9}$$

$$K_i = \int d^3x \, x_i T^{oo}(x) \tag{10}$$

$$J_i = \frac{1}{2} \sum_{mn} \varepsilon_{imn} \int d^3x [x^m T^{on}(x) - x^n T^{om}(x)] \tag{11}$$

In this realization relativistic invariance requires infinitely many degrees of freedom, local fields and antiparticles.[3] These features are, however, not necessary. It is possible to satisfy the commutation relations (4)-(8) by constructing operators $\vec{P}$, $\vec{X}$, $\vec{j}$ and M that satisfy the relations

$$[X_j,X_k] = [P_j,P_k] = 0 \quad , \tag{12}$$

$$[X_j,P_k] = i \delta_{jk} \quad , \tag{13}$$

$$[j_i,j_k] = i \varepsilon_{ikm} j_m \quad , \tag{14}$$

$$[j_i,\vec{P}] = [j_i,\vec{X}] = 0 \quad , \tag{15}$$

$$[\vec{j},M] = [\vec{X},M] = [\vec{P},M] = 0. \quad (16)$$

It follows that

$$H = \sqrt{\vec{P}^2+M^2} \quad , \quad (17)$$

$$\vec{K} = \frac{1}{2}(XH+HX)-\vec{j}\times\vec{P}(M+H)^{-1} \quad , \quad (18)$$

$$\vec{J} = \vec{X}\times\vec{P}+\vec{j} \quad , \quad (19)$$

satisfy the Lie algebra (4)-(8). This construction is possible for a fixed number of particles but it permits particle creation. For instance for the pion nucleon system we may write

$$M_{\pi N} = M^{o}_{\pi N}+v_{N\pi} \quad , \quad (20)$$

where $M^{o}_{\pi N}$ is the mass operator of the non-interacting system. The interaction term $v_{N\pi}$ must commute with $\vec{X}$ $\vec{P}$ and $\vec{j}$ and vanish for large separation of the nucleon and the pion. In the models considered here the operator $v_{N\pi}$ is a vertex interaction $N\pi\to\Delta$ in the P_{33} partial wave and a two-body potential in other partial waves.

The Hilbert space of states for the NN-system is

$$H = H_N \otimes H_N \oplus H_N \otimes H_\Delta \oplus H_N \otimes H_N \otimes H_\pi \quad , \quad (21)$$

where H_N, H_Δ and H_π are one-particle Hilbert spaces of the nucleon, the Δ isobar and the pion. The mass operator in this space has the general form

$$M = M^{o}+V_{o}+V'+V'' \quad , \quad (22)$$

where V_o is a two-baryon operator with matrix elements $NN\to NN$, $NN\rightleftarrows N\Delta$ and $N\Delta\to N\Delta$. The term V' is the interaction within the three-body channel and includes transitions to the $N\Delta$ channel via the $NN\pi$ vertex. The last term V" includes three-body interactions and transitions $NN\to NN\pi$ which are needed to describe off-resonance pion production.

The Betz-Lee model[4] is a particular realization of this scheme. It assumes V" = 0 and makes a nonrelativistic approximation for the $N\Delta$ and $NN\pi$ channels. All two-body interactions are parameterized in a separable form. After fitting the πN interaction to πN scattering the operator V_o is fitted to phase shifts for laboratory energies up to 800 MeV. The two nucleon mass operator so determined yields a many-body Hamiltonian of the form (1)-(3) in which all baryons are treated nonrelativistically. In the application to nuclear matter we further neglect the pion-nucleon interaction in the non-resonant partial waves.

The coupled cluster theory for homogeneous nuclear matter[5] can be easily generalized to accommodate our Hamiltonian. Let $|\Phi\rangle$ be the ground state of the non-interacting Fermi gas satisfying by definition

$$a(x)|\Phi\rangle = b(x)|\Phi\rangle = \alpha(k)|\Phi\rangle = 0 \quad (23)$$

where a(x) and b(x) are particle and hole annihilation operators. They are related to the baryon annihilation operator c(x) by

$$c(x) = a(x)+b^{\dagger}(x) \quad . \tag{24}$$

All spin and isospin variables are implied. There are no pions and no Δ's in $|\Phi>$. The eigenstate $|\Psi>$,

$$H|\Psi> = E\,|\Psi> \quad , \tag{25}$$

is written in the form

$$|\Psi> = \exp(S)|\Phi> \quad , \tag{26}$$

where by definition the operator S has the form

$$S := \sum_{n\geq 2} S_n + \sum_{n\geq 1}\sum_{m\geq 1} S_{n,m} \quad , \tag{27}$$

$$\begin{aligned} S_n := \frac{1}{(n!)^2} \int d^3x_1 \ldots \int d^3x_n \int d^3y_n \ldots \int d^3y_1 \\ a^{\dagger}(x_1)\ldots a^{\dagger}(x_n)\; b^{\dagger}(y_n)\ldots b^{\dagger}(y_1)\; S_n(x_n\ldots x_1 ; y_1\ldots y_n) \quad , \end{aligned} \tag{28}$$

$$\begin{aligned} S_{n,m} := \frac{1}{m!}\frac{1}{(n!)^2} \int d^3k_1 \ldots \int d^3k_m \int d^3x_1 \ldots \int d^3x_n \int d^3y_n \ldots \int d^3y_1 \\ \alpha^{\dagger}(k_1)\ldots\alpha^{\dagger}(k_m)\; a^{\dagger}(x_1)\ldots a^{\dagger}(x_n)\; b^{\dagger}(y_n)\ldots b^{\dagger}(y_1)\; S_{n,m}(k_1\ldots k_m, x_n\ldots x_1 ; y_1\ldots y_n) \quad . \end{aligned} \tag{29}$$

From (25) and (26) follows

$$E = <\Phi|e^{-S}\; H\; e^{S}|\Phi> \tag{30}$$

and the coupled cluster equations

$$<\Phi|b(y_1)\ldots b(y_n)\; a(x_n)\ldots a(x_1)\; e^{-S}\; H\; e^{S}|\Phi> = 0 \quad , \tag{31}$$

$$<\Phi|b(y_1)\ldots b(y_n)\; a(x_n)\; a(x_1)\; \alpha(k_1)\ldots\alpha(k_m)\; e^{-S}\; H\; e^{S}|\Phi> = 0 \tag{32}$$

for all n≥1, m≥1. It is easy to see that only S_2 contributes to the energy (30).

The following definitions are useful in bringing the coupled cluster equations into a form in which approximations can be formulated.

$$S(x:) := [a(x),S] \quad , \tag{33}$$

$$S(x,y:) := \{a(x),[a(y),S]\} \quad , \tag{34}$$

$$S(:x) := [b(x),S] \quad , \tag{35}$$

$$S(:x,y) := \{b(x),[b(y),S]\} \quad , \tag{36}$$

$$\begin{aligned} \chi(x,y) := \; & S(x,y:) + b^{\dagger}(x)\; S(y:) + S(x:)\; b^{\dagger}(y) \\ & + S(x:)S(y:) + b^{\dagger}(x)\; b^{\dagger}(y) \quad . \end{aligned} \tag{37}$$

The motivation for the definition (37) comes from the observation that

$$e^{-S}\; c(x)\; c(y)\; e^{S}|\Phi> = \chi(x,y)|\Phi> \quad . \tag{38}$$

Therefore $\chi(x,y)$ has the correlation structure of the complete eigenfunction and

$$W(x,y) := \frac{1}{2} \int d^3x' \int d^3y' (x,y|V|y',x') \, \chi(x',y') \tag{39}$$

is a well behaved operator even for singular potentials. The exact energy per particle $\mathcal{E}$ is a functional of W_2,

$$\mathcal{E} = \frac{3}{5}\frac{k_F^2}{2m} + \frac{1}{2} \int d^3p_1 \int d^3p_2 \int d^3p_1' \, W_2(p_1',p_2:p_2,p_1) \quad , \tag{40}$$

where $W_2(x,y)$ is that part of $W(x,y)$ that contains exactly two hole creation operators, and the notation defined in (36 is used. The coupled cluster equations (31) and (32) can be written in the form

$$[H_1,S] + \mathcal{V}^o + \mathcal{V}' + \mathcal{V}_\pi^{(+)} + \mathcal{V}_\pi^{(-)} = 0 \quad , \tag{41}$$

where

$$\begin{aligned} H_1 := & \int d^3k \, \alpha^\dagger(k) \sqrt{k^2+\mu^2} \, \alpha(k) + \int d^3p \sum_\tau a^\dagger(p,\tau) a(p,\tau)(m_\tau - m + \frac{p^2}{2m_\tau}) \\ & - \int d^3p \, b^\dagger(p) b(p) [\frac{p^2}{2m} + \int d^3p' \int d^3p'' \, W_2(p',p'':p'',p)] \quad , \end{aligned} \tag{42}$$

$$\mathcal{V}^o := \int d^3x \int d^3y \, a^\dagger(y) \, a^\dagger(x) \, W(x,y) \quad , \tag{43}$$

$$\begin{aligned} \mathcal{V}' := & \int d^3x \int d^3p \, a^\dagger(x) \{W(p,x:p) + S(:p) W(p,x)\} \\ & + \frac{1}{2} \int d^3p_1 \int d^3p_2 \, W(p_2,p_1:p_1,p_2) \\ & + \int d^3p \int d^3p' \, S(:p)[W(p',p:p') - W_2(p',p:p')] \\ & + \frac{1}{2} \int d^3p_1 \int d^3p_2 [S(:p_1)S(:p_2) + S(:p_1,p_2)] W(p_2,p_1) \quad , \end{aligned} \tag{44}$$

$$\begin{aligned} \mathcal{V}_\pi^{(+)} := & \int d^3x \int d^3k \int d^3z \, a^\dagger(x) \, \alpha^\dagger(k)(k,x|V_{\pi\Delta}|z) S(z:) \\ & + \int d^3k \int d^3z \int d^3p \, \alpha^\dagger(k)(k,p|V_{\pi\Delta}|z)[S(:p)S(z:) + S(z:p)] \quad , \end{aligned} \tag{45}$$

and

$$\mathcal{V}_\pi^{(-)} = \int d^3z \int d^3k \int d^3x \, a^\dagger(z)(z|V_{\pi\Delta}|x,k)\{S(k;x:) + S(k;)[S(x;) + b^\dagger(x)]\}. \tag{46}$$

The two-body potential V can be eliminated from the coupled cluster equations without approximation in favor of the reaction matrix G defined by

$$G := (1 + V \frac{Q}{e})^{-1} V \quad , \tag{47}$$

where the operator e is defined by

$$e \, S := [H_1,S] \tag{48}$$

and Q is the Pauli projection operator. We have then

$$W(x,y) = G\{\phi(x,y) - \frac{Q}{e}\,[U(x,y) + \,'(x,y:) + \mathcal{V}_\pi^{(+)}(x,y:) + \mathcal{V}_\pi^{(-)}(x,y:)]\} \tag{49}$$

and

$$S(x,y:) = -\frac{Q}{e}\,\{G\,\phi(x,y) + (1-G\,\frac{Q}{e})[U(x,y) + \mathcal{V}'(x,y) + \mathcal{V}_\pi^{(+)}(x,y:) + \mathcal{V}_\pi^{(-)}(x,y:)]\} \tag{50}$$

where $\phi(x,y)$ and $U(x,y)$ are defined by

$$\phi(x,y) := \chi(x,y) - S(x,y:) \tag{51}$$

and

$$U(x,y) := \mathcal{V}^{o}(x,y:) - W(x,y) \tag{52}$$

Note that $U_2(x,y) \equiv 0$ and $\phi_2(x,y) \equiv b^{\dagger}(x)\, b^{\dagger}(y)$. By inspection of Eqs. (49) and (50) we see that all components $m \neq 0$, and $n > 2$ are satisfied if S vanishes. We can therefore write the exact equation for W_2 and S_2 in the form derived approximately by Day[6]

$$W_2(x,y) = G\, b^{\dagger}(x)\, b^{\dagger}(y) - G\,\frac{Q}{e}\,\mathcal{M}\, S_2(x,y:) \tag{53}$$

$$S_2(x,y:) = -\frac{Q}{e}\,\{G\, b^{\dagger}(x)\, b^{\dagger}(y) + (1-G\,\frac{Q}{e})\,\mathcal{M}\, S_2(x,y:)\} \tag{54}$$

where $\mathcal{M}$ is a functional of S_2 which must be determined by approximate solution of the other equations.

The basis for the standard approximation is the observation that the integral over the hole momenta can be estimated by evaluating the integrand at a representative point and multiplying by the volume of the Fermi sphere. Formally terms can thus be classified by powers of the density. In the absence of pions the leading approximation is = 0 and the nonvanishing contributions of involve at least three-nucleon correlations. It is known that they are not small.[7]

Equations (53) and (54) are equivalent to

$$W_2(x,y) = V[b^{\dagger}(x)\, b^{\dagger}(y) + S_2(x,y:)] \tag{55}$$

$$S_2(x,y:) = -\frac{Q}{e}[W_2(x,y) + \mathcal{M}\, S_2(x,y:)] \quad . \tag{56}$$

The pion contributes to $\mathcal{M}$ in lowest order an off-shell Δ-self energy and an effective pion-exchange potential in the NΔ channel which includes NN interactions in the presence of the pion. The binding energy of nuclear matter has been calculated in that approximation using the Reid potential in the T = 0 partial waves.[8] As seen in Fig. 1 the saturation point falls on the line formed by other lowest order calculations. All the interesting pion effects involve three nucleons. They have not yet been calculated, but they could easily be sufficient to give a saturation point near the empirical region.

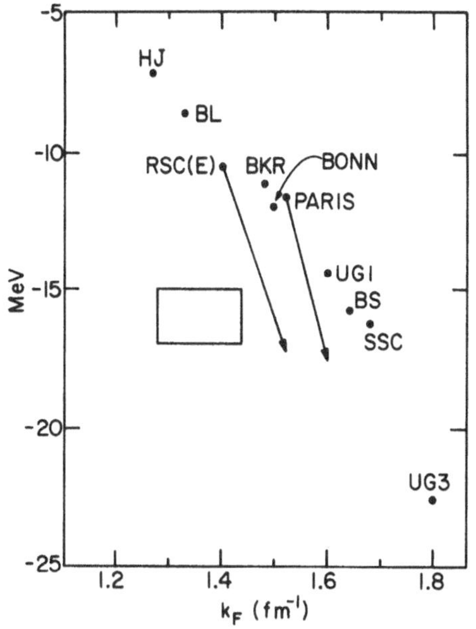

Fig. 1. Saturation points for various potentials in lowest order. The point marked BL is the result for the Betz-Lee model. The arrows give Day's results for the Reid and Paris potentials. See ref. 7.

References

1. M. Betz and F. Coester, Phys. Rev. C 21, 2505 (1980).
2. J. Schwinger, Phys. Rev. 127, 324 (1962).
3. Of course, the presence of some of the features required by relativistic field theories does not necessarily make a theory relativistically invariant. See for instance C. Mahaux, proceedings of this conference.
4. M. Betz and T.-S. H. Lee, Phys. Rev. C 23, 375 (1981).
5. F. Coester, Lectures in Theoretical Physics , K. T. Mahanthappa and W. E. Brittin ed. Vol. XIB, p. 157 (Gordon and Breach, New York, (1969); H. Kümmel, K. H. Lührmann and J. G. Zabolitzky, Phys. Rep. 36C, 1 (1978).
6. B. D. Day, in Proceedings of the International School of Physics "Enrico Fermi," Varenna, 1980.
7. B. D. Day, proceedings of this conference.
8. T.-S. H. Lee, private communication.

RECENT DEVELOPMENTS IN THE BONN POTENTIAL AND ITS APPLICATION IN NUCLEAR MATTER

K. Holinde
Institut für Theoretische Kernphysik der Universität Bonn
Nußallee 14-16, D-5300 Bonn, W.-Germany

It is argued that a consistent description of light and heavy nuclei requires an explicit treatment of the 2π-exchange contribution to the NN interaction providing the intermediate-range attraction. A corresponding model is presented, which contains also explicit 3π--exchange contributions (like e. g. non-iterative (π,ρ)-exchange). Such processes influence the medium-range part of the tensor force, which plays a decisive role in nuclear structure. The starting point of the underlying scheme is a field-theoretic Hamiltonian containing suitable meson-nucleon-nucleon and meson-nucleon-isobar vertices.

The consequences of such an explicit description for the behaviour of the 2π-exchange contribution in the medium is discussed, taking as an example the iterative Δ-isobar diagrams.

INTRODUCTION

In contrast to earlier belief there is now strong indication that a purely phenomenological nucleon-nucleon (NN) interaction cannot provide a consistent description of light and heavy nuclei. For example, the Reid soft-core (RSC) potential [1] strongly underbinds the triton and O^{16}, but slightly overbinds nuclear matter. Thus, a phenomenological potential automatically predicts relatively too much binding for higher densities. Of course, this implies that the predicted saturation density is always too high.

We stress here that this wrong tendency completely remains if we apply usual mesontheoretic potentials like e. g. the dispersion--theoretic Paris-potential [2] or one-boson-exchange (OBE) potentials. Of course, the numbers depend sensitively on the amount of tensor force: for example, one OBE-version of the Bonn potential (HM2 [3]), with a deuteron D-state probability p_D of 4.3 %, only slightly underbinds the triton and O^{16}, but strongly overbinds nuclear matter (by as much as 10 MeV).

In fact, the tensor force is probably considerably smaller than one thought before. Namely, strong ρ-exchange, some πNN vertex structure and nonlocality effects point to an appreciable suppression of the tensor force in the medium-range region, leading to values for p_D not larger than 5 %. Therefore, we consider the RSC--potential not to be realistic potential inspite of the fact that it fits the NN scattering phase shifts satisfactorily. Let me note that a partial omission of nonlocalities, which is usually performed in mesontheoretic r-space models in order to obtain an easily handable analytic expression, increases p_D by 10 - 20 %, leaving the quadrupole moment unaltered [4]. We feel this to be the reason why most r-space potential models have a larger medium-range tensor force than models which stay in momentum space. In other words, keeping the full nonlocality structure makes it possible to predict a large deuteron quadrupole moment (which seems to be required by experiment) and in the same time a small D-state probability.

Thus, realistic NN potentials ($p_D \simeq 4 - 5$ %), either phenomenological or derived from meson theory in the conventional way, give more or less correct results for light nuclei, but will surely strongly overbind nuclear matter, i. e. such a treatment obviously breaks down for higher densities. Although one might be tempted to consider this fact as a strong indication for the appearance of quark degrees of freedom in nuclear physics (which, after all, should show up at sufficiently high density) we strongly feel that one can get rid of this deficiency with the help of an extended, physically more consistent treatment than the conventional one, however, still within the mesontheoretic framework. In other words, we are deeply convinced that, at least for the binding energy problem in nuclear physics, i. e. for not too extreme situations, meson theory (to be considered as an effective theory, of course) is necessary, but also sufficient.

INTERMEDIATE-RANGE ATTRACTION

The problem of the conventional procedure (dispersion-theoretic or of one-boson-exchange type) lies in the inconsistent treatment of the 2π-exchange contribution to the NN interaction in the medium. It can be schematically split up into the following types of processes:

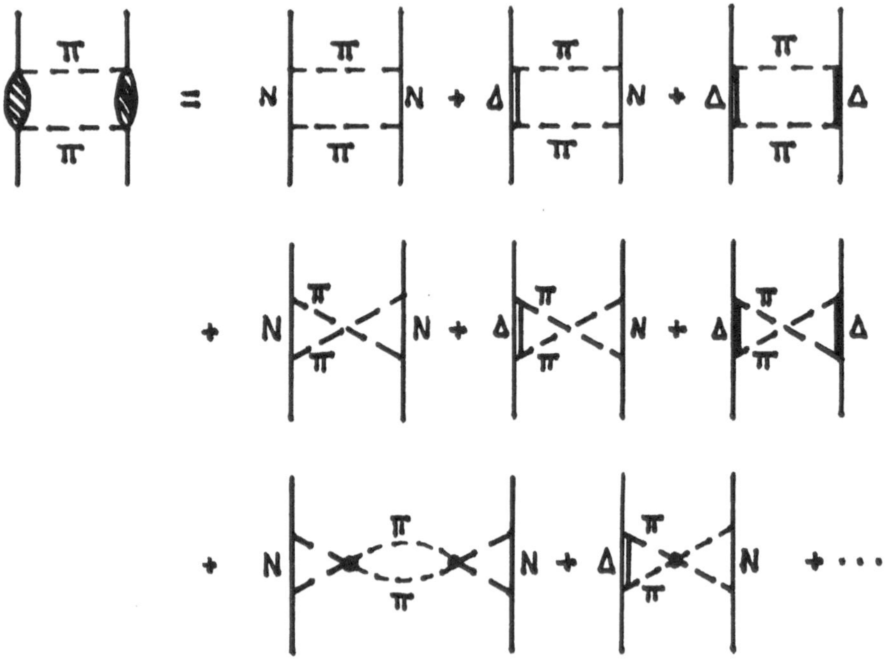

Fig. 1

Here N denotes an intermediate-nucleon state, whereas Δ represents an intermediate Δ-isobar (m_Δ = 1236 MeV, spin 3/2, isospin 3/2). The last two terms in this figure show typical rescattering contributions. The essential point is that the contributions of all diagrams are reduced in the medium, for example, part of the intermediate-nucleon states are now forbidden by the Pauli principle. In the conventional treatment, however (either dispersion-theoretic or of one-boson-exchange type), this modification is taken into account only in the first diagram treated correctly as a second iteration of OPEP. All the other diagrams are treated as part of the NN potential (providing the intermediate-range attraction); their reduction in the medium (growing strongly with the density) is completely neglected. Consequently, an overbinding should be expected within the usual scheme, at least for higher density systems like nuclear matter or, even more, neutron stars.

THE SCHEME

A realistic treatment of such modifications, especially for the crossed-box diagrams, suggests a much more explicit dynamical scheme, which starts from a field-theoretic Hamiltonian containing as interaction part not a potential, but nucleon-nucleon-meson and nucleon-isobar-meson vertices neglecting antiparticle contributions (which are small for pseudovector πNN coupling suggested by chiral symmetry arguments), for details see ref. [5]. Old-fashioned perturbation theory is used because it corresponds to standard non-relativistic many-body theory and, therefore, allows a direct comparison with the usual procedure.

Thus we start from a Hamiltonian

$$H = h_o^{(N)} + h_o^{(\Delta)} + t_o + W \tag{1}$$

where $h_o^{(N)}, h_o^{(\Delta)}$ and t_o describe the free relativistic energies (with bare masses) of nucleons, isobars and mesons, respectively. $W = W^{(N)} + W^{(\Delta)}$ is given by

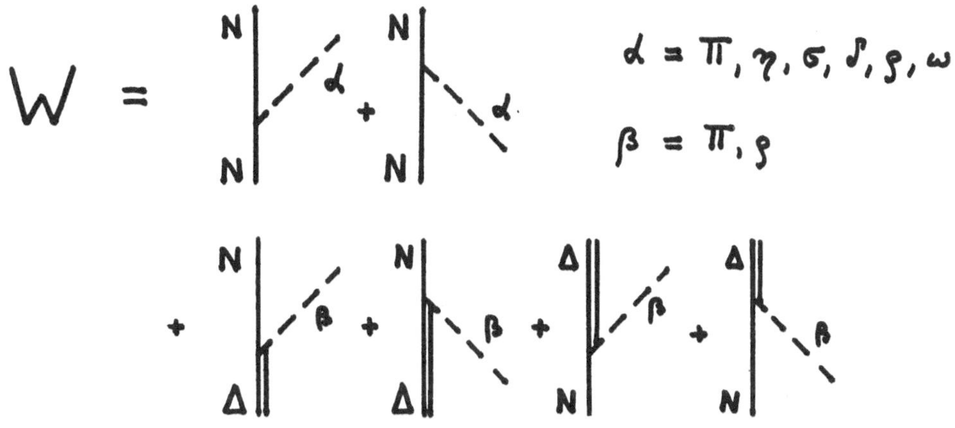

Fig. 2

In the two-body case, the corresponding perturbation series for the NN scattering amplitude can be partially summed by solving an integral equation of Lippmann-Schwinger type

$$T(z) = V_{eff}(z) + V_{eff}(z) \frac{1}{z-h_o^{(N)}} T(z) \tag{2}$$

The energy-dependent quasi-potential $V_{eff}(z)$ (being at least of second order in W) contains the (infinite) sum of all irreducible diagrams, i. e. those with at least one meson or one Δ-isobar present in each intermediate state. In the many-body case, we first introduce a single-particle potential like in the conventional scheme in order to improve convergence, i. e. we now write H as

$$H = h^{(N)} + h_o^{(\Delta)} + t_o + H'$$

$$H' = W - (h^{(N)} - h_o^{(N)}) \qquad (3)$$

and treat H' again in old-fashioned perturbation theory. The total energy E of the ground state of H can then be obtained from a series expansion quite analogous to standard many-body theory. However, the Bethe-Goldstone equation now becomes

$$G(\tilde{z}) = U_{eff}(\tilde{z}) + U_{eff}(\tilde{z}) \frac{Q}{\tilde{z}-h^{(N)}} G(\tilde{z}) \qquad (4)$$

with $U_{eff}(\tilde{z})$ representing the analogous diagrams in the medium compared to $V_{eff}(z)$ in free two-body scattering.

The essential point is that in this formalism the effective potential $U_{eff}(\tilde{z})$ itself is now modified in the medium in two respects: first, the Pauli principle suppresses part of the fourth- and higher-order diagrams; second, the propagators are modified $(z, h_o^{(N)} \to \tilde{z}, h^{(N)})$ called a dispersive effect.

THE MODEL

Of course, it is impossible to evaluate all diagrams occurring in $V_{eff}(z)$. Anyhow, only those should be taken into account which can be trusted: in view of the quark structure of hadrons, conventional meson theory starting from pointlike nucleons and mesons is only an effective theory and should break down for smaller distances. Especially we feel that there is no reason to take the numerical result of self-energy diagrams and vertex corrections too seriously, i. e. we do not trust any numbers growing out of an explicit renormalization of masses and coupling constants, although such a renormalization might be formally possible in simple (often unphysical) models. Thus our first prescription is to neglect, consistently in the two- and many-body problem, all diagrams corresponding to those processes in which one meson is emitted and absorbed by the same nucleon or Δ-isobar. Those processes are reasonably taken into account by using empirical masses and phenomenological vertex functions (form factors) in both the two- and the many-body system. This implies that we neglect possible modifications of masses and vertex functions in the medium, simply because we feel one cannot reliably calculate them without considering quark degrees of freedom. Anyhow, at least the vertex corrections should be small in nuclear matter since the form factors are rather short-ranged and should be affected by Pauli blocking to a small extent only.

The same argument restricts the (still infinite) sum of the diagrams further: diagrams of too high-order, i. e. of too short--range, should not be trusted and thus be omitted. Anyhow, their effect in the NN problem is probably suppressed by the strong short--range repulsion generated by ω-exchange and is effectively hidden in partly phenomenological form factors whose parameters are adjusted in order to fit the empirical NN scattering and deuteron data. Thus we propose the following model for $V_{eff}(z)$:

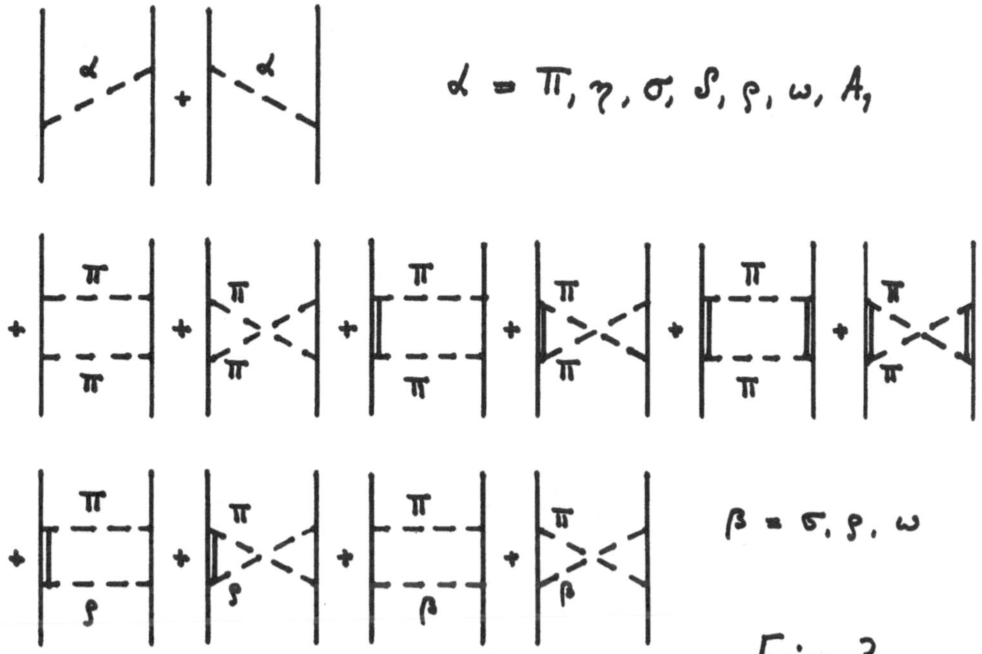

Fig. 3

Here, one diagram stands for all possible time-orderings. The four iterative time-orderings of box diagrams involving two-nucleon intermediate states have to be left out since these are generated by iteration of lower-order diagrams in the scattering equation (2). Compared with fig. 1, it is seen that this model describes only part of the 2π-exchange diagrams explicitly. The rest (mainly the rescattering part) is effectively described by single σ- and ρ-exchange, i. e. many-body corrections are suppressed here. It has to be adjusted in such a way that the total 2π-exchange contribution roughly agrees with what is known from dispersion theory. This should be a reasonable procedure [6].

Concerning 3π-exchange, we believe it worth while to include, apart from the dominant ω-exchange, explicit diagrams with (π,α) exchange, $\alpha = \sigma,\rho$. Such contributions might be important for the determination of the medium-range part of the tensor force, which, for the purpose of nuclear structure calculations, has to be described as accurately as possible. For the same reason, we include also (π,ω) and A_1 exchange.

NUCLEAR MATTER

We have recently completed the (analytical and numerical) evaluation of all these diagrams in fig. 3. Some meson parameters remain to be fixed by adjusting our 2π-exchange contribution to the dispersion--theoretical result and by fitting empirical NN data, before one can do sensible nuclear matter calculations. This has not been finished up to now. Nevertheless, I want to present some results we obtained some time ago, starting from a much simpler model

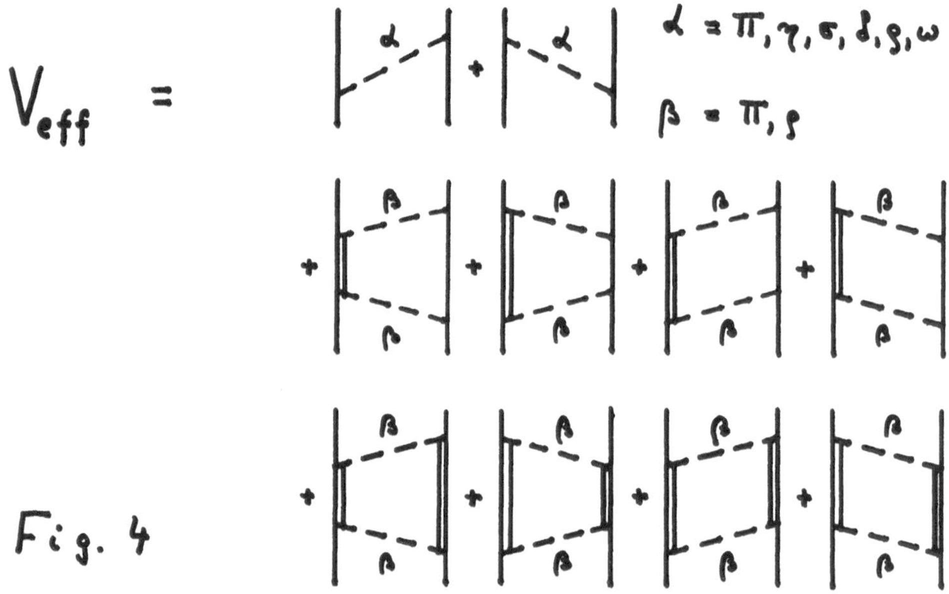

Fig. 4

Here, only the contribution of the iterative isobar diagrams are described explicitly, i. e. the many-body modifications of the corresponding crossed-box diagrams have been neglected, which, however, should be quite small because of the relatively short range of these contributions. Thus, concerning the size of the modifications, this model is probably quite reasonable.

The meson parameters have been adjusted and a reasonable fit of NN scattering and deuteron data has been obtained [7]. Nuclear matter calculations in lowest-order Brueckner theory have been performed [8] with the corresponding $U_{eff}(\tilde{z})$, see eq. (4), using both the standard and the continuous choice for the single-nucleon potential.

Results for the binding energy per particle, E/A, as function of the Fermi momentum k_F are given in fig. 5 (solid lines).

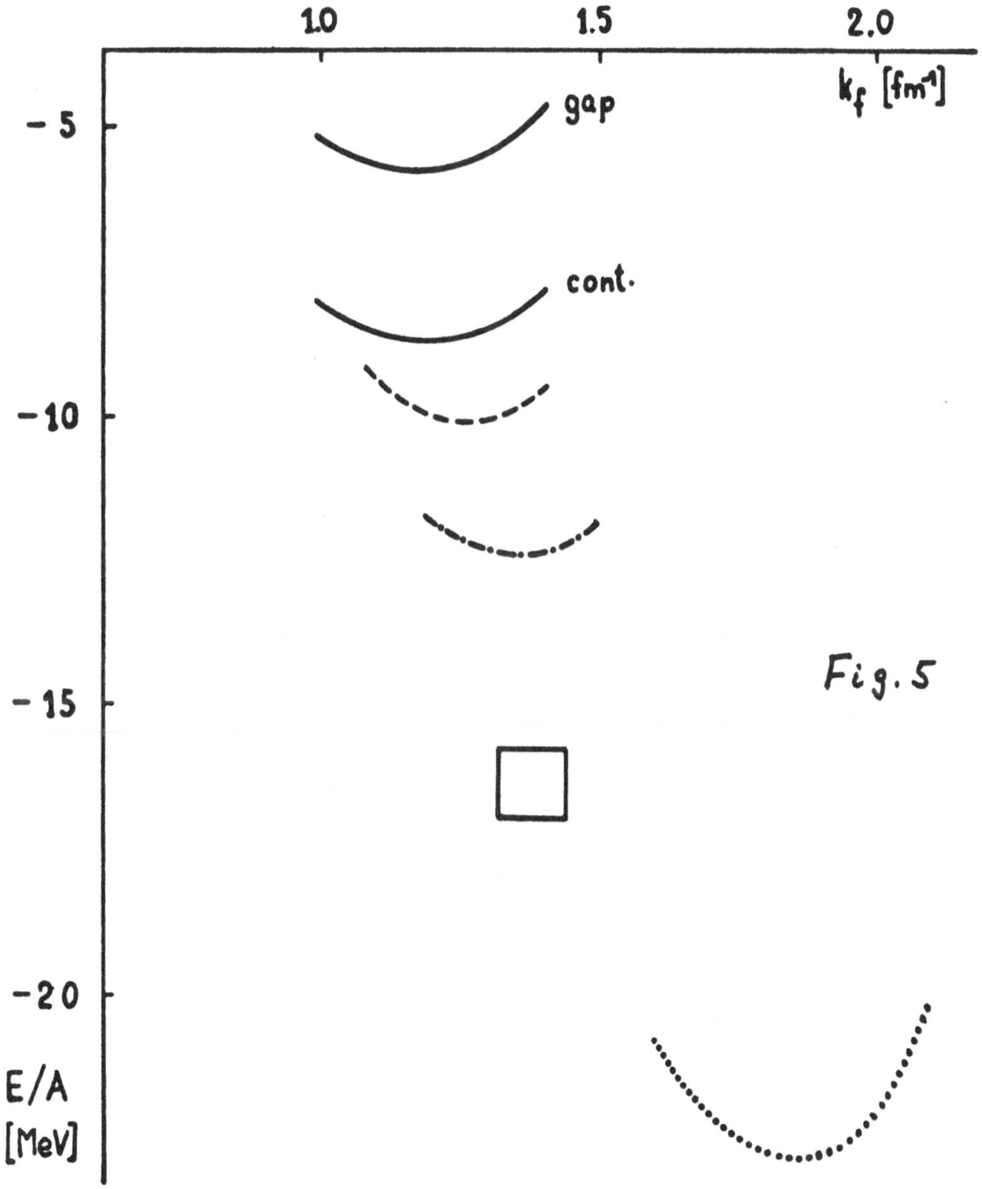

Fig. 5

If, for the continuous choice, we also introduce a Δ-potential under the admittedly crude assumption $U(\Delta) = U(N)$ we get the dashed line, yielding roughly 1 MeV more binding. The smallness of this value is due to the fact that $U(\Delta)$ appears only in the effective quasi--potential $U_{eff}(\tilde{z})$, not in the denominator of the Bethe-Goldstone equation. The propagators in $U_{eff}(\tilde{z})$ contain additional terms from mesons and the Δ-N mass difference, which suppress the sensitivity to a change in $U(\Delta)$.

A suppression of all many-body effects, i. e. taking $V_{eff}(z)$ instead of $U_{eff}(\tilde{z})$, results in the dotted line and demonstrates clearly the sizeable repulsive effect if an important part of the 2π-exchange contribution is treated correctly in the medium. If we leave out the dispersive effects in those propagators containing one meson, we end up with the dash-dot line. In fact, this case might be rather near to the true physical situation in the following sense. Due to the use of a field-theoretic Hamiltonian, additional diagrams of a new type occur, which describe the modification of the meson energy in the medium due to particle-hole excitations. Inspite of being of higher order in a hole-line expansion, such contributions should perhaps be built in from the beginning by introducing a suitable single-particle potential for the meson, too. In fact it turns out that the well-known change of the pion energy due to the medium is of the order of the change in the nucleon energies and they roughly cancel in the propagator of the quasipotential. Consequently, (mesonic) dispersive effects are probably rather small; however, the interplay between possibly different density-dependence of nucleon and meson energies in the medium might ultimately play a decisive role in obtaining a sufficiently low saturation density. Detailed calculations are in progress.

REFERENCES

[1] R.V. Reid, Ann. of Physics 50 (1968) 411

[2] M. Lacombe, B. Loiseau, J.M. Richard, R. Vinh Mau, J. Coté, P. Pires and R. de Tourreil, Phys. Rev. C 21 (1980) 861

[3] K. Holinde and R. Machleidt, Nucl. Phys. A 256 (1976) 479

[4] K. Holinde and H. Mundelius, Approximations in OBE-potentials and their effect on two-nucleon data, Nucl. Phys. A, in press

[5] K. Holinde, Nucl. Phys. A 328 (1979) 439; Physics Reports, in press

[6] J.W. Durso, A.D. Jackson and B.J. Verwest, Nucl. Phys. A 345 (1980) 471

[7] K. Holinde, R. Machleidt, M.R. Anastasio, A. Fäßler and H. Müther, Phys. Rev. C 18 (1978) 870

[8] R. Machleidt and K. Holinde, Nucl. Phys. A 350 (1980) 396

SEARCH FOR BASIC PROPERTIES OF THE NUCLEON-NUCLEON INTERACTION

A. Gersten
TRIUMF and Department of Physics, University of British Columbia
Vancouver, B.C., Canada V6T 2A6
and Department of Physics, Ben-Gurion University, Beer-Sheva, Israel

I. Introduction

Most commonly the N-N interaction is analysed via potential or pseudopotential models at low and intermediate energies and through Regge trajectory exchanges and various diffraction models at high energies. In this work the N-N amplitudes are obtained directly from phase-shift analysis.

From one point of view our aim is to check whether some features, evident in the direct amplitude analysis, are not omitted in the commonly used models. We also wish to check if some new relations exist for the N-N amplitudes which can lead to a simplified analysis of the N-N interaction. For example, it is possible that new relations or symmetries exist which will limit the number of independent amplitudes.

In recent years new N-N experiments were performed and phase-shift analyses up to 800 MeV lab energy are available.[1] It seems that by going to higher energies the real part of the amplitudes becomes better approximated by the first Born term. In this way one can get an insight into the meson exchange structure of the N-N amplitudes. We would like to emphasize a significant difference in the present and the potential model approach. The potential models fit the experimental data up to about 400 MeV lab energy, while the information we extract from our analysis comes mainly from the 400-800 MeV experiments.

In Sections II and III we give some new formulations of facts which are most probably known but not directly exposed in the literature.

II. The number of independent amplitudes

Let us consider first the N-N helicity amplitudes of isospin I=0 and I=1: $\langle\lambda_3\lambda_4|T^I(E)|\lambda_1\lambda_2\rangle$, where E is the total c.m. energy of one nucleon, λ is the helicity equal to $\pm\frac{1}{2}$. Parity conservation, time reversal invariance, and charge independence reduce the 16 possible helicity amplitudes (for each value of the isospin) to 5 independent ones.[2] This way one can construct 10 *linearly* independent amplitudes

$$\phi_1^I = \langle\tfrac{1}{2}\,\tfrac{1}{2}|T^I(E)|\tfrac{1}{2}\,\tfrac{1}{2}\rangle,\ \phi_2^I = \langle\tfrac{1}{2}\,\tfrac{1}{2}|T^I(E)|-\tfrac{1}{2}-\tfrac{1}{2}\rangle,\ \phi_3^I = \langle\tfrac{1}{2}-\tfrac{1}{2}|T^I(E)|\tfrac{1}{2}-\tfrac{1}{2}\rangle,$$
$$\phi_4^I = \langle\tfrac{1}{2}-\tfrac{1}{2}|T^I(E)|-\tfrac{1}{2}\,\tfrac{1}{2}\rangle,\ \phi_5^I = \langle\tfrac{1}{2}\,\tfrac{1}{2}|T^I(E)|\tfrac{1}{2}-\tfrac{1}{2}\rangle. \qquad (2.1)$$

From the Pauli principle the following *non-linear* dependence results (θ is the c.m. scattering angle):

$$\phi_i^I(\pi-\theta) = (-1)^{I+1}\,\phi_i^I(\theta), \quad i = 1,2\,;$$
$$\phi_3^I(\pi-\theta) = (-1)^I\,\phi_4^I(\theta);\quad \phi_5^I(\pi-\theta) = (-1)^I\,\phi_5^I(\theta)\,. \qquad (2.2)$$

Let us construct the following amplitudes:

$$\phi_i^{np}(\theta) = \tfrac{1}{2}[\phi_i^0(\theta) + \phi_i^1(\theta)], \quad i = 1,\ldots,5 , \tag{2.3}$$

which are the neutron-proton scattering amplitudes. From Eqs. (2.2) and (2.3) we obtain the following relations

$$\begin{aligned} \phi_i^I(\theta) &= \phi_i^{np}(\theta) + (-1)^{I+1}\,\phi_i^{np}(\pi-\theta), \quad i = 1,2 , \\ \phi_i^I(\theta) &= \phi_i^{np}(\theta) + (-1)^{I}\,\phi_i^{np}(\pi-\theta), \quad i = 3,4,5 . \end{aligned} \tag{2.4}$$

From Eq. (2.4) we deduce that all the information about the N-N amplitudes is contained in the five amplitudes $\phi_i^{np}(\theta)$ from which the ten linearly independent amplitudes $\phi_i^0(\theta)$ and $\phi_i^1(\theta)$ can be derived. Another relation of limited consequence is

$$\phi_i^{np}(-\theta) = \phi_i^{np}(\theta), \ i = 1,2,3,4; \quad \phi_5^{np}(-\theta) = -\phi_5^{np}(\theta) , \tag{2.5}$$

which may reduce the number of independent amplitudes to 4. For example, one can construct

$$\phi_{15}^{np}(\theta) = \tfrac{1}{2}[\phi_1^{np}(\theta) + \phi_5^{np}(\theta)] , \tag{2.6}$$

from which we obtain

$$\phi_1^{np}(\theta) = \phi_{15}^{np}(\theta) + \phi_{15}(-\theta); \quad \phi_5^{np}(\theta) = \phi_{15}^{np}(\theta) - \phi_{15}(-\theta) .$$

Equation (2.5) may have an application in the study of ambiguities of phase-shift analysis.[3] Using Eq. (2.4) for $\theta=0$ and replacing isospin I=1 by pp or nn scattering we have

$$\phi_i^{np}(\pi) = \phi_i^{pp}(0) - \phi_i^{np}(0), \ i = 1,2; \quad -\phi_4^{np}(\pi) = \phi_3^{pp}(0) - \phi_3^{np}(0) . \tag{2.7}$$

Taking the imaginary parts of Eq. (2.7) we obtain an extension of the optical theorem to backward angles

$$\begin{aligned} \frac{2\pi}{p}\,\mathrm{Im}\left[\phi_1^{np}(\pi) - \phi_4^{np}(\pi)\right] &= \sigma_{tot}^{pp} - \sigma_{tot}^{np} \\ -\frac{4\pi}{p}\,\mathrm{Im}\,\phi_2^{np}(\pi) &= \Delta\sigma_T^{pp} - \Delta\sigma_T^{np} \\ \frac{4\pi}{p}\,\mathrm{Im}\left[\phi_1^{np}(\pi) + \phi_4^{np}(\pi)\right] &= \Delta\sigma_L^{pp} - \Delta\sigma_L^{np}. \end{aligned} \tag{2.8}$$

III. The Fermi amplitudes

In order to investigate the meson exchange content of the scattering amplitude it will be advantageous to use the decomposition of the scattering matrix in terms of the Fermi invariants[2] S, V, T, A, P. At a given energy

$$T = a_S(t)S + a_V(t)V + a_T(t)T + a_A(t)A + a_P(t)P , \tag{3.1}$$

where t is the Mandelstam variable. We shall relate the Fermi amplitudes a_S, a_V, a_T, a_A, a_P to the helicity amplitudes. As the direct relations are well known we shall concentrate on inverting them. We shall write the direct relations in the following new way

$$\begin{aligned} &\alpha\phi_1 = A_1 + 4\beta A_2 + S_1 Z; \quad \alpha\phi_2 = -A_1 + \beta A_3 + (S_1+\beta S_2)Z ; \\ &\alpha(\phi_3-\phi_4) = \beta A_4 + (2S_1 + \beta S_3)Z ; \\ &\alpha(\phi_3+\phi_4) = 2S_1 + \beta S_3 + \beta A_4 Z; \quad \alpha\phi_5 = -E\sin\theta\, S_1/m , \end{aligned} \tag{3.2}$$

where $\beta = p^2/m^2$, $Z = \cos\theta$, p is the c.m. momentum, m is the nucleon mass, $\alpha = 2$ if the ϕ_i are normalized according to Eq. (2.8), and

$$A_1 = a_S + a_V - 3a_T - 3a_A; \quad A_2 = a_V - a_A; \quad A_3 = -a_S + 6a_T - a_P$$
$$S_1 = a_S + a_V + a_T + a_A; \quad S_2 = a_S + 2a_T + a_P$$
$$A_4 = -a_S + 2a_V + 2a_A + a_P = 4A_2 - 2A_1 - A_3$$
$$S_3 = a_S + 2a_V + 2a_A - a_P = 2S_1 - S_2 \,. \tag{3.3}$$

Comparing Eqs. (3.2) and (2.2) we see that the intermediate amplitudes A_i and S_i have the following symmetry properties resulting from the Pauli principle

$$A_i^I(\pi-\theta) = (-1)^{I+1} A_i^I(\theta); \quad S_i^I(\pi-\theta) = (-1)^I S_i^I(\theta) \,, \tag{3.4}$$

which are equivalent to Eq. (2.2). We shall find the inverse relations in two stages

$$S_1 = -\alpha m\phi_5/(E\sin\theta); \quad A_1 = S_1 Z + \alpha m^2(\phi_1-\phi_2-\phi_3+\phi_4)/(2E^2)$$
$$A_2 = m^2(\alpha\phi_1 - A_1 - S_1 Z)/(4p^2); \quad A_4 = \alpha m^2[\phi_3/(1+Z) - \phi_4/(1-Z)]/p^2$$
$$S_3 = m^2[\alpha\phi_3/(1+Z) + \alpha\phi_4/(1-Z) - 2S_1]/p^2 \tag{3.5}$$

and

$$\begin{pmatrix} a_S \\ a_V \\ a_T \\ a_A \\ a_P \end{pmatrix} = \frac{1}{8}\begin{pmatrix} 2 & -4 & -1 & 6 & -1 \\ 0 & 4 & 1 & 0 & 1 \\ -2 & 4 & -1 & 2 & -1 \\ 0 & -4 & 1 & 0 & 1 \\ 2 & -4 & 3 & 6 & -5 \end{pmatrix}\begin{pmatrix} A_1 \\ A_2 \\ A_4 \\ S_1 \\ S_3 \end{pmatrix} . \tag{3.6}$$

IV. The meson exchange content of the Fermi amplitudes

The meson exchange content of the Fermi amplitudes is complicated due to the presence of exchange diagrams. Let us denote by $\mathcal{D}$ a Feynman diagram of one-meson exchange (direct diagram) and by $\mathcal{E}$ the exchange diagram. The one-meson exchange contribution to p-p scattering is given by $\mathcal{D} + \mathcal{E}$, while in the case of n-p scattering we have $-\mathcal{D} + 2\mathcal{E}$ for isovector meson exchange and $\mathcal{D}$ for isoscalar meson exchange. The one-meson exchange contribution to the lowest order direct Feynman diagram acquires the following form

$$b_i(t) = \pm m^2 g_i^2/[4\pi E(t-\mu_i^2)] \,, \tag{4.1}$$

where the sign depends on the particular type of the interaction, g is the coupling constant and μ the meson mass. The exchange diagrams can be related to the direct ones by the matrix of the Fierz transformation. For instance, if we assume one-meson exchanges for p-p scattering, the Fermi amplitudes will have the following form:

$$\begin{pmatrix} a_S(t) \\ a_V(t) \\ a_T(t) \\ a_A(t) \\ a_P(t) \end{pmatrix} = \begin{pmatrix} b_S(t) \\ b_V(t) \\ b_T(t) \\ b_A(t) \\ b_P(t) \end{pmatrix} - \frac{1}{4}\begin{pmatrix} -1 & -4 & -6 & -4 & -1 \\ -1 & 2 & 0 & -2 & 1 \\ -1 & 0 & 2 & 0 & -1 \\ -1 & -2 & 0 & 2 & 1 \\ -1 & 4 & -6 & 4 & -1 \end{pmatrix}\begin{pmatrix} b_S(u) \\ b_V(u) \\ b_T(u) \\ b_A(u) \\ b_P(u) \end{pmatrix}, \tag{4.2}$$

where $t = -2p^2(1-\cos\theta)$, $u = -2p^2(1+\cos\theta)$.

Let us write Eq. (4.2) in the matrix form

$$\vec{a}(t) = \vec{b}(t) - F\,\vec{b}(u) \,, \tag{4.3}$$

where F is the matrix of the Fierz transformation. Let us assume for the moment that the Fermi amplitudes can also be represented in the form of Eq. (4.3). In this way we can try to resolve the meson exchange content of the Fermi amplitudes. From Eq. (4.3) we have

$$\vec{a}(u) = \vec{b}(u) - F\,\vec{b}(t)\ . \qquad (4.4)$$

The amplitudes $\vec{a}(t)$ and $\vec{a}(u)$ can be obtained from the phase-shift analysis. Thus Eqs. (4.3) and (4.4) together consist of 10 equations with 10 unknowns, $\vec{b}(t)$ and $\vec{b}(u)$. Unfortunately 5 of the 10 equations are dependent because of Eq. (3.4). Thus it is impossible to resolve the meson exchange content of the Fermi amplitudes in a simple way. Instead, we shall get information on the meson exchange content by generating particular combinations of the Fermi amplitudes.

We shall concentrate on two combinations of the isospin amplitudes:
(i) $T^{pp} + T^{np} = 1/2\ T^{0} + 3/2\ T^{1}$ - for this choice the isovector meson exchange is cancelled in the forward direction. *We shall call this combination the isoscalar amplitude.*
(ii) $T^{pp} - T^{np} = -1/2\ T^{0} + 1/2\ T^{1}$ - here the isoscalar meson exchange is cancelled in the forward direction. *We shall call this combination the isovector amplitude.* This is also the so-called charge exchange amplitude.

Instead of analysing the isovector amplitude in the forward direction one can analyse the np amplitude in the backward direction. Notice that the np amplitudes have exchange diagrams only for isovector exchanges.

V. Energy-independent results

Below we display two figures in which the results become stable with the increase of energy. Let us introduce the notation

$$\gamma_j^i = \tfrac{1}{2} E \cdot \mathrm{Re}[a_j(0)],\ j = S,V,T,A,P\ , \qquad (5.1)$$

with i=0 for isoscalar amplitudes and i=1 for isovector amplitudes. These quantities are simply related to the one-meson-exchange contributions at zero momentum transfer

$$\gamma \simeq \pm m_N^2\ g_j^2 F_j^2(0)/(4\pi\mu_j^2)\ , \qquad (5.2)$$

where μ_j is the meson mass, g_j the coupling constant and $F_j(0)$ the value of the form factor at zero momentum transfer. Our results are based on a recent phase-shift analysis of Arndt and VerWest[1] in which the higher partial waves (J>6) are represented by the one-pion-exchange (OPE) phase shifts with the pseudoscalar coupling constant $g^2/4\pi = 14.5$. In Fig. 1 the energy dependence of the γ_j^i is given. As one can see, except for γ_P^0, all γ_j^i become more or less stable above 500 MeV lab energy. There are four large quantities: γ_P^1 related to the pion exchange, γ_V^0 related to ω and ϕ exchange, γ_S^0 related to the ε exchange, and finally the energy-unstable γ_P^0. There is one surprise in the above results: there is no large contribution which can be associated with the ρ exchange.

Figure 2 is related to the πNN form factor. For this purpose we consider the backward np scattering for which all the Fermi amplitudes are dominated by the pion pole.

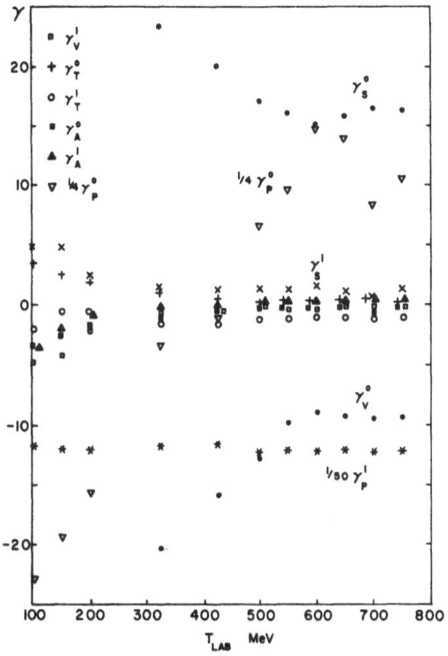

Fig. 1

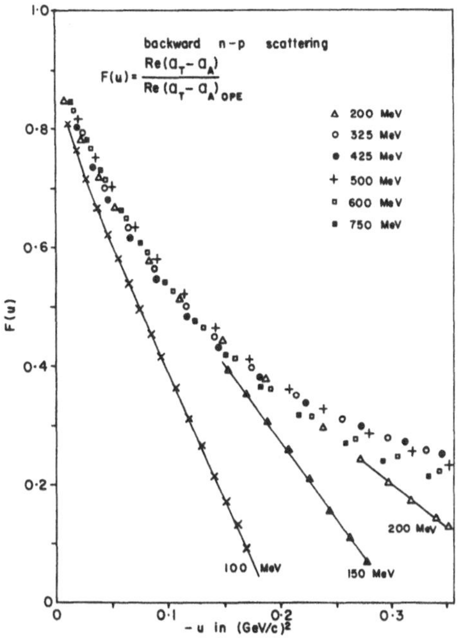

Fig. 2

We should avoid taking amplitudes which are large in the forward direction as such amplitudes may have still a residual large contribution in the backward direction. One test of an absence of this influence will be a consistent u dependence and energy independence of our result. As shown in Fig. 2 it seems that the combination $a_T^{np}(u) - a_A^{np}(u)$ satisfies this criterion. Above 200 MeV lab energy this curve is found to be independent of energy. One cannot interpret the result obtained in terms of the πNN form factor only, because of the contribution of the two-pion-exchange cut. However, this cut can be approximated by the conventional ρ-meson contribution.

Let β be the slope of the curve formed in Fig. 2 near the pion pole $[\beta \simeq 5(GeV/c)^{-2}]$. We can compare this value with that of some models of the πNN form factor if we include an approximation for the ρ exchange. Let us take, for example, the form factor derived on the basis of the cloudy bag model.[4] Comparing slopes near the pion pole we obtain in the non-relativistic approximation:

$$\beta \simeq R^2/(5+m_\pi^2R^2/7) + (g_\rho+f_\rho)^2/[g_\pi^2(m_\rho^2-m_\pi^2)] \simeq R^2/5+(g_\rho+f_\rho)^2/(g_\pi^2m_\rho^2)\,, \qquad (5.3)$$

where R is the radius of the bag, g_ρ and f_ρ are the electric and magnetic ρ coupling constants, respectively. If we substitute $g_\pi^2/4\pi = 14.5$, vector meson dominance model values $g_\rho^2/4\pi = 0.6$, $f_\rho/g_\rho = 3.7$ and $m_\rho = 765$ MeV, we obtain $R \simeq 0.8$ fm.

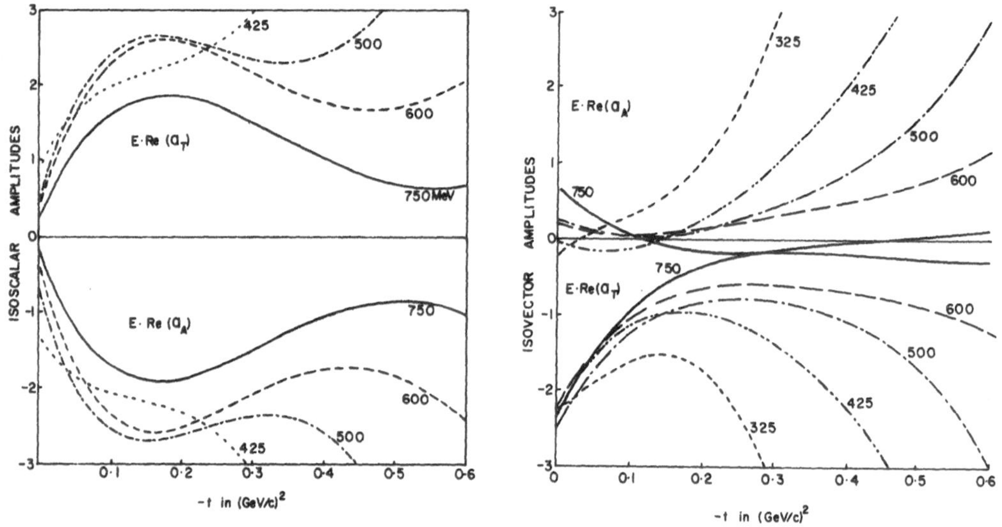

Fig. 3. The a_A and a_T amplitudes for different lab energies.

VI. The T-A and the V-S cancellations

The name "strong interaction" should be used with reservation when applied to the N-N interaction which is much weaker than the $\overline{N}$-N interaction. The reason seems to be the presence of cancellations in the N-N interaction. In Figs. 3 and 4 we depict the Fermi amplitudes a_T and a_A. The cancellation of the isoscalar amplitudes is rather striking. This cancellation is also evident in the imaginary part of the isovector

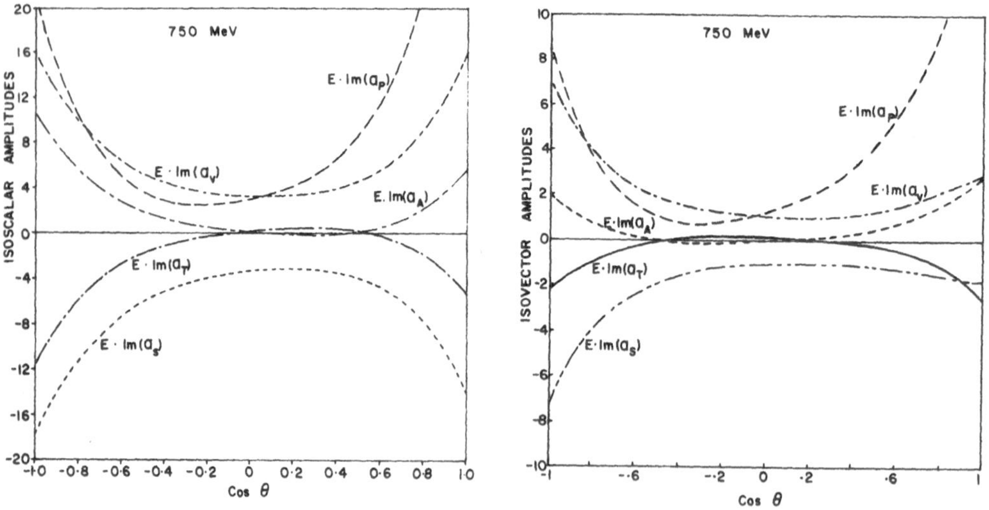

Fig. 4. The imaginary part of the a_S, a_V, a_T, a_A amplitudes for 750 MeV.

amplitudes, but seems to be absent in the forward direction of the real part. To our knowledge the origin of this T-A cancellation is not yet understood.

Let us compare the T-A cancellation with some of the consequences of isospin symmetry. As we approach the elastic threshold, isospin symmetry is strongly violated. As we increase the energy along the elastic cut the isopin symmetry "opens". The "opening" occurs much earlier in the imaginary part of the amplitudes than in the real part. A similar situation seems to occur for the T-A amplitudes, although the cancellation is not accurate enough to serve as a constraint on phase-shift analyses. In Fig. 5 the "opening" of the T-A cancellation is demonstrated for the imaginary part of the $a_T + a_A$ neutron-proton scattering amplitude.

In Figs. 6 and 4 the a_S and a_V amplitudes are depicted. The $a_S + a_V$ cancellation is less accurate than the $a_T + a_A$ cancellation and again is better satisfied for the imaginary parts and for the isoscalar amplitudes. The V-S cancellation was advocated before by A.E.S. Green and coworkers in a potential model.[5]

VII. Summary and discussion

We have shown in Sec. IV that it is impossible to extract ten independent amplitudes having a definite exchange meson properties because five of them must be dependent. By taking combinations of the Fermi amplitudes it is possible to obtain some insight into the meson exchange content of the N-N amplitudes. We defined two isospin combinations: the isoscalar and isovector amplitudes. By studying these amplitudes we can gain both qualitative and quantitative information about the N-N interaction. For example, from Eq. (5.2) and Fig. 1 we can get estimates for the coupling constants

$$g_\varepsilon^2/4\pi \simeq 16.4\ m_\varepsilon^2/[m_N^2\ F_\varepsilon^2(0)]\ ,$$
$$g_\delta^2/4\pi \simeq 1.2\ m_\delta^2/[m_N^2\ F_\delta^2(0)]\ ,$$
$$(1+f/g)_\omega (g_\omega^2/4\pi) F_\omega^2(0)/m_\omega^2 + (1+f/g)_\phi (g_\phi^2/4\pi) F_\phi^2(0)/m_\phi^2 \simeq 9.5/m_N^2 .$$

From Fig. 2 we can get information about the πNN form factor. The results are uncertain to some extent because of the difficulty of taking out the ρ-meson contribution which simulates the two-pion-cut contributions. With the parameters of the vector-meson dominance model one obtains a form factor consistent with a nucleon bag radius of about 0.8 fm.

A new feature of the discussed amplitudes is the T-A cancellation in the isoscalar amplitudes and the T-A cancellation in the imaginary part of the isovector amplitudes. A less accurate cancellation is the scalar-vector one. The origin of these cancellations is not yet clear.

The existence of the T-A and S-V cancellations suggests probability that new relations or symmetries may apply to the N-N interaction. In their present form the discussed cancellations are not accurate enough in order to be applied to reduce the number of independent amplitudes. One can look for further relationships. For example, related

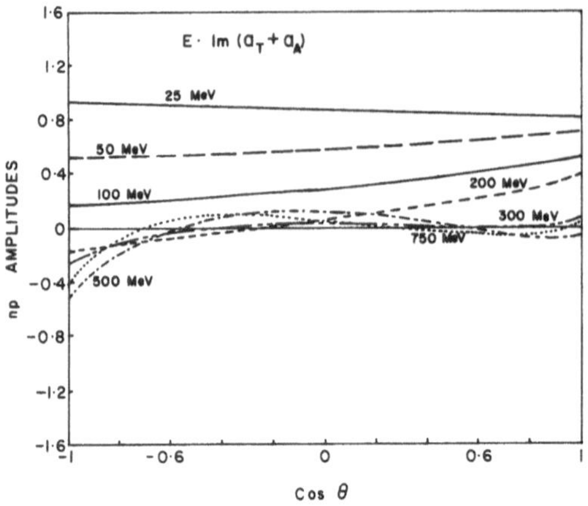

Fig. 5

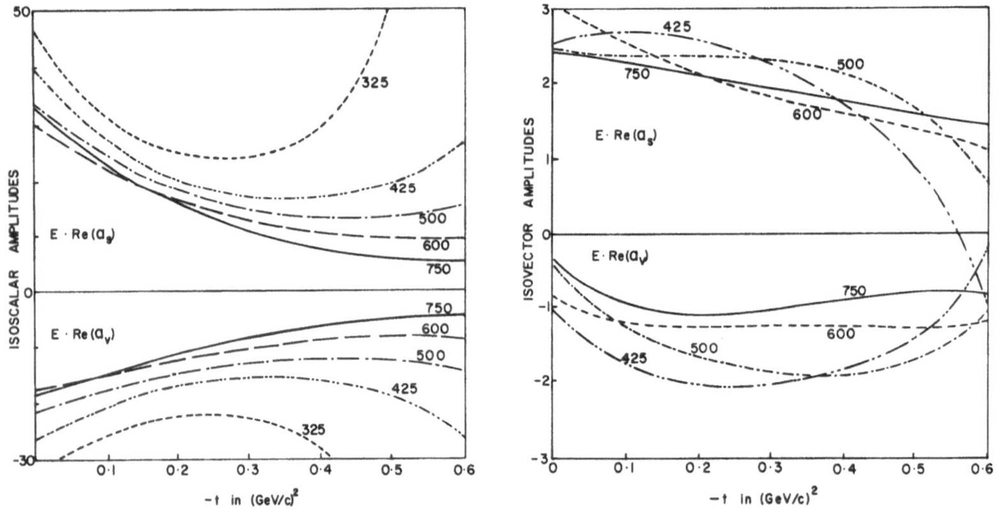

Fig. 6. The a_S and a_V amplitudes for different lab energies.

to the T-A cancellation is the following amplitude

$$a_T + (4m^2 a_A + t a_P)/(4E^2 - u) . \qquad (7.1)$$

If only direct Feynman diagrams and only meson exchanges of the so-called natural parity ($J^P = 0^+, 1^-, 2^+, \ldots$) are used, the amplitude (7.1) should vanish. Calculations show that this amplitude is rather small, but still not small enough to be a practical constraint. Another small amplitude (especially in the backward direction) is the real part of the following amplitude

$$a_T(t) + a_A(t) + \tfrac{1}{2} a_S(u) + \tfrac{1}{2} a_V(u) . \qquad (7.2)$$

The examples of the amplitudes (7.1) and (7.2) demonstrate that the T-A cancellation can be an approximation to a more accurate relationship.

Apart from the existence of the cancellations the depicted amplitudes show the following unusual features:

(i) An unusual behaviour of the real part of the isoscalar a_P amplitude. The change of sign to positive in Fig. 1 at higher energies, and the enormous magnitude, are somewhat puzzling. The maximum about 600 MeV suggests correlation with Δ-production or dibaryon resonances.

(ii) The isoscalar a_T amplitude is positive. This suggests that f_ω/g_ω is positive, contrary to the usually accepted value -0.12.

(iii) The isovector amplitudes a_V and a_T are relatively small, suggesting a rather small contribution of the ρ-meson.

(iv) The isovector a_T amplitude is not consistent with natural parity meson exchanges. The constant value observed at t=0 suggests a meson exchange of non-natural parity and spin $J \geqslant 2$. A possible candidate is the A_3 meson.

(v) The resulting mass of the isoscalar exchange should be smaller than the ω-meson mass.

It seems to us that the study of the Fermi amplitudes may lead to a better understanding of the properties of the N-N interaction. More calculations, especially those including an error analysis, will be of utmost value.

I would like to acknowledge valuable discussions with A.W. Thomas and J.M. Richard.

References

1. R. Arndt and B.J. VerWest, in Proc. of Ninth Int. Conference on the Few-Body Problem, Eugene, Oregon, August 1980;
   J. Bystricky, C. Lechanoine and F. Lehar, Saclay preprint DPhPE 79-01, October 1979.
2. M.L. Goldberger *et al.*, Phys. Rev. 120, 2250 (1960).
3. A. Gersten, Nucl. Phys. B12, 537 (1969).
4. G.A. Miller, A.W. Thomas and S. Théberge, Phys. Lett. 91B, 192 (1980) and TRIUMF preprint TRI-PP-80-8 (to be published in Phys. Rev. D).
5. A.E.S. Green and T. Sawada, Rev. Mod. Phys. 39, 594 (1967).

RELATIVISTIC QUANTUM FIELD APPROACH TO NUCLEAR MATTER AND NUCLEI

M. Jaminon and C. Mahaux,
Institut de Physique, Université de Liège,
Sart Tilman, 4000 Liège 1, Belgium

I. INTRODUCTION

The problem of calculating *relativistic corrections* to the properties of nuclear matter and nuclei comes up against a fundamental difficulty that Bethe[1] accurately described as follows : "*All calculations in nuclear matter theory are made using the nonrelativistic Schroedinger equation. There is a good reason for this. The nuclear forces assumed are phenomenological, hence it is unknown how they should be transformed relativistically. This will change only if and when a reliable nucleon interaction can be derived from meson theory*". Without recourse to the latter, one had been limited to the evaluation of the so-called "minimal" effect of relativity, which essentially consists in modifying the Schroedinger equation in order to fulfill the relativistic form of the unitarity condition. This minimal effect only adds approximately 0.5 MeV to the average binding energy per nucleon in nuclear matter at normal density.[2] Progress in the description of the nucleon-nucleon interaction in the framework of meson theory[3] makes it conceivable to investigate more closely some relativistic corrections. These have recently been claimed to be sizeable since for instance Anastasio, Celenza and Shakin concluded from their studies[4-6] that "*contrary to current thought nuclear matter should be treated as a relativistic system*".[7]

In the present survey, we discuss the nature and the size of *relativistic mean field corrections* which have recently been considered by various authors. The word *approach* in the title is meant to remind one that there does not yet exist a reliable relativistic quantum field *theory* of the many-body problem. In view of this, we shall find it instructive and useful to investigate first *simplified* dynamical models, and also to rely upon *phenomenological* approximations. Since nuclear theory essentially rests upon the shell model, most of our discussion will be devoted to relativistic corrections to the *single-particle* potential and wave functions. Here, the expression *relativistic corrections* refers to effects which directly or indirectly involve the *small components* of the Dirac spinor which represents the nucleon embedded in nuclear matter or in a nucleus.

2. SINGLE-PARTICLE WAVE EQUATION

The nonrelativistic Pauli-Schroedinger single-particle wave equation reads

$$[\frac{\vec{p}^2}{2m} + U^{(S)}(r;E) + U^{(S)}_{so}(r;E)\,\frac{\vec{\sigma}.\vec{L}}{r}]\,\Psi^{(S)}(\vec{r}) = \varepsilon\,\Psi^{(S)}(\vec{r}) \tag{2.1}$$

in the case of a spherical nucleus with equal number of protons and neutrons. We omitted nonlocality and the Coulomb field for simplicity, and set $\hbar = 1$. In the case of nuclear matter Eq. (2.1) reduces to

$$[\frac{k^2}{2m} + U^{(S)}(k)]\,\Psi^{(S)}(\vec{r}) = \varepsilon_k\,\Psi^{(S)}(r) \quad . \tag{2.2}$$

The nonrelativistic wave function $\Psi^{(S)}$ has *two* components, which correspond to the spin degree of freedom.

The relativistic Dirac single-particle wave equation reads

$$[\vec{\alpha}.\vec{p} + \gamma^0 m + \gamma^0 \mathcal{U}]\,\Psi^D(\vec{r}) = E\,\Psi^{(D)}(r) \quad , \tag{2.3}$$

where $E = \varepsilon + m$ is the total energy and the quantities α_j , γ^0 , $\mathcal{U}$ are 4×4 matrices. The relativistic wave function $\Psi^D(\vec{r})$ has *four* components, two *large* ones and two *small* ones. In the case of a *free* nucleon, the ratio of the small to the large components is of the order of $\beta = v/2c$. For a nucleon at the top of the Fermi surface, $\beta \stackrel{\sim}{\sim} 0.15$ Since $\beta^2 \stackrel{\sim}{\sim} 0.02$ this suggests that for *free* nucleons relativistic corrections are of the order of 2 % , in keeping with the finding of Brown et al.[2] We describe in the next section that this estimate may have to be drastically changed if the Lorentz components of the relativistic mean field $\mathcal{U}$ are large.

3. THE $\sigma + \omega$ MODEL

3.1. Definition

For invariance reasons, the general form of the relativistic field in symmetric nuclear matter is the following[8)]

$$\mathcal{U}(k) = 1\,U_s(k) + \gamma^0\,U_o(k) + \vec{\gamma}\,.\,\frac{\vec{k}}{m}\,U_v(k) + i\,\gamma^0\,\vec{\gamma}\,.\,\frac{\vec{k}}{m}\,U_t(k) \quad , \tag{3.1}$$

where 1 denotes the unit matrix; either one of the three quantities U_s , U_o or U_v may in fact be eliminated.[8)] The phenomenological "*σ+ω model*" consists in retaining only the first two terms on the right-hand side of Eq. (3.1), i.e. in assuming that the average field $\mathcal{U}$ is the sum of a *scalar* component U_s and of the fourth component U_o of a vector. Moreover, the dependence of U_o and U_s upon k is usually dropped. Below, we shall often for simplicity call U_o the *vector* potential. The expression "$\sigma+\omega$ model" derives from the fact that the form $\mathcal{U} = 1\,U_s + \gamma^0\,U_o$ is obtained from the Hartree approximation when one includes only the exchange of a scalar meson (σ) and of a vector meson (ω) (section 4.1). This Hartree model moreover suggests that $U_s < 0$, $U_o > 0$; we shall see that these inequalities are fulfilled if one fits the parameters of the $\sigma+\omega$ model to experimental data.

3.2. Relativistic effects

In the $\sigma+\omega$ model, the single-particle wave function $\Psi_k(\vec{r})$ in nuclear matter has the form $\Psi_k(\vec{r}) = \phi(\vec{k})\,\exp(i\vec{k}.\vec{r})$, with

$$\phi(\vec{k}) = Q_k \begin{pmatrix} (k^2 + m_*^2)^{\frac{1}{2}} + m_* \\ 0 \\ k_3 \\ k_1 + i\,k_2 \end{pmatrix} , \qquad (3.2)$$

where Q_k is a normalization constant, while the *effective mass* m_* is given by

$$m_* = m + U_s \quad . \qquad (3.3)$$

For small k, the ratio of the small to the large components is about

$$\phi_</\phi_> \approx k/2m_* \approx (v/2c)\,(m/m_*) \quad . \qquad (3.4)$$

Hence, relativistic effects are increased by the ratio m/m_* as compared to the case of a free nucleon. Equations (3.3) and (3.4) indicate that *even at low energy*, relativistic corrections may be important if $-U_s$ is not small compared to m.

In the framework of the $\sigma+\omega$ *model*, the magnitude of the quantities U_o and U_s can be estimated by fitting a few empirical observables. For instance, one may require that the *average binding energy per nucleon* B/A be approximately equal to -16 MeV at $k_F = 1.35$ fm^{-1}, and that the depth and the strength of the spin-orbit component of the standard *nonrelativistic optical-model potential* be reproduced (see section 5). This yields[9-11]

$$\langle U_s \rangle \approx -350 \text{ MeV} \quad , \quad \langle U_o \rangle \approx +270 \text{ MeV} \quad , \qquad (3.5)$$

where $\langle\rangle$ refers to an average between 0 and k_F over some possible momentum dependence of U_o and U_s. These values indicate that significant relativistic corrections exist in the phenomenological $\sigma+\omega$ model, since $-U_s$ is not negligible compared to the nucleon mass m.

As an example, let us consider the *kinetic energy* $K(k)$ of the nucleon described by $\Psi_k(\vec{r})$

$$K(k) = \phi^{\dagger}(\vec{k})\ \vec{\alpha}.\vec{k}\ \phi(\vec{k}) \tag{3.6}$$

$$\approx \frac{k^2}{2m}\ (1 - U_s^2/m_*^2) \quad . \tag{3.7}$$

This result will be illustrated in Fig. 3 below. Note that expression (3.7) vanishes for $U_s = -m/2 \approx -470$ MeV and becomes negative for large values of $|U_s|$. This is a warning that the relativistic mean field approximation should be treated with caution in the case of strong fields, which is precisely the case when relativistic corrections are large.

4. THEORETICAL APPROACHES

The phenomenological $\sigma+\omega$ model described in section 3 consists in making assumptions on the nature of the Lorentz components of the mean field $\mathcal{U}(k)$, i.e. in dropping the last two terms on the right-hand side of Eq. (3.1). In the present section, we discuss to what extent this phenomenological model is corroborated by more microscopic approaches. By increasing order of complexity, these are the relativistic Hartree, Hartree-Fock and Brueckner-Hartree-Fock approximations.

4.1. The Hartree mean field approximation

The Hartree mean field approximation consists in assuming that $\mathcal{U}$ is given by

$$\mathcal{U}^H = \gamma^0 \sum_{j<k_F} \int d^3r_2\ \Psi_j^{\dagger}(2)\ \mathcal{V}(1,2)\ \Psi_j(2) \quad , \tag{4.1}$$

where $\mathcal{V}(1,2)$ is the nucleon-nucleon quasipotential as calculated from OBEP. For instance, the contribution to $\mathcal{V}$ of the scalar σ meson is given by

$$\mathcal{V}(1,2) = -\gamma_0(1)\ \gamma^0(2)\ g_\sigma^2\ e^{-m_\sigma r_{12}}/r_{12} \quad . \tag{4.2}$$

The spinor ψ_j should be computed from Eq. (2.3) with $\mathcal{U}$ replaced by $\mathcal{U}^H$; this is emphasized by the words "mean field" attached to Hartree. If one only includes the mesons σ and ω, one gets $\mathcal{U}^H = 1\ U_s + \gamma^0\ U_o$ as assumed in the phenomenological $\sigma+\omega$ model, with[9)]

$$U_o = \rho\ g_\omega^2/m_\omega^2 \quad , \qquad U_s = -\rho_s\ g_\sigma^2/m_\sigma^2 \quad ; \tag{4.3}$$

here, ρ_s denotes the scalar density and is given by

$$\rho_s = 2\pi^{-2} \int_0^{k_F} k^2 \, dk \, m_*[k^2 + m_*^2]^{-\frac{1}{2}} \quad . \tag{4.4}$$

Since $m_* = m + U_s$, Eqs. (4.3) and (4.4) show that the relativistic Hartree mean field approximation in nuclear matter implies a self-consistent calculation even in nuclear matter, in contradistinction with the nonrelativistic Hartree case. This reflects the fact that the Hartree *mean field* approximation is *not* a first-order approximation in the coupling constants g_σ^2 and g_ω^2 . A first-order approximation would indeed consist in replacing in Eq. (4.1) the wave functions $\Psi_j(2)$ (Eq. (3.2)) by *free* plane wave spinors. Chin[12] has shown that the "Hartree mean field" approximation results from adding a selected set of higher-order Feynman diagrams to the first-order contribution. However, divergences appear which must be eliminated by introducing cubic and quartic terms in the Lagrangian for the free scalar meson.[12,13]

The Hartree mean field approximation has been used in finite nuclei by Miller and Green[14] and by Brockmann and Weise.[15,16] These authors obtained fair agreement with the empirical charge distribution and single-particle energies. Nuclear matter calculation based on Brockmann's parameters yields[11] $U_s^H = -329$ MeV and $U_o^H = +241$ MeV for $k_F = 1.35$ fm^{-1} . This is close to the estimate (3.5) derived from the phenomenological $\sigma+\omega$ model. It is noticeable that Brockmann's input values are identical to those contained in one of the OBEP nucleon-nucleon potentials constructed by the Bonn group.[17]

4.2. Hartree-Fock mean field approximation

The Hartree-Fock mean field approximation consists in assuming that $\mathcal{U}$ is given by

$$\mathcal{U}^{HF} = \mathcal{U}^H + \mathcal{U}^F \quad , \tag{4.5}$$

where $\mathcal{U}^H$ is given by Eq. (4.1), while

$$\mathcal{U}^F = -\gamma^0 \sum_{j<k_F} \int d^3r_2 \, \psi_j^\dagger(2) \, \mathcal{V}(1,2) \, \psi_j(1) \, \exp[i\vec{k}.\vec{r}_{12}] \quad . \tag{4.6}$$

The spinor ψ_j is now computed from Eq. (2.3) with $\mathcal{U}$ replaced by $\mathcal{U}^{HF}$, whence the words "mean field" attached to Hartree-Fock. If only the mesons σ and ω are included the Fock contribution $\mathcal{U}^F$ has the form[8]

$$\mathcal{U}^F = I \, U_s^F(k) + \gamma^0 \, U_o^F(k) + \vec{\gamma}.\frac{\vec{k}}{m} \, U_v^F(k) \quad . \tag{4.7}$$

The σ and the ω mesons *both* contribute to *all three* quantities U_s^F, U_o^F and U_v^F, which moreover depend upon k. Hartree-Fock mean field calculation with Brockmann's input parameters has been performed in the case of nuclear matter.[8] Since the calculated value $|U_s^H + U_s^F|$ is large, *relativistic corrections* are important in the Hartree-Fock approximation.

4.3. Brueckner-Hartree-Fock mean field approximation

The "relativistic" Brueckner-Hartree-Fock mean field approximation consists in assuming that $\mathcal{U}$ is given by[4-7]

$$\mathcal{U}^{BHF} = \overline{\mathcal{U}}^H + \overline{\mathcal{U}}^F \quad , \tag{4.8}$$

where $\overline{\mathcal{U}}^H$ and $\overline{\mathcal{U}}^F$ have the same expression as $\mathcal{U}^H$ and $\mathcal{U}^F$ (Eqs. (4.1) and (4.6)) with, however, V replaced by a quantity M which is a relativistic extension of Brueckner's reaction matrix, as defined by the Bonn group.[3]

Anastasio, Celenza and Shakin[5-7] evaluated $\mathcal{U}^{BHF}$ for some of the OBEP potentials constructed by the Bonn group. They write $\mathcal{U}^{BHF}$ in the same form as the right-hand side of Eq. (4.7). The calculated components of $\mathcal{U}^{BHF}$ are represented in Fig. 1. The corresponding average energy

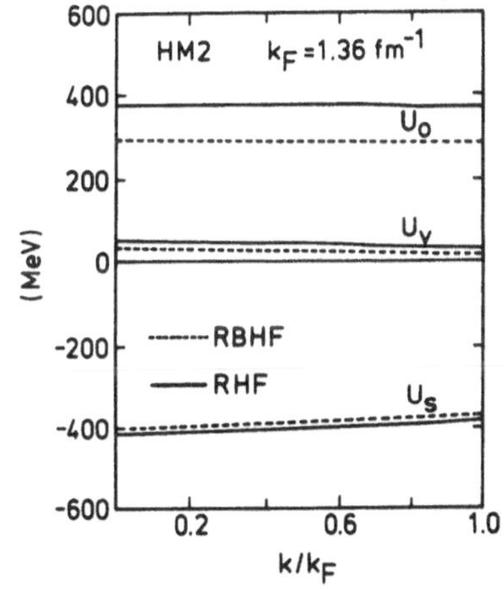

Fig. 1. From Ref. 7. Dependence upon k of the quantities U_s, U_o and U_v as calculated in the framework of the Brueckner-Hartree-Fock approximation (dashed lines) and of the Hartree-Fock approximation (full lines), for $k_F = 1.36$ fm^{-1}. The input is the HM2 OBEP potential of Holinde and Machleidt.[17]

per nucleon saturates near the empirical value. Note that the value of U_v is quite small, in keeping with the $\sigma+\omega$ model. Since moreover $|U_s^{BHF}|$ is quite large, relativistic corrections are important. This is illustrated in Fig. 2.

A number of intruiging features should be pointed out. Firstly, Anastasio et al. did not encounter the necessity of eliminating the divergences found by Chin[12] and mentioned in section 4.1. Secondly, T/A is *not* identical to the kinetic energy per nucleon of a relativistic *free* Fermi gas of nucleons, in contrast to what one would expect from a Rayleigh-Schroedinger-Brueckner approach. These two features possibly reflect the fact that Anastasio et al. actually start from Green's func-

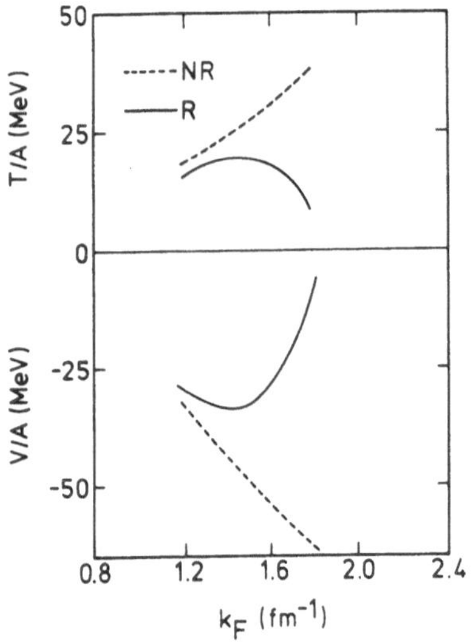

Fig. 2. From Ref. 7. The upper part of the figure shows the average kinetic energy per nucleon and the lower part the average potential energy per nucleon. The dashed curves correspond to the nonrelativistic Brueckner-Hartree-Fock approximation used by the Bonn group[17] and the solid lines to the relativistic Brueckner-Hartree-Fock mean field approximation.

tion theory rather than from perturbation theory. A third intruiging feature is that the average kinetic energy per nucleon becomes *negative* for large k_F (see Eq. (3.7)). The proper interpretation of the Dirac equation is then quite delicate and the use of an approximation of independent particles (or quasiparticles) moving in a mean field appears questionable.

5. OPTICAL-MODEL POTENTIAL

It is possible to eliminate the small components of ψ^D from Eq. (2.3). The resulting equation for the large components $\Psi_>$ reads

$$\left(\frac{\vec{p}^{\,2}}{2m} + U_e + U_{so}\,\frac{\vec{\sigma}.\vec{L}}{r}\right)\Psi_> = \frac{E^2 - m^2}{2m}\,\Psi_> \quad , \tag{5.1}$$

with[8])

$$U_e(\varepsilon) \approx (U_s + U_o)\left[1 + \frac{U_s - U_o}{2m}\right] + U_o\,\frac{\varepsilon}{m} \quad , \tag{5.2}$$

$$U_{so} \approx (2m)^{-2}\,\frac{d}{dr}\,(U_o - U_s) \quad . \tag{5.3}$$

Equations (2.1) and (5.1) show that the quantities U_e and U_{so} can be compared with the empirical real part of the central and of the spin-orbit potentials. The Hartree and the Hartree-Fock mean field approximations with σ and ω mesons both yield good agreement with the experimental values if the coupling strengths g_σ^2 and g_ω^2 are adjusted to reproduce the empirical saturation properties of nuclear matter.[8,11] At intermediate energy, the Schroedinger equivalent potential U_e has a wine-bottle bottom shape.[18] This appears to be compatible with analyses of recent proton elastic scattering data.[19]

Anastasio, Celenza and Shakin[20] proposed another way than Eq. (5.1) for defining a Schroedinger equivalent potential. Their potential is nonlocal even if U_s and U_o are local. In the case of nuclear matter, it can be shown[21)] that their potential is identical to U_e as defined by Eq. (5.2).

6. SMALL COMPONENTS OF THE SINGLE-PARTICLE WAVE FUNCTION

We emphasized in section 3.2 that the relativistic mean field effects discussed here are associated with an enhancement of the small components of the single-particle wave function. One possible test would thus consist in investigating experimental observables which involve a matrix element of an operator which couples the small and the large components of single-particle Dirac spinors.

Noble[22)] calculated the static magnetic moments of ^{15}O and ^{15}N and found fair agreement with experiment, thus showing that conclusions drawn earlier by Miller[23)] were excessively pessimistic. Noble also looked at the beta-decay asymmetry and found that the $\sigma+\omega$ model disagrees with experiment by about one standard deviation. In the two cases discussed by Noble, meson exchange corrections are unfortunately quite important and blur the conclusions.

Eisenberg[24)] investigated the ratio of the induced pseudoscalar coupling constant to the axial vector couping constant. This ratio can in principle be extracted from the study of muon capture. The present experimental error bars are unfortunately too large to draw any definite conclusion.

7. DISCUSSION

We have surveyed a few features which point to the possibility that relativistic mean field corrections may be significant in nuclear matter once the density becomes larger than about $0.15\ fm^{-3}$, even for low energy nucleons. For lower densities, such as encountered in finite nuclei, some relativistic effects may also be relevant. They concern in particular the size of the small components of the Dirac spinor, the energy dependence of the optical-model potential and its shape at intermediate energy. It should be kept in mind, however, that the relativistic mean field approximations are not based on a firmly established theory when mesons other than σ and ω are taken into account in the nucleon-nucleon interaction. Nevertheless, the outcome of the preliminary investigations surveyed here are sufficiently provocative to call for further studies.

We are grateful to P. Rochus for many stimulating discussions and to C.M. Shakin for an early communication of the results of Refs. 4-7 and 20.

REFERENCES

1. H.A. Bethe, Ann.Rev.Nucl.Sci. 21, 93 (1971)
2. G.E. Brown, A.D. Jackson and T.T.S. Kuo, Nucl.Phys. A133, 481 (1969)
3. K. Erkelenz, Phys.Reports 13, 191 (1974)
4. M.R. Anastasio, L.S. Celenza and C.M. Shakin, Phys.Rev. C23, 569 (1981)
5. M.R. Anastasio, L.S. Celenza and C.M. Shakin, Phys.Rev.Lett. 45, 2096 (1980)
6. M.R. Anastasio, L.S. Celenza and C.M. Shakin, A relativistic model of interacting nucleons and mesons, Phys.Rev. C (in press)
7. M.R. Anastasio, L.S. Celenza and C.M. Shakin, Relativistic effects in the Bethe-Brueckner theory of nuclear matter, Phys.Rev.C(in press
8. M. Jaminon, C. Mahaux and P. Rochus, submitted for publication
9. J.D. Walecka, Ann.Phys. 83, 491 (1974)
10. J.V. Noble, Phys.Rev. C17, 2151 (1978)
11. M. Jaminon, C. Mahaux and P. Rochus, Phys.Rev. C22, 2027 (1980)
12. S.A. Chin, Ann.Phys.(N.Y.) 108, 301 (1977)
13. L.N. Savushkin, Sov.J.Nucl.Phys. 30, 340 (1979)
14. L.D. Miller and A.E.S. Green, Phys.Rev. C5, 241 (1972)
15. R. Brockmann and W. Weise, Phys.Rev. C16, 1282 (1977)
16. R. Brockmann, Phys.Rev. C18, 1510 (1978)
17. K. Holinde and R. Machleidt, Nucl.Phys. A256, 479 (1976)
18. M. Jaminon, C. Mahaux and P. Rochus, Phys.Rev.Lett. 43, 1097 (1979)
19. H.O. Meyer et al., Phys. Rev. C23, 616 (1981)
20. M.R. Anastasio, L.S. Celenza and C.M. Shakin, Dirac phenomenology in nuclear structure and reactions, Phys.Rev. C (in press)
21. M. Jaminon and C. Mahaux, to be published
22. J.V. Noble, Phys.Rev. C20, 1188 (1979)
23. L.D. Miller, Ann.Phys. 91, 40 (1975)
24. J.M. Eisenberg, Observational Tests of Models for a Relativistic Nucleon bound in Scalar and Vector Potentials, Tel Aviv University preprint, 1980.

PATH INTEGRALS, BOSON EXPANSIONS AND MEAN FIELD APPROXIMATIONS

J. P. Blaizot
Department of Physics
University of Illinois at Urbana-Champaign
Urbana, Illinois 61801 USA
and
H. Orland
Service de Physique Theorique
CEN Saclay
BP02-91190 Gif-sur-Yvette, FRANCE

Path integrals or functional integrals have proved to be a useful theoretical tool to solve problems in statistical mechanics or quantum field theory and they have been recently applied to the (nuclear) many-body problem.[1)] In particular path integrals have been used to analyse and calculate some of the corrections to the mean field approximations.[2)] In this contribution, we show how path integrals defined on overcomplete sets of the Hilbert space can provide a deep understanding of the relations between various approximation schemes or various phenomenological models. We emphasize the special role of the mean field approximation and show the connection between the functional integral and the well known boson expansion methods. Detailed presentation of this work is given elsewhere,[3-5)] so that only the main results will be discussed here.

The central concept in this work is that of continuous overcomplete set of the Hilbert space.[6)] This is defined as a subset of vectors of the Hilbert space $\mathcal{H}$ from which one can extract a complete basis. Thus the decomposition of unity in $\mathcal{H}$ can be written:

$$\mathbb{1} = \int d\mu(z)\ |z\rangle\langle z| \qquad (1)$$

where the state labels z (generally a set of complex numbers) vary continuously. A familiar example of overcomplete set is that of the coherent states of the harmonic oscillator. These are defined by:

$$|z\rangle = e^{zc^+}|0\rangle \qquad (2)$$

where c^+ is the creation operator of a quantum of the oscillator. The coherent state (2) is the eigenstate of the operator c with eigenvalue z:

$$c|z\rangle = z|z\rangle. \qquad (3)$$

In terms of the state (2) the closure relation (1) takes the form:

$$\mathbb{1} = \int \frac{dz^* dz}{2\pi i}\, e^{-z^* z}|z\rangle\langle z| \qquad (4)$$

and the integration is extended to the whole complex plane of the variable z.

An overcomplete set of interest in nuclear physics is that of the Slater determinants. Let $|\phi_o\rangle$ be a given Slater determinant. As well known,[7] any Slater determinant not orthogonal to $|\phi_o\rangle$ can be written:

$$|\underset{\sim}{Z}\rangle = \exp\{ \sum_{ph} Z_{ph}\, a_p^\dagger a_h \} \; |\phi_o\rangle \tag{5}$$

where:

$$a_h^\dagger |\phi_o\rangle = a_p \; |\phi_o\rangle = 0 \; . \tag{6}$$

Using simple group theoretical arguments,[8] one easily obtains the measure $\mu(z)$ necessary to write the decomposition of the identity in terms of the states (5). One has:

$$\mathbb{1} = \int \prod_{ph} \frac{dZ^*_{ph} dZ_{ph}}{2\pi i} \; [\mathrm{Det}(1 + \underset{\sim}{Z}^\dagger \underset{\sim}{Z})]^{n-1} \; |\underset{\sim}{Z}\rangle\langle \underset{\sim}{Z}| \tag{7}$$

where n is the total number of single particle states available. The change of variables:

$$\beta_{ph} = \sum_{h'} Z_{ph'} (1+Z^\dagger Z)^{-1/2}_{h'h} \tag{8}$$

transforms (7) into a simpler form:

$$\mathbb{1} = \int \prod_{ph} \frac{d\beta^*_{ph} d\beta_{ph}}{2\pi i} \; |\underset{\sim}{\beta}\rangle\langle \underset{\sim}{\beta}| \quad . \tag{9}$$

The parameters β_{ph} may be identified[5] with the coefficients of the unitary transformation which carries $|\phi_o\rangle$ into $|\underset{\sim}{Z}\rangle$. It also turns out that the β_{ph}'s are the "classical" analog of the Holstein-Primakoff representation of spin system generalized to SU(n).[9,10] A further change of variables given explicitly by:

$$\alpha = \sqrt{1-\beta^\dagger\beta}\; U \qquad \beta' = \beta U \tag{10}$$

where U is an arbitrary unitary matrix brings (9) into:

$$\mathbb{1} \sim \int d\alpha d\alpha^\dagger d\beta' d\beta'^\dagger \; \delta(\alpha^\dagger\alpha + \beta^\dagger\beta - 1) \; |\alpha\beta\rangle\langle\alpha\beta| \; . \tag{11}$$

The transformation (10) is the one which relates the generalized Holstein-Primakoff representation to the generalized Schwinger representation.[10] Note that the parameters α and β are the expansion coefficients of single particle wave function on a fixed basis and the expression (11) may as well be written:

$$\mathbb{1} \sim \int \prod_{k=1}^{N} \prod_{x} \frac{d\phi^*_k(x) d\phi_k(x)}{2\pi i} \; \prod_{\ell} \delta(\langle\phi_k|\phi_\ell\rangle - \delta_{k\ell}) \; |\phi\rangle\langle\phi|$$

$$|\phi\rangle = \prod_k a_k^\dagger \; |0\rangle \quad . \tag{12}$$

A functional integral representation of the matrix element $\langle \underset{\sim}{Z}^b | e^{-iHt} | \underset{\sim}{Z}^a\rangle$ can be given, using a standard procedure. One gets the following expression:

$$\langle Z^b | e^{-iHt} | Z^a\rangle = \int_{|Z(t_i)\rangle = |Z^a\rangle}^{\langle Z(t_f)| = \langle Z^b|} \mathcal{D}(Z^*, Z) \; e^{iS[Z^*, Z]} \tag{13}$$

where:

$$\mathcal{D}(Z^*,Z) = \prod_{0<\tau<t} d\mu(Z(\tau))\langle Z(\tau)|Z(\tau)\rangle$$

$$S[Z^*Z] = \int_0^t dt \frac{\langle Z(\tau)|i\partial_t - H|Z(\tau)\rangle}{\langle Z(\tau)|Z(\tau)\rangle} - i\log\langle Z^b|Z(t)\rangle. \tag{14}$$

A few comments on the expression (13) are in order. a) First of all, the mathematical steps involved in the derivation of (13) are not totally justified and as a result it is not guaranted that a naive calculation of the integral (13) (if it were possible!) would reproduce the result of the Schrödinger equation, independently of the choice of the overcomplete set. b) It may be verified easily that the action $S[Z^*Z]$ given by eq. 14 is the one which has to minimizedin order to recover the time dependent Schrödinger equation. Let us assume indeed that the overcomplete set $\{|Z\rangle\}$ is the Hilbert spaceitself. Then, minimizing S with respect to Z and Z^* yield:

$$\frac{\delta S}{\delta Z^*(t)} = 0 \Rightarrow i\partial_t|Z^{(-)}\rangle = H|Z^{(-)}\rangle, |Z^{(-)}(0)\rangle = |Z^a\rangle$$

$$\frac{\delta S}{\delta Z(t)} = 0 \Rightarrow -i\langle Z^{(+)}|\overleftarrow{\partial}_t = \langle Z^{(+)}|H, \langle Z^{(+)}(t)| = \langle Z^b| \quad . \tag{15}$$

The matrix element $\langle Z^b|e^{-iHt}|Z^a\rangle$ is related to the value of the action (14) calculated with the solutions $Z^{(-)}$ of the eq. 15 by:

$$\langle Z^b|e^{-iHt}|Z^a\rangle = e^{iS_{st}} = \langle Z^b|Z^{(-)}(t)\rangle . \tag{16}$$

The expression (16) is exact when $\{|Z\rangle\}$ is the set of all the vectors of the Hilbert space. Approximate expressions may be obtained by considering smaller sets. The mean field approximation to the S-matrix is one of these, obtained by restricting $\{|Z\rangle\}$ to be the set of all Slater determinants. (See the contribution by S. E. Koonin at this conference). c) The state labels Z play the role of classical coordinates which describe the evolution of the system in a generalized phase space. These classical coordinates are effectively quantized as boson degrees of freedom in the functional integral formalism. Thus the functional integral on overcomplete sets represents a boson theory, equivalent to the original fermion theory. This feature is very much reminiscent of the boson expansion method, and is further discussed below. d) The use of overcomplete sets provide a very powerful way for realizing changes of variables and introducing collective coordinates. In many works, collective coordinates are introduced by a transformation of the form:

$$(X_1,X_2,\ldots,X_A) \to (Q_1\ldots Q_j;\ \xi_{j+1}\ldots\xi_A) \tag{17}$$

where the X_i's represents the coordinates of the nucleons, $Q_1\ldots Q_j$ are the collective coordinates and the ξ_i's are the intrinsic coordinates. In the functional integral formulation, one introduces many (∞) new coordinates, the overcompleteness of the description being taken care of by the integration measure. For example, in the representation given by eq. 12, the coordinates are the $\phi_i(X)$, so that the change of variable effectively realized is of the form:

$$(X_1,X_2,\ldots,X_A) \rightarrow (\phi_1(X_1),\phi_2(X_2),\ldots,\phi_A(X_A)) \quad . \tag{18}$$

The method has therefore great flexibility.

As mentioned above, path integrals on overcomplete sets and boson expansion method have common features, which we now examine more closely. The basic idea of boson representations is to map injectively the Fermion Fock space into a subspace, called the physical subspace, of a large boson space. The mapping is such that it preserves the matrix elements and that the boson image of the fermion operators have no matrix elements between physical and unphysical states. It follows that the matrix elements of the evolution operator may be calculated either in the Fermion space or in the boson space:

$$<\phi_b|e^{-iHt}|\phi_a> = {}_B<\phi_b|e^{-iH_Bt}|\phi_a>_B \tag{19}$$

where the subscript $_B$ denotes boson images. A simple boson representation is obtained by "quantizing" the solutions of the Schrödinger equation. Let us consider the case of a single particle first. The wave function may be written:

$$\phi(X) = <X|\phi> = <X|\phi_c> \tag{20}$$

where $|\phi_c>$ is the boson coherent state:

$$|\phi_c> = \exp\int dx\phi(X)\psi_B^\dagger(X)\ |0>_B \tag{21}$$

$\psi_B^\dagger(X)$ is a boson creation operator, and $|0>_B$ is the boson vacuum. Thus the "physical" state $|\phi>$ may be considered as the one-particle component of the state $|\phi_c>$. The projector on the physical subspace may be written:

$$P = \int_{-\pi}^{\pi} \frac{d\theta}{2\pi}\, e^{i\theta} e^{-i\theta\hat{N}} \qquad \hat{N} = \int dx\psi_B^\dagger(x)\psi_B(x) \tag{22}$$

and commutes with the boson image of the Hamiltonian

$$H = \int dxdx'\psi^\dagger(X)H(X,X')\psi(X') \tag{23}$$

obtained simply by replacing in (23) $\psi^\dagger$ and ψ by boson field operator. As a very simple illustration of the formalism let us consider the calculation of the partition function. One has:

$$\begin{aligned} \mathrm{Tr}e^{-\beta H} &= \int d\theta e^{i\theta}\prod_\alpha\left(e^{-\beta(\varepsilon_\alpha + i\frac{\theta}{\beta})}-1\right)^{-1} \\ &= \sum_\alpha e^{-\beta\varepsilon_\alpha} \end{aligned} \tag{24}$$

where ε_α are the eigenvalues of H. One recognizes in the intermediate step the partition function of a system of bosons with an imaginary chemical potential $-i\theta/\beta$.

The generalization of the preceding representation to the N-particle system is straightforward.[3)] It constitutes a natural extension of the work of ref. 10. Let us label the N particles by a label i (i = 1,2,...,N) and let $\alpha,\beta,\gamma\ldots$denote a complete set of single particle states. The following states:

$$|\psi\rangle_B = \sum_P (-)^P C_1^\dagger(\alpha_{p_1}) C_2^\dagger(\alpha_{p_2}) \ldots C_N^\dagger(\alpha_{p_N}) \, |0\rangle_B \tag{25}$$

where $(\alpha_{p_1} \ldots \alpha_{p_N})$ is a permutation of $(\alpha_1, \ldots, \alpha_N)$ and the operators $C_i^{(\dagger)}(\alpha_j)$ are boson operators:

$$[C_j(\beta), C_i^\dagger(\alpha)] = \delta_{\alpha\beta} \, \delta_{ij} \tag{26}$$

are in one-to-one correspondence with the fermion-states:

$$|\psi\rangle_F = a_{\alpha_1}^\dagger a_{\alpha_2}^\dagger \ldots a_{\alpha_N}^\dagger \; |0\rangle_F \; . \tag{27}$$

They span the physical subspace of the boson representation. The boson image of the Hamiltonian is:

$$H_B = \sum_{i=1}^{N} \sum_{\alpha\beta} T_{\alpha\beta} C_i^\dagger(\alpha) C_i(\beta) + \frac{1}{2} \sum_{i,j=1}^{N} \sum_{\alpha\beta\gamma\delta} (\alpha\beta|V|\gamma\delta) C_i^\dagger(\alpha) C_j^\dagger(\beta) C_j(\delta) C_i(\gamma) \; . \tag{28}$$

One can now proceed to the calculation of the matrix element of the evolution operator using a functional integral. To this end, one first uses the identity (19) to replace the calculation in the fermion space by a calculation in the boson space. Now, in the boson space, one can use as overcomplete set, the set of coherent states:

$$|\underset{\sim}{Z}\rangle = \exp \Big\{ \sum_{i=1}^{N} \sum_{\alpha} Z_i(\alpha) C_i^\dagger(\alpha) \Big\} \, |0\rangle_B \; . \tag{29}$$

Note that these states do not belong to the physical subspace. The projection on the physical subspace is realized at the end points in the functional integral:

$${}_B\langle\phi^b| e^{-iH_B t} |\phi^a\rangle_B = \int d\mu(\underset{\sim}{Z}^b) d\mu(\underset{\sim}{Z}^a) {}_B\langle\phi^b|\underset{\sim}{Z}^b\rangle\langle\underset{\sim}{Z}^b| e^{-iH_B t} |\underset{\sim}{Z}^a\rangle\langle\underset{\sim}{Z}^a|\phi^a\rangle_B \tag{30}$$

$$\langle\underset{\sim}{Z}^b| e^{-iH_B t} |\underset{\sim}{Z}^a\rangle = \int_{\underset{\sim}{Z}(0)=\underset{\sim}{Z}^a}^{\underset{\sim}{Z}^*(t)=\underset{\sim}{Z}^{b*}} \mathcal{D}(\underset{\sim}{Z}^*, \underset{\sim}{Z}) \; e^{\, i \int_0^t \{ i\underset{\sim}{Z}^* \dot{\underset{\sim}{Z}} - H_B(\underset{\sim}{Z}^*, \underset{\sim}{Z}) \} dt - i\underset{\sim}{Z}^{b*} \cdot \underset{\sim}{Z}(t)} \tag{31}$$

where $H_B(Z^*, Z) = \dfrac{\langle Z|H_B|Z\rangle}{\langle Z|Z\rangle}$ is obtained simply by replacing in the expression (28) the operators $C_i^\dagger(\alpha)$ and $C_j(\beta)$ by C-numbers $Z_i^*(\alpha)$ and $Z_j(\beta)$ respectively.

One can show that the functional integral (31) together with the eq. (30) retrieves the perturbation expansion for the fermion problem, which gives some confidence in the validity of the formalism. It is also worth-mentioning that there are many ways of projecting onto the physical subspace. In the present case we have treated this constraint globally, projecting only at the end points of the functional integral. It is also possible to require that the "path" lie entirely within the physical subspace. This is obtained by inserting a projector onto this subspace at each time in the construction of the path integral. One then ends up with an expresion very similar to (31) with two noticeable differences.

i) The measure contains a δ-function: $((Z_k|Z_\ell) - \delta_{k\ell})$.

ii) In the hamiltonian $H_B(Z^*,Z)$, the direct matrix element of the two-body interaction is replaced by an antisymmetrized matrix element:

$$(\alpha\beta|V|\gamma\delta) \rightarrow \langle\alpha\beta|V|\gamma\delta\rangle = (\alpha\beta|V|\gamma\delta) - (\alpha\beta|V|\delta\gamma) \ .$$

The structure of this functional integral is identical to that obtained by using as overcomplete set, the set of the Slater determinants and the measure (12). The presence of the δ-function makes it complicated and it has to be used with great care.

Let us now show that calculating the functional integral (31) in the saddle point approximation retrieves the mean field equations. The saddle points are given by the equations:

$$\frac{\delta S}{\delta Z_k^*(\alpha,\tau)} = 0 \quad \Rightarrow \quad i\overset{\circ}{Z}_k(\alpha,\tau) = \sum_\beta h_{\alpha\beta} Z_k(\beta,\tau) \qquad Z_k(\alpha,0) = z_k^a(\alpha)$$

$$\frac{\delta S}{\delta Z_k(\alpha,\tau)} = 0 \quad \Rightarrow \quad -i\overset{\circ}{Z}{}_k^*(\alpha,\tau) = \sum_\beta Z_k^*(\beta,\tau) h_{\beta\alpha} \qquad Z_k^*(\alpha,t) = z_k^{b^*}(\alpha) \tag{32}$$

where h is the familiar Hartree Hamiltonian:

$$h_{\alpha\beta} = T_{\alpha\beta} + \sum_{i=1}^{N} \sum_{\gamma\delta} (\alpha\gamma|V|\beta\delta) Z_i^*(\gamma) Z_i(\delta) \ . \tag{33}$$

Had one done the same approximation on the functional integral defined over Slater determinants one would have obtained the time dependent Hartree-Fock approximation, i.e. a single particle hamiltonian similar to (33) but with the exchange terms included.

The time dependent Hartree approximation can also be obtained in a different way. Consider the exact equation of motion:

$$i \frac{d}{dt} C_k(\alpha) = \sum_\beta \hat{h}_{\alpha\beta} C_k(\beta) \tag{34}$$

with:

$$\hat{h}_{\alpha\beta} = T_{\alpha\beta} + \sum_{i=1}^{N} \sum_{\gamma\delta} (\alpha\gamma|V|\beta\delta) C_i^\dagger(\gamma) C_i(\delta) \ . \tag{35}$$

By replacing the operators $C_i^\dagger(\gamma)$, $C_i(\delta)$ by C-numbers $Z_i^*(\gamma)$, $Z_i(\delta)$ respectively, one recovers the eq. (32).

This replacement of the boson operators by C-numbers, or the calculation of the functional integral by the saddle point method, correspond to a "classical" limit to the many-body problem. This classical limit has to be understood in the following way. The state labels of the overcomplete set satisfy, in this limit, equations of motion which can be interpreted as equations of motions of classical mechanics, defining a continuous "path" in the Hilbert space. The "quantum" effects are recovered, as in the standard Feynman integral formalism, by allowing fluctuations around these classical paths. The functional integral provides a natural framework to calculate these effects, which lead to the so-called "quantization" of the time dependent mean field equations. As we have seen, boson representations are an alternative to path integrals to realize this quantization. In most cases both technics are equivalent, at least at the formal level: Since the state labels of the overcomplete set are quantized as bosons, to each choice of overcomplete set or of its parameterization corresponds a boson representation. For example it was shown at the beginning of this paper how various parameterization of Slater determinants lead naturally to some well-known boson representations. The functional integral formalism has also some intimate connection with the generator coordinate method, but we will not discuss it here.

In conclusion, we have shown that path integrals on overcomplete sets provide a unifying understanding of various approaches to the nuclear many-body problem. Most of the important methods of approximations developed to describe nuclear collective motion can be simply interpreted in the functional integral formalism and it is not unreasonable to believe that this method may lead to interesting new approximation schemes.

References

1. See for example, H. Reinhard, Nucl. Phys. A298 (1977) 77. S. Levit, Phys. Rev. C 21 (1980) 1594.
2. H. Reinhard, Nucl. Phys. A346 (1980) 1.
3. J. P. Blaizot and H. Orland, J. Phys. Lett. 41 (1980) 53, J. Phys. Lett. 41 (1980) 523.
4. J. P. Blaizot and H. Orland, to be published in Phys. Lett. B.
5. J. P. Blaizot and H. Orland, submitted to Phys. Rev. C.
6. J. R. Kla der, Ann. Phys. 11 (1960) 123.
7. D. J. Thouless, Nucl. Phys. 21 (1960) 225.
8. H. Kuratsiyi and T. Suzuki, Phys. Lett. 92B (1980) 19.
9. D. Jansen, F. Dönan, S. Frauendorf and R. V. Jolos, Nucl. Phys. A172 (1971) 145.
10. J. P. Blaizot and E. R. Marshalek, Nucl. Phys. A309 (1978) 422,453.

APPLICATIONS OF GENERALIZED HOLSTEIN-PRIMAKOFF TRANSFORMATIONS TO PROBLEMS OF NUCLEAR COLLECTIVE MOTION

Abraham Klein and C. T. Li
Department of Physics, University of Pennsylvania
Philadelphia, Pennsylvania 19104/USA

and

Michel Vallieres
Department of Physics and Atmospheric Sciences
Drexel University
Philadelphia, Pennsylvania 19104/USA

1. Introduction

The Holstein-Primakoff (HP) transformation was introduced originally into the theory of ferromagnetic spin waves in order to map a problem of coupled spins into a problem of coupled oscillators [1]. That such a mapping is possible is seen most simply by considering the irreducible representations of the familiar angular momentum or SU(2) algebra as represented by the three operators $J_\pm$, J_z satisfying the commutation relations

$$[J_+,J_-] = 2J_z \quad , \tag{1.1}$$

$$[J_z,J_\pm] = \pm J_\pm \quad , \tag{1.2}$$

where $J_\pm$ are hermitian conjugates. In an irreducible representation labeled by $|j,m\rangle$, we have the non-vanishing matrix elements

$$\langle jm|J_z|jm\rangle = m, \ -j\leq m \leq j \ , \tag{1.3}$$

$$\langle j,m+1|J_+|jm\rangle = [(j-m)(j+m+1)]^{\frac{1}{2}} \ . \tag{1.4}$$

With a change of notation

$$m = -j+n \quad , \tag{1.5}$$

these formulas can be written

$$\langle n|J_z|n\rangle = -j+n, \quad 0 \leq n \leq 2j \quad , \tag{1.6}$$

$$\langle n|J_+|n+1\rangle = [(n+1)(2j-n)]^{\frac{1}{2}}. \tag{1.7}$$

We have suggestively removed the symbol j from the basis vectors because it permits an interpretation of the formulas (1.6) and (1.7) in an oscillator basis (boson basis)

$$|n\rangle = \frac{(b^+)^n}{\sqrt{n!}}\,|0\rangle \quad , \tag{1.8}$$

with J_z, $J_\pm$ given by the operators

$$J_z = -j + b^\dagger b \quad , \tag{1.9}$$

$$J_+ = (J_-)^\dagger = b^\dagger[2j-b^\dagger b]^{\frac{1}{2}} \ . \tag{1.10}$$

These are the HP formulas which are thus seen to realize the irreducible representation (1.6,7) in a subspace of a boson space, since the transformation is restricted to the values $n \leq 2j$. Using the boson commutation relations, $[b,b^\dagger] = 1$, it can be verified directly that the operators (1.9,10) satisfy the commutators (1.1,2).

In the ferromagnetic problem, where one has a large number of kinematically independent spins, the total mapping is a direct product of individual mappings which, upon introduction into the Hamiltonian, effect the desired reformulation.

The HP transformation first appeared in nuclear physics |2| in connection with the study of a highly simplified shell model problem known as the MGL model[3]. In this model fermions are assigned to one of two orbits of equal degeneracy, separated by an energy difference ε. Let $\alpha_{m\sigma}^\dagger$ create a fermion in the sublevel m, $1 \leq m \leq N$, $\sigma = \pm$ for upper and lower level respectively. The bilinear operators

$$J_+ = \sum_m \alpha_{m_+}^\dagger \alpha_{m_-} = (J_-)^\dagger, \tag{1.11}$$

$$J_z = \tfrac{1}{2}(n_+ - n_-) \quad , \tag{1.12}$$

$$n_\pm = \sum_m \alpha_{m_\pm}^\dagger \alpha_{m_\pm} \quad , \tag{1.13}$$

which satisfy the SU(2) algebra (1.1,2) are often called quasi-spin operators because the real angular momentum content, identified with the "magnetic quantum number" m has been recoupled in the combinations (1.11-13) to total angular momentum zero. For this reason, in the typical Hamiltonian studied,

$$H_{MGL} = \varepsilon J_z + (f/2N)\ [J_+^2 + J_-^2] \quad , \tag{1.14}$$

the second term which is largely a two-fermion interaction is characterized as a monopole-monopole interaction. The problem usually studied is one in which there are N fermions present and thus the lower level is fully occupied for $f = 0$. This is the state with $J_z = -\tfrac{1}{2}N$ and under the mapping (1.9,10) becomes the boson vacuum with $n = 0$. The mapping is a convenient starting point both for exact diagonalization and for the study of various approximations. Selected aspects of a generalized version of this problem are considered in Sec. 3.

Quasi-spin appears in a more essential context in nuclear physics in connection with the study of pairing interactions or nuclear superconductivity. For a single level, creation operators $\alpha_m^\dagger$, the operators

$$A^\dagger = \sum_{m>0} s_m \alpha_m^\dagger \alpha_{-m}^\dagger = (A)^\dagger \quad , \tag{1.15}$$

where s_m is a phase factor satisfying $s_{-m} = -s_m$, and $\tfrac{1}{2}\,(N-\Omega)$, where

$2\Omega = 2j+1$, and

$$N = \sum_m \alpha_m^{\dagger} \alpha_m \tag{1.16}$$

can be identified with $J_{\pm}$, J_z, respectively, since we have the algebra

$$[A^{\dagger},A] = N-\Omega, \tag{1.17}$$

$$[N,A^{\dagger}] = 2A^{\dagger} . \tag{1.18}$$

If we have a set of non-degenerate levels labeled by j_a, the conventional pairing Hamiltonian

$$H_{pairing} = \sum_a \varepsilon_a N_a - \sum_{ab} G_{ab} A_a^{\dagger} A_b \quad , \tag{1.19}$$

where ε_a are the single particle energies and G_{ab} the matrix elements of the pairing interaction, is conveniently studied with the help of a direct product of HP transformation [4,5].

During the past decade, understanding has spread that the concepts involved in the HP mapping can usefully be extended to more general Lie algebras, most simply and directly to the unitary algebras [6,7] but also to the algebras of orthogonal groups [8,9]. In this paper, we shall restrict attention to two recent applications carried out by us of a generalized HP transformation applicable to the systematic representations of SU(n).

2. Relation between the Interacting Boson Model (IBM) and the Bohr-Mottelson Model (BMM).

The past five years has witnessed the onslaught of a new phenomenological description of "vibrations and rotations" in medium to heavy nuclei away from closed shells, precisely the range of phenomena that was previously the preserve of the Bohr-Mottelson Model (BMM). We shall first describe the essential content of this model [10] which utilizes the machinery of the Lie algebra SU(6) and then ask for the relationship of this model to the BMM. It is in response to the latter question that the generalized HP transformation will play a role.

In its simplest version, to which we shall restrict our attention, the IBM postulates that the collective phenomena under discussion can be understood as the manifestation of two elementary excitations and of their interactions: (i) The s-boson carries no angular momentum but adds two neutrons or two protons to the system. These s-pairs describe nuclear superconductivity. (ii) The d-boson, d_μ (with five components $\mu = -2,\dots 2$) carries angular momentum two and also adds two neutrons or two protons to the system. The d-pairs are mainly responsible for producing quadrupole deformations. There is a (different) binding energy associated with each pair, and the pairs may interact. Up to now calculations have been confined to hamiltonians with two body interactions only. As an interesting example, though hardly the general case, we consider the hamiltonian

$$H_{IBM} = \varepsilon \sum_\mu d_\mu^\dagger d_\mu - \kappa \sum_\mu Q_\mu (-1)^\mu Q_{-\mu} , \qquad (2.1)$$

where Q_μ is a quadrupole operator

$$Q_\mu = s^\dagger d_{-\mu} (-1)^\mu + d_\mu^\dagger s + \chi (d^\dagger d)^{(2)}_\mu , \qquad (2.2)$$

and ε, κ, χ are parameters. The hamiltonian (2.1) conserves the total number of bosons, N,

$$N = s^\dagger s + \sum_\mu d_\mu^\dagger d_\mu , \qquad (2.3)$$

which is taken to be the number of "active fermions". Since we require any generalized IBM hamiltonian to commute with N, it follows that such a hamiltonian will be a polynomial in the 36 operators $d_\mu^\dagger d_\nu$, $d_\mu^\dagger s$, $s^\dagger d_\mu$, and $s^\dagger s$ which close under commutation and thereby generate the Lie algebra U(6), or since $s^\dagger s$ can be eliminated by means of (2.3), the Lie algebra SU(6).

The simplest and perhaps most fundamental way to describe the IBM - at least for our purposes - is to assert that a certain class of nuclear states can be set into one-to-one correspondence with states in the finite dimensional vector space

$$|[\mu_i], n_s> = [n_d! n_s!]^{-\frac{1}{2}} \left(\prod_{i=1}^{n_d} d_{\mu_i}^\dagger\right) (s^\dagger)^{n_s} |0>, \qquad (2.4)$$

where $N = n_d + n_s$. For each N, this is an invariant space under the action of any hamiltonian allowed by the model. In fact, the states (2.4) constitute the carrier space of the symmetric representation of SU(6), completely labeled by the value of N. We are invited to diagonalize H_{IBM} in this space. Thus, we may say that certain states of a given even nucleus can be described in a sub-space of a six-dimensional coupled oscillator.

By contrast, the BMM model [11] may be said to establish a correspondence between the aforementioned "collective" states of even nuclei and a subspace of a 5-dimensional anharmonic oscillator described by a boson, b_μ, which carries angular momentum two but does not change the number of nucleons. Thus there is no requirement that the number n_b of b bosons,

$$n_b = \sum_\mu b_\mu^\dagger b_\mu \qquad (2.5)$$

be conserved. The hamiltonian may be taken to be any rotationally invariant function of the $b_\mu^\dagger$, b_λ, but it is usually chosen to be a polynomial. This hamiltonian is, for a given nucleus, to be diagonalized in a basis

$$|[\mu_i]\rangle = (n_d!)^{-\frac{1}{2}} \prod_{i=1}^{n_d} b_{\mu_i}^{\dagger} |0\rangle_N \quad , \tag{2.6}$$

where $|0\rangle$ means a vacuum for quadrupole bosons, and one usually does not consider any theoretical upper bound on n_d, only a practical one, in carrying out a diagonalization.

The possible relationship between IBM and BMM can now be considered. As we shall see, it can be made precise provided we sharpen the definition of the BMM just given in two ways: (i) We assume that the state $|0\rangle_N$ may be written as

$$|0\rangle_N = (N!)^{-\frac{1}{2}}(s^{\dagger})^N \; |0\rangle, \tag{2.7}$$

where $s^{\dagger}$ and $|0\rangle$ are to be identified with similar quantities in the IBM description. (ii) We bound n_d by $n_d \leq N$. Thus we arrive at the finite basis

$$|[\mu_i], N\rangle = (N!n_d!)^{-\frac{1}{2}} \prod_1^{n_d} (b_{\mu_i}^{\dagger})(s^{\dagger})^N \; |0\rangle \; ,$$

$$n_d \leq N \quad . \tag{2.8}$$

If we accept (2.8) together with associated statements about the hamiltonian as the "definition" of BMM just as we took (2.4) as the corresponding characterization of the IBM, then we can state the relationship between them in the form of a Theorem: The basis vectors (2.4) and (2.8) are in one to one correspondence. As expressed by Eqs. (2.9-11) below, that correspondence relates any given IBM hamiltonian to its equivalent BMM hamiltonian and conversely.

The proof is based on the availability of a generalized HP transformation. The IBM formulation is a special illustration of the general circumstance that the generators of the algebra U(n) (or SU(n)) may be realized as bilinear, "number conserving" operators constructed from n-bosons a_λ, namely $a_\lambda^{\dagger} a_\nu$, λ, $\nu=1...n$. For n=2 the HP transformation allows us to realize SU(2) in terms of n-1 bosons, namely a single one. This yields all the irreducible representations of SU(2). For SU(n), a corresponding realization in terms of n-1 bosons is possible only for the symmetric representation. For the IBM the mapping chosen is

$$d_\mu^{\dagger} d_\nu = b_\mu^{\dagger} b_\nu \quad , \tag{2.9}$$

$$d_\mu^{\dagger} s = (s^{\dagger} d_\mu)^{\dagger} = b_\mu^{\dagger} \sqrt{N - \sum_\lambda b_\lambda^{\dagger} b_\lambda} \; , \tag{2.10}$$

$$s^{\dagger} s = N - \sum_\mu d_\mu^{\dagger} d_\mu \equiv N - \sum_\mu b_\mu^{\dagger} b_\mu \quad . \tag{2.11}$$

It may be verified directly that both sets of operators satisfy the same commutation relations. Next, by substituting the last equality from

(2.11) into (2.10) and inverting the radical, we obtain the formula

$$b_\mu^\dagger = d_\mu^\dagger \; s \; (N-\sum_\lambda d_\lambda^\dagger d_\lambda)^{-\frac{1}{2}} \; . \qquad (2.12)$$

Substituting (2.12) into (2.8), it is a trivial exercise that this becomes a member of the set (2.4). The process is easily inverted, requiring only a rearrangement after use of the formula

$$(n+1)(s^\dagger)^n \,|0\rangle = s(s^\dagger)^{n+1} \,|0\rangle \quad . \qquad (2.13)$$

The substitution of (2.9-11) into the IBM hamiltonian, for instance (2.1), yields the equivalent BMM hamiltonian. However this is not of polynomial type. Conversely the substitution of (2.12) into a BMM hamiltonian of polynomial form would yield an IBM hamiltonian of non-polynomial type. Though we are guaranteed equivalence of physical content of the two formulations, there remain questions of elucidation of this content which go somewhat beyond the limits of the present subject. We pursue these matters only in the briefest outline.

Starting with (2.1), as an illustration, we deal with the radicals in the transformed BMM form by introducing (non-Hermitian) canonical coordinates,

$$b_\mu^\dagger = 2^{-\frac{1}{2}}(x_\mu - ip_\mu^\dagger) \; , \qquad (2.14)$$

$$(-1)^\mu \, b_{-\mu} = 2^{-\frac{1}{2}}(x_\mu + ip_\mu^\dagger) \; , \qquad (2.15)$$

where

$$x_\mu^\dagger = (-1)^\mu x_{-\mu}, \; p_\mu^\dagger = (-1)^\mu p_{-\mu} \quad , \qquad (2.16)$$

$$[x_\mu, \, p_\nu] = [x_\mu^\dagger, p_\nu^\dagger] = i\delta_{\mu\nu} \; . \qquad (2.17)$$

If we set $p_\mu = 0$, the hamiltonian becomes a function only of two shape parameters β, $\hat{\gamma}$ according to the standard-looking formulas

$$\sum_\mu x_\mu^\dagger x_\mu \equiv N\bar{\beta}^2 \qquad (2.18)$$

$$[(x \otimes x)^{(2)} \otimes x]^{(0)} \equiv -(2/7)^{\frac{1}{2}} \, N^{3/2} \, \bar{\beta}^3 \cos 3\gamma \; , \qquad (2.19)$$

though $\bar{\beta}$ is here bounded by $\sqrt{2}$, arising from the finiteness of the vector space (the radical of the HP transformation). Therefore a further scale change is made to a variable β which ranges over the entire positive real axis,

$$\beta = \sqrt{2} \; \bar{\beta}/(1+\bar{\beta}^2)^{\frac{1}{2}} \; . \qquad (2.20)$$

This yields a potential energy function

$$\lim_{N\to\infty} (H/N\varepsilon) \equiv v(\beta,\gamma) = \frac{(\tfrac{1}{2}-2F)\ 2\beta^2}{1+\beta^2} + \frac{4F\chi(2/7)^{\frac{1}{2}}\ \beta^3 \cos 3\gamma}{(1+\beta^2)^2}$$
$$+ F\ \{1-(\chi^2/14)\}\frac{4\beta^4}{(1+\beta^2)^2} \qquad (2.21)$$

where $F = (\kappa N/\varepsilon)$. This result has been obtained previously by a completely different method [12,13], which, however is not adequate to give the full BMM hamiltonian. Except for the denominators, which reflect the finiteness of the vector space and therefore the finite number of bound states, this potential energy function shares all the properties of the most general such function given by a BMM hamiltonian of fourth power in the b_λ, $b_\mu^\dagger$ bosons.

We shall emphasize but one more point. Early discussions of the IBM [10] emphasized limiting dynamical properties, namely SU(5), associated with vibrations, SU(3), associated with strong deformation and O(6), associated with a γ-independent deformation. Of these, only the SU(5), which occurs when not only N but also n_d is conserved and $\varepsilon >> \kappa$ in Eq. (2.1), had been previously well understood [14]. Why had the other symmetries been overlooked? The answer is clearly given by the HP mapping. As an example, the generators of the SU(3) subgroup are the operators $(d^\dagger d)_\mu{}^{(1)}$ (the angular momentum operator) and the operators Q_μ of (2.2) for the special value $\chi = -(\sqrt{7}/2)$. Under the HP mapping the latter operator acquires the characteristic square root. Thus this SU(3) can be formed from the BMM bosons, but <u>not</u> as a polynomial. A corresponding analysis holds for the O(6), though in the latter case its O(5) subgroup was known previously [15] because its generators contain no square roots. Thus the HP transformation shows us that the symmetries evidenced by the IBM are also implicit in the BMM model, but only in the form of non-linear realizations.

3. Derivation of a Boson Model from a Shell Model

In the previous section, we have utilized the generalized HP correspondence to establish the physical equivalence between the IBM and the BMM. A fundamental problem of nuclear collective motion is to derive either of these models from the underlying shell model. This is a formidable task which we can hardly address as a small add-on to the present considerations. We can, however, show how this problem is attacked and solved for a highly simplified class of models, a set of generalized MGL models to which we can apply the same tools as previously developed.

In a typical one of these models there are n non-degenerate levels each with sublevel degeneracy N. Let $\alpha_{mr}^\dagger$ be the creation operator for a fermion in the mth sublevel of level r. The bilinear operators

$$A_r^{\,s} = (A_s^{\,r})^\dagger = \sum_m \alpha_{mr}^\dagger \alpha_{ms}, \quad r, s=1,\ldots,n \tag{3.1}$$

are a set of generators of the Lie algebra, U(n). The operators

$$J_o^{(r)} = \tfrac{1}{2}(A_{r+1}^{\,r+1} - A_1^{\,1}) \quad , \tag{3.2}$$

$$J_+^{(r)} = (J_-^{(r)})^\dagger = A_{r+1}^{\,1} \quad r=1,\ldots,n-1 \tag{3.3}$$

and the remaining $A_{r+1}^{\,s+1}$, $r \neq s$, generate the algebra SU(n). We study the shell model Hamiltonian

$$H_{MGL} = \varepsilon\{\sum_{r=1}^{n-1} \eta_r J_o^{(r)} - (f/2N) \sum_r [(J_+^{(r)})^2 + (J_-^{(r)})^2]\} \ , \tag{3.4}$$

which generalizes (1.14). Here $\varepsilon\eta_r$, with $\eta_{r+1} > \eta_r$ and $\Sigma\, \eta_r = 1$, measures the unperturbed energy difference between level r+1 and the ground level 1, and ($\varepsilon f/2N$) is an overall coupling strength for the interaction which connects the ground level with any other level, but the latter do not make direct transitions to one another. This problem was studied some time ago for n=3 [16,17], and more recently, the features which we shall discuss were first studied by coherent state methods [18]. Again we consider the problem with the ground level fully occupied, with N particles - when $f \to 0$.

To apply the HP mapping, we introduce n-1 mutually commuting bosons b_r, r=1,...,n-1 and write

$$A_{r+1}^{\,s+1} = b_r^\dagger b_s, \qquad r, s=1,\ldots,n-1 \ , \tag{3.5}$$

$$A_{r+1}^{\,1} = (A_1^{\,r+1})^\dagger = b_r^\dagger \theta(N) \quad , \tag{3.6}$$

$$A_1^{\,1} = \theta(N)\theta(N) \quad , \tag{3.7}$$

where

$$\theta(N) = (N-\hat{n})^{\frac{1}{2}} \quad , \tag{3.8}$$

$$\hat{n} = \sum_{r=1}^{n-1} b_r^\dagger b_r \quad . \tag{3.9}$$

The substitution of (3.5-9) into (3.4) maps the shell model problem onto a subspace of (n-1) bosons. We shall be interested in approximate methods of solving the resulting problem defined by the hamiltonian which we thus obtain,

$$H = H_o + H_1 \ , \tag{3.10}$$

$$H_o = -\tfrac{1}{2}\varepsilon(N-\hat{n}) + \tfrac{1}{2}\,\varepsilon\sum_r \eta_r b_r^\dagger b_r \ , \tag{3.11}$$

$$H_1 = -\tfrac{1}{2}f\varepsilon \sum_r\{b_r^\dagger b_r^\dagger[1-((\hat{n}+1)/N)]^{\frac{1}{2}}\,[1-(\hat{n}/N)]^{\frac{1}{2}} + h.c.\}. \tag{3.12}$$

This problem can be studied by different techniques in three interesting regimes [19]:

(i) Weak coupling, $|f/\eta_1| \ll 1$. In this regime the ground state is almost the unperturbed state and all expectation values of $b_r^\dagger b_r$ are small compared to N. Under these conditions the square root may be expanded, and we obtain a standard problem of coupled oscillators. Here it is understood that we are concerned with low-lying energy levels, for otherwise the condition imposed on the average occupation numbers is certainly incorrect.

(ii) Strong coupling and semi-classical regimes. $|f/\eta_1| \gg 1$. As the relevant ratio increases we encounter a "phase transition" (see below). We may study this possibility and prepare to study both the strong coupling limit and (the distinct) semi-classical limit by introducing canonical variables, according to the formula

$$b_r = (b_r^\dagger)^\dagger = (\tfrac{1}{2}J)^{\frac{1}{2}} (p_r - i x_r) \quad , \quad (N=2J) \quad , \tag{3.12}$$

which implies the commutation relation

$$[x_r, p_s] = (i/J)\, \delta_{rs} \quad , \tag{3.13}$$

and shows that the parameter (1/J) plays the role of $\hbar$. Now if we assume provisionally that both of the combinations

$$\underline{x}^2 = \sum_r x_r^2 , \quad \underline{p}^2 = \sum_r p_r^2 \tag{3.14}$$

are large compared to unity (of order J), but that we can expand zero point fluctuations, then, an expansion to order J^{-1} rids (3.10-12) of the square roots and yields a hamiltonian of the form

$$h \equiv (H/\varepsilon J) = t + v \quad , \tag{3.15}$$

$$t = \tfrac{1}{4} \sum_r (\eta_r + 1) p_r^2 + \tfrac{1}{2}\bar{f}\, \underline{p}^2 - \tfrac{1}{8} f (\underline{p}^2)^2 \quad , \tag{3.16}$$

$$v = -1 - (n/4J) + \tfrac{1}{4} \sum_r (\eta_r + 1)\, x_r^2 - \tfrac{1}{2}\bar{f}\, \underline{x}^2 + \tfrac{1}{8} f\, (\underline{x}^2)^2 \quad , \tag{3.17}$$

where $\bar{f} = f\,[1 + (n/4J)]$.

Equations (3.15-3.17) have been utilized to study two regimes. For the strong-coupling low energy regime, we have $x_r \sim 1$, but $p_r \sim J^{-1}$ (the product is of the same order as the commutator in this quantum domain). Then the leading term in h is the potential energy v. For f small enough v has a "spherical" minimum at $x_r = 0$. But for

$$[2\bar{f}/(\eta_1 + 1)] > 1 \quad , \tag{3.18}$$

we obtain degenerate minima at the solutions of

$$f x_1^2 = 2\bar{f} - (\eta_1 + 1) \quad . \tag{3.19}$$

As f increases further new local minima intervene. To study the energy levels in this regime, it should also suffice to keep only the quadratic terms in t - this is the regime usually called adiabatic.

The other domain for which (3.15-17) may be utilized is the semi-classical one for which h must be used as it stands. In this case a complete study has been carried out by means of the WKB approximation [20], but we cannot enter here into any of the details.

4. Final Remarks.

In this paper we have emphasized an exact realization by means of bosons of the symmetric representation of SU(n). Exact realizations which have been utilized in comparable ways exist for models based on the O(5) Lie algebra [21] and on the SO(8) algebra [22]. This work is clearly at a young stage, but represents an interesting prelude to the real physical problem, where at best only approximate mappings can be utilized at present [9].

REFERENCES

1. T. Holstein and H. Primakoff, Phys. Rev. 58, 1098 (1940).
2. S. C. Pang, A. Klein, and R. M. Dreizler, Ann. Phys. (N.Y.) 49, 477 (1968).
3. H. J. Lipkin, N. Meshkov, and A. J. Glick, Nucl. Phys. 62, 188 (1965).
4. M. Kleber, Phys. Letters 30B, 588 (1969).
5. A. Klein and M. Vallieres, U. of Pennsylvania preprint UPR-0163T (1980).
6. D. Janssen, R. V. Jolos and F. Donau, Nucl. Phys. A224, 93 (1974).
7. S. Okubo, J. Math. Phys. 16, 528 (1975).
8. E. R. Marshalek, Nucl. Phys. A347, 253 (1980).
9. T. Otsuka, A. Arima, and F. Iachello, Nucl. Phys. A309, 1 (1978).
10. Interacting Bosons in Nuclear Physics, ed. F. Iachello (Plenum Press, New York, 1979).
11. A. Bohr and B. R. Mottelson, Nuclear Structure, Vol. II (W. A. Benjamin, Reading, Mass., 1975).
12. J. N. Ginocchio and M. W. Kirson, Phys. Rev. Lett. 44, 1744 (1980).
13. A. Dieperink, O. Scholten, and F. Iachello, Phys. Rev. Lett. 44, 1747 (1980).
14. T. K. Das, R. M. Dreizler, and A. Klein, Phys. Rev. C2, 632 (1970).
15. L. Wilets and M. Jean, Phys. Rev. 102, 788 (1956).
16. N. Meshkov, Phys. Rev. C3, 2214 (1971).
17. S. Y. Li, A. Klein and R. M. Dreizler, J. Math. Phys. 11, 975 (1970).
18. R. Gilmore and D. H. Feng, Phys. Lett. 85B, 155 (1979).
19. A. Klein, Phys. Lett. 95B, 327 (1980).

20. A. Klein and C. T. Li, U. of Pennsylvania report UPR-0166T (1980).

21. A. Klein, H. Rafelski, and J. Rafelski, Nucl. Phys. A (to be published) (UPR-0159T).

22. A. Arima, N. Yoshida, and J. Ginocchio, Los Alamos Report LA-UR-80-2596 (1980).

THEORY OF METAL SURFACES: VARIATIONAL CALCULATION USING A CORRELATED WAVE FUNCTION

Xin Sun, Tiecheng Li, and Chia-Wei Woo
Department of Physics
University of California, San Diego
La Jolla, CA 92093, USA

Density profiles, surface energies, and Gibbs surfaces are calculated for metal surfaces over a wide range of electron densities using a variational approach. The trial wave function contains a determinant of model single particle orbitals which are solutions of a variational surface potential, and a pair correlation factor. Integral equations relating the wave function and the distribution functions previously derived by Chakravarty and Woo for homogeneous electron liquids and by Rajan and Woo for inhomogeneous Coulomb systems are employed. Divergences appear in the integral equations and the energy expression, but are proved to cancel exactly. Surface energies obtained with just one variational parameter are high compared to those of Lang and Kohn using the density functional formalism, but are close to those extrapolated from experimental data on liquid metals.

We report here a successful attempt to treat the metal surface with a correlated wave function. Space limitation dictates that only a brief summary will be given. Details of the theory and calculations will be published elsewhere.

The conventional many-body theoretic treatment of the metal surface is by using the density functional formalism invented by Kohn and co-workers.[1] We have sought an alternative which will (i) provide us with wave functions, (ii) have the variational principle applied directly on the wave function, (iii) be free of a density gradient expansion, (iv) treat the homogeneous bulk and the inhomogeneous system in the same framework, and (v) lend itself to extensions beyond the jellium (uniform charge background) approximation without resorting to a first-order perturbation theory. (i) will be of practical importance in future work on adsorption. For example, the availability of wave functions enables us to calculate vertices that describe the coupling between adsorbed atoms (and molecules) and the elementary excitations in the adsorbent surface. Elimination of these vertices by a suitable transformation will then dress the adsorbed particles as well as renormalize adatom-adatom interactions. (ii)-(iv) refer to internal consistencies in the theoretical formalism. (v) places our many-body problem in touch with the real world. The alternative that we seek lies in the method of correlated basis functions invented by Feenberg and coworkers.[2]

We have approached the development of such a theory of metal surface with much caution. First, the long range correlation effects in a Coulomb system can give rise to unpleasant divergences. And secondly, inhomogeneities, especially those as sharp as that near the surface of a metal, can render many of the standard techniques (e.g. Monte Carlo) and well tested approximations (e.g. the local approximation) impotent.

Our first breakthrough came four years ago[3] with the derivation and solution of an integral equation for the homogeneous electron liquid. The results were found to be in good agreement with those of Singwi et al.[4] throughout the range of metallic densities. The same integral equation (in the configuration representation) was applied to model nuclear matters and found, after some initial difficulties,[5] to agree[6] with FHNC/4 calculations.[7] Then we moved on to inhomogeneous electron liquids, deriving and solving in the process coupled integral equations,[8,9] which were subsequently applied to calculating correlation energies for metallic hydrogen.[9] The results checked out well against those obtained with a standard perturbation theory[10] and the density functional formalism.[11] Only then have we become sufficiently emboldened to face the strongly inhomogeneous metal surface.

The problem is defined by the Hamiltonian

$$H = \sum_i \frac{-\hbar^2}{2m}\nabla_i^2 + \sum_{i<j} \frac{e^2}{|\vec{r}_i - \vec{r}_j|} - \sum_i \int d\vec{R}\, \frac{e^2 n_+(\vec{R})}{|\vec{r}_i - \vec{R}|} + \frac{1}{2}\int d\vec{R}\, d\vec{R}'\, \frac{n_+(\vec{R})\, n_+(\vec{R}')}{|\vec{R} - \vec{R}'|}, \tag{1}$$

where $n_+(\vec{R})$ denotes an input static distribution of neutralizing charges. In the jellium model, $n_+(\vec{R})$ is a semi-infinite, uniform positive charge background truncated near the metal surface. Its exact boundary coincides with the Gibbs surface of the electron distribution $n(\vec{r}) \equiv n(z)$. Deep inside the metal, both $n_+(\vec{R})$ and $n(\vec{r})$ approach the bulk density ρ.

Our trial wave function for the ground state ψ takes the form of a Feenberg-Jastrow factor $F \equiv \prod_{i<j} \exp \frac{1}{2} u(r_{ij})$ multiplied to a determinant of single particle orbitals $\varphi_\alpha(\vec{r})$. We choose

$$u(r) = \frac{-2e^2}{\hbar\,\omega_p}\,\frac{1 - e^{-br}}{r}, \tag{2}$$

where $\omega_p \equiv (4\pi\rho e^2)/m$ denotes the plasmon frequency, and b is a variational parameter taken as that determined in Ref. 3, so as to guarantee that deep in the metal the asymptotic behavior of the wave function is consistent with that for the bulk. Note that the Coulomb force requires $u(r)$ to be long-ranged and slowly varying. The former property is the source of certain divergences that appear in our theory. The latter turns out to be responsible for the divergences to cancel exactly. For the single particle orbitals, we take the eigenfunctions of an electron moving in some model field characterized by a potential $v(\vec{r})$:

$$\left[\frac{-\hbar^2}{2m}\nabla^2 + v(\vec{r})\right]\varphi_\alpha(\vec{r}) = \varepsilon_\alpha \varphi_\alpha(\vec{r}) \; . \tag{3}$$

$v(\vec{r})$ will be parameterized. In fact, this is how the variational parameters enter our calculations. $v(\vec{r})$ is undoubtedly related to the effective surface potential, and ours represents a parametric approach to determining it self-consistently. A more formal analysis of its significance will be given elsewhere.

The energy expectation value is given by

$$\begin{aligned} E &\equiv \langle\psi|H|\psi\rangle/\langle\psi|\psi\rangle = \{T_1+T_2+T_3+T_4\} + V_{es} + V_c \\ &\equiv \Big\{\sum_\alpha \varepsilon_\alpha - \int n(\vec{r})\,v(\vec{r})\,d\vec{r} + \frac{\hbar^2}{8m}\int P(1,2)[\nabla_1 u(r_{12})]^2 d\vec{r}_1\,d\vec{r}_2 \\ &\quad + \frac{\hbar^2}{8m}\int P(1,2,3)[\nabla_1 u(r_{12})\cdot\nabla_1 u(r_{13})]\,d\vec{r}_1\,d\vec{r}_2\,d\vec{r}_3\Big\} \\ &\quad + \frac{e^2}{2}\int\frac{[n(\vec{r}_1)-n_+(\vec{r}_1)][n(\vec{r}_2)-n_+(\vec{r}_2)]}{|\vec{r}_1-\vec{r}_2|}d\vec{r}_1\,d\vec{r}_2 + \frac{e^2}{2}\int\frac{n(\vec{r}_1)n(\vec{r}_2)}{|\vec{r}_1-\vec{r}_2|}[g(1,2)-1]\,d\vec{r}_1\,d\vec{r}_2 \,. \end{aligned} \tag{4}$$

$P(1,2) \equiv n(\vec{r}_1)\,n(\vec{r}_2)\,g(1,2)$ and $P(1,2,3) \equiv n(\vec{r}_1)\,n(\vec{r}_2)\,n(\vec{r}_3)\,g(1,2,3)$ are the two- and three-particle distribution functions, respectively. Our task comes down to solving for every chosen $v(\vec{r})$ the eigenvalue equation (3), calculating the density profile $n_0(\vec{r}) \equiv n(z|\lambda=0) \equiv \sum_\alpha |\varphi_\alpha(\vec{r})|^2$ of the non-interacting system, determining then the density profile $n(\vec{r}) \equiv n(z|\lambda=1)$ for the interacting system and the distribution functions $g(1,2)$ and $g(1,2,3)$, evaluating the energy E, and finally evaluating the surface energy per unit area σ by carefully subtracting off the bulk energy E_B. σ is then to be minimized with respect to $v(r)$. Such a series of calculation is to be done for a range of average densities $\rho \equiv k_F^3/3\pi^2$, or r_s, where $\frac{4\pi}{3}(r_s a_B)^3 = \rho^{-1}$, a_B denoting Bohr's radius. We should also calculate the electronic work function for every r_s, thus:

$$\Phi \equiv \Delta\phi - \bar{\mu} \equiv [\phi(\infty) - \phi(-\infty)] - \bar{\mu} \;, \tag{5}$$

where
$$\phi(z) \equiv \phi(\vec{r}) \equiv \int \frac{n(\vec{r}\,') - n_+(\vec{r}\,')}{|\vec{r}-\vec{r}\,'|}\,d\vec{r}\,' \tag{6}$$

and $\bar{\mu}$ denotes the bulk chemical potential which we take consistently from Ref. 3.

In this first attempt, we take $v(\vec{r})$ to be a step function. It vanishes for negative z, rises sharply to $v_0 \equiv (\hbar^2\beta)/2m$ at $z = 0$, and stays at v_0 for positive z. v_0 is thus the only variational parameter.

The position of the potential discontinuity, $z = 0$, does not coincide with the Gibbs surface. The jellium ends on the Gibbs surface, just short of the potential discontinuity, i.e., at $z = -\Delta L$ where

$$\Delta L = \frac{3\pi}{8k_F} - \frac{3}{8}\left(1 - \frac{k_F^2}{\beta}\right)^{1/2}\left(\beta^{-1/2} + \frac{\beta^{1/2}}{k_F^2}\right) - \frac{3\beta^{1/2}}{8k_F^2}\left(1 - \frac{k_F^2}{\beta}\right)^{3/2} + \left(\frac{3\beta}{4k_F^3} - \frac{3}{2k_F}\right)\sin^{-1}\left[\left(\frac{k_F^2}{\beta}\right)^{1/2}\right] . \tag{7}$$

At the other end: $z = -L \to -\infty$, where we let the potential $v(\vec{r})$ rise sharply to ∞, the jellium ends on a second Gibbs surface which lies just short of $z = -L$ by $\Delta L'$, where $\Delta L' = (3\pi)/8k_F$. While ΔL enters prominently into the surface calculation because the cleaving of the bulk is carried out at the first Gibbs surface, $\Delta L'$ has no role to play since it appears in an energy term that cancels out an equivalent contribution to the bulk energy.

Also, it can be shown exactly that occupation of the single particle spectrum terminates at k_M which differs from k_F by only a term of order L^{-1}. The counting problem is tricky and must be handled with meticulous care. We find

$$n(z|0) = \begin{cases} 3\rho \int_0^1 (1 - x^2)\sin^2[k_F xz - \sin^{-1}(\beta^{-1/2}k_F x)]\,dx, & z \le 0 , \\ 3\rho \int_0^1 (1 - x^2)\beta^{-1}k_F^2 x^2 \exp[-2(\beta - k_F^2 x^2)^{1/2} z]\,dx, & z > 0 . \end{cases} \tag{8}$$

It approaches the average density ρ as $z \to -\infty$.

The integral equation[9,12]

$$\ln[n(z_1|\lambda)/n(z_1|0)] = \int_0^\lambda d\lambda' \int d\vec{r}_2\, n(z_2|\lambda')\, u(r_{12})\, g(1,2) + \frac{1}{2}\int_0^\lambda d\lambda' \int d\vec{r}_2 d\vec{r}_3\, n(z_2|\lambda')\, n(z_3|\lambda')\, u(r_{23})[g(1,2,3) - g(2,3)] \tag{9}$$

then gives us $n(z|1) \equiv n(\vec{r}) \equiv n(z)$. There are two problems in this equation. First, the equations that govern $g(1,2)$ and $g(1,2,3)$ relate them to still higher-order distribution functions. This we resolve by taking $g(1,2)$ to be that obtained in Ref. 3 for the bulk (realizing that it is not strongly density dependent) and by approximating $g(1,2,3)$ in the convolution approximation (which preserves proper sequential relations):

$$g(1,2,3) \approx 1 + h_{12} + h_{13} + h_{23} + h_{12}h_{13} + h_{12}h_{23} + h_{13}h_{23} + \int d\vec{r}_4 \, n(z_4)\, h_{14}h_{24}h_{34} , \qquad (10)$$

with $h_{ij} \equiv g(i,j) - 1$. Second, the resulting expression on the right side of Eq. (9) contains divergences. This we overcome by noting that the divergences are spurious: they cancel out diagrammatically, leaving a finite integral which is complicated but manageable, for $u(r)$ varying slowly relative to the extent of the correlation hole, as in the present case. This finite integral, when neglected, causes errors which are less than 10% in the surface energy. Apart from it, Eq. (9) reads:

$$\ell n[n(z_1|\lambda)/n(z_1|0)] = \frac{1}{2}\int_0^{\lambda} d\lambda' \int d\vec{r}_2 \, n(z_2|\lambda') \left[u(r_{12})h_{12} + \int d\vec{r}_3 \, n(z_3|\lambda')\, u(r_{23})h_{13}h_{23}\right] . \qquad (11)$$

In the limit $z \to -\infty$, $n(z|\lambda) \to \rho$, retaining the proper asymptotic (bulk) behavior. This is a crucial feature in any surface calculation.

In the space provided, it is impossible to display the energy expressions in all their details. Using subscripts corresponding to those in Eq. (4), we find

$$\sigma_1 = \frac{\hbar^2}{2\pi^2 m}\left[\frac{\pi}{16} k_F^4 - \frac{2}{15}\Delta L \, k_F^5 + \int_0^{k_F} \gamma(k)\, k(k^2 - k_F^2)\, dk\right] , \qquad (12)$$

where $\gamma(k)$ is a phase shift given by $\tan^{-1}[k(\beta - k^2)^{-1/2}]$,

$$\sigma_2 = -\frac{\hbar^2\beta}{2m}\int_0^{\infty} n(z)\, dz , \qquad (13)$$

$$\sigma_3 + \sigma_4 = \frac{\hbar^2}{8m L_x L_y}\int d\vec{r}_1 d\vec{r}_2 d\vec{r}_3 [\nabla_1 u(r_{12})\cdot\nabla_1 u(r_{13})]\, g(r_{12}) g(r_{13}) \left[n(z_1)n(z_2)n(z_3) - \frac{1}{2}\rho^3\right] \qquad (14)$$

apart from a small finite integral that remains after cancelling two spurious divergences, $L_x L_y$ being the total surface area,

$$\sigma_{es} = 2\pi e^2 \int_{-\infty}^{\infty} dz \left\{\int_{-\infty}^{z} dz' [n(z') - \rho\theta(-z')]\right\}^2 , \qquad (15)$$

and

$$\sigma_c = \frac{e^2}{2L_xL_y}\int \frac{n(z_1)\,n(z_2) - \frac{1}{2}\rho^2}{|\vec{r}_1 - \vec{r}_2|}\,[g(r_{12}) - 1]\,d\vec{r}_1\,d\vec{r}_2 \;. \tag{16}$$

Minimizing

$$\sigma \equiv \{\sigma_1 + \sigma_2 + \sigma_3 + \sigma_4\} + \sigma_{es} + \sigma_c \tag{17}$$

with respect to the one variational parameter v_0, we obtain the results shown in the two figures below.

Figure 1 shows $n(z)$ at $r_s = 5.65$, as compared to $n(z|0)$ and the density profile obtained by interpolating the results of Lang and Kohn[1] at $r_s = 5$ and 6 using the density functional method. Results at smaller r_s show larger Friedel oscillations than Ref. 1.

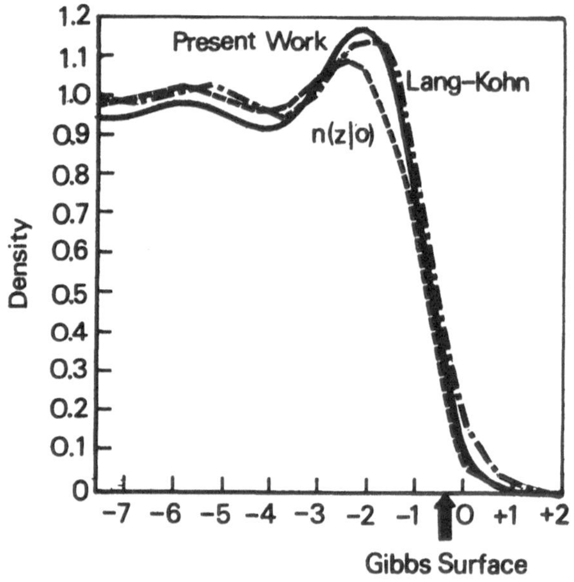

Fig. 1. Density Profile

Figure 2 shows the surface energy as a function of r_s, as compared to that of Lang and Kohn[1] before first order perturbation corrections were included to account for the discrete nature of the background charges. "Experimental" data obtained by extrapolating surface tensions of liquid metals to zero temperature are shown for comparison. The close agreement between our results and the data is suggestive but in no way conclusive since we have not accounted for the discreteness of the background. Variational principle dictates that the use of a more flexible $v(\vec{r})$ will necessarily lower our σ.

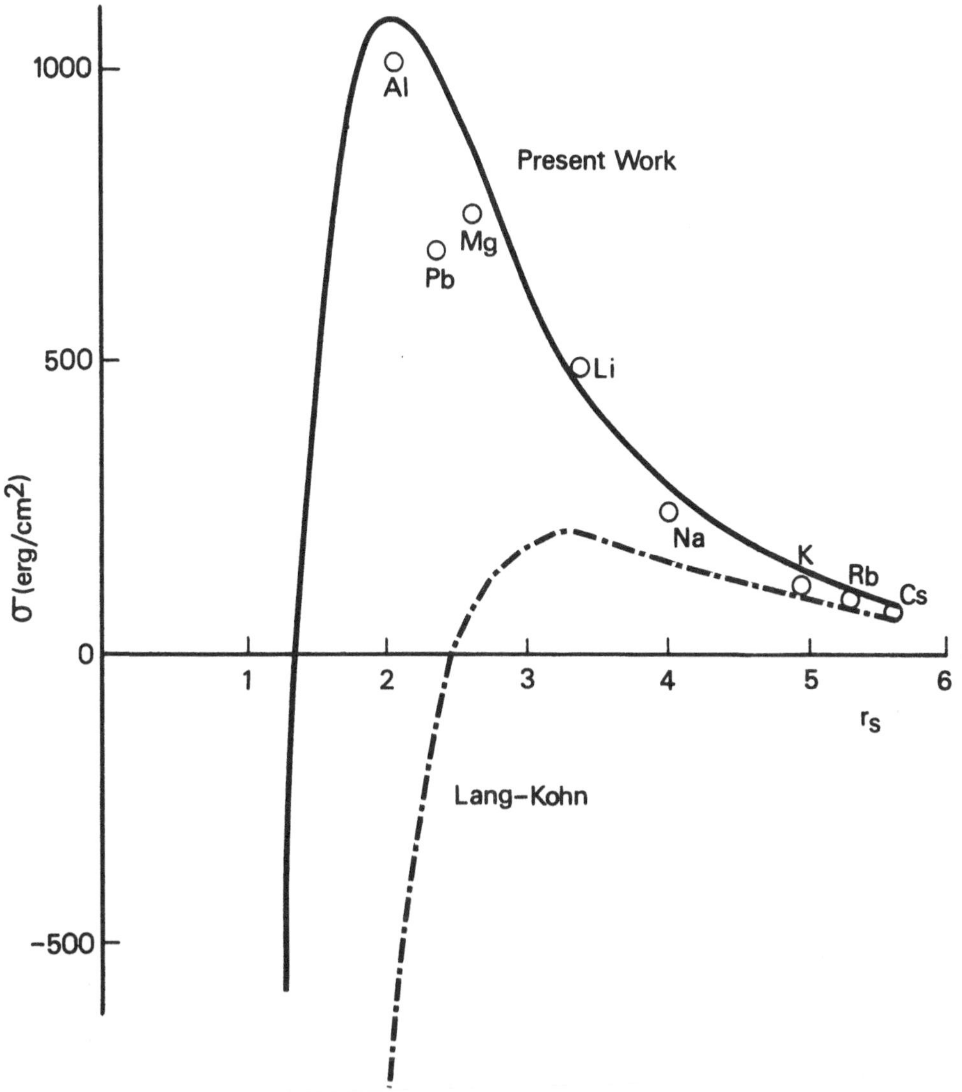

Fig. 2. Surface Energy

The calculation of work functions depends sensitively upon the density profile, in particular the magnitude of the overshoot. We shall report on those results elsewhere.

We are very much encouraged by the results obtained so far, and have begun to refine our calculations by adopting a two-parameter $v(\vec{r})$. Of the five goals stated in the introductory paragraphs, we feel that headways have been made in (i) - (iv).

This work is supported in part by the U.S. National Science Foundation through grant No. DMR80-08816, and by Fudan University (Shanghai) and the Institute of Physics (Beijing), China, which have supported X. Sun and T. Li on research leave in the U.S. since 1979.

References:

1. P. Hohenberg and W. Kohn, Phys. Rev. 136, B864 (1964); W. Kohn and L. J. Sham, Phys. Rev. 140, A1133 (1965); N. D. Lang and W. Kohn, Phys. Rev. B1, 4555 (1970); N. D. Lang, Solid State Phys. 28, 225 (1973).
2. E. Feenberg, Theory of Quantum Fluids (Academic Press, 1969); J. W. Clark, unpublished notes (1966); C.-W. Woo, in Physics of Liquid and Solid Helium (John Wiley-Interscience, 1976).
3. S. Chakravarty and C.-W. Woo, Phys. Rev. B13, 4815 (1976).
4. K. S. Singwi, M. P. Tosi, R. H. Land, and A. Sjölander, Phys. Rev. 176, 589 (1968); P. Vashishta and K. S. Singwi, Phys. Rev. B6, 875 (1972).
5. Y. R. Lin-Liu, S. Chakravarty, and C.-W. Woo, Phys. Rev. C18, 516 (1978).
6. T. Li, E. Zhao, X. Zhu, and C.-W. Woo, Phys. Rev. B21, 2745 (1980).
7. J. G. Zabolitzky, Phys. Lett. 64B, 233 (1976); Phys. Rev. A16, 1258 (1977).
8. V. T. Rajan, C.-W. Woo, and F. Y. Wu, J. Math. Phys. 19, 892 (1978).
9. V. T. Rajan and C.-W. Woo, Phys. Rev. B18, 4048 (1978).
10. L. G. Caron, Phys. Rev. B9, 5025 (1974).
11. G. A. Neece, F. J. Rogers, and W. G. Hoover, J. Comp. Phys. 7, 621 (1971).
12. L. Senbetu and C.-W. Woo, Phys. Rev. B18, 3251(1978).

MANY-BODY EFFECTS IN THE OPTICAL PROPERTIES OF QUASI-TWO DIMENSIONAL SYSTEMS

Amitabha Bagchi
Xerox Corporation
800 Phillips Road
Webster, N.Y. 14580/U.S.A.

The importance of many-body effects in two and quasi-two dimensional systems has been studied for many years [1] in such contexts as electrons in the inversion layer at semiconductor-oxide interfaces and electron layers deposited on the surface of liquid helium. For the quasi-two dimensional electron gas as found in inversion layers, attention has focused on electron-electron interaction with regard to its influence on band energy levels [2], longitudinal and transverse dielectric functions [3], inter-subband optical absorption [4] and exciton effects [5]. Dahl and Sham [6] and Equiluz and Maradudin [7] have studied carefully the question of optical reflectance from an inversion layer, taking fully into account depolarization effects whose importance in shifting the optical absorption energy was stressed earlier.[8] In the present work, we wish to focus attention on a different quasi-two dimensional system, viz., that of chemisorbed atoms on a metal surface. The object of our interest is the role of many body effects in the optical properties of such a system. As we shall see, there are important similarities and differences between this system and that of an inversion layer.

Experimentally when one studies a chemisorbed system by optical means, the quantity measured in the fractional change of reflectance, $\Delta R/R$, caused by adsorption, where R and $R+\Delta R$ are reflectances of the pure substrate and the substrate with adsorbates respectively. The measurement can be made, for a given angle of incidence of light, at various frequencies of photon energies, and two independent polarizations for each frequency. The theoretical aim is to correlate structures found in these experiments with inter-band transition energies. The problem is quite complicated because the dielectric response of the system is non-local and the translational symmetry is broken normal to the surface. Simplifications arise, however, from the long-wavelength nature of light, and from Maxwell's equations which dictate that the transverse E-field and the normal D-field should remain approximately constant as one goes across the surface region. It is thus possible to set up the problem of reflection from a chemisorbed system as a perturbation on reflection from the background substrate.[9,10] Treating the background within the Fresnel model and using only the first Born Approximation, one finds [10]

$$(\Delta R/R)_s = 4\left(\frac{\omega}{c}\right)\cos\theta_i \, \mathrm{Im}\left[\frac{\delta\Lambda_y(\omega)}{\epsilon_b(\omega)-1}\right] , \qquad (1a)$$

$$(\Delta R/R)_p = 4\left(\frac{\omega}{c}\right)\cos\theta_i \, \mathrm{Im}\left\{\frac{[\epsilon_b(\omega) - \sin^2\theta_i]\delta\Lambda_x(\omega) + \epsilon_b^2(\omega)\sin^2\theta_i\,\delta\Lambda_z(\omega)}{[1-\epsilon_b(\omega)][\sin^2\theta_i - \epsilon_b(\omega)\cos^2\theta_i]}\right\}, \qquad (1b)$$

where s and p stand for s-polarized (E-field perpendicular to the plane of incidence) and p-polarized (E-field parallel to the plane of incidence) light respectively, $\hbar\omega$ is the photon energy, θ_i the angle of incidence of light, $\epsilon_b(\omega)$ the dielectric function (assumed local) of the bulk substrate, and $\delta\Lambda_x(\omega)$ and $\delta\Lambda_z(\omega)$ are two length parameters defined as

$$\delta\Lambda_x(\omega) = \iint_{-\infty}^{\infty} dz\, dz' \left[\epsilon_{xx}^{(a)}(\vec{Q}=0,\omega;z,z') - \epsilon_{xx}^{(c)}(\vec{Q}=0,\omega;z,z')\right] \qquad (2a)$$

$$\delta\Lambda_z(\omega) = \iint_{-\infty}^{\infty} dz\, dz' \left[\epsilon_{zz}^{-1(a)}(\vec{Q}=0,\omega;z,z') - \epsilon_{zz}^{-1(c)}(\vec{Q}=0,\omega;z,z')\right] . \qquad (2b)$$

Here $\epsilon^{(a)}(Q,\omega;z,z')$ and $\epsilon^{(c)}(Q,\omega;z,z')$ stand for the non-local dielectric tensors of the adsorbate-covered and the clean substrate respectively. $\vec{Q}$ is a wavevector in the xy plane parallel to the surface. Many-body effects come into consideration through $\overleftrightarrow{\epsilon}(\vec{Q},\omega;z,z')$ and $\overleftrightarrow{\epsilon}^{-1}(\vec{Q},\omega;z,z')$. It is interesting to note that because of depolarization effects as discussed by Chen et al.[8], the response of the system to electric fields normal to the surface is $\overleftrightarrow{\epsilon}^{-1}$ rather than $\overleftrightarrow{\epsilon}$. The formula of Eqs. (1) in the first Born approximation in effect treats the surface region as decoupled from the bulk. It is possible, however, to sum the Born series exactly[11,12] and thus arrive at more exact but more complicated formulas for differential reflectance. We shall not discuss the formulas in detail here as they are discussed by Barrera elsewhere in the volume. Some work has also been done on the change of reflectance by treating the background metal within a better approximation[13] (viz., the Semi-Classical Infinite Barrier Model) than the Fresnel model.

Many-body effects determine the properties of the dielectric function and thus determine the differential reflectance. Several authors[14-19] have studied the transverse dielectric response of a semi-infinite metal within the Random Phase Approximation by treating the metal as an electron gas either within the Semi-classical Infinite Barrier model[14-16] or within some quantum-mechanical models [17-19]. The situation, however, is quite different when adsorbed atoms exist on the metal surface. Typical experiments on differental reflectance involve the adsorption of either light gases such as H and O or rare gases such as Ar and Kr on transition metals like W or Ni. Various coverages upon adsorption are used. To obtain the dielectric response of, say, a monolayer of adsorbed atoms it is necessary to know a number of things about which information is sketchy. We need to know whether the adsorbed atoms form an ordered array or lattice on the surface. It is necessary to know the band structure, at least as projected on the surface Brillouin Zone, and have some information about wave functions so as to be able to calculate optical matrix elements. Very limited information is available so far in this regard [20-22] and no serious attempt has been made to calculate $\overleftrightarrow{\varepsilon}^{(\alpha)}(0,\omega;z,z')$ directly or to invert it numerically in order to obtain $\overleftrightarrow{\varepsilon}^{(\alpha)-1}(0,\omega;z,z')$. What has been tried instead [23] is to treat the chemisorbed layer as a quasi-two dimensional system, quite distinct from the substrate, with its own dielectric response function. This cannot be a good approximation for strongly chemisorbed systems as it ignores or minimizes the importance of electron tunneling between the adsorbate and the metal. It should, however, be a reasonable approximation for weak adsorbates such as rare gases, although even there, chemical bonding effects may be quite significant in the excited state.

Let us consider the weak chemisorption of rare gas atoms on a metal surface, and treat the chemisorbed layer as a quasi-two dimensional electron system. Let us imagine the adsorbed atoms to form an ordered lattice on the surface. Electron states on the atoms will form two-dimensional bands, much like the electronic subbands of inversion layers. There is considerable experimental evidence, mainly from photoemission, for the existence of such bands. The dielectric response of the chemisorbed layer, which will be reflected in differential reflectance, will involve inter-band transitions between occupied and unoccupied bands. It can be shown [23] that, quite in keeping with expectations, structures in differential reflectance ought to be correlated with inter-band transition energies at specific symmetry points or over regions of the surface Brillouin Zone.

What are the major effects of the electron interaction expected to be in the optical properties of the chemisorbed layer? As in the case of the inversion layer, two major effects are expected. The single-particle or Hartree band energies should be renormalized. And there is the possibility of exciton formation, i.e., the existence of a bound state of an excited electron and the remaining hole, below the edge of the inter-band transition energy. For the chemisorbed system, with limited knowledge of the system wavefunctions, a quantitative estimate of energy renormalization effects is not very meaningful. It is, however, interesting to speculate on whether excitons are formed in the chemisorbed system, and if they are, how they can be detected and their origin established unambiguously.

An exciton is a highly correlated or bound state of an excited electron with the hole it leaves behind. In the strong coupling or Frenkel case, the electron and hole stay on the same atom but hop from atom to atom. In the weak coupling or Wannier case, the electron or hole states are highly de-localized, i.e., band-like rather than atomic, yet they form a bound state. The latter situation closely parallels a hydrogen atom, and one looks commonly for some kind of Balmer series in the optical absorption spectra. Whether a particular solid will have Frenkel-type or Wannier-type excitons depends on the constituent atoms and the band structure. In rare gas solids, optical absorption reveals well defined Balmer lines except for the most deeply bound exciton which is presumably Frenkel-like.

In this work we wish to report on characteristic features of Wannier excitons in the quasi-two dimensional system of a chemisorbed layer. We therefore imagine that electrons in the chemisorbed layer occupy two dimensional parabolic bands with an isotropic effective mass M_h up to the Fermi momentum p_F. An electron from this band can be excited to an unoccupied band, also assumed to be parabolic, leaving a hole behind. The electron and hole bands have the dispersion relations (setting h=1)

$$\epsilon_e(\vec{K}) = \Delta + K^2/2m_e \tag{3a}$$

$$\epsilon_h(\vec{K}) = K^2/2m_h \ . \tag{3b}$$

The scattering of the electron and the hole can be described by the Bethe-Salpeter equation [24, 25] which is pictured diagrammatically in Fig. 1, where γ is the total electron-hole interaction while I is the irreducible interaction. The Bethe-Salpeter equation can be represented as

$$\Gamma(p,p';q) = I(p,p';q) + \sum_{p''} I(p,p'';q)\, G(p''-q/2)\, G(p''+q/2)\, \Gamma(p'',p';q) \tag{4}$$

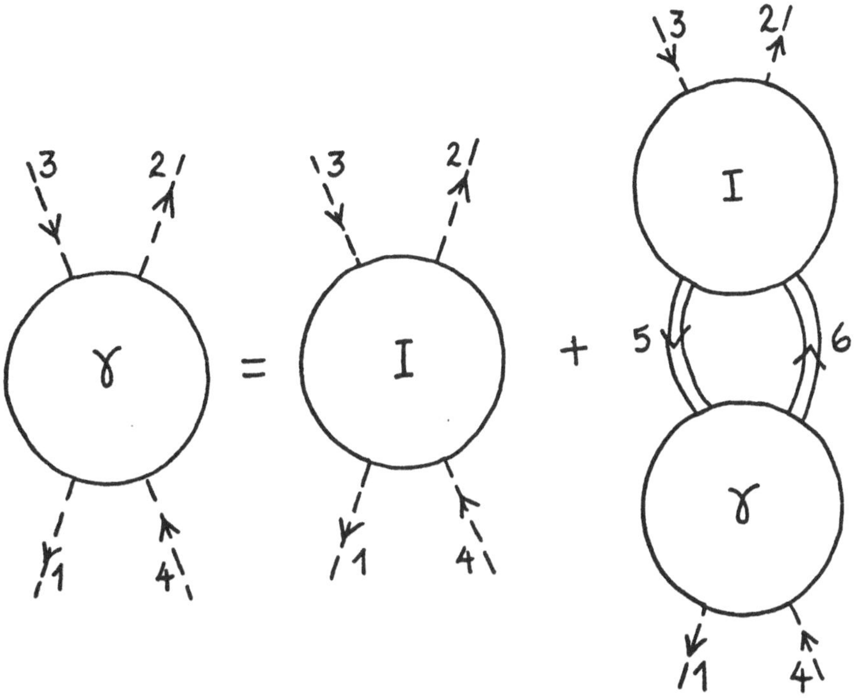

FIG. 1 The diagram for the electron-hole interaction in the Bethe-Salpeter equation

where G stands for the single-particle Green's function, either of the electron or the hole. Collective or bound states appear as poles in Γ, and the equation determining them becomes homogeneous as the irreducible interaction term can be ignored near it. Following Kohn's derivation [25] and straight forward algebra relevant to this case, we find that an exciton will exist with momentum $\vec{q}$ and energy q_0 provided the following homogeneous equation,

$$\left[\frac{(\vec{p}+\vec{q}/2)^2}{2m_e} - \frac{(\vec{p}-\vec{q}/2)^2}{2m_h} + \Delta - q_0\right]\Phi(\vec{p}) = -\sum_{|\vec{p}''-\vec{q}/2|\leq p_F} I(\vec{p},\vec{p}'';\vec{q})\,\Phi(\vec{p}'') \tag{5}$$

has a solution. For optical processes, we can set $\vec{q} = 0$, and define

$$1/m^* = 1/m_e - 1/m_h \tag{6a}$$

$$E = \Delta - q_0 \; . \tag{6b}$$

Note that $E > 0$ means that the exciton is bound. The final version of the Bethe-Salpeter equation is then

$$(p^2/2m^* + E)\Phi(\vec{p}) = -\sum_{|\vec{p}'|\leq p_F} I(\vec{p},\vec{p}';\vec{q}=0)\,\Phi(\vec{p}') \, . \tag{7}$$

All momenta here are two-dimensional.

It is important at this stage to distinguish the Wannier exciton we are studying here from previous work and from the two-dimensional hydrogen atom. Since both the valence and the conduction bands disperse upward, m^* is not the usual effective mass. If the hole band dispersed downward, as happens commonly in indirect-gap semiconductors, m^* would be the usual effective mass. Furthermore, since the valence band is occupied up to the Fermi energy, it imposes important restriction on the integral equation for exciton of Eq. (7). For the two-dimensional hydrogen atom, $I(\vec{p},\vec{p}';\vec{q}=0)$ would be the Fourier transform of the Coulomb potential and the sum over $\vec{p}$ would be unrestricted. It is easy to show [26] that for the two-dimensional hydrogen atom, energy levels fall on a Bohr series with only odd integral values of the principal quantum number n. Such a series is not expected for our excitons. In fact, Del Sole and Tosatti [27] in their study of surface state excitons in semiconductors with a logarithmic electron-hole interaction (Coulomb interaction as projected in the two-dimensional layer) found quite different behavior of the bound states from the Bohr series. Vinter [5] considered excitons in inversion layers by assuming the subbands to be parallel, i.e., having the same dispersion, so that $m^* \to \infty$. His results naturally would be quite different from ours.

To analyze the Bethe-Salpeter equation, we take $I(\vec{p},\vec{p}',\vec{q}=0)$ to be the screened Coulomb interaction in two dimensions. This would be exactly true for triplet excitons, where the electronic excitation involves a spin flip. Singlet excitons will have an additional contribution to $I(\vec{p},\vec{p}')$ from the possibility of direct annihilation of the electron and the hole. We imagine a screened Coulomb interaction in real space between the electron and hole of the form $e^{-\Lambda\rho}(-e^2/\rho)$, so that

$$I(\vec{p},\vec{p}';\vec{q}=0) = -\frac{2\pi e^2}{[\Lambda^2+(\vec{p}-\vec{p}')^2]^{1/2}} \quad . \tag{8}$$

It is possible to make an expansion of this expression in Bessel functions [28], so that, assuming a solution of the form

$$\Phi(\vec{p}) = \Phi_{m'}(p)\exp(im'\varphi) \tag{9}$$

one arrives at the "radial" integral equation

$$(p^2/2m^*+E)\,\Phi_{m'}(p) = e^2\int_0^{p_F} p'dp'\int_0^\infty d\xi\; J_{m'}(\xi p)\,J_{m'}(\xi p')\,e^{-\Lambda\xi}\,\Phi_{m'}(p'). \tag{10a}$$

The ξ-integral can be worked out [29] and finally yields

$$(p^2/2m^* + E)\Phi_{m'}(p) = \frac{e^2}{\pi}\int_0^{p_F}\sqrt{\frac{p'}{p}}\,dp'\,Q_{m'-1/2}\left(\frac{p^2+p'^2+\Lambda^2}{2pp'}\right)\Phi_{m'}(p'), \qquad (10b)$$

where $Q_{m'-1/2}(x)$ is the associated Legendre polynominal. For a more compact notation, we define

$$E/(p_F^2/2m^*) = \beta \quad ; \qquad x = p/p_F$$

and the mean spacing r_o between electrons,

$$1/\pi r_o^2 = p_F^2/2\pi \quad .$$

Finally, concentrating on the S type (m' = 0) states we obtain

$$\Phi_0(x) = \frac{\sqrt{2}}{\pi}\,\frac{r_s}{x^2+\beta}\int_0^1 dx'\sqrt{\frac{x'}{x}}\,Q_{-1/2}\left(\frac{x^2+x'^2+\tilde{\Lambda}^2}{2xx'}\right)\Phi_0(x') \qquad (11)$$

where $r_s = r_o/a^*_o$, $a^*_o = 1/m^* e^2$ is the effective Bohr radius, and $\tilde{\Lambda} = \Lambda/P_f$

We have solved the integral equation approximately by numerical means. The results for the first S-state energies, which have the greatest reliability, are shown in Fig. 2, where we have plotted the binding

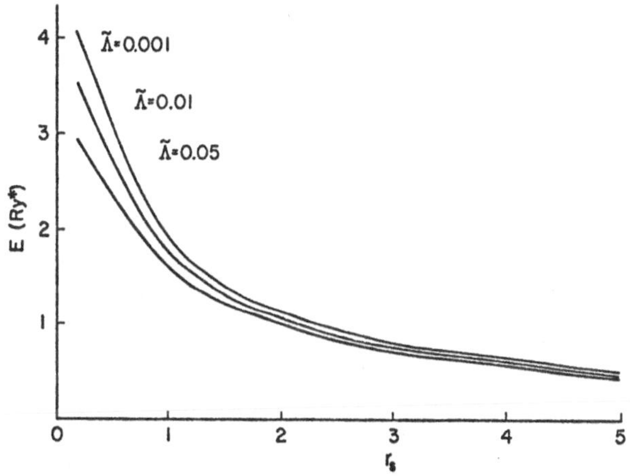

FIG. 2 The exciton energies E (in units of $Ry^* = e^4m^*/2$) Plotted against the ratio of the mean spacing of electrons to the effective Bohr radius, for various values of the damping parameter $\tilde{\Lambda}$.

energy versus r_s for various choices of the parameter Λ . The binding energy is expressed in the units of $Ry^* = e^4m^*/2$. The general trends shown in Fig. 2 are that the binding energy increases as the screening parameter decreases, which is expected as the electron-hole interaction becomes stronger. Similarly, the binding energy increases as r_s

decreases. Since the coverage is proportional to r_s^{-2}, we can say equivalently that the binding energy of the exciton increases as the coverage increases.

In conclusion, the possibility of observing excitons by optical means in a chemisorbed layer remains speculative. If present, the exciton will show up as a peak below the edge of inter-band absorption. We suggest that a study of the location of the peak as a function of coverage is one way of identifying its origin.

References

1. An excellent review is contained in Proceedings of the International Conference on the Electronic Properties of Two-Dimensional Systems (Berchtesgaden, Germany, 19-22 August 1977), reprinted in Surf. Sci. 73 (1978).
2. T. Ando, Surf. Sci. 73, 1 (1978).
3. A. K. Rajagopal, Phys. Rev. B15, 4264 (1977).
4. T. Ando, Z. Physik B26, 263 (1977).
5. B. Vinter, Phys. Rev. B15, 3947 (1977).
6. D. A. Dahl and L. J. Sham, Phys. Rev. B16, 651 (1977).
7. A. Eguiluz and A. A. Maradudin, Ann. Phys. (N.Y.) 113, 29 (1978).
8. W. B. Chen, Y. B. Chen and E. Burstein, Surf. Sci. 58, 263 (1976).
9. A. Bagchi and A. K. Rajagopal, Solid State Commun. 31, 127 (1979).
10. A. Bagchi, R. G. Barrera and A. K. Rajagopal, Phys. Rev. B20, 4824 (1979).
11. J. E. Sipe, Phys. Rev. B22, 1589 (1980).
12. R. G. Barrera and A. Bagchi, unpublished.
13. B. B. Dasgupta and R. Fuchs, preprint.
14. K. L. Kliewer, Surface Photoexcitation, in Photoemission from Surfaces (B. Feuerbacher, B. Fitton and R. F. Willis ed., Wiley, New York, 1977), Chapter 3.
15. G. Mukhopadhyay and S. Lundqvist, Physica Scripta 17, 69 (1978).
16. P. Apell, Physica Scripta 17, 535 (1978).
17. P. J. Feibelman, Phys. Rev. B12, 1319 (1975).
18. A. Bagchi, Phys. Rev. B15, 3060 (1975).
19. R. G. Barrera and A. Bagchi, Phys. Rev. B20, 3186 (1979).
20. N. V. Smith and L. F. Mattheiss, Phys. Rev. Lett. 37, 1494 (1976).
21. A. Liebsch, Phys. Rev. B17, 1653 (1978).
22. C. S. Wang and A. J. Freeman, Phys. Rev. B19, 4930 (1979).
23. M. H. Lee and A. Bagchi, Phys. Rev. B22, 1687 (1980).
24. P. Nozières, Theory of Interacting Fermi Systems (Benjamin, New York, 1964), Chapter 6.
25. W. Kohn in Les Houches Lectures (Gordon and Breach, N.Y., 1967).
26. M. H. Lee, Ph.D. Thesis (University of Maryland, College Park, 1980), unpublished.
27. R. Del Sole and E. Tosatti, Solid State Commun. 22, 307 (1977).
28. J. D. Jackson, Classical Electrodynamics (Wiley, New York, 1962), Problem 3.12.
29. Gradshteyn and Ryzhik, Tables of Integrals, Series and Products (Academic Press, N.Y., 1965), Section 3.147.5.

NON-LOCAL EFFECTS IN THE ELECTROMAGNETIC PROPERTIES OF INTERFACES

Rubén G. Barrera
Instituto de Física, UNAM

México

In the present work we report some of the results that we have obtained in relation with problems concerning the non-local response of metallic interfaces to an electromagnetic probe.

First we define the meaning of non-locality through the integral relationship

$$\bar{j}_\omega(\bar{r}) = \int \overleftrightarrow{\sigma}(\bar{r},\bar{r}';\omega)\cdot\bar{E}_\omega(\bar{r}')\,d^3r' \qquad (1)$$

between the electric current density $\bar{j}(\bar{r})$ and the electric field $\bar{E}(\bar{r})$. The sub-index ω indicates the Fourier component of frequency ω and $\overleftrightarrow{\sigma}(\bar{r},\bar{r}';\omega)$ is known as the non-local conductivity.

The fact that $\overleftrightarrow{\sigma}$ depends on both $\bar{r}$ and $\bar{r}'$ means that if we set locally a delta function field at $\bar{r}_0$ in the direction $\hat{e}_0$ it will generate an electric current distribution

$$\bar{j}(\bar{r}) \propto \overleftrightarrow{\sigma}(\bar{r},\bar{r}_0;\omega)\cdot\hat{e}_0 \qquad (2)$$

which, in general, will cover a finite region around $\bar{r}_0$. The typical size of this region is known as the non-locality range.

By electromagnetic properties we mean here the solution of Maxwell's E-quations for the metallic system in the absence of external charge and current density distributions. But with the integral relationship between the current density and the electric field (Eq. 1) Maxwell's Equations become a set of integro-differential equations. This problem has an exact solution in the case of an infinite homogeneous system in which $\overleftrightarrow{\sigma}$ depends on $|\bar{r}-\bar{r}'|$ and the solution of the integro-differential equations can be obtained through the Fourier-integral transform $\overleftrightarrow{\sigma}(k, \omega)$. In this case one finds several transverse electromagnetic modes and the appearance of longitudinal modes all with the same frequency ω but with different wave vector $\bar{k}$. For any given system the number of different modes with frequency ω depends on the functional form of $\overleftrightarrow{\sigma}(k, \omega)$.

When the symmetry of the system is broken in one direction by the presence of a flat interface, the Fourier-integral transformation along this direction is not useful anymore and one has to deal with a true one-dimensional integral equation.

Nevertheless one can try to solve the problem assuming that the system at each side of the interface is infinite and homogeneous and then match the solutions at the boundary. One finds that the two boundary conditions obtained from Maxwell's Equations are not sufficient to determine the amplitudes of all the different modes which are able to propagate in a non-local medium thus Additional Boundary Conditions (ABC) are required. The ABC represent the additional information required about the specific structure of the surface and therefore a unique determination of them is not possible, in general.

A different approach to the interface problem is the microscopic one. In this case one starts from a many-body Hamiltonian for a metal with a plane surface out of which one calculates the non-local conductivity and then one looks for solutions to Maxwell's Equations. The work we have done is along these lines.

Using a half-space jellium model for the metal and a finite step square potential for the surface we have calculated[1] the non-local conductivity tensor within the Random Phase Approximation (RPA). We then compare our calculation with the bulk non-local conductivity calculated within the same approximation and we found that both functions approach each other after a certain distance ℓ_{op} (optical surface region) from the surface. We found that the optical surface region is strongly frequency dependent and in the optical frequency region is, for all metals, much larger than the density-profile healing distance a . Since at optical frequencies $k\, l_{op} \gg 1$ $(k = \omega/c)$, it is not possible to simply match boundary conditions at the surface and at a plane located at a distance l_{op} inside the metal and then connect them through a first order expansion in $k l_{op}$ as is usually done in a local theory where $ka \ll 1$. Thus the solution of Maxwell's Equations becomes rather involved because the structure of the integral equation will depend on the specific functional form of the kernel $\overleftrightarrow{\sigma}(\bar{r}, \bar{r}'; \omega)$.

Nevertheless using the same microscopic model as above and a very crude approximation in solving the integral equation we were able to interpret the vanishing of the photo-electron yield along the normal direction to the (100) face of tungsten, beyond the plasma frequency, as an effect due to the behavior or the refracted electric field[2,3].

We were able also to generate, in a systematic way, approximate solutions to Maxwell's Equations through a perturbative approach[4]. We simply split the non-local response functions into two parts, one of which is assumed to have an exact solution for the fields and the other one is to be taken as a perturbation. Then we solve the integral equation in the first Born approximation assuming only that the tangential components of the electric field and the normal components of the displacement field vary slowly within the range of non-locality (long-wavelength approximation). We were able to write closed-form expressions for the optical coefficients (reflection amplitudes and ellipsometric coefficients) as a function of frequency, angle of incidence and appropriate integrals over the non-local response functions.

Lately we have extended our solution beyond the first Born approximation to all orders in perturbation theory[5] finding a new general dispersion relation of surface plasmons which takes into account possible band structure effects near the surface.

We have applied this formalism to the calculation of the optical coefficients of quasi-two-dimensional systems[6]. Starting with a simple model Hamiltonian with parameters adjusted to an inversion layer and to a metallic monolayer we compare the optical coefficients and the electric field calculated with a non-local response versus the ones calculated with several local approximations derived from the same Hamiltonian. Our results show that the theoretical interpretation of experimental data requires detailed non-local calculation. This is specially important in the case of the ellipsometric coefficient which measure the phase change after reflection.

Using the same formalism we have also studied[7] the role of the local-field within a classical model for an ordered overlayer of a weakly adsorbed species on a local metal substrate. Results for Ar on Al at two coverages strongly indicate the importance of the local-field in such optical studies. Lately we have derived[8] expressions for the potential of an oscillating dipole near the surface of a metal described by a non-local dielectric response. Calculations have been carried out using a hydrodynamic non-local dielectric function and the effects of the non-local surface plasmon dispersion as well as the coupling to bulk plasmons are apparent. At present we are using these results in order to extend our classical model for an adsorbed overlayer on a local metal substrate[7] to one which takes into account the non-local nature of the substrate.

REFERENCES

1. R. G. Barrera and A. Bagchi, Phys. Rev. B20, 3186(1979).
2. A. Bagchi, N. Kar and R. G. Barrera, Phys. Rev. Lett. 40, 803(1978).
3. A. Bagchi and R. G. Barrera, Rev. Bras. Fis. 9, 167(1979).
4. A. Bagchi, R. G. Barrera and A. K. Rajagopal, Phys. Rev. B20, 4824(1979).
5. R. G. Barrera and A. Bagchi, (to be published, 1981).
6. L. Mochán and R. G. Barrera, Phys. Rev. B (to appear, 1981).
7. A. Bagchi, R. G. Barrera and B. B. Dasgupta, Phys. Rev. Lett. 44, 1475(1980).
8. R. Fuchs and R. G. Barrera, (to be published, 1981).

CALCULATION OF THE COEFFICIENTS OF THE NUCLEAR MASS FORMULA

S. Rosati

Istituto di Fisica dell'Università, Pisa, Italy

Istituto Nazionale di Fisica Nucleare, Sezione di Pisa, Italy

Abstract. The variational theory is well suited for calculating the nuclear mass formula for a given nucleon-nucleon potential. An independent check of the available variational calculations on light nuclei is then possible; however, accurate estimates of the symmetry-energy contributions are required.

In the last few years, the variational approach to infinite homogeneous and translationally invariant systems has been extensively studied. The status of the research in the field is well discussed in some recent review papers[1-3]. On the contrary, only a limited number of applications to finite and/or non homogeneous systems is at present available. Particular interest has been devoted to the variational study both of the surface effects[4] in finite Fermi systems and of the ground states of light nuclei[5-7].

A relevant problem is the calculation of the coefficients of the nucear mass formula by means of the variational theory. A satisfactory NN potential must give correct values for the nuclear binding energies and, consequently, for the coefficients of the nuclear mass formula. Let us write the binding energy of a nucleus with A= N+Z in the form

$$\frac{B(N,Z) + E_C}{A} = -a_1 - a_2 A^{-1/3} - J\,\alpha^2 + \varepsilon \,, \qquad (1)$$

where E_C is the coulomb energy and $\alpha = (N-Z)/A$. a_1, a_2 and J are the volume-,surface- and symmetry-energy coefficients, respectively. The quantity ε involves corrections of higher order in the expansion parameters α and $A^{-1/3}$. Till now, most of the efforts have been devoted to calculating a_1, i.e. the energy per nucleon in infinite symmetrical nuclear matter without coulomb repulsion. No difficulties are met to extend the theory to infinite asymmetrical nuclear matter in order to evaluate the coefficient J. Such an extension has been presented by **ROSATI** and **FANTONI**[8] for the case of state-independent Jastrow correlated wave functions. The estimate J= 31 MeV, as obtained in ref.(8) for densities close to the experimental density and using the so-called OMY potential[9], agrees well with the value J= 32 MeV calculated[10] for the Reid soft core potential in the lowest order of the Brueckner theory. Moreover, the variational approach with semirealistic potentials (including also tensorial components) gives[11] to a large extent nearly the same result as the OMY potential. It should be noticed that the available semiphenomenological mass formulas give estimates for J spread over a rather large range of values since the symmetry-energy contributions are small. As a consequence, it should be convenient to fix the symmetry-energy coefficient of the semiphenomenological mass formulas at the value J= 31 MeV.

The calculation of a_2 is a rather difficult task. An approximate and easy procedure for estimating the surface-energy coefficient has been outlined in ref.(8). The presence of a boundary surface is taken into account by a proper modification of the density of single particle states; the energy per particle of the (infinite) nuclear matter is then re-evaluated and the surface energy term is extracted. For the OMY potential, the value a_2= 21 MeV has been obtained: the agreement with the values 20÷21 MeV suggested by the most accurate semiphenomenological mass formulas[12] is excellent, but one must not take it with too large a confidence due to the approximations involved. Nevertheless, a_2 can be calculated in this way for different ρ density-values and the dependence on the density is expected to be reasonable. The volume and exchange contributions to the coulomb energy are given by the expression

$$(E_C)_{vol.} + (E_C)_{exch.} = \frac{3}{5}\left(\frac{4\pi\rho}{3A}\right)^{1/3} Z^2 e^2 (1 + \eta_C A^{-2/3}) , \qquad (2)$$

with

$$\eta_C = \frac{5}{8\pi}\left(\frac{4\pi\rho}{3}\right)^{2/3} \int d\vec{r}\, \frac{g(r) - 1}{r} , \qquad (3)$$

where g(r) denotes the radial distribution function.
In correspondence to OMY potential, the values of a_1, a_2, J and E_C, as given by eq.(3), can be calculated for different ρ values: by using eq.(1), the binding energy of a generic nucleus is easily obtained if the correction term ε is disregarded (for spherical nuclei $|\varepsilon| \simeq 0.5$-1. MeV). This has been accomplished in ref.(13), where the results for the ground states of ^{16}O and ^{40}Ca are given. A quite independent check is therefore realized for the estimates obtained by means of a direct variational calculation on a specific nucleus (see, for example, refs. (5,6)).
The procedure outlined above requires to be improved under a few respects. First of all, a local density approximation can be chosen to estimate accurately E_C, then the calculation of a_2 can be performed by using a proper variational FHNC treatment. At present, work on these topics is in progress.

References

(1) J.W. CLARK, in: Progress in Particle and Nuclear Physics, vol. 2, edited by D. WILKINSON, Pergamon, Oford (1979)
(2) V.R. PANDHARIPANDE and R.B. WIRINGA, Rev. Mod. Phys. 51 (1979) 821
(3) S. ROSATI,in: "From Nuclei to Particles", 1980, Varenna summer school, in press
(4) C.-W. WOO and L. SENBETU, Nucl. Phys. A328 (1979) 309
(5) L.R. MEAD and J.W. CLARK, Phys. Lett. 90B (1980) 331
(6) R. GUARDIOLA, Nucl. Phys. A328 (1979) 490
A. FAESSLER, R. GUARDIOLA, H. MÜTHER and A. POLLS, to be published
(7) S. FANTONI and S. ROSATI, Nucl. Phys. A328 (1979) 478
(8) S.ROSATI and S. FANTONI, Nuovo Cim. 58A (1980) 327
(9) T.OHMURA, M.MORITA and M. YAMADA, Progr. Theor. Phys. 15 (1956) 222
(10) O. SJÖBERG, Nucl. Phys. A222 (1974) 161
(11) V.R. PANDHARIPANDE, private communication
(12) Atomic Data and Nuclear Data Tables,, 17 (1976) 411
(13) S. ROSATI,in: "From Collective States to Quarks in Nuclei", Bologna 1980, and "Lecture Notes in Physics", to be published

COUPLED CLUSTERS AND COULOMB CORRELATIONS

R.F. Bishop
Department of Mathematics
University of Manchester Institute of Science and Technology
P.O. Box 88, Manchester M60 1QD, England

1. INTRODUCTION

In this paper I intend to give a summary of work carried out in the last few years on developments of the coupled-cluster formalism, and its application to infinite systems of either bosons or fermions. In particular I shall concentrate wholly on systems interacting via the two-body Coulomb force, and hence mainly, but not entirely, on problems involving the long-range behaviour of many-body systems. Much of this work on the one-component Bose and Fermi plasmas has been carried out in collaboration with K.H. Lührmann.

2. COUPLED-CLUSTER FORMALISM

I present first a brief outline of the main elements of the coupled-cluster formalism needed here. A full review of the formalism has recently been given,[1] although this deals almost exclusively with applications in nuclear physics, and hence largely with problems involving short-range correlations, rather than with the long-range correlations of the sort induced by the Coulomb force and which largely concern us here. The interested reader is directed to the article by Lührmann[2] for a formulation that perhaps best stresses its physical content, and to another article with the present author that sets the formalism firmly in the context of the one-component electron plasma.[3]

In terms of a suitable model, or uncorrelated, N-body wavefunction $|\Phi\rangle$, the (usual linked-cluster) ansatz for the exact Coulomb-correlated N-body ground-state (g.s.) wavefunction $|\Psi\rangle$,

$$|\Psi\rangle = e^{S}|\Phi\rangle \, , \tag{1}$$

is made, and we consider $|\Psi\rangle$ normalized to $|\Phi\rangle$ by $\langle\Phi|\Psi\rangle = 1$. I deal here with model Fermi states of Slater determinant form,

$$|\Phi_F\rangle = a^{\dagger}_{\nu_1} \ldots a^{\dagger}_{\nu_N}|0\rangle, \tag{2}$$

with $|0\rangle$ the vacuum state, and where the operators $a^{\dagger}_{\nu_i}$ are a set of fermion creation operators for the orthonormalized single-particle (s.p.) states $|\nu_i\rangle$. For bosons, the antisymmetrized product of s.p. states is replaced by the (symmetric) single-state condensate,

$$|\Phi_B\rangle = (N!)^{-\frac{1}{2}}(b^{\dagger}_{0})^{N}|0\rangle \, , \tag{3}$$

where the operator $b^{\dagger}_{0}$ creates a boson in state o, and, more generally, the operators $b^{\dagger}_{\alpha_i}$ create bosons in a complete orthonormal s.p. set $|\alpha_i\rangle$.

Although it is evidently possible to consider more general s.p. states, it is important for later discussions to realize that for all later results reported, I deal exclusively with plane-wave s.p. states. Thus $|\Phi_F\rangle$ represents the usual filled Fermi sea, and $|\Phi_B\rangle$ the usual completely occupied zero-momentum condensate: both isotropic, homogeneous states of zero total momentum. The correlation operator S is decomposed into n-body ($n \le N$) components,

$$S = \sum_{n=2}^{N} S_n , \tag{4}$$

$$S_n = \begin{cases} (n!)^{-1} \sum\limits_{\rho_1\ldots\rho_n} b^{\dagger}_{\rho_1}\cdots b^{\dagger}_{\rho_n} S_n(\rho_1\ldots\rho_n)(N^{-1/2}\, b_o)^n ; \text{ bosons} \\ (n!)^{-2} \sum\limits_{\substack{\rho_1\ldots\rho_n \\ \nu_1\ldots\nu_n}} a^{\dagger}_{\rho_1}\cdots a^{\dagger}_{\rho_n} \langle\rho_1\ldots\rho_n|S_n|\nu_1\ldots\nu_n\rangle a_{\nu_n}\cdots a_{\nu_1} ; \text{ fermions} \end{cases} \tag{5}$$

The notation used in Eq.(5), and henceforth, reflects the linked-cluster aspect of the expansion, viz. s.p. labels ν_i indicate states normally occupied in $|\Phi\rangle$ (i.e. states inside the filled Fermi sea for fermions or the zero-momentum state, $\nu\equiv 0$, for bosons); and s.p. labels ρ_i indicate normally unoccupied states. Where necessary later, s.p. labels α_i run over all s.p. states (i.e. a complete set). It is important to realize that the sum in Eq.(4) omits the term $n=1$ only as a consequence of our implied assumption that the exact g.s. wavefunction $|\Psi\rangle$ is also an eigenstate of total momentum (with eigenvalue zero).

Physically, S_n represents the true correlation operator for an n-body subsystem that remains after all the factorizable (or unlinked) correlations have been removed from the (complete) n-body subsystem amplitude operator Ψ_n, defined by its matrix elements,

$$\begin{aligned} \langle\alpha_1\ldots\alpha_n|\Psi_n|\nu_1\ldots\nu_n\rangle_A &\equiv \langle\Phi|a^{\dagger}_{\nu_1}\cdots a^{\dagger}_{\nu_n} a_{\alpha_n}\cdots a_{\alpha_1}|\Psi\rangle, \text{ fermions} \\ \Psi_n(\alpha_1\ldots\alpha_n) &\equiv \langle\Phi|(N^{-1/2}\, b^{\dagger}_o)^n b_{\alpha_n}\cdots b_{\alpha_1}|\Psi\rangle, \text{ bosons} \end{aligned} \tag{6}$$

where, for fermions, the subscript A on a ket state indicates an explicitly antisymmetric state

$$|\nu_1\ldots\nu_n\rangle_A \equiv a^{\dagger}_{\nu_1}\cdots a^{\dagger}_{\nu_n}|0\rangle .$$

Thus, for bosons as example, the 2- and 3-body subsystem amplitudes Ψ_2 and Ψ_3 can be expressed as,

$$\begin{aligned} \Psi_2(\alpha_1\alpha_2) &= (N^{1/2}\delta_{\alpha_1,o})(N^{1/2}\delta_{\alpha_2,o}) + S_2(\alpha_1\alpha_2) \\ \Psi_3(\alpha_1\alpha_2\alpha_3) &= (N^{1/2}\delta_{\alpha_1,o})(N^{1/2}\delta_{\alpha_2,o})(N^{1/2}\delta_{\alpha_3,o}) \\ &\quad + S_{123}[S_2(\alpha_1\alpha_2)(N^{1/2}\delta_{\alpha_3,o})] + S_3(\alpha_1\alpha_2\alpha_3), \end{aligned} \tag{7}$$

in the thermodynamic limit ($N \to \infty$; volume $\Omega \to \infty$, $\rho = N/\Omega$ finite), and where S_{123} generates the sum of all terms obtained by cyclic permutation of the labels α_1, α_2 and α_3.

An equivalent physical description of S_n (for fermions) is that its matrix elements give the exact amplitudes that describe the excitation of n particle-hole pairs; where particles and holes refer respectively to states normally unoccupied and normally occupied (in the model wavefunction $|\Phi\rangle$). For bosons the role of the hole states is played by the condensate. It seems intuitively apparent that in order for our ansatz (1) to be useful, the physical system under consideration ought to share at least qualitatively the features built into the model state $|\Phi\rangle$. More explicitly we expect our choices $|\Phi_F\rangle$ and $|\Phi_B\rangle$ to have relevance respectively only to real fermion systems in states where some semblance of the sharp Fermi surface still remains, and to real Bose systems which contain a finite fraction of the particles in a zero-momentum condensate. This would seem to rule out from the outset for fermions, for example, an accurate description of "abnormal" or "super" phases, or indeed of anything but the usual "liquid" or "Fermi fluid" phase. Later, I give some indication that this intuitive feeling may well be false; or at least that the coupled-cluster formalism may be much more powerful than this too pessimistically narrow interpretation would seem to allow.

Formally, the g.s. coupled-cluster formalism now proceeds by decomposing the N-body Schrödinger equation

$$H|\Psi\rangle = E|\Psi\rangle \qquad (8)$$

into a coupled set of equations for the matrix elements of the correlation operators S_n. Formally this may be achieved by taking the overlap of Eq.(8) with the states $\langle\Phi|(a^\dagger_{\nu_1}\ldots a^\dagger_{\nu_n} a_{\alpha_n}\ldots a_{\alpha_1})$ for $n = 0,1,\ldots N$, to get a set of coupled equations for the elements $\langle|\Psi_n|\rangle$. Finally the amplitudes Ψ_n are decomposed in terms of the correlation amplitudes of the S_n, which has the effect of eliminating all macroscopic terms (i.e. those, like E, which are proportional to N) from the essentially microscopic subsystem equations. This wholly algebraic procedure results in a coupled set of equations for the elements of S_n, in which the $i^{\underline{th}}$ equation for S_i is coupled to both S_{i+1} and S_{i+2} (as well as to all S_j with $j<i$), for a Hamiltonian H involving two-body potentials only. For the technical details of the derivation the reader is referred to Refs.[1-3]. Clearly, in order to be useful this exact coupled hierarchy needs to be truncated; and as an obvious initial step I discuss the so-called SUBn approximation scheme in which I set $S_i = 0$ for all $i>n$, and the remaining equations are treated as accurately as possible.

I now apply the coupled-cluster formalism to one-component Coulomb systems, stressing mainly the qualitative nature of the results in order more clearly to demonstrate the power of the formalism. To this end I spend more time on the mathematically much simpler boson equations, and indicate only more briefly their fermion counterparts.

3. APPLICATION TO ONE-COMPONENT COULOMB SYSTEMS

The Coulomb potential with a uniform, rigid and neutralizing background present is,

$$v(q) = \frac{4\pi e^2}{\Omega q^2}\left(1 - \delta_{\vec{q},0}\right) . \tag{9}$$

The density ρ may be expressed either in terms of the usual dimensionless coupling constant r_s, which is the average interparticle spacing in units of the Bohr radius a_o, or in terms of a (for bosons, fictitious) Fermi wavenumber k_F applicable to an unpolarized spin-½ system,

$$\rho = (4\pi r_s^3 a_o^3/3)^{-1} = k_F^3/3\pi^2 . \tag{10}$$

Henceforth, the g.s. energy per particle is expressed in Rydberg units by

$$E/N = \varepsilon(e^2/2a_o) , \tag{11}$$

and any dimensionless momentum variables that appear have been scaled against the Fermi momentum $\hbar k_F$, defined by Eq.(10).

3.1 Charged Bose system

For spin-zero bosons, the exact two-body equation for $S_2(q) \equiv S_2(\vec{q},-\vec{q})$ may readily be found by the method sketched in Sect.2:

$$\frac{\hbar^2 q^2}{m} S_2(q) + T_{RPA} + T_{CP} + T_{LAD} + \sum_{\vec{q}'} v(q')[2N^{\frac{1}{2}}S_3(\vec{q},\vec{q}',-\vec{q}-\vec{q}') + \tfrac{1}{2}NS_4(\vec{q},-\vec{q},\vec{q}',-\vec{q}')] = 0,$$

$$T_{RPA} = Nv(q)[1+S_2(q)]^2; \quad T_{CP} = -4\frac{E}{N}S_2(q); \quad T_{LAD} = \sum_{\vec{q}'} v(\vec{q}-\vec{q}')S_2(q'), \tag{12}$$

and where the g.s. energy per particle is given by,

$$E/N = \tfrac{1}{2}Nv(0) + \tfrac{1}{2}\sum_{\vec{q}} v(q)S_2(q) . \tag{13}$$

The SUB2 approximation is obtained from Eq.(12) by putting S_3 and S_4 to zero. The remaining first four terms in Eq.(12) represent respectively (i) the kinetic energy (KE) contribution, (ii) the terms that generate the ring or bubble diagrams of the random-phase approximation (RPA), (iii) the terms that generate the self-consistent (s.c.) energy insertions on the zero-momentum condensate lines, i.e. the s.c. condensate potential (CP), and (iv) the terms that scatter the two particles outside the condensate and hence generate the two-particle ladder (LAD) diagrams. Inserting the potential from Eq.(9) into the SUB2 equation gives, in dimensionless variables,

$$q^2S_2(q) + 4\alpha r_s(3\pi q^2)^{-1}[1+S_2(q)]^2 - 2(\alpha r_s)^2\varepsilon S_2(q) + \frac{\alpha r_s}{\pi q}\int_0^\infty dq'q'\ln\left|\frac{q'+q}{q'-q}\right|S_2(q') = 0; \tag{14}$$

$$\varepsilon = 2(\pi\alpha r_s)^{-1}\int_0^\infty dq\, S_2(q), \tag{15}$$

where $\alpha \equiv (9\pi/4)^{-1/3}$.

Although Eqs.(14) and (15) are readily amenable to numerical solution, it is more instructive here to examine them in the high density $(r_s \to 0)$ and low density $(r_s \to \infty)$ limits. In the high density limit it is readily shown that to leading order for the energy only the KE and RPA terms contribute. The resulting quadratic equation is trivially solved to give

$$\varepsilon \xrightarrow[r_s \to 0]{} \varepsilon_{RPA} = Q\, r_s^{-3/4}$$
$$Q = -2\pi^{-1}(6^{1/4})\int_0^\infty dx[1+x^4-x^2(x^4+2)^{1/2}] \approx -0.8031 \ , \tag{16}$$

which is the exact result first obtained by Foldy.[4] It is not difficult to show that the next term in the high-density expansion is a constant,

$$\varepsilon \xrightarrow[r_s \to 0]{} Q\, r_s^{-3/4} + R \ , \tag{17}$$

and that both the CP and LAD terms now contribute to R. I find,

$$R_{SUB2} = R_{CP} + R_{LAD}; \quad R_{CP} = 16/15\pi, \ R_{LAD} = 32/45\pi \ . \tag{18}$$

By inspecting Eq.(12) and the equivalent equations for S_3 and S_4 it can however be shown that the coupling terms to S_3 and S_4 in the <u>exact</u> Eq.(12) also contribute to the constant R (although not to Q). I have also calculated the constant R given by the exact two-body Eq.(12), keeping the complete coupling to three- and four-body clusters. I find that, to this order, the 3-body correlation amplitude S_3 needed in Eq.(12) may be replaced by,

$$-N^{1/2}S_3(\vec{q}_1,\vec{q}_2,\vec{q}_3) \to [\omega(q_1)+\omega(q_2)+\omega(q_3)]^{-1}S_{123}[\{v(q_1)(1+S_2(q_1)) + v(q_2)(1+S_2(q_2))\}S_2(q_3)], \tag{19}$$

for $\vec{q}_1+\vec{q}_2+\vec{q}_3 = 0$, and where the effective s.p. energy is,

$$\omega(q) = \hbar^2q^2/2m + Nv(q)(1+S_2(q)); \tag{20}$$

and a similar replacement may be made for S_4 by examining the 4-body equation. Equations (12) and (13) then lead to

$$R = R_{SUB2} + R_3 + R_4 \approx 0.0280 \tag{21}$$

where both the contributions R_3 and R_4 from the coupling terms to S_3 and S_4 in Eq.(12) are finite. The final result of Eq.(21) is <u>exact</u>, and an extremely tedious rearrangement of the integral expressions shows it to be in precise agreement with the first correct result reported, of Brueckner.[5] It is worth pointing out that, by contrast with most competing methods, each of the terms in Eqs.(18) and (21) is finite, and no cancellation of spurious logarithmic singularities occurs. This particular point highlights a more general advantage of the coupled-cluster formalism — namely that terms which tend to cancel each other are automatically grouped together.

Turning now to the more revealing low-density limit, naively one would not expect the SUB2 approximation to give any reasonable result at all in this strong-coupling regime, since one imagines that the n-body clusters even with $n >> 2$ are still very important. Indeed one believes the real Coulomb system to undergo a phase transition to a Wigner solid[6] in this limit, and the solid may be regarded as an archetypal system where the N-body correlations dominate. At any rate it is clear that the low-density Coulomb systems provide one of the most stringent tests for our formalism.

It is readily seen from Eqs.(14) and (15) that in the SUB2 approximation,

$$\varepsilon_{SUB2} \xrightarrow[r_s\to\infty]{} -Ar_s^{-1} + Br_s^{-3/2} + O(r_s^{2}), \tag{22}$$

where the KE term contributes only to the constant B in leading order. In this limit the terms RPA, CP and LAD are all necessary for a quantitative evaluation of the constant A, but they play distinctly different qualitative roles. Thus it is vital to keep the RPA terms to get the correct analytic behaviour because, as expected, these terms continue to be crucial for the long-range $(q \to 0)$ screening of the Coulomb potential. Similarly the CP plays a crucial role now in the short-range $(q \to \infty)$ limit. Whereas the inclusion of the LAD term quantitatively changes the constants A and B in Eq.(22), it may safely be omitted without changing the analytic form of Eq. (22). Dropping the LAD term from Eq.(14), I find

$$\text{SUB2 - LAD approxn.:}\quad A = (32/3\pi^2)^{1/3} \approx 1.03; \quad B = 3^{1/2}\pi/8 \approx 0.68 . \tag{23}$$

Use of the virial theorem verifies our expectation that the leading term in Eq.(22) is purely potential energy; and furthermore shows that the much more interesting second term is exactly one half each kinetic and potential energies — which at the very least is strongly reminiscent of simple harmonic motion and of the behaviour expected of a solid.

Indeed, as first pointed out by Wigner,[6] the energy of the system is minimised in this low density limit by the particles crystallizing into a regular periodic lattice, which leads to an electrostatic potential energy proportional to r_s^{-1}. Whereas in a fluid phase the particles are free to occupy the whole volume, which by the uncertainty principle leads to a kinetic energy proportional to r_s^{-2}, in the Wigner solid phase the particles are constrained to oscillate about the fixed lattice sites and hence to have a greater kinetic energy. Elementary considerations of simple harmonic motion lead to a kinetic energy proportional to $r_s^{-3/2}$. The exact expansion in the Wigner solid phase is a power series in $r_s^{-1/2}$ where the terms of order r_s^{-2} and higher are due to anharmonicities in the zero-point motion. Based on a b.c.c. lattice, Carr et al.[7] give,

$$\varepsilon_{exact} \xrightarrow[r_s\to\infty]{} -1.792\, r_s^{-1} + 2.65\, r_s^{-3/2} - 0.73\, r_s^{-2} + \ldots . \tag{24}$$

Clearly our approximation in Eqs.(22) and (23) has the correct analytic form for the energy of a solid, although the values of the coefficients are considerably underestimated. What is more important however is that even the lowest SUB2 approximation in the coupled-cluster scheme gives a low-density energy which cannot possibly represent what is normally understood by a fluid phase, since the particles are definitely not free to occupy the whole volume.

Thus it is clear that our intrinsically fluid-like and everywhere translationally-invariant approach can provide a good description of both fluid and solid phases; and we can understand the possibility of this by similar reasoning to that behind the familiar "floating crystal model" of Feenberg.[8] On the other hand, we must still face the fact that three- and more-body effects can only be treated in SUB2 approximation in an average sense, and cannot possibly represent the detailed internal structure of a lattice wavefunction. While I have presented explicit evidence that this is not a severe limitation at least as far as the qualitative behaviour of the g.s. energy is concerned (and probably also for matrix elements of other few-body operators), the fact remains that for a charged Bose system in the low-density limit, the third and higher order correlations are still very strong. Thus accurate values of the coefficients A and B cannot be expected. I note here that numerical calculations of the complete SUB2 equation including the LAD term do not change this overall picture. In fact our approximate value of A from Eq.(23) is lowered by about 20% by including the LAD term, thereby increasing the discrepancy with the Wigner value. (I note also however that an evaluation of the two-body radial distribution function within the SUB2 approximation gives a positive-definite function at all densities only so long as the LAD term is included.)

Finally I note that although the SUB2 approximation works superbly over the entire density regime for the Bose Coulomb system, the g.s. energy is quantitatively unsatisfactory in the low-density limit. It is clear that higher-order clusters must be incorporated; but due to the relative simplicity of the Bose coupled-cluster equations this is quite practicable, as indeed I have already indicated in the high-density limit.

3.2 Charged Fermi system

The SUB2 equations for fermions, although conceptually similar to those for bosons, are mathematically vastly more complex due both to the many more terms required by antisymmetrization, and to the state- (i.e. momentum-) dependence induced by the hole states inside the Fermi sea in comparison with the unique zero-momentum condensate for bosons. In particular for fermions the matrix elements $S_2(\vec{k}_1,\vec{k}_2;\vec{q}) \equiv \langle\vec{k}_1+\vec{q},\vec{k}_2-\vec{q}|S_2|\vec{k}_1,\vec{k}_2\rangle$ depend not only on a momentum transfer $\vec{q}$ as for bosons but also on the two hole momenta $\vec{k}_1$ and $\vec{k}_2$. The complete SUB2 equation for charged Fermi systems has been discussed in detail,[3] and it is clear that a numerical solution of this non-linear integral equation for a function of three 3-vectors, while perhaps just feasible, is not to be undertaken lightly! Accordingly we have again examined various limits and approximation schemes for handling the coupled-cluster Fermi equations, and I now briefly report on these.

In the high-density limit, the RPA again gives the leading contribution to the correlation energy, ε_c, i.e. the g.s. energy relative to the (uncorrelated) Hartree-

Fock energy. In Ref.[3] the nonlinear integral equation for S_2 in RPA was solved exactly and in some detail, both confirming the well-known results of Gell-Mann and Brueckner,[9] and giving for the first time exact analytic forms for the four-point function S_2 and the once-integrated three-point particle-hole vertex function. The Tamm-Dancoff approximation (TDA) to the ring summation was also formulated, and the analogous exact solutions in TDA were also presented for the electron gas for the first time.

Turning to the intermediate-coupling ($1 \lesssim r_s \lesssim 5$) metallic-density regime, we no longer expect the RPA plus second-order exchange to be a good approximation, although it gives the first two terms in the high-density expansion for ε_c exactly. Thus, quite apart from ignoring (a) the simple exchange effects necessary to antisymmetrize RPA, we have ignored even in SUB2 approximation: (b) all of the combined particle-particle and hole-hole ladder terms, some at least of which are important for the correct short-range behaviour; (c) the generalized self-energy correction terms which self-consistently generate both the particle potential and, much more importantly the hole potential (which now for fermions plays the same crucial role as the CP for bosons); (d) classes of higher ring-exchange terms; and (e) a class of additional exchange terms which includes the particle-hole ladder terms. In order systematically to deal with these effects I have proposed and implemented a further approximation that enables us to study these terms much more readily.

Based on a comparison with Bose systems, the fermion equations should be much simpler if they could be "state-averaged"; and the basic approximation is thus to average over the initial hole momenta $\vec{k}_1$ and $\vec{k}_2$ in $S_2(\vec{k}_1,\vec{k}_2;\vec{q})$ but keeping the important exact property that final states $(\vec{k}_1+\vec{q})$, $(\vec{k}_2-\vec{q})$ lie outside the Fermi sea (i.e., the Pauli principle is exactly implemented). In this way the exact $S_2(\vec{k}_1,\vec{k}_2;\vec{q})$ is replaced by an averaged $\bar{S}_2(q)$, and the resulting coupled-cluster equation considered still then itself has to be state-averaged. Although the procedure for this latter step is not unique, this works to our advantage, for two reasons: (a) the averaging can be made on physically-motivated grounds rather than being imposed arbitrarily; and (b) since we know exact results for S_2 in at least one limit, namely the RPA and TDA results for $r_s \to 0$, the errors induced by the various averaging schemes can be checked. As an illustration: carrying out the above scheme in RPA leads to an equation for S_2 which involves only KE and RPA terms. After the replacement $S_2 \to \bar{S}_2$ has been made the only state-dependence left is in the KE term, which for fermions is proportional to $[|\vec{k}_1+\vec{q}|^2+|\vec{k}_2-\vec{q}|^2-k_1{}^2-k_2{}^2]S_2 \equiv eS_2$. As two obvious averaging schemes one could imagine (a) replacing $e \to \langle e\rangle$; or (b) the intuitively and physically more appealing idea of first dividing through by e and then averaging the "energy denominator"; $e^{-1} \to \langle e^{-1}\rangle$. I have shown that the former procedure leads precisely to the mean-spherical approximation (which Zabolitzky[10] discusses in this context, and which his state-independent, variational, Fermi hypernetted-chain (FHNC) formalism leads to in this $r_s \to 0$ limit), which gives an ε_c in error by 8.4% at $r_s \to 0$. The latter procedure on the other hand is exact at

$r_s \to 0$, and by comparison with the exact RPA results I show that it is for no density in error by more than 2%. There seems to be no reason why this result should not hold for all other terms in the Fermi SUB2 equation, which we may therefore now with confidence systematically include.

Based partly on experience with the Bose equations, our best coupled-cluster (CC) calculations to date, for ε_c in the metallic regime, include from the S_2 equation the following terms, all treated simultaneously and fully self-consistently: (i) KE; (ii) RPA; (iii) hole potential (HP); (iv) all particle-particle ladder (LAD) terms included in SUB2 and furthermore (motivated by experience with nuclear matter) a much broader class of generalized ladder terms obtained by taking into account part of the coupling terms to S_3 and S_4, and which involves replacing the bare potential by a self-consistent G-matrix (obtained from the full S_2 solution itself); (v) a class of particle-hole ladder terms called PHA in Ref.[3]; and (vi) exchange (EX) terms to keep the resulting S_2 explicitly antisymmetric. The results of this CC calculation are shown in the Table below, where for comparison I also show both the essentially exact (unpublished) results of Ceperley who used an approximate Green's function Monte Carlo (GFMC) method; and, as representatives of the best variational results, the FHNC results of both Zabolitzky[10] (FHNC-Z) and Lantto[11] (FHNC-L).

Table: The Fermi correlation energy ε_c for the unpolarized electron gas

r_s	ε_c(CC)	ε_c(GFMC)	ε_c(FHNC-Z)	ε_c(FHNC-L)
$\to 0$	$0.0622 \ln r_s$	$(0.0622 \ln r_s)$	$0.0570 \ln r_s$	?
1	-0.123	-0.121	-0.114	-0.140
2	-0.0917	-0.0902	-0.0859	-0.098
3	-0.0751	...	-0.0710	-0.079
4	-0.0644	...	-0.0612	-0.067
5	-0.0568	-0.0563	-0.0541	-0.058

Turning finally to the low-density limit, the situation for fermions is much more favourable than for bosons since the Pauli principle very effectively hinders electrons from clustering in groups of more than two, thus forcing the higher correlations to be smaller. Although in the exact Wigner low-density limit the effects of quantum statistics vanish, with the fermion and boson solid both described by the same asymptotic expansion (24) (and the different statistics reflected only in differing terms which vanish exponentially with $r_s \to \infty$) this is by no means true in our translationally-invariant CC description. In the case of electron system, exchange terms do not vanish and the convergence of the CC hierarchy is thereby considerably improved from the Bose case. Thus for the analogue of the result (23) for bosons, I find $A \approx 1.58$ for electrons in a "state-averaged" RPA+HP scheme — which is in much better agreement with the Wigner solid value of 1.79.

4. FINAL REMARKS

I intend further to explore the low-density regime with the full SUB2 approximation for electrons since it provides a scheme that offers what is essentially the first unified framework in which to calculate at all densities the g.s. properties (at least) of the charged quantum fluids/solids. Although I have stressed only g.s. energy calculations it is important to realise that recent extensions of the CC formalism permit calculations both of excited states and of the density matrices. For excited states, Emrich has given a very elegant formulation in which he derives a coupled set of eigenvalue equations for the energies and amplitudes of the excitations. As a first step I have applied this formalism in its lowest level of approximation to the electron plasma as $r_s \to 0$. As input this requires the exact g.s. (RPA) S_2 already found.[3] To this level of approximation I find the usual plasmon "bound-state" plus the one-particle-one-hole scattering continuum. Of particular interest at the next step will be the usefulness of the excited states to pin down further the low-density solid aspects of the g.s. calculations. Thus presumably the real electron system has plasmon excitations (with a finite energy gap at low momenta q) at high densities, and a phonon spectrum (with an acoustic branch linear in q at small q) characteristic of solids at low densities. It will be of great interest to see whether this behaviour is also seen in our calculations; and, if so, whether it can be used to obtain the critical density.

It is also intended further to examine the one- and two-body density matrices at metallic and low densities, since these can provide much more sensitive tests of various theories than the g.s. energy. For example, while the approximation RPA+EX gives quite good values for the g.s. energy at metallic densities, the density matrices can be badly wrong — even giving negative values for the two-body radial distribution function at small separations. All preliminary investigations indicate that the CC calculations also give extremely good Coulomb distribution functions.

Finally, work is also in progress to extend these results to such multi-component plasmas as the hydrogen plasma, simple metals, and the electron-hole droplets observed in various semiconductors (with a particular aim to study the excitonic phase).

References

[1] H. Kümmel, K.H. Lührmann and J.G. Zabolitzky, Phys. Reports 36C (1978) 1; and earlier references contained therein.

[2] K.H. Lührmann, Ann.Phys.(NY) 103 (1977) 253.

[3] R.F. Bishop and K.H. Lührmann, Phys.Rev. B 17 (1978) 3757.

[4] L.L. Foldy, Phys.Rev. 124 (1961) 649; 125 (1962) 2208.

[5] K.A. Brueckner, Phys.Rev. 156 (1967) 204.

[6] E.P. Wigner, Phys.Rev. 46 (1934) 1002.

[7] W.J. Carr, Jr., R.A. Coldwell-Horsfall and A.E. Fein, Phys.Rev. 124 (1961) 747.

[8] E. Feenberg, J. Low Temp. Phys. 16 (1974) 125.

[9] M. Gell-Mann and K.A. Brueckner, Phys.Rev. 106 (1957) 364.

[10] J.G. Zabolitzky, Phys.Rev. B 22 (1980) 2353.

[11] L.J. Lantto, Phys.Rev. B 22 (1980) 1380.

Calculation of Gaps of Superconductors by Coupled Cluster Methods (CCM)

K. Emrich

Ruhr-Universität Bochum, Institut für Theoretische Physik II, D-4630 Bochum 1

I will discuss a generalization of the CC-method which makes it possible to calculate the gap of a superconductor.

The CC-method - invented by Coester and Kümmel more than twenty years ago - uses the exp(S)-form of the correlated ground state wave function

$$|\Psi_g\rangle = \exp S|\Phi\rangle \quad , \; S = \sum_n S_n \tag{1a}$$

where S_n is the linked n-particle-n-hole ground state amplitude.

For a Fermi system $|\Phi\rangle$ is usually given by the finite or infinite product

$$|\Phi\rangle = \prod_\nu a_\nu^+ |vac\rangle \tag{1b}$$

The states ν are then called the occupied states, the unoccupied states are usually denoted by ρ .

The definition (1b) can be replaced by a more general one - demanding that

$$a_\nu^+|\Phi\rangle = 0 = a_\rho|\Phi\rangle \quad , \; \{\nu\} \cap \{\rho\} = 0 \tag{1c}$$

The Schrödinger equation is written in the form

$$\exp(-S) H \exp S|\Phi\rangle = E_g|\Phi\rangle \tag{2}$$

and then projected with the states

$$\langle\Phi| a_{\nu_1}^+ \dots a_{\nu_n}^+ a_{\rho_n} \dots a_{\rho_1} \tag{2a}$$

One obtains then a coupled system of equation which must be truncated in a suited way.

The CC-method has been applied with great success in nuclear physics as well as quantum chemistry. For a review article see ref. 1.

The gap Δ_0 of a superconductor is given as the minimum of excitation energies i.e.

$$\Delta_0 = \mathop{Min}_{\vec{q}} \varepsilon_{\vec{q}}$$

In order to obtain it one has to generalize the method in two respects:

1) One has to calculate excitation energies. The excited state $|\Psi_q\rangle$ can be written in the following way

$$|\psi_q\rangle = R^q|\psi_g\rangle = R^q \exp S|\Phi\rangle \tag{3}$$

where

$$R^q = \sum_n R^q_n \tag{3a}$$

R^q_n is a linked n-particle-n-hole excited state amplitude. The Schrödinger equation is written in the form [2]

$$\exp(-S)[H,R^q]\exp S|\Phi\rangle = \varepsilon_q R^q|\Phi\rangle \quad , \; \varepsilon_q = E_q - E_g \tag{4}$$

and projected from the left with the state (2a).

Already the lowest order (non trivial) approximation yields the RPA-terms and all (generalized) exchange terms. It fulfills the Pauli principle [3]. A test at the exactly solvable Lipkin model shows that for particle numbers $N \gtrsim 10$ the method is far superiour to the configuration interaction method, Schrödinger-perturbation theory etc.

ii) For the ground state superconductors we use (1a) where $|\Phi\rangle$ is given by the BCS-state, i.e.

$$|\Phi\rangle = |\Phi_{BCS}\rangle = \prod_k (u_k + v_k a^+_k a^+_{-k})|vac\rangle = \prod_k \beta_{-k}\alpha_k|vac\rangle \tag{5}$$

where α_k and β_k are given by the Bogoljubov-transformation

$$\begin{aligned} u_k a^+_{k\uparrow} - v_k a_{-k\downarrow} &= \alpha^+_k \\ u_k a^+_{-k\downarrow} + v_k a_{k\uparrow} &= \beta^+_{-k} \end{aligned} \tag{6}$$

Instead of (1c) we have the relations

$$\beta_{-k}|\Phi_{BCS}\rangle = 0 = \alpha_k|\Phi_{BCS}\rangle \tag{7}$$

In order to obtain (1c) we introduce the operators $\tilde{\alpha}_{k\downarrow}$ and $\tilde{\alpha}_{k\uparrow}$ by

$$\beta_{-k} = \tilde{\alpha}^+_{k\downarrow} \quad , \quad \alpha_k = \tilde{\alpha}_{k\uparrow} \tag{8}$$

Then we project the Schrödinger equations (2) and (4) with

$$\langle\Phi|\tilde{\alpha}^+_{k_1\downarrow}\cdots\tilde{\alpha}^+_{k_n\downarrow}\tilde{\alpha}_{k_n\uparrow}\cdots\tilde{\alpha}_{k_1\uparrow} \tag{9}$$

We see that the new equations - denoted in the following by Bogoljubov-CC-equations - are obtained from the usual CC-equations by the replacements

$$\rho \rightarrow \text{spin}\uparrow = \text{part.}\uparrow + \text{hole}\downarrow \tag{10a}$$

$$\nu \rightarrow \text{spin}\downarrow = \text{part.}\downarrow + \text{hole}\uparrow \; . \tag{10b}$$

Of course the Hamiltonian, too, has to be transformed. We will see that it has a special simple form if the $\tilde{\alpha}$ -operators are used. We will discuss at first pure electronic systems (with an effective force e.g. simulating the pairing) and later on electronic-ionic-systems.

Electronic Systems

We assume that the Hamiltonian is given by

$$H = T_E + V_E \quad , V_E = V_{Coul.} + V_R \tag{11}$$

(local potential)

$$V_E \underset{q \to 0}{\longrightarrow} \frac{const.}{q^2} + V_R(q) \quad , \quad V_E(0) = V_R(0) \tag{12}$$

(Coul. part = 0 because of charge conservation)

We introduce in the usual way the thermodynamical potential K and decompose it in a one body part H_1 and a two-body part H_2

$$K = H - \mu N = \mathcal{U} + H_1 + H_2 \tag{13}$$

The one body-part H_1 yields the ordinary gap equation. H_1 and the constant $\mathcal{U}$ are given in ref. 4, p. 326-33o, for example.

In the α, β notation H_2 is a complicated operator containing products of 4 creation operators, 3 creation operators X 1 annihilation operator, 2 creation operators X 2 annihilation operators etc. In the $\tilde{\alpha}$-notation H_2 becomes rather simple. H_1 and H_2 are given by ($\underset{\sim}{N}$ stands for the normal product)

$$H_1 = \sum_k \sum_{\sigma_1 \sigma_2} \tilde{\alpha}^+_{k\sigma_1} \tilde{\alpha}_{k\sigma_2} \tag{14a}$$

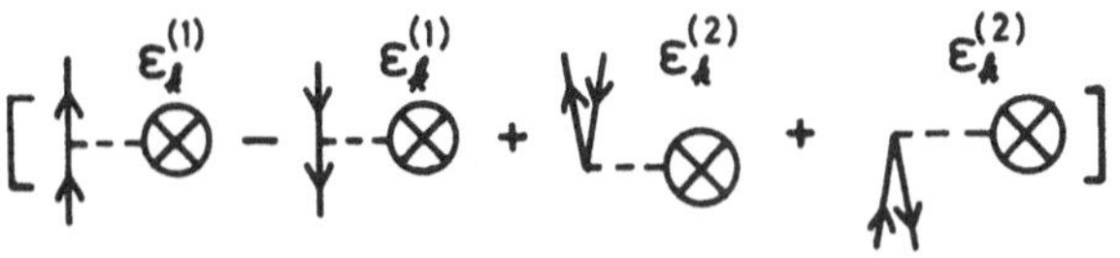

$$H_2 = \frac{1}{2} \sum_{k_1 k_1'} \sum_{\sigma_1 \sigma_1'} \underset{\sim}{N}\, \tilde{\alpha}^+_{k_1\sigma_1} \tilde{\alpha}^+_{k_2\sigma_2} \tilde{\alpha}_{k_2'\sigma_2'} \tilde{\alpha}_{k_1'\sigma_1} \tag{14b}$$

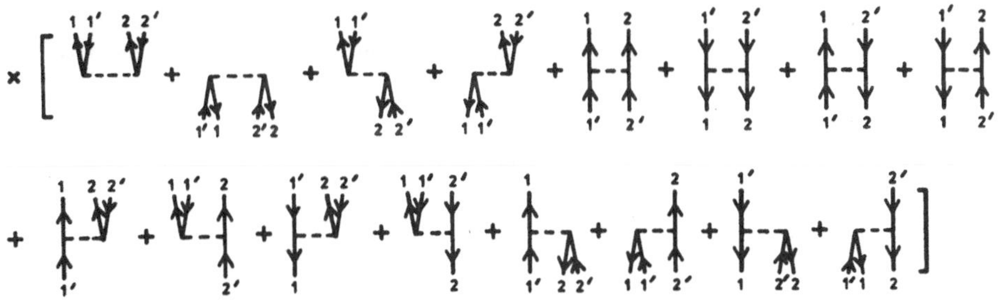

where

The vertices a and b are given by

$$b_{kk'} = u_k v_{k'} + v_k u_{k'} \qquad , \quad a_{kk'} = u_k u_{k'} - v_k v_{k'} \tag{15}$$

If the function v_k has a sharp edge at $k = k_F$ i.e.

$$v_k \rightarrow \Theta(k_F - |k|) = \begin{cases} 1 & |k| \leq k_F \\ 0 & |k| > k_F \end{cases} \tag{16}$$

then the expression (14b) for H_2 is identical with the standard form of a local interaction term. From this it follows that the Bogoljubov-CC eqs. do contain the standard CC-eqs. as a special case. Moreover, it follows that - up to the vertices a and b, the one-particle energies $\varepsilon_k^{(1)}$ and $\varepsilon_k^{(2)}$ and the replacement (1o) - the equations are identical. E.g. the equation for "two pair cluster" - written in terms of diagrams - is given by

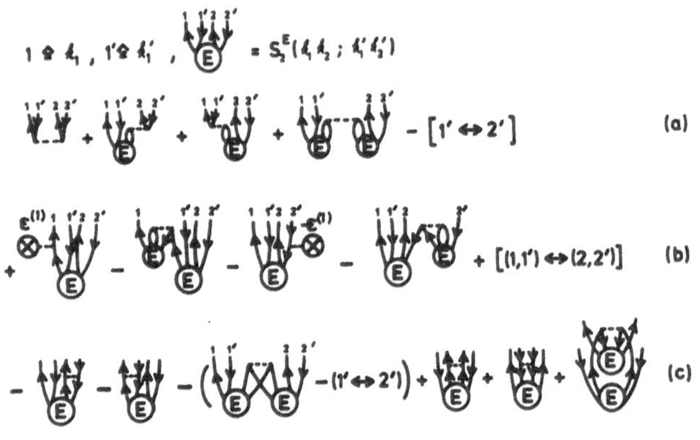

= 0 (17)

Note that the arrows denote spin up and spin down states resp. The spin up spin down pairs do contain particle-hole as well as particle-particle - as well as hole-hole-pairs.

If $k_1 \rightarrow k_1'$ the terms of 17a become divergent. Because the sum of these divergent terms must be finite, they have to cancel each other. This is reached by the identity

$$\sum_k b_{kk} S_2^E(k k_1; k k_1)_A = - b_{k_1 k_1} \tag{18}$$

For arbitrary $\vec{q} = k_1 - k_1'$ it holds

$$\sum_{\substack{k k' \\ k - k' = \vec{q}}} b_{kk'} S_2^E(k k_1; k' k_1')_A = - b_{k_1 k_1'} + I(k_1 k_1') \quad \text{(\underline{screening equation})} \tag{19}$$

where $I(k_1 k_1')$ is of the order of q. Eq. (19) means that the divergent term $V(q)$ is screened by the other divergent terms of (17a).

Using (19) it follows that the terms of (17a, b) cancel to a large part. If (17c) is neglected one obtains - without further approximations -

$$S_2^E(k_1 k_2; k_1' k_2')_A = - A_{k_1' k_2'} V_E(|k_1 - k_1'|) \frac{I(k_1 k_1')\, I(k_2 k_2')}{\tilde{\varepsilon}(k_1) + \tilde{\varepsilon}(k_1') + \tilde{\varepsilon}(k_2) + \tilde{\varepsilon}(k_2')} \tag{2o}$$

($A_{k_1' k_2'}$ = Antisymmetr. operator)

where
$$\tilde{\varepsilon}(k) = (T_k - \mu)(u_k^2 - v_k^2) + \sum_k V_E(|k-k'|)[v_k^2 - b_{kk'} I(k\,k')] \tag{21}$$

From (19) and (2o) one gets a rather simple integral equation for $I(k\,k')$. Even if the "Brueckner" terms of (17c) are taken into amount one obtains an equation for I which can be solved without too large numerical effort.

If only the "direct" terms of (17a) and the kinetic energy terms of (17b) are taken into account one gets an equation which is analogous to the Random Phase Approximation of the standard CCM. Therefore, I denote this approximation by σRPA.
It can be easily shown that

$$\sigma RPA \xrightarrow[v_k \to \Theta(k_F - |k|)]{} RPA \tag{22}$$

If V_E is given by the Coulombpotential, one just obtains the non linear integral equation of Ref.5. Eq. (22) illustrates the fact that the BogCC eqs. contain the CC eqs.

One can also compare the method with the hypernetted chain method. Using for the kinetic energy terms of the RPA (in the limit (22)) the averaging procedure of the "Mean Spherical Approximation (MSA)" of Ref. 6 one easily rederives Zabolitzky's result (58c) for the "unsymmetric structure function" $\langle \phi | S_{-q} S_{q} | \psi_g \rangle$.

The Bog-CC-equation for the function v_k reads ($\varepsilon_k^{(2)}$ is decomposed into an interaction part and a "kinetic energy part" $2(T_k - \mu) u_k v_k$)

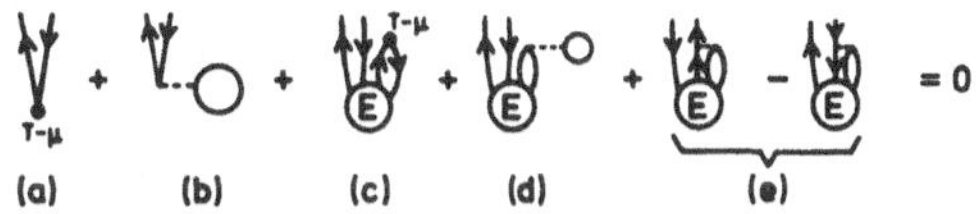

(23)

If the S_2^E terms were zero one would obtain the old well-known gap-equation - however with an unscreened potential V_E. (23d) evidently screens the term (23b). Also the (relatively small) terms (23e) can be interpreted as "higher screening terms". The term (23c), however is quite new. The μ-parts of (23a) and (23c) exactly cancel because of the identity (18). (Note that μ occurs also in eq. 21). The T-part of (23c) is a large term. It originates from a combined effect, i.e. screening (large S_2^E) and pairing; in a normal system this term would be zero because the matrix elements $\langle k_g | T | k_v \rangle$ are zero because of momentum conservation.

Of course (23) is <u>not</u> a new kind of gap equation; it only defines v_k and therewith $|\Phi_{BCS}\rangle$. The gap is obtained as the minimum of excitation energies. The excitation energies are given in first approximation by the equation

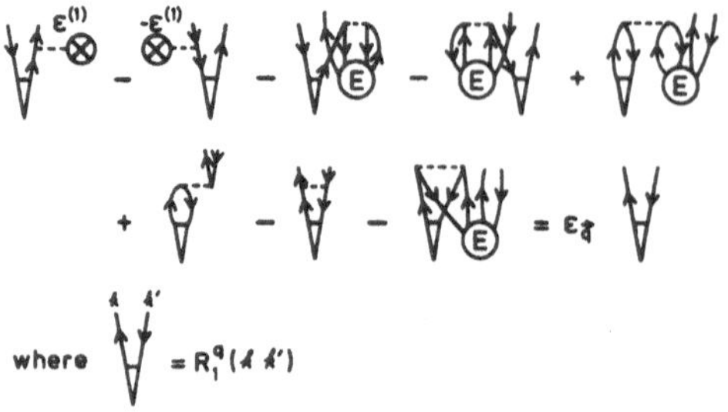

(24)

Higher clusters R_m^q are connected with the break up of n-pairs and can be neglected at least if $q \to 0$. The eigenvector $R_1^q(k\,k')$ depends on 2 scalar variables (for $q = 0$ on one). Of course, one needs the groundstate clusters $I(k\,k')$ and v_k depending on 3 resp. 1 scalar variable(s).

Electronic-Ionic-Systems

We start from the Hamiltonian

$$H = T_I + V_I + V_{EI} + T_E + V_{Coul} \tag{25}$$

Where V_I is the ion-ion potential and V_{EI} the pseudo potential of the electrons. We transform to normal coordinates Q_{qj} and P_{qj} and express these operators in terms of the Boson creation and annihilation operators [7,8] b^+_{qj} and b_{qj} , i.e.

$$Q_{qj} = \sqrt{\frac{\hbar}{2\omega_j(q)}}\,(b^+_{-qj} + b_{qj}) \quad , \quad P_{qj} = i\sqrt{\frac{\hbar\,\omega_j(q)}{2}}\,(b^+_{-qj} - b_{qj}) \tag{26}$$

The transformation yields [7]

$$H = H_{Ph} + H_{EPh} + V_{Coul} + (T_E + U_{EI}) \tag{27}$$

where

$$H_{Ph} = \sum_{q,j} \hbar\,\omega_j(q)\,(b^+_{qj}\,b_{qj} + \tfrac{1}{2}) \tag{27a}$$

$$H_{EPh} = \sum_q \sum_{kk'} a^+_k\,a_{k'}\,Q_{qj}\,v_{kk'}(\vec{q}) \tag{27b}$$

U_{EI} is the electron potential created by the ions fixed on their lattice sites. The matrixelement $v_{kk'}(\vec{q})$ is essentially given by the Fourier transform of $\vec{\nabla} V_{EI}$ (also see below) which is proportional to q^{-1} if $q \to 0$. Because of the factor $q^{-1/2}$ occouring in Q_{qj} the matrix elements of H_{EPh} tend to $q^{-3/2}$ if $q \to 0$.

Note that e.g. the discussion of Ref. 7 is not quite complete. H_{Ph} and therewith $\omega_j(q)$ can not be obtained from the "bare" ion-ion force because the metallic electrons make covalent bonds between the ions. Of course, this effect has to be taken into account. A nice discussion is given in the appendix of Ref. 8. It comes out that (27) is correct for the model of harmonic lattice vibrations plus the "adiabatic approximation". The latter means that - as to the calculation of the dynamical matrix or $\omega_j(q)$ - the electrons can be treated as they were adiabatically following the movements of the ions. This is certainly a good approximation. In the following it is assumed that V_{EI} (H_{EPh}) as well as $\omega_j(q)$ are known from elsewhere (experiment and/or calculation).

In the model of harmonic lattice vibration the exact ground state is given by

$$|\Psi_g\rangle = \exp\,(S^E + S^{EPh})\,|\,\Phi_{BCS}\,\Phi_{Bos}\rangle \tag{28}$$

$(S^{Ph}$ is zero because $b_{qj}|\Phi\rangle = 0)$

Because H_{EPh} is divergent for $q \rightarrow 0$ it becomes screened by S_2^{EPh} - similarly as $V_{Coul.}$ by S_2^E.

Eq. (24) is modified by the replacement $T_E \rightarrow (T_E + U_{EI})$ and the following new terms:

"self energy terms"

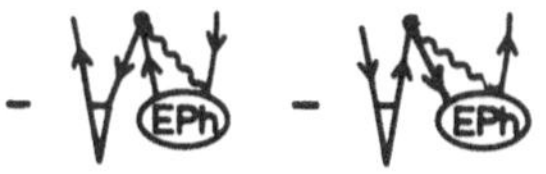

"RPA like terms"

In eq. (24) one can average over the directions $\hat{q}$ and calculate an averaged excitation energy ε_q instead of $\varepsilon_{\vec{q}}$. At least for cubic and similar cristals the same approximation can be made for S_2^{EPh}; the averaged quantity then depends on 3 scalar variables. Because $I(kk')$ - yielding S_2^E - equally depends on 3 scalar variables, the ground state problem is manageable and so is the (modified) equation (24).

Higher clusters are neglegible, at least if the excitation spectrum looks like that of Fig. 1a. If it were looking like such of Fig. 1b it cannot be excluded that

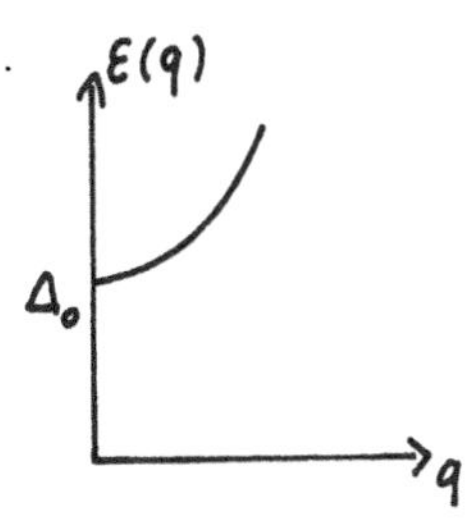

Fig. 1a

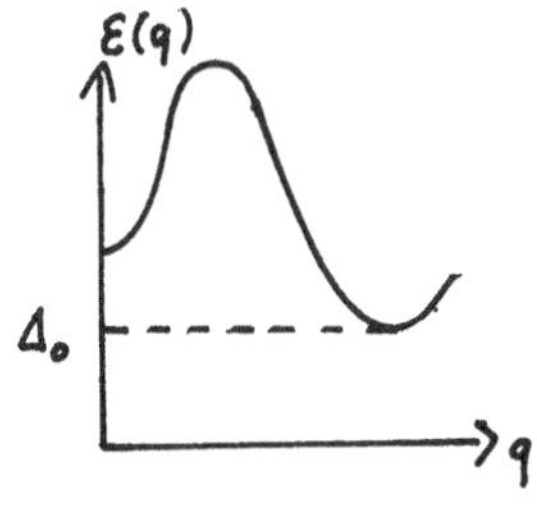

Fig. 1b

R_n^q-clusters (describing the break-up of two pairs) etc. are coming into play.

If one has different kinds of ions the matrix elements of H_{EPh} are given by [7,8]

$$(2\omega_j(\vec{q}))^{-1/2} v_{kk'}(\vec{q}) = \sum_{\varkappa} (2\omega_j(\vec{q}))^{-1/2} \sqrt{\frac{n}{M_\varkappa}}\ \vec{e}(\varkappa;\vec{q}j)\cdot\vec{q}\ V_{EI}(q) \tag{29}$$

where: n = number density of unit cells, $M_\varkappa$ = ion mass
$\vec{e}(\varkappa;\vec{q}j)$ = polarization operator (eigenvector of the dynamical matrix)

The sum goes over the ions in the unit cell. For complicated cristals $\omega_j(q)$ is difficult to calculate (or to measure).

Instead of calculating the quantities $v_{kk'}(\vec{q})$, $\omega_j(\vec{q})$ and U_{EI} one can perform a variation of these "parameters" - of course in resonable limits - and then look for "islands" with large gaps in the parameter space. This islands are - roughly speaking - islands of large critical temperature T_c, too, because there is a relation between the gap Δ_0, and the critical tempeature T_c, i.e. [4]

$$k_B T_c = \frac{\Delta_0}{1.76} \pm 25\% \quad , k_B = \text{Boltzmann constant} \tag{3o}$$

The error is an estimate obtained from comparison with experiment.

References

/1/ H. Kümmel, K.H. Lührmann and J.G. Zabolitzky, Phys. Rep. <u>36C</u> 11 (1978)

/2/ K. Emrich, Nucl. Phys. A 351 (1981) 379

/3/ K. Emrich, Nucl. Phys. A 351 (1981) 397

/4/ A.L. Fetter and J.D. Walecka, Quantum Theory of Many Particle Systems, Mc Graw Hill, 1971

/5/ R. Bishop and K.H. Lührmann, Phys. Rev.B17 (1978), 3757

/6/ J.G. Zabolitzky, Phys. Rev. B22, 2353 (198o)

/7/ C. Kittel, Quantum Theory of Solids, J. Wiley & Sons, 1963

/8/ A.A. Maradudin et al., Theory of Lattice Dynamics in the Harmonic Approximation, Sol. State Phys. Suppl. 3, 1971

CORRELATED BCS THEORY*

S. Fantoni+
Department of Physics and
Materials Research Laboratory
University of Illinois at Urbana-Champaign
Urbana, Illinois 61801/USA

Abstract

A variational theory of the BCS pair-condensed state of strongly-interacting Fermi system is briefly outlined. An FHNC scheme is devised to calculate the radial distribution function and the one-and the two-body density matrices.

*Supported by NSF-DMR-77-2389 and DOE-DE-AC02-76ER01198
+Permanent Address: Istituto di Fisica dell'Universita, Pisa, Italy

1. Introduction

In this contribution a method is presented to study pair-condensed states of strongly-interacting Fermi systems, by means of the variational approach, based on trial correlated wave functions. Most of the theories, which have been developed for explaining the superfluid and superconducting phenomena, rely on the assumption that the particles interact through an effective weak interaction. The noticeable advances made in the last few years on the many-body theories encourage to study the relationship of such effective interactions with the realistic interactions, through microscopic calculations.

The form of the trial correlated state vector, for the paired condensed states of Fermi fluids, is suggested by the variational calculations successfully carried out for the normal phase[1,2] of these systems, namely

$$|\psi_s\rangle = \hat{F}|\Phi_s\rangle \quad . \tag{1}$$

The model superstate $|\Phi_s\rangle$ adequately describes the weak-interacting system. For instance, in the case of S-wave pairing, $|\Phi_s\rangle = |BCS\rangle = \prod_{\underline{k}}(u_{\underline{k}} + v_{\underline{k}} a^+_{\underline{k}\uparrow}\, a^+_{-\underline{k}\downarrow})|0\rangle$, where $|0\rangle$ is the vacuum state and a^+_m is the creation operator for a fermion in the state $|m\rangle$. The operator $\hat{F}$ induces on $|\Phi_s\rangle$ the correlations due to the presence of the strong interaction among the particles. Since, in general, $|\Phi_s\rangle$ is not eigenvector of the particle number operator $\hat{N}_{op} = \sum_m a^+_m a_m$, it is convenient to rewrite eq. (1) in the form

$$|\psi_s\rangle = \sum_N \sum_{(m_N)} \hat{F}_N\, |\Phi_N^{(m)}\rangle\langle \Phi_N^{(m)}|\Phi_s\rangle \ , \tag{2}$$

where the label m_N specifies a set of N single particle states corresponding to $|\Phi_N^{(m)}\rangle$. In coordinate representation $\hat{F}_N|\Phi_N^{(m)}\rangle$ is assumed to be of the form

$$\langle \underline{x}_1 \ldots \underline{x}_N|\hat{F}_N|\Phi_N^{(m)}\rangle = F_N(\underline{x}_1 \ldots x_N)\,\mathcal{A}\{\phi_{m_1}(\underline{x}_1)\ldots\phi_{m_N}(\underline{x}_N)\} \ , \tag{3}$$

where $\phi_m(\underline{x})$ is a normalized plane wave orbital and $F_N(\underline{x}_1 \ldots, \underline{x}_N)$ is the correlation factor involving the particles $\underline{x}_1, \ldots, \underline{x}_N$. State-dependent correlations should be included in F_N in order to do realistic calculations on liquid ^3He and nuclear matter as well. In dealing with such a correlated superstate, suitable cluster expansions of the interesting quantities, as for instance the distribution functions, must be devised. Remarkable progress has been done in this direction, in the case of a simple Jastrow ansatz for F_N, i.e. $F_N = \prod_{i>j=1}^{N} f(r_{ij})$. This progress seems to be promising in view of realistic choices for F_N. Yang and Clark[3] used a truncated AHT expansion[4] in calculating the energy shift due to the transition induced by S-wave pairing in neutron matter. A generalized version of the Wu-Feenberg approximation has been adopted by Paulick and Campbell in their analysis of the BCS and Balian-Werthamer pairings on liquid ^{3}He. The approximations used in the above approaches resulted to be questionable in variational calculations on the normal phase of Fermi fluids. More recently, the pairing phenomena have been extensively analyzed by Krotschek and Clark,[6] who used the machinery of the CBF theory to derive criteria of instability and to calculate the condensation energy. In what follows the main results of a full cluster expansion of the radial distribution function and the one-body and two-body density matrices for the correlated |BCS> superstate are presented. The details of the derivation are given in ref. (7) and will not be reported here.

2. Radial Distribution Function and Density Matrices

The scheme for deriving the cluster expansion of the radial distribution function $g(r_{12})$ strictly follows the procedure[2] derived for the Jastrow-Slater ansatz. The main difference is that the squared Slater determinant is substituted here by the following quantity

$$\Delta_p\,(\underline{r}_1,\ldots,\underline{r}_p) = \sum_{s_1\ldots s_p}\ \sum_{\substack{m_1\ldots m_p\\ n_1\ldots n_p}} \phi^*_{n_p}(\underline{x}_p)\ldots\phi^*_{n_1}(\underline{x}_1)\phi_{m_1}(\underline{x}_1)\ldots\phi_{m_p}(\underline{x}_p)\cdot \langle BCS|a^+_{n_p}\ldots a^+_{n_1}\ a_{m_1}\ldots a_{m_p}|BCS\rangle \quad , \tag{4}$$

where $\underline{x}_i \equiv (\underline{r}_i, s_i)$, s_i being the spin variable. The radial distribution function $g(r_{12})$ turns out to be expressed by a series of reducible linked cluster terms involving the dynamical correlation $h(r) = f^2(r)-1$ and the complex function $L(r) = l_v(r) + i\ l_u(r)$, where

$$l_v(r) = \frac{\nu}{(2\pi)^3\rho_0}\int d\underline{k}\ v^2(k)\ \exp\{i\ \underline{k}\cdot\underline{r}\} \quad ,$$

$$l_u(r) = \frac{\nu}{(2\pi)^3\rho_0}\int d\underline{k}\ u\,(k)\ v(k)\ \exp\{i\ \underline{k}\cdot\underline{r}\} \quad , \tag{5}$$

where $\rho_0 = \frac{\nu}{(2\pi)^3} \int d\underline{k}\, v^2(k)$ is the density of the uncorrelated |BCS> state vector and ν is the degeneracy of the system. The series can be summed by means of a FHNC technique, in which the complex function L(r) plays the role of the statistical correlation. As a result, g(r) is given by

$$g(r) = 1 + N_{dd}(r) + X_{dd}(r) + 2\,\frac{cd}{c}\,(N_{de}(r) + X_{de}(r)\ ) + \\ + (\frac{cd}{c})^2\,(N_{ee}(r) + X_{ee}(r)\ )\ . \tag{6}$$

The evaluation of the functions N_{nm} and X_{nm} requires the solution of five coupled integral equations, and two algebraic equations are needed for calculating the vertex corrections c and c_d. These integral equations have a structure which closely resembles that of the FHNC equations derived for the normal Fermi fluids and the computational effort required for their solution, at least in FHNC/0 approximation, is very modest.

The vertex correction c is related to the density ρ of the system through the equation $\rho = c\rho_0$. This means that the mean value of the particle number operator $<\hat{N}_{op}>$ depends, in general, both on the BCS amplitudes and on the correlation factor f(r). However, the dependence of $<\hat{N}_{op}>$ on f(r) is rather weak if $v^2(k) - \Theta(k_f-k)$ is a small quantity.[(8)]

The procedure used to derive the cluster expansion of $g(r_{12})$ has proved very efficient to analyze also the successive density matrices, which furnish a complete description of a quantum fluid. The cluster terms of the corresponding series are linked and reducible, and can be rearranged in such a way that, ultimately, result to be expressed in terms of irreducible quantities. In particular, the two-body density matrix $n^{(2)}_{\sigma_1,\sigma_2}(\underline{r}_1, \underline{r}_{1'}, \underline{r}_2, \underline{r}_{2'})$ shows the BCS - pairing behaviour, i.e. $n^{(2)}_{\sigma_1,\sigma_2}(\underline{r}_1, \underline{r}_{1'}, \underline{r}_2, \underline{r}_{2'}) \rightarrow \delta_{\sigma_1,-\sigma_2}\chi_F(r_{12})\chi_F(r_{1'2'})$ when $|\underline{r} - \underline{r}'|$ approaches infinity. Moreover it can be easily verified that such a long-range ordering disappears when v_k^2 is kept as the Fermi gas θ-function. The one-body density matrix $n^{(1)}(r)$ and the pairing function $\chi_F(r)$ have the following simple structures:

$$n(r) = -\nu\, \rho_0 c_\xi^2 \;\mathrm{Re}\{L(r)+N_{\xi\xi cc}(r) + E_{\xi\xi cc}(r)\} \exp\{N_{\xi\xi}(r) + E_{\xi\xi}(r)\},$$

$$\chi_F(r) = \nu\rho_0 c_\xi^2 f(r)\; \mathrm{Im}\{L(r) + N_{\xi\xi cc}(r) + E_{\xi\xi cc}(r)\} \exp\{N_{\xi\xi}(r)+E_{\xi\xi}(r)\}. \tag{7}$$

The vertex correction c_ξ is the sum of the cluster terms in which the external point is involved by dynamical correlations $\xi(r) = f(r) - 1$ only, and the functions $E_{\xi\xi cc}(r)$ and $E_{\xi\xi}(r)$ are related to elementary cluster terms. Four coupled integral equations are to be solved for evaluating $N_{\xi\xi cc}(r)$ and $N_{\xi\xi}(r)$.
Owing to the fact that the diagonal part of $n^{(1)}(r)$ gives the density ρ of the system, the following useful sum rules hold

$$n(0) = c\rho_0 \;;\; c_\xi^2 \exp\{N_{\xi\xi}(0) + E_{\xi\xi}(0)\} = 1. \tag{8}$$

Finally, the Fourier transform of $n^{(1)}(r)$ and $\chi_F(r)$ are respectively related to the one-particle momentum distribution and to the shift of the pair distribution from the value of uncorrelated pairs.

The method presented provides for accurate microscopic calculations on the BCS-pair condensed state of infinite Fermi systems, if the Jastrow ansatz is assumed to correlate the $|BCS\rangle$ state vector. The improvement of this method with respect to existing theories, based on trial correlated wave functions, rely on the fact that no truncations of the cluster series have been done and no approximations have been introduced on the BCS amplitudes. In addition, the method maintains the simplicity of the standard FHNC theory and allows for the calculation of the one- and two-body density matrices. In some way, it complements the method of ref. (6), in which the expansion of the expectation values on the superstate have been carried out up to the first order in $v_k^2 - \theta(k_F-k)$ and the second in $u_k v_k$. Even if this approximation is in principle justified by the fact that the condensation energy is expected to be much smaller than the absolute energies, it could be interesting to use the approach of ref. (6) to give inputs to the present one.

The simplicity of the method is encouraging for further developments, which are necessary in order to carry out realistic calculations on liquid ^{3}He and nuclear matter as well. On one side the variational calculations[1,2] performed on normal Fermi fluids may suggest how state dependent correlations can be taken into account in F_N, on the other side the theories developed for the effective weak interactions acting among the particles may suggest how to improve the model state vector $|\Phi_s\rangle$.

References

1. V. R. Pandharipande and R. B. Wiringa, Rev. Mod. Phys. 51 (1979) 821; J. W. Clark, in "Progress in Nuclear and Particle Physics" ed. D. H. Wilkinson (Pergamon, Oxford, 1979) Vol. 2.
2. S. Rosati, in "From Nuclei to Particles", Verenna Summer School (1980), in press.
3. C. H. Yang and J. W. Clark, Nucl. Phys. A174 (1971) 49; C. H. Yang, Ph.D. Thesis, Washington University (1971), unpublished.
4. J. B. Aviles Jr., Ann. of Phys. 5 (1958) 251; C. D. Hartogh and H. A. Tolhoek, Physica 24 (1958) 721, 875, 896.
5. T. C. Paulick and C. E. Campbell, Phys. Rev. B16 (1977), 2000.
6. E. Krotscheck and J. W. Clark, Nucl. Phys. A338 (1980), 77.
7. S. Fantoni, Nucl. Phys. in press.
8. S. Fantoni, in "The Many Body Problem, Jastrow versus Brueckner Theory" Granada Summer School. Lecture Notes in Physics (Springer-Verlag) (1980), in press.

Microscopic Parameters for Superconductivity[†]

J.P. Carbotte

Physics Department, McMaster University
Hamilton, Ontario L8S 4M1
Canada

Abstract

A discussion is given of the parameters that enter the superconducting state and of our present knowledge of how they relate to the size of the critical temperature.

I want to talk about the microscopic parameters that determine the superconducting state and discuss how they relate to the size of the critical temperature T_c as well as one other property.

Superconductivity results from a competition between the Coulomb repulsions between electrons which inhibit superconductivity and a phonon mediated attraction favouring the formation of Cooper pairs. To describe the polarization of the system of ions we need to know:

1) the lattice dynamics i.e. the phonon dispersion curves $\omega_\lambda(\underline{k})$ and polarization vectors $\underline{\varepsilon}_\lambda(\underline{k})$ with λ a branch index and $\underline{k}$ momentum restricted to the first Brillouin zone.
2) the electron-ion matrix element entering the electron-phonon vertex.
3) the dynamics of the system of electrons i.e. their wave functions and Fermi surface properties.

At first sight this may seem discouraging since it means that the description of the superconducting state requires a detailed knowledge of not only the electron and lattice dynamics but also of the coupling between them. In general, this can be very complicated. Fortunately it turns out that all of this detailed information can be condensed into a single function, the electron-phonon spectral density $\alpha^2(\omega)F(\omega)$ which depends on phonon energy ω. This spectral density can be thought of as a phonon frequency distribution in which each phonon mode is appropriately weighted by the strength of the electron-phonon interaction.

Denote by $g_{\underline{kk}'\lambda}$ the electron-phonon vertex (Fig. 1) for scattering for an electron from state $\underline{k}$ to $\underline{k}'$ due to the absorption or emis-

---

[†]Research supported in part by the Natural Sciences and Engineering Research Council

sion of a phonon $\omega_\lambda(\underline{k}'-\underline{k})$.

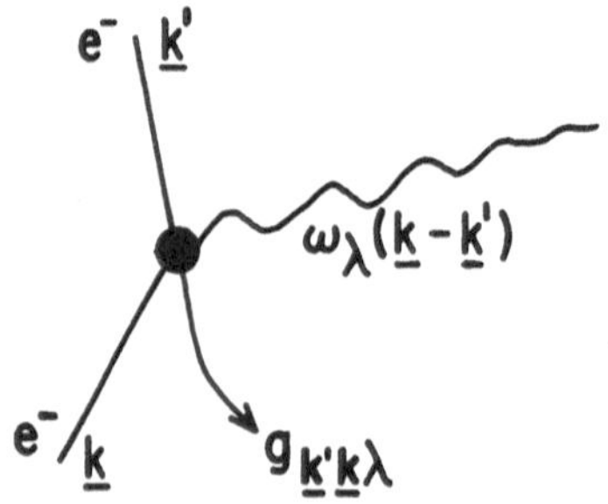

Fig. 1

Schematic representation of electron-phonon interaction

Then the spectral density $\alpha^2(\omega)F(\omega)$ is given by

$$\alpha^2(\omega)F(\omega) = \frac{\sum_{\underline{k}\underline{k}'\lambda} |g_{\underline{k}'\underline{k}\lambda}|^2 \delta(\varepsilon_{\underline{k}'}-\varepsilon_F)\delta(\varepsilon_{\underline{k}}-\varepsilon_F)\delta(\omega-\omega_\lambda(\underline{k}'-\underline{k}))}{\sum_{\underline{k}} \delta(\varepsilon_{\underline{k}}-\varepsilon_F)} \quad (1)$$

where ε_F is the Fermi energy so that $\sum_{\underline{k}} \delta(\varepsilon_{\underline{k}}-\varepsilon_F)$ gives the density of electronic states at the Fermi energy $N(\varepsilon_F)$. Note that the electronic transitions described in (1) are from the Fermi energy $\varepsilon_k = \varepsilon_F$ to the Fermi energy $\varepsilon_{k'} = \varepsilon_F$ and that the phase space on the phonon index $\underline{k}-\underline{k}'$ is controlled by the electronic transitions $\varepsilon_{\underline{k}} \to \varepsilon_{\underline{k}'}$ with each process weighted by $|g_{\underline{k}'\underline{k}\lambda}|^2$.

It is important to realize that $\alpha^2(\omega)F(\omega)$ is defined by (1) however complex the phonon and electronic structure of the material of interest may be. Further, the microscopic equations for superconductivity refer to the material parameters only through $\alpha^2(\omega)F(\omega)$ and the Coulomb repulsion parameter μ^* - a constant.

The imaginary frequency axis representation of the Eliashberg equations are[1,2,3]

$$\tilde{\Delta}(\omega_n) = \pi T \sum_m \{\lambda(\omega_n-\omega_m)-\mu^*\} \frac{\tilde{\Delta}(\omega_m)}{\sqrt{\tilde{\omega}_m^2+\tilde{\Delta}^2(\omega_m)}} \quad (2)$$

$$\tilde{\omega}_n = \omega_n + \pi T \sum_m \lambda(\omega_n-\omega_m) \frac{\tilde{\omega}_m}{\sqrt{\tilde{\omega}_m^2+\tilde{\Delta}^2(\omega_m)}} \quad (3)$$

where T is the temperature, $i\omega_n = i(2n+1)\pi T$, $n = 0, \pm 1, \cdots$ the Matsubara frequencies and

$$\lambda(\omega_n-\omega_m) = 2\int_0^\infty \frac{d\omega\, \alpha^2(\omega)F(\omega)\omega}{\omega^2+(\omega_n-\omega_m)^2} . \quad (4)$$

The electron Green's function can be written in terms of the $\tilde{\Delta}(\omega_n)$'s which are closely related to the gap function and the renormalized frequencies $\tilde{\omega}_n$.

The critical temperature T_c is the temperature at which (2) and (3) stop having a non trivial solution. Numerical solution of these equations gives the functional relationship $\mathcal{F}$ between T_c and the material parameters $\alpha^2(\omega)F(\omega)$ and μ^*.

$$T_c = \mathcal{F}(\alpha^2(\omega)F(\omega),\mu^*) \ . \tag{5}$$

General statements about the exact functional relationship $\mathcal{F}$ can be made.

Consider any $\alpha^2(\omega)F(\omega)$; characterize it by a shape and an overall strength defined by A (the area under $\alpha^2(\omega)F(\omega)$). It can be shown that for <u>any</u> shape whatsoever the inequality[4]

$$K_B T_c \leqq C(\mu^*)A \tag{6}$$

holds with $C(\mu^*)$ some number dependent only on μ^* (the Coulomb repulsion pseudopotential). Its calue is determined from equations (2) and (3). The variation of C with μ^* is given in Fig. 2 and the relationship (6) is tested by placing on the same figure the ratio $k_B T_c/A$ for the many superconductors for which it is known. They all fall below the solid curve, as they should. Note that many fall near the optimum ratio.

It can be shown that the equality in equation (6) holds when the shape of $\alpha^2(\omega)F(\omega)$ is a delta function at an optimum phonon energy $\omega_E = d(\mu^*)A$ with $d(\mu^*)$ another number also determined from (2) and (3). Thus, for

$$\alpha^2(\omega)F(\omega) = A\delta(\omega-\omega_E) \tag{7}$$

we have

$$k_B T_c = C(\mu^*)A \ . \tag{8}$$

The result tells us that the optimum phonon energy for T_c is at $\omega_E = D(\mu^*)A$ and also that there is no limit on the size of T_c imposed by the mathematical structure of the theory of superconductivity itself, i.e. by the form of the functional relationship $\mathcal{F}$. If the strength of the electron-phonon interaction measured by A is large and the phonons fall near ω_E the critical temperature T_c will also be large. This reduces the search for large T_c materials to a search for stable materials with a large electron-phonon interaction (characterized by a large A) and at the same time with phonons of energy $d(\mu^*)A$. Any limit on T_c will then involve limits on our ability to

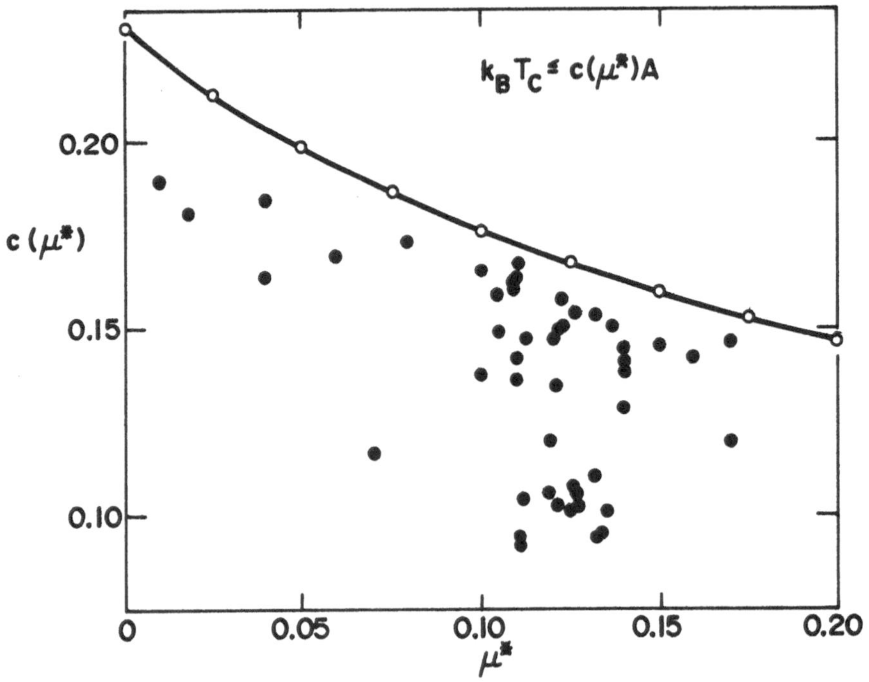

Fig. 2. The function $C(\mu^*)$ vs μ^* and the ratio k_BT_c/A for a number of superconductors.

achieve the above conditions.

A detailed discussion of the functional derivative of T_c with respect to $\alpha^2(\omega)F(\omega)$ is useful at this point. By definition[1,3]

$$\frac{\delta T_c}{\delta\alpha^2(\omega_0)F(\omega_0)} = \lim_{\varepsilon\to 0} \frac{F(\alpha'^2(\omega)F(\omega),\mu^*)-F(\alpha^2(\omega)F(\omega),\alpha^*)}{\varepsilon} \tag{9}$$

where

$$\alpha'^2(\omega)F(\omega) = \alpha^2(\omega)F(\omega) + \varepsilon\delta(\omega-\omega) \ . \tag{10}$$

Thus $\delta T_c/\delta\alpha^2(\omega_0F(\omega_0)$ tells us how T_c changes when an infinitesimal amount of spectral weight is added at energy ω_0. In other words the functional derivative measures the effectiveness in T_c of the various phonon energies. Results for several materials are given in Fig. 3 We note that very low and very high energy phonons are not effective in T_c since $\delta T_c/\delta\alpha^2(\omega)F(\omega)$ goes like ω as $\omega \to 0$ and like $1/\omega$ for $\omega \to \infty$. All curves have a broad maximum around $\omega/k_BT_c \sim 7$. We see, therefore, that an optimum energy exists for the phonons and that the best shape for $\alpha^2(\omega)F(\omega)$ is a delta function at this energy, although, because the maximum is so broad, frequencies anywhere near the maximum are almost as desirable.

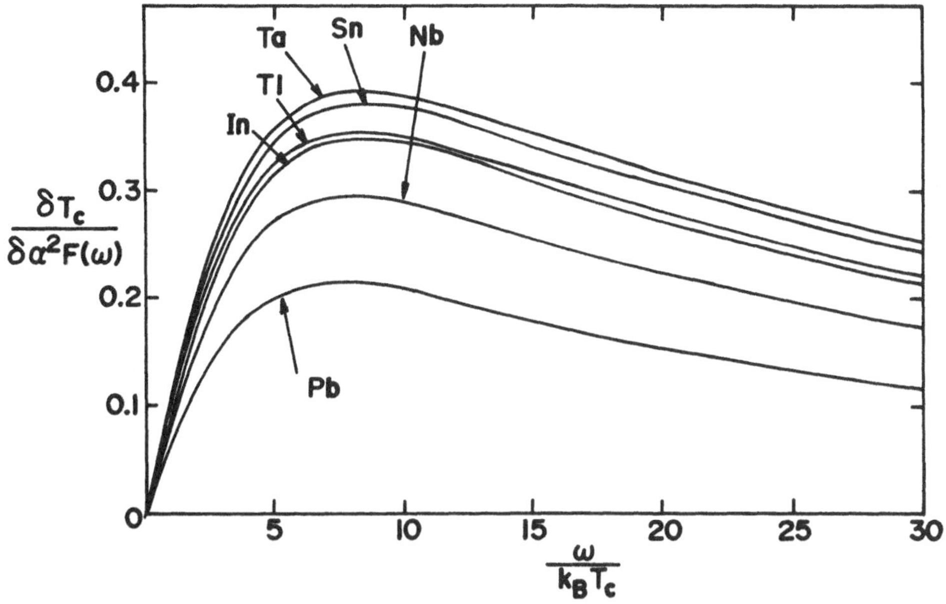

Fig. 3 $\delta T_c/\delta\alpha^2 F(\omega)$ for several superconductors.

A simple qualitative physical explanation of the origin of the maximum in $\delta T_c/\delta\alpha^2(\omega)F(\omega)$ may be helpful. Fig. 4 shows an electron polarizing the system of ions. It is however travelling with the Fermi velocity v_F. The typical distance of relevance to superconductivity is the coherence distance ρ_0. The time the electron stays within this distance is $t = \rho_0/v_F$. To get maximum polarization of the ions, having displacements $u = a\sin(\omega t)$ where ω is the oscillator energy and a is an amplitude, within time t we require[10]

$$\omega t = \frac{\pi}{2} \tag{11}$$

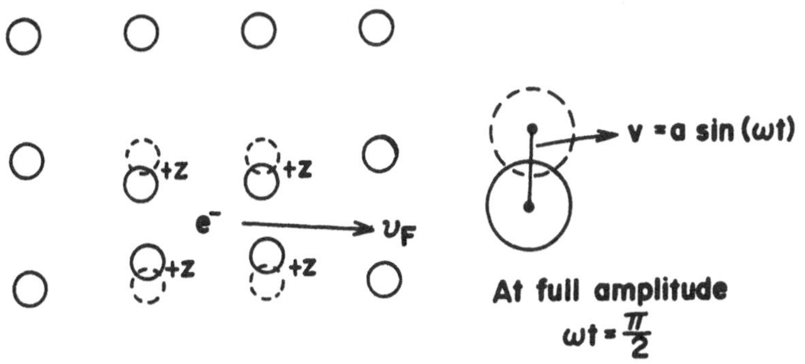

Fig. 4 Lattice polarization

or

$$\omega_{optimum} = \frac{\pi}{2t} = \frac{\pi}{2\rho_0} v_F \sim \Delta \sim k_B T_c \tag{12}$$

with the proportionality constant in this last relation equal to about 7.

The spectral density $\alpha^2(\omega)F(\omega)$, of course, enters in different ways in different quantities. As an example consider the zero temperature gap edge Δ_0, which can be obtained from an analytic continuation of the low temperature solution of equations (2) and (3). Its functional derivative $\delta\Delta_0/\delta\alpha^2F(\omega)$ has been calculated by Mitrović et. al.(5) for several systems. Fig. 5 shows results and a comparison with $\delta T_c/\delta\alpha^2F(\omega)$ for the case of Nb. We note that the functional derivative of the gap edge peaks at lower energies than that for T_c. An interesting comparison in this case is Nb with the alloy $Nb_{.75}Zr_{.25}$(6). For Nb A ≓ 7.247 to be compared with 7.425 for $Nb_{.75}Zr_{.25}$; a 2.5% increase in area under $\alpha^2(\omega)F(\omega)$. However, a considerable amount of phonon softening occurs in the region around 5 to 15 meV so that λ increases from 1.009 to 1.311. According to figure 5 this shift will have a larger effect on the gap edge Δ_0 than on the critical temperature T_c. In fact T_c only goes from 9.22 K to 10.8 K, but the dimensionless ratio $2\Delta/k_BT_c$ is greatly increased. It goes from 3.88 for Nb to 4.19 for $Nb_{.75}Zr_{.25}$ which can be traced to phonon softening.

The theory described to this point applies only to the case when the electronic density of states $N(\varepsilon)$ does not vary on the scale of the Debye energy about the Fermi energy ε_F. It is treated as a constant equal to $N(\varepsilon_F)$ and taken out of the integrals during the course of the derivation of equations (2) and (3).

To date, high values of the critical temperature T_c have been found only in the A-15 compounds. For example T_c = 16.8 K for V_3Ga, 17.2 for V_3Si, 18.3 for Nb_3Sn and 23.2 for Nb_3Ge. These materials show many anomalous properties which have often been described in terms of sharp structure in $N(\varepsilon)$[7,8]. For example, the one dimensional linear chain Labbé - Friedel model has been widely used with Fermi energy ε_F a few hundred degrees above a square root singularity in the electronic density of states[7,8].

Recent band structure calculations are not yet completely consistent[9,10] with one another but one such calculation, the self-consistent pseudopotential computations of Ho et.al.[10], gives a sharp peak in $N(\varepsilon)$ of width around 70 meV with the Fermi energy falling in this peak.

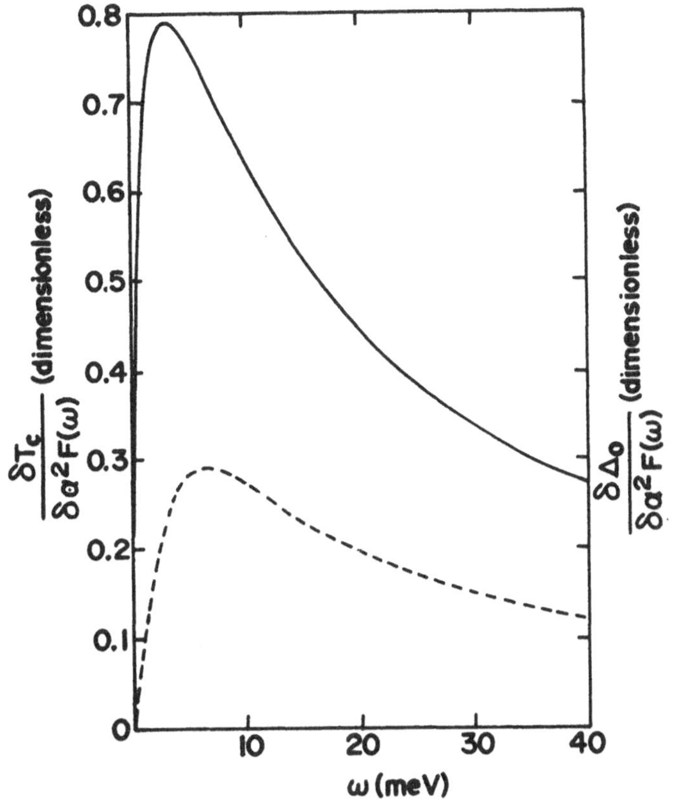

Fig. 5 Comparison of $\delta T_c/\delta\alpha^2F(\omega)$ (--- dashed line) and $\delta\Delta_0/\delta\alpha^2F(\omega)$ (—— solid line).

Gosh and Strongin[11] recently reviewed experimental evidence on radiation damage that bears directly on the question of the nature of $N(\varepsilon)$. The normal state electronic specific heat coefficient γ can be measured directly or inferred from upper critical magnetic field data H_{c2}. γ is directly proportional to $N(\varepsilon_F)$. Measurements of $N(\varepsilon_F)$ in this way reveal a strong dependence on disorder which can be interpreted in terms of the washing out, by scattering, of a sharp peak in $N(\varepsilon)$. In V_3Si a width of 100°K and a height of three times the background is a possibility.

To take account of sharp structure in $N(\varepsilon)$ the Eliashberg equations (2) and (3) need to be generalized. The new set applicable for a general $N(\varepsilon)$ with particle-hole symmetry assumed to still remain are[12]

$$\tilde{\Delta}(\omega_n) = \pi T \sum_m [\lambda(\omega_m-\omega_n)-\mu^*] \frac{\tilde{\Delta}(\omega_m)}{|\tilde{\omega}_m|} \tilde{N}(|\tilde{\omega}_m|) \qquad (13)$$

$$\tilde{\omega}_n = \omega_n + \pi T \sum_m \lambda(\omega_m - \omega_n)\,\mathrm{sgn}(\omega_m)\tilde{N}(|\omega_m|) \qquad (14)$$

with

$$\tilde{N}(|\tilde{\omega}_m|) = \int_{-\infty}^{+\infty} d\varepsilon \, \frac{|\tilde{\omega}_m|}{\varepsilon^2 + |\tilde{\omega}_m|^2} \frac{N(\varepsilon)}{\pi N_0} \qquad (15)$$

In (15) N_0 is some constant to be discussed shortly and the final integration can only be carried out once an explicit form is specified for $N(\varepsilon)$. It is noted that (13) and (14) have been written in their linearlized form applicable only near T_c.

To understand the role variations in $N(\varepsilon)$ can play in determining T_c we first work out the functional derivative of T_c with $N(\varepsilon)$. Results are given in Fig. 6. It is seen that $\delta T_c/\delta N(\varepsilon)$ is peaked at the Fermi energy and drops to half its value for $\omega = 6k_BT_c$ and $8k_BT_c$ for Nb and Nb_3Sn respectively. This sets the scale of energy on which variations in $N(\varepsilon)$ are important for the critical temperature.

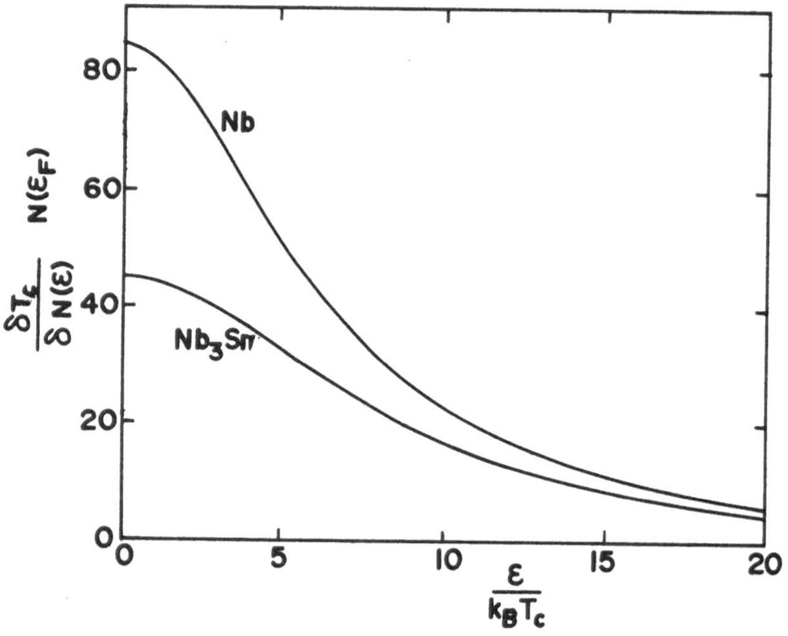

Fig. 6 Functional derivative $\delta T_c/\delta N(\varepsilon)$ vs ε.

It is enlightening to study a simple model. We will take for $N(\varepsilon)$ a Lorentzian form of strength g, width $\bar{a}$, centered about ε_F and superimposed on a constant background,

$$N(\varepsilon) = N_0\left[1 + \frac{g}{\pi} \frac{\bar{a}}{\bar{a}^2 + (\varepsilon - \varepsilon_F)^2}\right] \qquad (16)$$

At theFermi energy $N(\varepsilon_F) = N_0(1+d)$ with $d = g/\pi\bar{a}$ giving the height of the Lorentzian peak above the background. For this case substitution of (16) into (15) gives

$$N(|\tilde{\omega}_m|) = [1 + \frac{g}{\pi}\frac{1}{\bar{a}+|\tilde{\omega}_m|}] \tag{17}$$

To illustrate the point we wish to make we have solved (13) and (14) for several values of the parameters $\bar{a}$ and g. In all cases μ^* is fixed at a value of .175 and $\alpha^2(\omega)F(\omega)$ is chosen to be that obtained by Shen[(14)] (in the case of Nb_3Sn) from inversion of tunneling data with inversion procedure based on the assumption of a constant electronic density of states. Because this may not be realistic, we assume only that the shape obtained is representative of reality for the A-15. Its absolute value is fixed as follows. Shen's[(14)] $\lambda = 1.7$ is reduced to .57 so that $\bar{a}\to\infty$ (completely damaged example) the T_c is as observed equal to about 4°K. For a fixed choice of $\bar{a} = a_0$ g is then chosen to give a T_c of 18°K. Having done this $\bar{a}$ is varied to get T_c as a function for any value of the Lorentzian width. This allows us to understand the increase in T_c that can be achieved by sharpening up the peak in the electronic density of states.

Decreasing $\bar{a}$ makes $N(\varepsilon_F) = N_0(1 + \frac{g}{M\bar{a}})$ as large as we please yet T_c is found to be quite finite. In Table 1 we give results for the saturated value of $T_c(\bar{a}\to 0)$ for various choices of a_0 and g. All these numbers are very reasonable and we can conclude that making the density of states extremely sharp does not make T_c infinite. This is not unexpected since the superconducting transition depends on the density of states not just at ε_F but also around ε_F in a range of order ± the Debye energy. Formula (17) shows explicitly the saturation since for $\bar{a}$ much less than the lowest Matsubara energy $= \pi T_c$ it can be dropped from the equation and hence drops out of the theory altogether. Saturation occurs so that making the peak in $N(\varepsilon)$ sharper and sharper will not increase T_c indefinitely.

Table 1

a_0 (units of ω_D)	g(meV)	d	T_c^s(°K)
.5	132	2.9	24.
1	205	2.3	27.
2	395	2.2	31.
3	575	2.1	33.

References

(1) G. Bergmann and D. Rainer, Z. Phys. 263, 59 (1973).

(2) D. Rainer and G. Bergmann, Jour. Low Temp. Phys. 14, 50 (1974).

(3) J.M. Daams and J.P. Carbotte, Can. J. Phys. 56, 1248 (1978).

(4) C.R. Leavens and J.P. Carbotte, Ferroelectrics 16, 295 (1977).

(5) B. Mitrović, C.R. Leavens and J.P. Carbotte, Phys. Rev. 21, 5048 (1970).

(6) B. Mitrović and J.P. Carbotte, Jour. Low Temp. Phys. (in press).

(7) M. Weger and I.B. Goldberg, Solid State Physics, ed. H. Ehrenreich, F. Seitz and D. Turnbull, Academic Press, New York 28, 1 (1973).

(8) L.R. Testardi, Physical Acoustics, ed. W.P. Mason and R.N. Thurston, Academic Press, New York 10, 193 (1973).

(9) B.M. Klein, L.L. Boyer, D.A. Papaconstantopoulos and L.F. Mattheiss, Phys. Rev. B18, 6411 (1978).

(10) K.M. Ho, M.L. Cohen and W.E. Pickett, Phys. Rev. Lett. 41, 815 (1978).

(11) A.K. Ghosh and M. Strongin in Superconductivity in d- and f-Band Metals, Ed. H. Suhl and M.B. Maple, Academic Press, 305 (1980).

(12) S.G. Lie and J.P. Carbotte, Solid State Comm. 35, 127 (1980).

(13) S.G. Lie and J.P. Carbotte, Solid State Comm. 26, 511 (1978).

(14) L.Y.L. Shen, Phys. Rev. Lett. 29, 1082 (1972).

PHONONS IN SUPERCONDUCTING Pd-H(D) ALLOYS USING THE COHERENT POTENTIAL APPROXIMATION WITH OFF-DIAGONAL DISORDER.†

L.E. Sansores and J. Tagüeña-Martínez
Instituto de Investigaciones en Materiales
Universidad Nacional Autónoma de México
Apdo. Postal 70-360 México 20, D.F.

I. INTRODUCTION

The Pd-H(D) system has been studied for a long time but there are still a large number of open and interesting problems. One of them is the occurrence of superconductivity for large hydrogen or deuterium concentrations, as it was discovered by Skoskiewicz[1] in 1972.

The pure palladium has a fcc structure. When hydrogen is introduced in Pd it goes to interstitial sites. Pure Pd is a paramagnetic material due to an unfilled d band, susceptibility measurements[2] show that for H/Pd$\geq$.63 (atoms of H/atoms of Pd) the alloy becomes diamagnetic. Papaconstantopoulos et al[3] have calculated the electronic structure of Pd-H(D) using the Slater-Koster Hamiltonian, they have noticed that as hydrogen is introduced the d band is filled first and afterwards the s band begins to fill.

With respect to the superconducting properties. The transition temperature is a function of H/Pd and it shows an inverse isotope effect, that is, the stoichiometric PdH has a Tc=8.0 K, while for PdD Tc=9.8 K[4]. Calorimetric[5] and critical field[6] measurements of the superconducting state have been done by several authors.

Coherent neutron scattering[7] on $PdD_{.63}$ show that the Pd-H(D) phonon spectra can be considered as the sum of two contributions, one associated with the Pd ions and which is almost the same of pure Pd and another associated with the deuterium ions. Moreover incoherent neutron scattering on $PdH_{.63}$ shows this last peak is shifted by a factor of 1.2 which indicates anharmonicity.

Phonons, besides being one of the conceptually simplest elementary excitations, are essential for superconductivity. From the phonon spectra a great number of properties can be calculated. The lack of good phonon spectra for Pd-H(D) encouraged us to extend the coherent potential approximation (CPA) including off-diagonal disorder[8] to interstitial alloys which we shall cal C.P.A.I. The C.P.A. is considered the best approximation available to study alloys.

II. THEORY

We shall consider the system as formed by two intercalated sublattices: number 1 for the Pd and number 2 for the H in the intercalated positions.

We shall begin writing the harmonic Hamiltonian

$$H(t)=\frac{1}{2}\sum_{\substack{n,\gamma\\ \ell}}\frac{P^{\gamma^2}_{n}(\ell,t)}{M^{\gamma}_{n}(\ell)}+\frac{1}{2}\sum_{\substack{nn'\\ \gamma\gamma'\\ \ell\ell'}}U^{\gamma}_{n}(\ell,t)\,\phi^{\gamma\gamma'}_{nn'}(\ell,\ell')U^{\gamma'}_{n'}(\ell',t) \qquad (1)$$

where ℓ refers to the atom position in the γ sublattice and n takes the x,y,z directions. We will work with the following Green's functions

$$G^{\gamma\gamma'}_{nn'}(\ell,\ell';t) = \frac{2\pi}{\hbar} <<U^{\gamma}_{n}(\ell,t);\ U^{\gamma'}_{n'}(\ell',0)>> \tag{2a}$$

$$H^{\gamma\gamma'}_{nn'}(\ell,\ell';t) = \frac{2\pi i}{\hbar} <<P^{\gamma}_{n}(\ell,t);\ U^{\gamma'}_{n'}(\ell',0)>> \tag{2b}$$

where <<.....>> means the configurational average. The imaginary part of (2a) is related with the neutron scattering cross section while the imaginary part of (2b) is related with the phonon density of states. Since it is very easy to go from one to the other, thus we shall formulate the theory in terms of 2a, as it is normaly done.

Using the Fourier transform

$$G^{\gamma\gamma'}_{nn'}(\ell,\ell';t) = \int_{-\infty}^{\infty} G^{\gamma\gamma'}_{nn'}(\ell,\ell';\omega)\, e^{i\omega t}\, dt\ , \tag{3}$$

the equation of motion is

$$M^{\gamma}_{n}(\ell)\omega^2 G^{\gamma\gamma'}_{nn'}(\ell,\ell';\omega) = \delta_{nn'}\delta_{\gamma\gamma'}\delta(\ell-\ell') + \sum_{\substack{\gamma''n''\\ \ell''}} \phi^{\gamma\gamma''}_{nn''}(\ell,\ell'')G^{\gamma''\gamma'}_{n''n'}(\ell'',\ell';\omega) \tag{4}$$

which can be written in terms of the locator

$$g^{\gamma}_{n}(\ell;\omega) = \frac{1}{M^{\gamma}_{n}(\ell)\omega^2} \tag{5}$$

as

$$G^{\gamma\gamma'}_{nn'}(\ell,\ell';\omega) = g^{\gamma}_{n}(\ell;\omega)\,\delta_{nn'}\,\delta_{\gamma\gamma'}\,\delta(\ell-\ell') + g^{\gamma}_{n}(\ell;\omega) \sum_{\substack{\gamma''n''\\ \ell''}} \phi^{\gamma\gamma''}_{nn''}(\ell,\ell'')\, G^{\gamma''\gamma'}_{n''n'}(\ell'',\ell';\omega). \tag{6}$$

In each sublattice we will have a different host and different impurity. We will call A and A' to the host and impurity atoms respectively of sublattice 1. And B and B' the host and impurity atoms of sublattice 2. So now we can define

$$\phi^{11(22)}_{(i,j)} = \alpha^{11(22)}_{(i,j)} \quad \text{if sites i and j are A(B) type}$$

$$= \beta^{11(22)}_{(i,j)} \quad \text{if sites i and j are A'(B') type} \tag{7}$$

$$= \eta^{11(22)}_{(i,j)} = \zeta^{11(22)}_{(i,j)} \quad \text{if site i is A(B) type and j is A'(B') type or viceversa}$$

$$\phi^{12}_{(i,j)} = \alpha^{12}(i,j) \qquad \text{if i is A type and j is B type or viceversa}$$

$$= \zeta^{12}(i,j) \qquad \text{if i is A type and j is B' type or viceversa}$$

$$= \eta^{12}(i,j) \qquad \text{if i is A' type and j is B type or viceversa}$$

$$= \beta^{12}(i,j) \qquad \text{if i is A' type and j is B' type or viceversa}$$

and the ocupation indices $x^{\gamma}(i)$ and $y^{\gamma}(i)$

$$\begin{array}{lllll} x^1(i)=1 & y^1(i)=0 & x^2(i)=0 & y^2(i)=0 & \text{if i is an A site} \\ x^1(i)=0 & y^1(i)=1 & x^2(i)=0 & y^2(i)=0 & \text{if i is an A' site} \\ x^1(i)=0 & y^1(i)=0 & x^2(i)=1 & y^2(i)=0 & \text{if i is a B site} \\ x^1(i)=0 & y^1(i)=0 & x^2(i)=0 & y^2(i)=1 & \text{if i is a B' site} \end{array} \tag{8}$$

that have the following properties

$$\begin{array}{lll} x^{\gamma}(i)y^{\gamma'}(i)=0 & x^{\gamma}(i)x^{\gamma'}(i)=x^{\gamma}(i)\delta_{\gamma\gamma'} & \\ y^{\gamma}(i)y^{\gamma'}(i)=y^{\gamma}(i)\delta_{\gamma\gamma'} & <x^1(i)>=C_A & <x^2(i)>=C_B \end{array} \tag{9}$$

Now we can rewrite (6) as

$$\begin{aligned} G^{\gamma\gamma'}_{nn'}(\ell,\ell';\omega) &= g^{\gamma}_n(\ell;\omega)\delta_{nn'}\delta_{\gamma\gamma'}\delta(\ell-\ell') + g^{\gamma}_n(\ell;\omega) \sum_{m\gamma''\ell'} [x^{\gamma}(\ell)\alpha^{\gamma\gamma''}_{nm}(\ell,\ell'')x^{\gamma''}(\ell'') \\ &+ y^{\gamma}(\ell)\beta^{\gamma\gamma''}_{nm}(\ell,\ell'')y^{\gamma''}(\ell'') + x^{\gamma}(\ell)\zeta^{\gamma\gamma''}_{nm}(\ell,\ell'')y^{\gamma''}(\ell'') + \\ &+ y^{\gamma}(\ell)\ {}^{\gamma\gamma''}_{nm}(\ell,\ell'')x^{\gamma''}(\ell'') + x^{\gamma}(\ell)\eta^{\gamma\gamma''}_{nm}(\ell,\ell'')y^{\gamma''}(\ell'') + \\ &+ y^{\gamma}(\ell)\eta^{\gamma\gamma''}_{nm}(\ell,\ell'')x^{\gamma''}(\ell'')]\ G^{\gamma''\gamma'}_{m\,n'}(\ell'',\ell';\omega). \end{aligned} \tag{10a}$$

To weight appropriately the Green's function we pre and post multiply the last equation by the occupation indices. This gives us a set of four equations which we can write as one in a matrix form

$$\begin{aligned} \underline{G}_{nn'}(\ell,\ell';\omega) &= \underline{g}_n(\ell;\omega)\delta_{nn'}\delta_{\gamma\gamma'}\delta(\ell-\ell') + \\ &+ \underline{g}_n(\ell;\omega) \sum_{n''\ell''} \underline{t}_{nn''}(\ell,\ell'')\ \underline{G}_{n''n'}(\ell'',\ell';\omega) \end{aligned} \tag{10b}$$

where

$$\underline{G}_{nn'} = \begin{pmatrix} G^{AA} & G^{AB} & G^{AA'} & G^{AB'} \\ G^{BA} & G^{BB} & G^{BA'} & G^{BB'} \\ G^{A'A} & G^{A'B} & G^{A'A'} & G^{A'B'} \\ G^{B'A} & G^{B'B} & G^{B'A'} & G^{B'B'} \end{pmatrix}_{nn'} \quad . \tag{10c}$$

The CPAI

We can proceed to do the configurational average of the Green's function <G> following Blackman et al[8].

To start we define the fully renormalized locator

$$\underline{\gamma} = \langle \underline{G}(\ell,\ell) \rangle \tag{11}$$

so <G> must have the form

$$\langle \underline{G} \rangle = \gamma + \underline{\gamma t \gamma} + \underline{\gamma t \gamma t \gamma} + \ldots$$

where the first and last γ of the second and subsequent terms refer to different sites and all γ's of a term refer to different sites.

As it has been shown by Blackman et al the configurational average is easily done using functionals in the k representation. The final equations are

$$\underline{U}_\circ[\gamma] = \underline{\gamma}^{-1}(\underline{G}[\gamma]t)_\circ \tag{12a}$$

$$\underline{\gamma} = \langle \underline{g}(\underline{1} - \underline{U}_\circ[\gamma]\underline{g})^{-1} \rangle \tag{12b}$$

and

$$\underline{G}_k[\gamma] = [\underline{1} + \underline{\gamma}(\underline{U}_\circ[\gamma] - \underline{t}_k)]^{-1}\underline{\gamma} \tag{12c}$$

In our system the sublattice 2, which shall be in the interstitial positions, is either occupied by a H ion or empty. We consider the vacancies as impurities and since vacancies do not vibrate $g^{B'}=0$. The equation 12b gives

$$\gamma = \begin{pmatrix} \gamma^{A} & 0 & 0 & 0 \\ 0 & \gamma^{B} & 0 & 0 \\ 0 & 0 & \gamma^{A'} & 0 \\ 0 & 0 & 0 & 0 \end{pmatrix} \tag{13a}$$

where

$$\gamma A = \frac{C_A}{M_A\omega^2 - U_{11}} \quad , \quad \gamma B = \frac{C_B}{M_B\omega^2 - U_{22}} \quad , \quad \gamma A' = \frac{C_{A'}}{M_{A'}\omega^2 - U_{33}} \quad . \qquad (13b)$$

Using equation 11,12c and 13 we get a set of six equations which have to be solved simultaneosly.

III Results

For the substitutional alloy the set of six equations reduces to three. We have applied this method to the Ni_xPd_{1-x}alloy[9] with satisfactory results.

For an interstitial alloy as PdH(D) sublattice A can be considered as full with the host atoms and again the six equations reduce to 3

$$\gamma^A = \frac{1}{3N} \sum_{k,j} \frac{B - \alpha^{22}_{BB}}{D_j(k)} \quad , \quad \gamma^B = \frac{1}{3N} \sum_{k,j} \frac{A - \alpha^{11}_{AA}}{D_j(k)} \qquad (14)$$

$$0 = \frac{1}{3N} \sum_{k,j} \frac{U_{12} - \alpha^{12}_{AB}}{D_j(k)} \quad ,$$

where now $A = M_A\omega^2$, $B = M_B\omega^2 + \frac{1 - C_B}{\gamma^B}$

and

$$D_j(k) = (AB - U^2_{21}) - (A\alpha^{22}_{BB} + B\alpha^{11}_{AA} - 2U_{21}\alpha^{12}_{AB}) + (\alpha^{11}_{AA}\alpha^{22}_{BB} - \alpha^{21\,2}_{BA}) \qquad (15)$$

and the densities of states are

$$\rho_A = -\frac{2}{\pi}\,\mathrm{Im}\gamma^A, \quad \rho_B = -\frac{2}{\pi}\,\mathrm{Im}\gamma^B, \quad \rho_T = \rho_A + \rho_B \quad . \qquad (16)$$

In fig. 1 the PdD total density of states for different concentrations is shown. For these calculations we have used the force constants obtained from the neutron scattering experiments[7]. As it was expected we can see that the low frequency part is basically the pure Pd phonon spectra slightly modified.

Figure 2 shows the deuterium partial density of states. It can be seen that the contribution to the total density is only at high energies. Back to figure 1 it can be seen that as the vacancies concentration increases the area under the deuterium part decreases as well and that the valley between both peaks disappears. Figure 3 shows the total phonon density of PdH for H/Pd=.9.

Figure 4 show the configurational averaged Green's functions for the [00A] direction L branch for different concentrations. This has been calculated as

$$G_j(k;\omega) = G^{AA} + 2G^{AB} + G^{BB} \qquad (17)$$

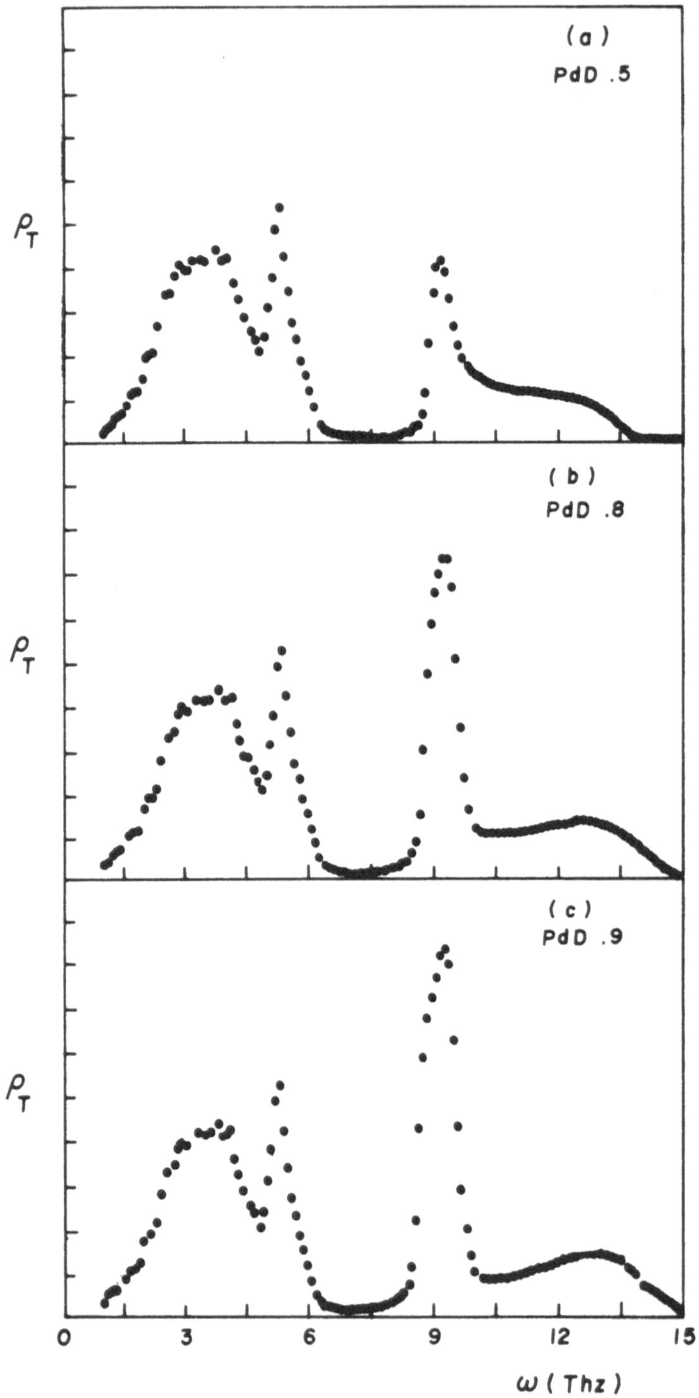

Figure 1 Total phonon density of PdD_x for different concentrations. a)x=.5, b)x=.8, c)x=.9

which gives the response function if the thermal neutron cross section and Debye-Waller factor are taken into account. They show that as the deuterium concentration increases the width increases which means that the lifetime of the excitations decreases. Also new excitations apperar, obviosly due to the H(D)-Pd coupling. On the other hand high frequency excitations lifetime increases with the deuterium concentration since the number of impurities decreases.

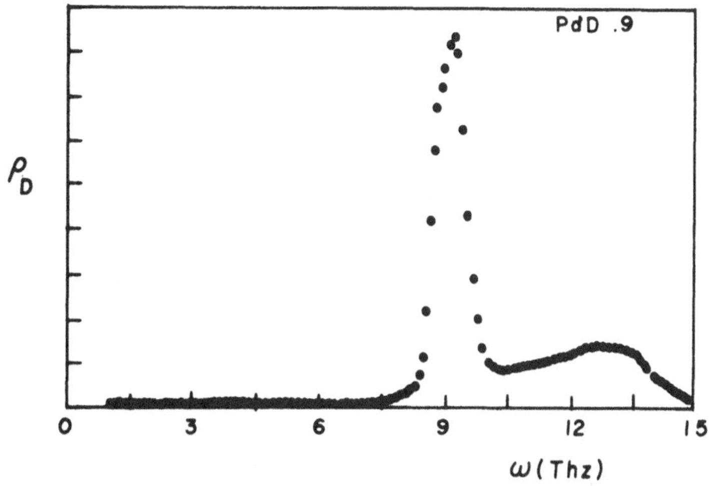

Figure 2 Deuterium partial density of $PdD_{.9}$

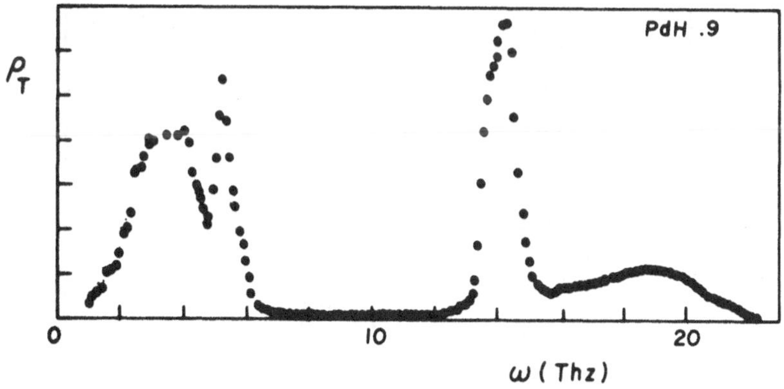

Figure 3 Total phonon density of $PdH_{.9}$

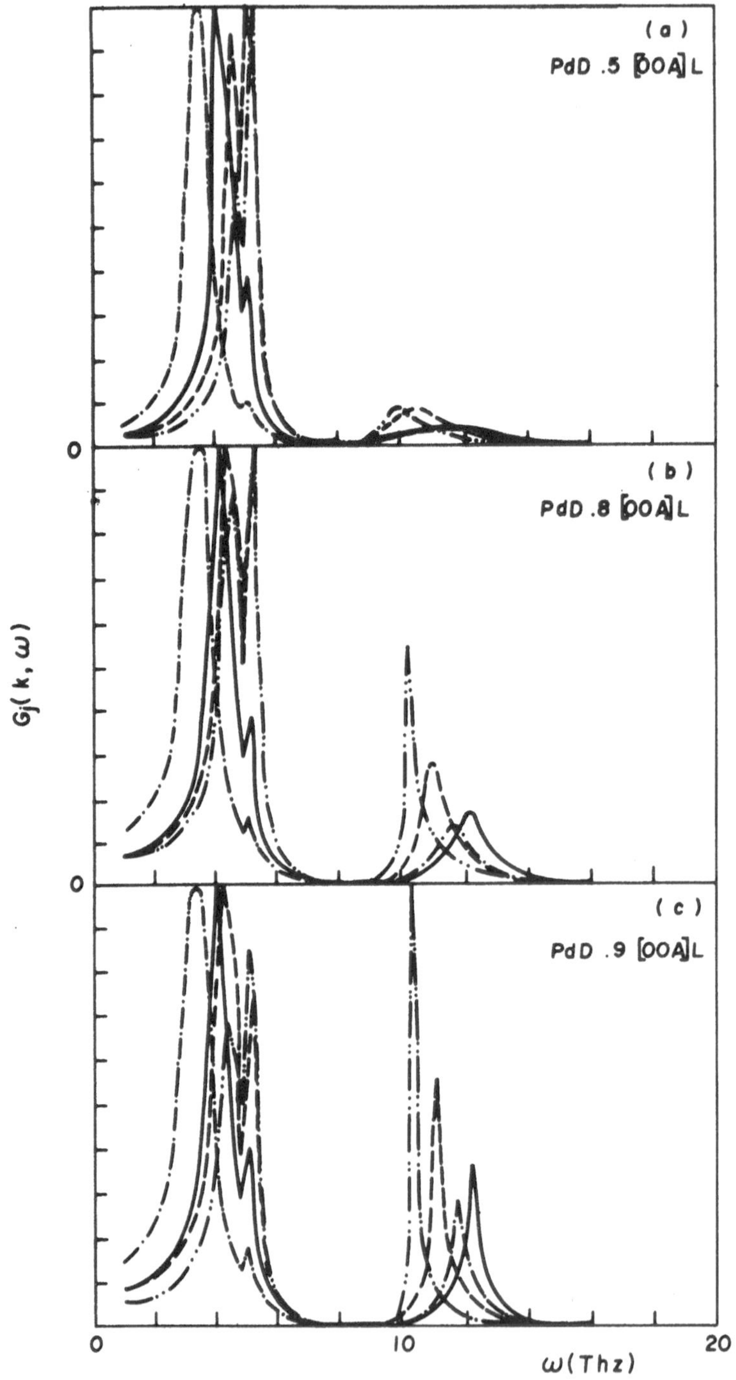

Figure 4 Green function for PdD_x in the [00A] direction L branch. --- A=.4, —— A=.6 ,- - A=.8, -··- A=1. a)x=.5, b)x=.8, c)x=.9

RECENT PROGRESS IN THE UNDERSTANDING OF STRONGLY COUPLED COULOMB SYSTEMS

G. Kalman

Department of Physics
Boston College
Chestnut Hill, MA 02167

After a dormancy of a decade or so, the study of the statistical behaviors of Coulomb-systems is again in the forefront of many-body physics. The emphasis now, however, on <u>strong coupling</u>, i.e. the situation where the potential energy is comparable to or greater than the kinetic energy. The ratio of the two energies can be characterized either by the parameter Γ, $\Gamma = \frac{Z^2e^2\beta}{d}$ or by the parameter γ, $\gamma = \frac{\kappa^3}{4\pi n}$ ($\kappa^2 = 4\pi e^2 n\beta Z$ is to Debye wave number, <u>n</u> the density, <u>d</u> the interparticle distance, $\frac{4\pi}{3} d^3 n = 1$, and $\beta = 1/kT$), with the simple relationship $\gamma = \sqrt{3}\Gamma^{3/2}$. Evidently, strong coupling is described by $\Gamma \gtrsim 1$, or by $\gamma \gtrsim 1$. The stimulus for the new work in the field has come from three sources: (1) New, high accuracy computer "experiments" on Coulomb-systems, both by Monte Carlo method, done by the Livermore group (H. deWitt and collaborators)[1] and by molecular dynamics, done by the Orsay group (J.-P. Hansen and collaborators)[2]. (2) Actual new experimental work, mostly X-ray scattering and experiments done on electron film on the surface of liquid helium. (3) The development of new, non-perturbative theoretical methods appropriate for the strong coupling situation. The latter fall into three categories: (a) static approaches - mainly HNC and its variants; (b) mean field theories, which, in principle, should provide a unified approach to the analysis both of static and of dynamic properties but which turned out to be wholly inadequate for the latter; and (c) genuine dynamical theories which attempt to properly account for the dynamical correlations of the system.

In this review I will survey the main results of the computer experiment and I will discuss to what extent these results we understand on theoretical grounds, and I will describe a rather successful method, developed by the Boston group (K.I. Golden and G. Kalman,[3] with contributions from M. Silevith, P. Bakshi, P. Carini, D. Merlini and R. Calinon)[4] which aims at describing dynamical correlational effects through the application of the hierarchy of fluctuation-dissipation(FD) relations.

While real Coulomb-systems are of great complexity, composed of at least two oppositely charged species, containing also bound configurations, and exhibiting both quantum and finite temperature effects, for the purpose of computers and theoretical studies, three models have gained prominence:

- the "jellium", a degenerate zero-temperature electrongas in a neutralizing positive background;

The Superconducting Properties

To calculate the superconducting properties we need the $\alpha^2\rho(\omega)$ function. We have just calculated $\rho(\omega)$ and shown that it is perfectly reasonable to separate the spectrum in two parts, one related with the Pd and another with the D(H). Papaconstantopoulos[3] has calculated α^2 under this assumption giving one constant value for the Pd and another for the D(H). So we can construct $\alpha^2\rho_T(\omega)$ as

$$\alpha^2\rho_T(\omega)=\alpha^2{}_{Pd}\rho_{Pd}(\omega)+\alpha_H^2\ \rho_H(\omega) \quad . \tag{18}$$

Since the total phonon density can be divided in two parts an alternative way to construc $\rho(\omega)$ is to take the stoichiometric PdH phonon density of states calculated experimentally and to scale the D(H) part according to the D(H) concentrations and the use eq. (18). The results from the CAPI and the experimental density of states should agree. Table I shows some properties of the superconducting state calculated with both methods. The agreemet is excelent.

TABLE I

	$PdD_{.8}$		$PdD_{.9}$		PdD
	(a)	CPAI	(a)	CPAI	(a)
λ	0.441	0.445	0.499	0.509	0.625
A(mev)	6.38	6.33	6.63	7.79	10.76
μ*(fitted)	0.097	.098	0.079	0.081	0.106
T_c(K)	2.4		5.4		9.8
$\partial T_c/\partial\mu^*$(K)	-2.91	-2.88	-4.91	-4.82	-6.40
$H_c(0)$ (G)	305	306	726	726	1420
$\Delta C(T_c)$(mJ/mole/K)	5.3	5.32	13.3	13.01	27.5
$\Delta C(T_c)/\gamma T_c$	1.44	1.43	1.48	1.48	1.59
$\gamma(T_c/H_c(0))^2$	.167	.167	.165	.166	.163
$\partial H_c/\partial T\|_{T_c}$	-2560	-2583	-2736	-2756	-3010
$(T_c\frac{\partial H_c}{\partial T})/H_c(0)$	1.74	1.75	1.76	1.77	1.79

(a) Sansores, L.E., Tagüeña-Martínez, J., Sánchez, A.M.: J.Low Temp.Phys.43

References

† Work supported in part by the CONACYT under contract PCAIEUA-800455
1.Skoskiewicz, T.: Phys. Status Sol. (a) 11, K123 (1972)
2.Miller, R.J., Satterthwaite, C.B.: Phys. Rev. Letters 34, 144 (1975)
3.Papaconstantopoulos, D.A., Klein, B.M., Faulkner, J.S., Boyer, L.L.: Phys. Rev. B 18, 2784 (1978)
4.Schirber, J.E., Northup, Jr., C.J.M.: Phys. Rev. B 10, 3818 (1974)
5.Mackliet, C.A., Gillespie, D.J. Shindler, A.I.: J. Phys. Chem. Solids 37, 379 (1979)
6.McLachlan, D.S., Doyle, T.B., Burger, J.P.: Proc. XIV Internat. Conf. Low Temperature Phys., Vol. 2, p.44, Otaniemi (1975)
7.Rowe, J.M.,Rush, J.J., Smith, H.G., Mostoller, M., Flotow, H.E.: Phys. Rev. Letters 33, 1297 (1974)
8.Blackman, J.A., Esterling, D.M., Berk, N.F.: Phys. Rev. B 4, 2412 (1971)
9.Sansores, L.E., Tagüeña-Martínez, J., to be published

- the "tcp", i.e. two-component plasma, a classical gas - mixture of positively and negatively charged particles;
- the "ocp", i.e. one-component plasma, a classical ion-gas is a neutralizing negative background of (degenerate) electrons.

We will be only concerned with the ocp. The requirement that the electrons be degenerate, while the ions are classical, and that the system is still in a gaseous (or liquid) state for a given ion species delineate a region in the T-n plane where the ocp approximation is reasonable. For He-ions Fig. 1 shows this region. Stellar and planetary interiors and laser compressed plasmas are typically within this domain.

RESULTS OF COMPUTER EXPERIMENTS

The equation of state has been fitted by an analytic formula by deWitt[5] and most recent by deWitt, Slattery and Doolen.[6] The excess internal energy per particle, E is given by the formula

$$\beta E(\Gamma) = a\Gamma + b\Gamma^{1/4} + c + e\Gamma^{-1/4} \qquad (1)$$

for $1<\Gamma<160$. The constants are

$$\begin{aligned} a &= -0.89752 \\ b &= +0.94544 \\ c &= -0.80049 \\ e &= +0.17954 \end{aligned} \qquad (2)$$

The value of a should be compared with the Madelung - constant for a bcc digital:

$$a_{MADELUNG} = -0.89593$$

The close agreement is striking, and indicates strongly developed order in the liquid phase. The source of the $\Gamma^{1/4}$term is, however, not easy to see.

The equation of state (1) can be contrasted with the equation of state valid for $\Gamma<<1$, obtained by perturbation method and recently extended[7] to order γ^4

$$\begin{aligned} \beta E(\Gamma) \equiv \beta E(\gamma) = \frac{\gamma}{2} &- (\frac{\gamma}{2})^2 \ (\ln\gamma + c_1) \\ &- (\frac{\gamma}{2})^3 \ (3\ln\gamma + c_2) \\ &- (\frac{\gamma}{2})^4 \ (a_3\ln\gamma - \frac{1}{2}(\ln\gamma)^2 + c_3) \end{aligned} \qquad (3)$$

with

$$\begin{aligned} c_1 &= \ln 3 + 2 \times 0.577 - \frac{4}{3} \\ c_2 &= 2.022 \\ c_3 &= 2 \\ a_3 &= -(2 \times 0.577 + \ln 3 - \frac{15}{18}) \end{aligned} \qquad (4)$$

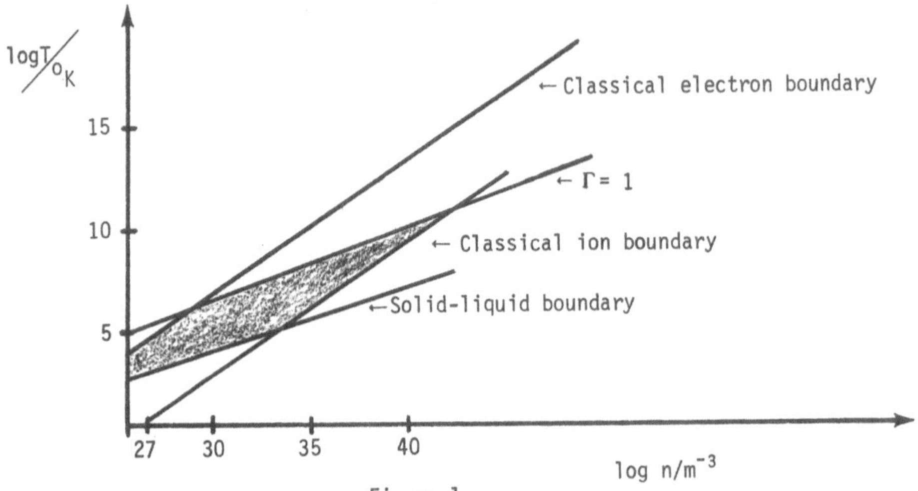

Figure 1

Domain of validity of the OCP approximation

At a high enough Γ value, say at Γ_s, a transition to a solid phase takes place. While Hansen and Pollock[8] quote

$$\Gamma_s = 155 \pm 10$$

more recent evaluation of the data by deWitt and colaborators [1] give

$$\Gamma_s = 171 \pm 3 \tag{5}$$

The behavior of the two body function $G(\underline{r})$ and that of the related static structure factor $S_{\underline{k}}$ (we recall that $S_{\underline{k}} = 1 + n \int e^{i\underline{k}\cdot\underline{r}}(G(r) - 1)\underline{dr}$) are illustrated for $\Gamma = 1, 10, 100$ in Figures 2 and 3. The most striking feature both of $G(r)$ and $S_{\underline{k}}$ is the onset of oscillations ("short range order") for approximately $\Gamma > 3$.

The dynamical properties of the ocp can be analyzed through the dynamized structure factor $S(\underline{k}\omega)$.[9] For $\Gamma = 2$ this quantity is shown in Fig. 4. Of greatest interest is the behavior of the characteristic plasmon frequency as a function of k and Γ. The plasma dispersion curve, as extracted from the analysis of the position of the plasma peak, is given in Fig. 5. As it can be seen, there exists a critical Γ value, Γ_c, somewhere $2 < \Gamma_c < 9.7$, where the slope of the $\omega(k)$ curve for $k \to 0$ changes from positive to negative. The understanding of the behavior is of great theoretical interest.

THEORETICAL RESULTS: HNC.

The best reproduction of the equation of state (1) generated by the computer experiments is provided by HNC calculations. The "classical" HNC gives fairly good quantitative results, but the analytic fitting of the data fails to reproduce the $\Gamma^{1/4}$ dependence. Instead, one finds

$$\beta E(\Gamma) = \bar{a}\,\Gamma + \bar{b}\,\Gamma^{1/2} + \bar{d}\,\ln\Gamma + \bar{c}$$

Quite recently Rosenfeld[10] and Rosenfeld and deWitt[10] have developed an improved HNC approximation in which the bridge-graph (ignored in the "classical" HNC) are included and approximated by their had-core value. As a result, an excellent fit

$$\beta E(\Gamma) = \bar{a}\Gamma + \bar{b}\Gamma^{1/4} + \bar{c} + \bar{e}\Gamma^{-1/4} + \bar{f}\Gamma^{-1/2} \tag{6}$$

with (1) is obtained with

$$\begin{aligned} \bar{a} &= -0.90000 \\ \bar{b} &= +0.97098 \\ \bar{c} &= -0.50000 \\ \bar{e} &= +0.18033 \\ \bar{f} &= +0.01003 \end{aligned} \tag{7}$$

The agreement with (2) is very good indeed.

THEORETICAL RESULTS: MEAN FIELD THEORIES

The principal idea of all mean field theories is that particle interactions can be described in terms of a "static effective potential" which is to be determined self-consistently[11]. In other words, in place of the bare Coulomb-potential

$$\phi_{\underline{k}} = \frac{4\pi e^2 Z^2}{k^2} \tag{8}$$

the effective potential

$$\psi_{\underline{k}} = (1+u_{\underline{k}})\,\phi_{\underline{k}} \tag{9}$$

is to be used for the computation of the dielectric response function; $u_{\underline{k}}$ is the still to be determined "screening function". Consider now the dielectric response function $\varepsilon(\underline{k}\omega)$ or the polarizability $\alpha(\underline{k}\omega)$, $\varepsilon(\underline{k}\omega) = 1 + \alpha(\underline{k}\omega)$. The $\phi_{\underline{k}} \to \psi_{\underline{k}}$ replacement results in

$$\alpha(\underline{k}\omega) = \alpha_{RPA}(\underline{k}\omega)\,\{\,1 + v(\underline{k}\omega)\,\} \tag{10}$$

were the "coupling function" $v(\underline{k}\omega)$ is related to the screening funciton $u_{\underline{k}}$ through

$$v(\underline{k}\omega) = -\frac{u_{\underline{k}}\alpha_{RPA}(\underline{k}\omega)}{1+u_{\underline{k}}\alpha_{RPA}(\underline{k}\omega)} \tag{11}$$

We can also note the relationship between the direct corelation function $c_{\underline{k}}$, the pair correlation function $g_{\underline{k}}$ and the effective potential $\psi_{\underline{k}}$

$$c_{\underline{k}} \equiv \frac{g_{\underline{k}}}{1 + ng_{\underline{k}}} = -\frac{\beta}{n}\,\psi_{\underline{k}} \tag{12}$$

This latter relationship indicates the kinship of the ocp mean field theories and the NELKIN-RAGANATHAN[11] mean field theory of neutral fluids.

The different mean field theories are distinguished from each other by the self-consistency expression they generate for $\psi_{\underline{k}}$. Thus the SINGWI-TOSI-LAND-SJOLANDER(STLS)[12] theory arrives at

$$u_{\underline{k}} = \frac{1}{V}\sum_{\underline{q}}\frac{\underline{k}\cdot\underline{q}}{q^2}\,g_{\underline{k}-\underline{q}} \tag{13}$$

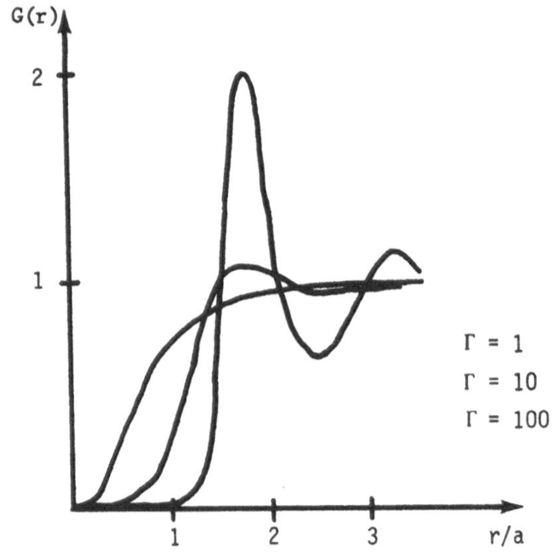

Figure 2

Two body function G(r)

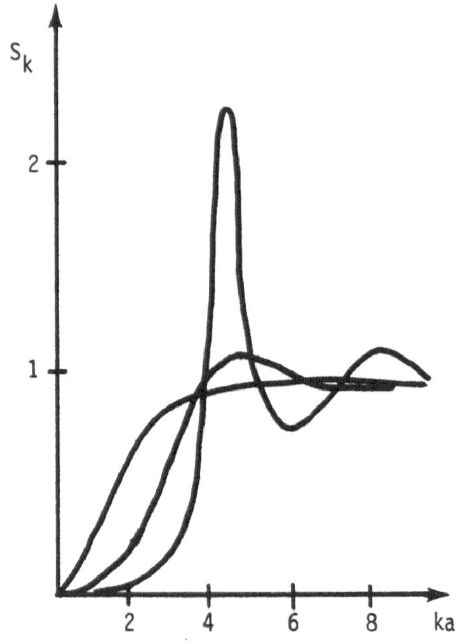

Figure 3

Static structure factor

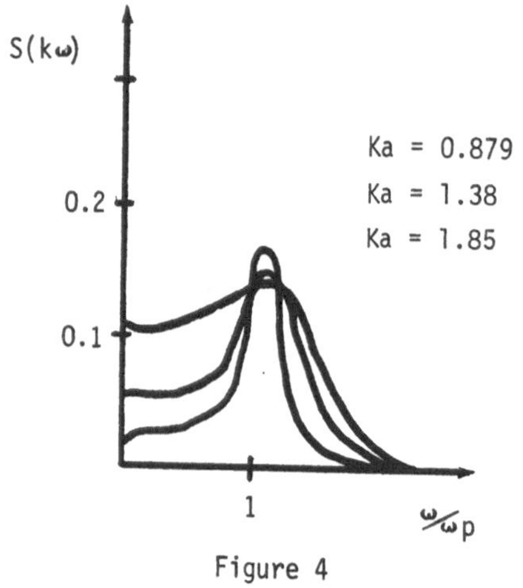

Figure 4

Dynamical structure factor for Γ=7

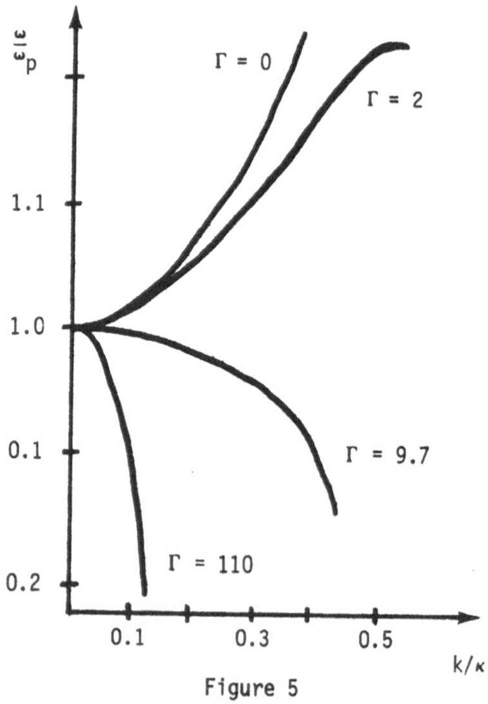

Figure 5

Plasmon dispersion

This theory has its problems with the satisfaction of sum rules and the g(r) calculated from it has poor $r \to 0$ behavior. A better mean field theory is due to Ichimaru and Totsuji,[13] which leads to

$$u_{\underline{k}} = \frac{1}{V} \sum_{\underline{q}} \frac{\underline{k}\cdot\underline{q}}{q^2} g_{\underline{k}-\underline{q}} (1+ng_{\underline{q}}) \tag{14}$$

with good sum rule properties, but with basically the same poor $r \to 0$ behavior. Indeed it can be shown quite generally,[13] that while it is required that for $r \to 0$ $g(r) \to e^{-\beta\psi(r)}-1$, mean field theories in general lead to

$$g(r) \to -1+Ar^3 \tag{15}$$

differing only in the value of the constant A. The mean field theories give reasonable results for the equation of state for small Γ ($\Gamma \lesssim 3$). For higher Γ values HNC is certainly superior. As to dynamical properties, even though the change of shape of the dispersion curve can qualitatively be described, by any of them details of the dynamical processes cannot be explained through a theory which entirely fails to include dynamical correlations.

FLUCTUATION-DISSIPATION RELATIONS

The theory to be outlined, which goes substantially beyond the mean field theories is based on exploiting generalized fluctuation-dissipation (FD) relations. We start with the survey of the latter.

Standard response functions (designated by χ) which we will refer to as "response functions of the first kind" relate perturbed density averages to perturbing densities ($\hat{\rho}$). Schematically, to first order

$$\langle\rho\rangle^{(1)} \sim \underset{1}{\chi}\, \hat{\rho} \tag{16}$$

for the linear response function, and to second order

$$\langle\rho\rangle^{(2)} \sim \underset{2}{\chi}\, \hat{\rho}\, \hat{\rho} \tag{17}$$

for the quadratic response function. Response functions of the first kind satisfy standard FD relations (FD relations of the first kind). Again, schematically

$$\underset{1}{\chi} \sim \langle\rho(1)\rho(2)\rangle^{(0)} \sim S(\underline{k}\omega) \tag{18a}$$

$$\underset{2}{\chi} \sim \langle\rho(1)\rho(2)\rho(3)\rangle^{(0)} \sim S(\underline{k}_1\omega_1;\, \underline{k}_2\omega_2) \tag{18b}$$

$S(\underline{k}\omega)$ is the dynamical structure function, while $S(\underline{k}_1\omega_1;\, \underline{k}_2\omega_2)$ is the three-point dynamical structure function (the derivation of (18b) is given in Ref. 14).
Response function of the second kind (Ξ) can now be defined through

$$\langle\rho(1)\rho(2)\rangle^{(1)} \sim \underset{1}{\Xi}\, \hat{\rho} \tag{19}$$

$$\langle\rho(1)\rho(2)\rangle^{(2)} \sim \underset{2}{\Xi}\, \hat{\rho}\, \hat{\rho} \tag{20}$$

They satisfy FD relations of the second kind

$$\underset{1}{\Xi} \sim \langle\rho(1)\rho(2)\rho(3)\rangle^{(0)} \tag{21}$$

$$\underset{2}{\Xi} \sim \langle\rho(1)\rho(2)\rho(3)\rho(4)\rangle^{(0)} \tag{22}$$

Comparison of (18a,b) and (21), (22) shows the connections

$$\begin{array}{ccccc} \chi_1 & & \chi_2 & & \chi_3 \\ & \swarrow\!\!\nearrow & & \swarrow\!\!\nearrow & \\ \Xi_1 & & \Xi_2 & & \end{array} \tag{23}$$

As an example, we display one of the above schematic relations explicity

$$S(\underline{q}\mu;\ \underline{p}\nu) = \frac{kqp}{2\pi e^3 n\beta^2} \ln \left\{ \frac{1}{\varepsilon(\underline{q}\mu)\ \varepsilon(\underline{p}\nu)\varepsilon^*(\underline{k}\omega)} \cdot \left[\frac{\alpha_2(\underline{q}\mu;\underline{p}\nu)}{\nu\mu} - \frac{\alpha_2(-\underline{k}-\omega;\underline{q}\mu)}{\mu\omega} - \frac{\alpha_2(\underline{p}\nu;-\underline{k}-\omega)}{\omega\nu} \right] \right\}$$

$$\omega = \mu+\nu$$

$$\underline{k} = \underline{p}+\underline{q} \tag{24}$$

VELOCITY AVERAGE APPROXIMATION AND GKS THEORY

The GKS theory[3] starts from the recognition of the fact that the shortcomings of the mean field theories originate from approximating the two body non-equilibrium function G(12) through

$$G(12) = \frac{F(1)\ F(2)}{\langle F(1)\rangle\langle F(2)\rangle}\ (1 + g(12)) \tag{25}$$

Here F(1), F(2) are the non-equilibrium one-body function, while g(12) is the equilibrium pair corelation function; $\langle\ \rangle$ here denote velocity averages. In order to restore the genuine non-equilibrium correlations, (25) is proposed to be replaced by

$$G(12) = \frac{F(1)\ F(2)}{\langle F(1)\rangle\langle F(2)\rangle}\ \langle G(12)\rangle \tag{26}$$

$\langle G(12)\rangle$ is easily shown to be related to the two-point function

$$\langle G(12)\rangle = \langle\rho(1)\ \rho(2)\rangle - \delta(12)\ \langle F(1)\rangle \tag{27}$$

Using now the FD relations listed above, $\langle\rho(1)\ \rho(2)\rangle^{(1)}$ can be expressed in terms of α_2, which in terms, allows one to express the linear $\alpha(\underline{k}\omega)$ as follows:

$$\alpha(\underline{k}\omega) = \alpha_{RPA}(\underline{k}\omega)\ \{1+v(\underline{k}\omega)\}$$

$$v(\underline{k}\omega) = -\frac{k^2}{\kappa^2}\ \frac{1}{N} \sum_{\underset{\sim}{p},\underset{\sim}{q}} \frac{\underline{k}\cdot\underline{p}}{p^2} \int d\mu \int d\nu\ \ \delta(\omega-\mu-\nu)\ \delta_{\underline{k}-\underline{p}-\underline{q}}\ (\delta_-(\mu)+\delta_-(\nu))\ \frac{\alpha_2(\underline{p}\mu;\underline{q}\nu)}{\alpha_{2RPA}(\underline{p}o;\underline{q}o)}\ \frac{1}{\varepsilon(\underline{p}\mu)\varepsilon(\underline{q}\nu)} \tag{28}$$

In order to render the above relation self-consistent, the decomposition of α_2 in terms linear α-s is necessary. This can be done for $k/\omega \to 0$. In this case and in $\gamma \to 0$ limit the Vlasov $\alpha_2(\underline{p}\mu,\underline{q}\nu)$ can be decomposed into product expressions containing $\alpha(\underline{p}\mu)\ \alpha(\underline{q}\nu)$ only. Following a philosophy similar to the one applied in approximating the three-body function through the cluster of two-body functions, which is exact for $\gamma \to 0$, we adopt the $\gamma \to 0$ structure as a valid structural approximation for arbitary γ. Even though the general expression is fairly complicated, the expression intergrated out according to (28) becomes quite simple:

$$v(\underline{k}\omega) = v_{stat}(\underline{k}\omega)+v_{dyn}(\underline{k}\omega)$$

$$v_{stat}(\underline{k}\omega) = \frac{1}{\omega^2}\ \frac{1}{N} \sum_{\underline{q}} \chi^2\ \{\ S_{\underline{k}-\underline{q}} - S_{\underline{q}}\}$$

$$v_{dyn}(\underline{k}\omega) = -\frac{k^2}{\omega^2}\frac{1}{N}\sum_{\underline{q}}(1-6\chi^2+8\chi^4) \int d\mu\ \delta_-(\mu)\ \hat{\alpha}(\underline{q}\mu)\ \hat{\alpha}(\underline{q},\omega-\mu) \tag{29}$$

with

$$\chi = \frac{\underline{k}\cdot\underline{q}}{kq}, \quad \hat{\alpha} = \frac{\alpha}{\varepsilon}$$

(Here and in the sequel ω and k are measured in units of ω_o, κ respectively.)

PLASMON DISPERSION

The result (29) has been applied[15] to the study of the γ-dependence of the plasmon dispersion curve. In order to make the self-consistency calculation feasible, we adopt a "two-pole" approximation for $\hat{\alpha}(\underline{k}\omega)$:

$$\hat{\alpha}(\underline{k}\omega) = -\frac{1}{\omega-\omega_{\underline{k}}+i\nu_{\underline{k}}}\ \frac{1}{\omega+\omega_{\underline{k}}+i\nu_{\underline{k}}} \tag{30}$$

$$\omega_{\underline{k}} = 1 + A(\gamma)k^2$$

$$\nu_{\underline{k}} = B(\gamma)k^2 \tag{31}$$

The rather lengthy calculation results in complicated algebraic equations for the coefficients A, B:

$$A = a(A,B)$$

$$B = b(A,B) \tag{32}$$

which can be solved numerically. The results we shown in Fig. 7, while the comparison with the results of the computer experiments are shown in Fig. 6. The agreement is satisfactory. More refined models for $\hat{\alpha}(\underline{k}\omega)$ should provide an even better result.

OTHER DYNAMICAL APPROACHES

In addition to the approach outlined, many other methods have been attempted to analyze the dynamical properties of strongly coupled ocp - s. Roughly speaking, they fall into two groups: (a) application of the "renormalized kinetic theory" to plasmas; (6) ad hoc "Ansaty"-es for the dynamical structure function. Gould and Mazenko;[16] Baus and collaborators[16] and Gross[16] should be mentioned under (a), and Lovesey, Abramo and Parinello,[17] Singh[17] and Takeno and Yosida[17] under (b). Especially good results have been obtained by Baus and collaborators[18] for transport coefficients. The study of the dispersion properties within this framework is, however, problematic, in the $\gamma \to 0$ limit in particular.

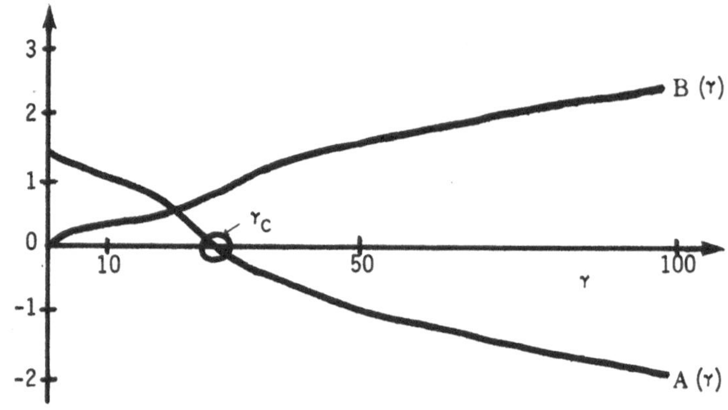

Figure 6

Composion of theoretical and experimental results for plasmon dispersion

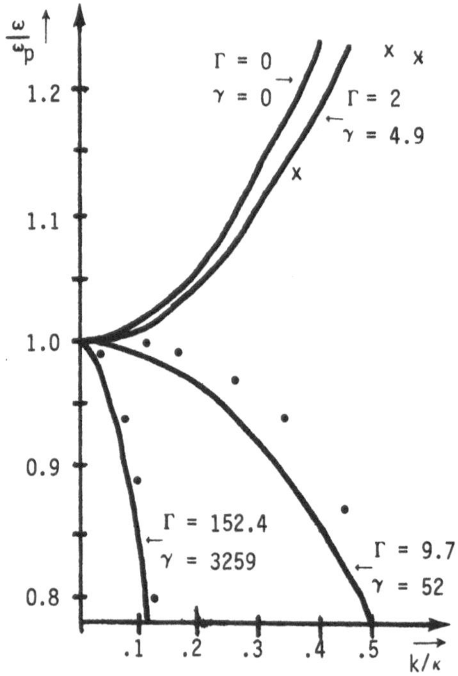

Figure 7

Plasmon dispersion coefficients, A(γ) and B(γ)

REFERENCES

1. H.E. DeWitt, in "Strongly Coupled Plasmas," ed. by G. Kalman, Plenum Press, N.Y. (1978), p. 81.

2. (a) J.P. Hansen, Phys. Rev. A8, 3096 (1973).
   (b) E.L. Pollock and J.P. Hansen, Phys. Rev. A8, 3110 (1973).
   (c) S. Galam and J.P. Hansen, Phys. Rev. A14, 816 (1976).
   (d) J.P. Hansen, I.R. McDonald, and E.L. Pollock, Phys. Rev. A11, 1025 (1975), Phys. Rev. Lett. 32, 277 (1974).
   (e) J.P. Hansen and I.R. McDonald, Phys. Rev. Lett. 41, 1379 (1978).
   (f) M. Baus and J.P. Hansen, Phys. Reports, 59, 1 (1980).

3. (a) K.I. Golden, in "Strongly Coupled Plasmas," (see Ref. 1), p. 223.
   (b) G. Kalman, in "Strongly Coupled Plasmas," (see Ref. 1), p. 141.

4. (a) K.I. Golden, G. Kalman and M.B. Silevitch, Phys. Rev. Lett. 33, 1544 (1974).
   (b) P. Bakshi, in "Strongly Coupled Plasmas," (see Ref. 1), p. 533.
   (c) R. Calinon, K.I. Golden, G. Kalman, and D. Merlini, Phys. Rev. A20, 329 (1979).
   (d) P. Bakshi, R. Calinon, K.I. Golden, G. Kalman and D. Merlini, Phys. Rev. A20, 336 (1979).
   (e) P. Carini, G. Kalman and K.I. Golden, Phys. Lett. 78A, 450 (1980).

5. H.E. DeWitt, Phys. Rev. A14, 816 (1976).

6. W.L. Slattery, C.D. Doolen and H.E. DeWitt, Phys. Rev. A21, 2087 (1980).

7. (a) E.G.D. Cohen and T.J. Murphy, Phys. Fluids 8, 1109 (1965).
   (b) R.L. Guernsey, Phys. Fluids, 21, 2162 (1978).

8. Ref. 2(a).

9. Ref. 2(d), 2(e).

10. (a) Y. Rosenfeld, Phys. Rev. Lett. 44, 146 (1980).
    (b) H.E. DeWitt and Y. Rosenfeld, Phys. Lett. 75A, 79 (1979).

11. (a) G. Kalman in "Strongly Coupled Plasmas," (see Ref. 1), p. 141.
    (b) M. Nelkin and S. Ranganathan, Phys. Rev. 164, 222 (1967).

12. (a) K.S. SIngwi, M.P. Tosi, R.H. Land, and A.S. Sjolander, Phys. Rev. 176, 589 (1968).
    (b) K.S. Singwi, A. Sjolander, M.P. Tosi, P. Vashishta and K.S. Singwi, Phys. Rev. B6, 875 (1972).

13. (a) S. Ichimaru in "Strongly Coupled Plasmas" (see Ref. 1), p. 187.
    (b) H. Totsuji and S. Ichimaru, Progr. Theor. Phys. 50, 753 (1973); 52, 42 (1974).

14. K.I. Golden, G. Kalman and M.B. Silevitch, J. Stat. Phys. 6, 87 (1972).

15. Ref. 4(e).

16. (a) H. Gould and G.F. Mazenko, Phys. Rev. A15, 1279 (1977).
    (b) M. Baus, Physica, 79A, 377 (1975), 88A, 319, 336, 591 (1977).
    (c) M. Baus and J. Wallenborn, J. Stat. Phys. 16, 91 (1977).
    (d) M. Baus, Phys. Rev. A15, 790 (1977).
    (e) E.P. Gross, J. Stat. Phys. 11, 503 (1974).

17. (a) S.W. Lovesey, J. Phys. 64, 3057 (1971).
    (b) M.L. Abramo and M.P. Tosi, Nuovo Cim. 21B, 363 (1974).
    (c) K. Singh, Preprint 1980.
    (d) S. Takeno and F. Yoshida, Progr. Theor. Phys. 62, 883 (1979).

18. J. Wallenborn and M. Baus, Lett. 61A, 35 (1977); Phys. Rev. A18, 1737 (1978).

THE PERTURBATION THEORY APPROACH TO THE GROUND STATE ENERGY IN AN INFINITE FERMION SYSTEM

George A. Baker, Jr.
Theoretical Division, Los Alamos Scientific Laboratory
University of California, Los Alamos, NM 87545/USA

The fundamental question which we wish to address is, "What is an effective procedure for finding the lowest, Fermionic eigenvalue of a system described by the Hamiltonian

$$H = \sum_{i=1}^{N} \frac{p_i^{\,2}}{2m} + \sum_{i<j}^{N} V(|\vec{r}_i - \vec{r}_j|), \tag{1}$$

when N becomes indefinitely large?"

If V is purely repulsive, then rather good methods are available[1] over a fair range of densities and potentials, but if V is partially attractive, then the problem is much more difficult, and, of course, more interesting. There are quite a number of ways to approach the question raised above, but we will confine our discussion to perturbation methods, and their progress toward the answer to our question. In particular we will specialize in the Rayleigh-Schrödinger perturbation theory. By way of a brief review, we start with the Schrödinger equation for the Hamiltonian(1),

$$\left[-\frac{\hbar^2}{2m}\sum_{j=1}^{N} \nabla_j^2 + \sum_{i<j}^{N} v(r_{ij})\right]\Psi = E\Psi, \tag{2}$$

where $r_{ij} = |\vec{r}_i - \vec{r}_j|$, and we define the useful operators

$$H_o = -\frac{\hbar^2}{2m}\sum_{j=1}^{N} \nabla_j^2, \quad h = \sum_{i<j}^{N} v(r_{ij}), \tag{3}$$

so that eq. (2) may be written more compactly as

$$(H_o + \lambda h)\Psi = E(\lambda)\Psi. \tag{4}$$

The formulas of the Rayleigh-Schrödinger perturbation theory can then be written in terms of the energy E, and wave matrix Ω,

$$\begin{aligned} E(\lambda) &= E_o + E_1\lambda + E_2\lambda^2 + \cdots , \\ \Omega(\lambda) &= I + \Omega_1\lambda + \Omega_2\lambda^2 + \cdots , \\ H_o\Phi &= E(0)\Phi, \qquad \Psi = \Omega(\lambda)\Phi, \end{aligned} \tag{5}$$

as

$$E_n = \langle\Phi|h \;\; \Omega_{n-1}|\Phi\rangle, \tag{6}$$

$$\Omega_n = (I - P_o)(E_o - H_o)^{-1}(h\Omega_{n-1} - \sum_{j=1}^{n-1} E_{n-j}\,\Omega_j),$$

where P_o is the projection operator for the state Φ.

The Fermi-statistics constraint is imposed by selecting the initial wave-function as a Slater determinant,

$$\Phi = (N!)^{-\frac{1}{2}}\,\Gamma^{-\frac{1}{2}N} \det \begin{vmatrix} \exp(i\vec{k}_1\cdot\vec{r}_1) & \cdots & \exp(i\vec{k}_N\cdot\vec{r}_1) \\ \vdots & \ddots & \vdots \\ \exp(i\vec{k}_1\cdot\vec{r}_N) & \cdots & \exp(i\vec{k}_N\cdot\vec{r}_N) \end{vmatrix}, \tag{7}$$

where the $\vec{k}_i$ are the lowest N eigenstates of H_o in a box of volume Γ with periodic boundary conditions. The corresponding momentum representation of the potential is

$$\langle\vec{\nu}\vec{\mu}|v|\vec{\lambda}\vec{\eta}\rangle = \Gamma^{-2}\int_{\text{Box}}\cdots\int d^3R\,d^3r\;v(r)$$

$$\times \exp\left[i(\vec{\lambda} + \vec{\eta} - \vec{\nu} - \vec{\mu})\cdot\vec{R} + \tfrac{1}{2}\,i\vec{r}\cdot(\vec{\lambda} + \vec{\mu} - \vec{\nu} - \vec{\eta})\right]$$

$$= (2\pi)^3\Gamma^{-1}\,\delta_{\vec{\lambda}+\vec{\eta},\,\vec{\nu}+\vec{\mu}}\;\tilde{v}(\tfrac{1}{2}(\vec{\lambda} + \vec{\mu} - \vec{\nu} - \vec{\eta})), \tag{8}$$

where δ is Kronecker's delta and $\tilde{v}$ is the momentum transform of v(r). The basic formula for the graphical expansion of the Rayleigh-Schrödinger perturbation theory is

$$h\Phi = \tfrac{1}{2}(N!)^{-\frac{1}{2}}\Gamma^{-\frac{1}{2}N}\sum_{\vec{m}\neq\vec{n}}\sum_{\vec{\nu},\vec{\mu}} \langle\vec{\nu}\vec{\mu}|v|\vec{m}\vec{n}\rangle\;\Phi(\vec{m}\to\vec{\nu};\;\vec{n}\to\vec{\mu}) \tag{9}$$

where $\vec{m}$ and $\vec{n}$ run over the indices in the Slater determinant Φ and $\vec{\nu}$ and $\vec{\mu}$ run over all momentum states. The $\Phi(m\to\nu)$ type notation means that $k_{\vec{\nu}}$ replaces $k_{\vec{m}}$ in eq. (7). This eq. expands $h\Phi$ as a sum of Slater determinants.

By use of the Hugenholtz factorization theorem, we can establish the linked cluster theorem, i.e., only connected diagrams contribute to the energy. This result is important because before the work of Brueckner[3], it was thought that perturbation theory was useless because every connected part contributed a factor of N so that

$$E(\lambda) = N\lambda + N^2\lambda^2 + N^3\lambda^3 + \cdots, \tag{10}$$

instead of the correct result,

$$E(\lambda) = N\lambda + N\lambda^2 + N\lambda^3 + \cdots. \tag{11}$$

There was a "physical" explanation of eq. (10). It was that for an attractive, square-well potential, there was nuclear collapse[4] no matter how weak the attraction, so of course the perturbation theory was non-sense!

So what happened to this explanation? The answer is that the physics has not disappeared, the radius of convergence of the series (11) for $E(\lambda)/N$ is zero. Thus it need not give the results for a pure attraction while there is still the possibility that it can work for a repulsion in an asymptotic sense. It can be shown[1] that the divergence of (11) is no worse than

$$|E_n/N| \leq M(n!)A^n, \tag{12}$$

so that the series is uniquely summable to the correct physical answer for a simple repulsive force. Thus, for at least some many-fermion, ground-state energy problems, perturbation theory can lead in principal to the correct physical answer.

There are further problems to be considered on our search for the ground state energy of a many-fermion system with attractive forces. In particular in applications like nuclear matter which are self-bound, we know that the system must be a liquid. Thus the saturation point lies on the liquid, coexistence curve in the $(k_F - \lambda)$ plane. It is reasonable on general grounds, and borne out in model calculations[5], that the coexistence curve is a line of analytic singularities so the usual procedure of looking for a minimum is in principle only half correct. That is to say the approach from high density at fixed potential (or to weaker potential at fixed density) is fine; the part as one passes the minimum and finds increasing energy as the density is lowered further is wrong. Here we know that a two phase system occurs. The gas phase (absolute temperature in this problem is zero) is the vacuum, and the liquid is a self-bound drop. Thus the energy remains at its minimum value and does not increase. The perturbation approach, when properly used, has the advantage of approaching the saturation point along a physically correct path exclusively in the one phase region. On the contrary a purely low density rearrangement tends to pass through the two-phase region and must be viewed with great caution.

An additional problem is that of the hard-core. We know, at least for sufficiently low density, that an infinitely strong, repulsive-core potential leads to only a finite shift in energy. The classical solution to this problem in perturbation theory is to make a change of variables. For example, the function f(r) defined as

$$\eta f(r) = 1 - e^{-\lambda v(r)} = \lambda v(r) - \frac{\lambda^2 v^2(r)}{2!} + \cdots . \tag{13}$$

has the property $\eta f(r) = 1$ for $v(r) = +\infty$, and $\eta f(r) \simeq \lambda v(r)$ for small $\lambda v(r)$. Thus if we replace

$$\lambda v(r) = - \ell n[1 - \eta f(r)] = \eta f(r) + \frac{1}{2}\eta^2 f^2(r) + \frac{1}{3}\eta^3 f^3(r) + \cdots . \tag{14}$$

in the $\lambda v(r)$ expansion we generate a new expansion in terms of $\eta f(r)$ which is not immediately singular for a repulsive hard core.

Brueckner had the idea to follow Watson's theory of multiple scattering and sum up in a K-matrix all the ladder diagrams (diagrams with only two hole-lines). These diagrams constitute all the leading order terms in a low density expansion and give good results for repulsive forces at low densities.

However, it was soon noticed by Emery[6] that for a central repulsion with an attractive tail that no matter how weak the attraction, for a sufficiently high value of the relative angular momentum, the K-matrix equations processed a singularity. What then is to be done about this problem? First, of course, it may be that the singularity comes from Cooper pairs and that it implies that the true ground state is a "superfluid" and not a "normal" ground state. In this case the perturbation theory which we are using

$$\sum_{k=0}^{\infty} (\lim_{N\to\infty} E_k^{(N)}/N)\lambda^k \tag{15}$$

does not work. On the other hand, if such a situation does not arise physically, we need to fix our formalism to allow progress. In the context of perturbation theory several ways have been suggested. (a) Brueckner and Gammel[7] simply modified the intermediate state denominator as

$$\frac{1}{(k'')^2 - k_o^2} \to \frac{1}{(k'')^2 - k_o^2 + \Delta} \tag{16}$$

where even a very small Δ serves to eliminate the problem. (It also changes the answer, in principal, a little bit.) (b) Baker and Kahane[8] solved the problem by using a different change of variables, i.e.,

$$K_\ell(k) = R_\ell(k)/\left[1 + (\tfrac{1}{2}\tau_1 - \tilde{a})R_\ell(k)\right] \tag{17}$$

where $R_\ell(k)$ (the "R matrix") is much better behaved. The Emery singularities now occur in $K_\ell(k)$ if $R_\ell(k) < 0$ since $\tau_1 \to \infty$ as $k \to k_F$. However, as any power of τ_1 is integrable, K can be expressed as a series expansion in R with finite coefficients. Baker and Kahane found that R itself is satisfactory for typical potentials. (c) Brandow's choice[9] for intermediate-state energy denominators corresponds to a re-arrangement of the perturbation series. It is to make the hole-line energies self-consistent in the intermediate state propagators, but to use the kinetic energy alone for the particle energies. This procedure is superficially attractive, but has a fatal flaw! Its advantages are: first, by the Hugenholtz factorization theorem the energy corrections to the hole lines, summed over all time orders, are on the energy shell and don't depend on the excitation of the Fermi sea so that they are easy to compute. Secondly, for a potential with a net attraction, there is a large energy gap at the Fermi surface, thus eliminating the Emery singularity problem. Finally, the resulting large denominators make the higher order terms smaller and thus less important. The principal disadvantage of this scheme is that it gives the *wrong* answer[10]! That is to say, the answer obtained in this manner using the "obvious" choice of the integration contour over intermediate state energies omits certain residue corrections which are included by the correct choice of integration contour. These details are explained at length by Baker and Gammel.[10] An additional aesthetic drawback to the method is that it splits certain finite contributions into the difference of two infinite ones.

In summary then, the progress thus far by perturbation methods is: (1) The straightforward·V expansion with appropriate resummations such as the K-matrix one, or the complete hole-line rearrangement with a symmetrical treatment of hole and particle energies appears to give satisfactory results for problems with purely repulsive forces. (2) The K-matrix and hole-line type rearrangements, as they stand, are not adequate for potentials with an attractive tail because of the Emery singularities. (3) A good change of variables, such as the R-matrix expansion[8] would appear to allow reasonable computational progress towards our goal. Other changes of variables could profitably be explored. (4) Modern computers would appear to make much more extended work in this area possible than in the past.

This work was performed under the auspices of the U.S. D.O.E.

References

1. G. A. Baker, Jr., Rev. Mod. Phys. 43, 479 (1971).
2. K. A. Brueckner, in *The Many Body Problem*, C. Dewitt, ed. (Dunod, Paris, 1959) pg. 47.
3. K. A. Brueckner, Phys. Rev. 100, 36 (1955).
4. J. M. Blatt and V. F. Weisskopf, *Theoretical Nuclear Physics*, (Wiley, New York, 1952).
5. G. A. Baker, Jr. and D. Kim, J. Phys. A 13, L103 (1980).
6. V. J. Emery, Nucl. Phys. 12, 69 (1959).
7. K. A. Brueckner and J. L. Gammel, Phys. Rev. 109, 1023 (1958).
8. G. A. Baker, Jr. and J. Kahane, J. Math. Phys. 10, 1647 (1969).
9. B. H. Brandow, Phys. Rev. 152, 863 (1966).
10. G. A. Baker, Jr. and J. L. Gammel, Phys. Rev. C 6, 403 (1972).

BRUECKNER-BETHE CALCULATIONS OF NUCLEAR MATTER

B. D. Day

Physics Division
Argonne National Laboratory*

Argonne, IL 60439, USA

I. Introduction

The calculations described here are based on the following model. The nucleus is treated as a collection of point nucleons that obey the nonrelativistic Schrödinger equation and interact through a 2-body potential. The potential has a one-pion-exchange tail, and in some cases additional constraints based on theory are imposed. Typical potentials of interest are Hamada-Johnston,[1] Reid,[2] Paris,[3] and Bonn.[4] The basic question is whether these potentials can account for the saturation properties of nuclei and nuclear matter. To answer this question for nuclear matter, we must calculate the energy per particle as a function of density, find the minimum, or saturation point, and compare with the saturation point deduced empirically.

This model of the nucleus is clearly incomplete. It neglects 3-body forces, which are surely present even if small. Also, the degrees of freedom of mesons, antinucleons, isobars, and quarks are buried in the 2-body potential. One of the most fascinating aspects of nuclear physics is to try to identify those phenomena where non-nucleonic degrees of freedom show themselves. But to do this, we must first see how far we can go with nucleons alone. This means that we need accurate approximate solutions of the many-body Schrödinger equation. In this paper the Brueckner-Bethe approach to this problem is described.

II. Outline of the Brueckner-Bethe Method.

The Brueckner-Bethe method can be described in terms of partial summations of perturbation theory,[5] but a much more powerful formulation is that of the coupled cluster equations[6-8] invented by Coester and Kümmel. I will describe the latter method for nuclear matter. It can also be used for nuclei.

The Hamiltonian is the sum of single-particle kinetic energies and 2-body potentials:

$$H = \sum_{i=1}^{A} T_i + \sum_{i<j}^{A} V_{ij} \quad . \tag{1}$$

Denoting the non-interacting Fermi-gas state by Φ, we write the exact ground state Ψ as

---

*This work was performed under the auspices of the U. S. Dept. of Energy.

$$\Psi = e^{S}\Phi, \quad S = \sum_{n=2}^{\infty} S_n \quad . \tag{2}$$

The operator S_n applied to Φ produces n particle-hole pairs and is represented by the amplitude

$$(k_1 \ldots k_n | S_n | p_1 \ldots p_n), \quad k_i > k_F, \quad p_i < k_F \quad , \tag{3}$$

where k_F is the Fermi momentum. In nuclear matter we have $S_1 = 0$ from momentum conservation. The exponential form eq. (2) is convenient because it causes the S_n amplitudes (3) to be linked. In perturbation theory the amplitude (3) is given by the sum of all linked diagrams that produce n particle-hole pairs.

The many-body Schrödinger equation $H\Psi = E\Psi$ takes the form $He^{S}\Phi = Ee^{S}\Phi$. By manipulating this equation, one derives formulas for E and for the S_n. The exact ground-state energy E is given by

$$E = T_\phi + \frac{1}{2} \sum_{p_1, p_2 < k_F} (p_1 p_2 | V | \psi_{p_1 p_2}) \tag{4}$$

$$\psi_{p_1 p_2}(\vec{r}_1, \vec{r}_2) = e^{i\vec{p}_1 \cdot \vec{r}_1} e^{i\vec{p}_2 \cdot \vec{r}_2} + (\vec{r}_1 \vec{r}_2 | S_2 | \vec{p}_1 \vec{p}_2) \quad . \tag{5}$$

If S_2 is known, this determines the correlated 2-body wavefunction $\psi_{p_1 p_2}$, which in turn determines the exact ground state energy.

The equations for S_2 and S_3 can be written schematically as

$$e_2 (k_1 k_2 | S_2 | p_1 p_2) = F_2 (V, S_2, S_3, S_4) \tag{6}$$

$$e_3 (k_1 k_2 k_3 | S_3 | p_1 p_2 p_3) = F_3 (V, S_2, S_3, S_4, S_5) \tag{7}$$

where the nonlinear functionals F_2 and F_3 involve linked products of matrix elements of V and the S_n. The general equation for S_n involves all the S_k for $k = 2,3,\ldots n+2$. Eqs. (6), (7) are the first two of a set of infinitely many coupled nonlinear equations that are equivalent to the many-body Schrödinger equation. These are the coupled cluster equations.

The quantity e_2 in eq. (6) is given by

$$e_2 = T(k_1) + T(k_2) - E(p_1) - E(p_2) \tag{8}$$

$$E(p_1) = T(p_1) + U(p_1); \quad U(p_1) = \sum_{p_2 < k_F} (p_1 p_2 | V | \psi_{p_1 p_2}) \tag{9}$$

This is the conventional choice of single-particle spectrum: kinetic energy above the Fermi sea and Hartree-Fock energy in the Fermi sea. This single-particle spectrum is <u>not</u> an observable quantity. In particular, it need not be related to the excitation spectrum of the system. It is at our disposal and should be chosen to define a starting approximation that leads to good convergence.

For various reasons a spectrum that is continuous at the Fermi surface has often been advocated.[9,10] If one defines a single-particle potential U(k) for $k>k_F$, then e_2 is changed by $\Delta e_2 = U(k_1)+U(k_2)$, and eq. (6) becomes

$$(e_2+\Delta e_2)S_2 = F_2(V,S_2,S_3,S_4)+\Delta e_2 S_2 \qquad (10)$$

In lowest order we neglect S_3 and S_4 and solve for S_2. If U(k) is cleverly chosen, the term $\Delta e_2 S_2$ on the right of (10) might partially cancel the terms involving S_3 and S_4. Neglecting these terms but keeping $\Delta e_2 S_2$ on the left of (10) would then give a more accurate lowest-order approximation than simply using $U(k) = 0$.

The choice of spectrum can also be used to formulate a consistency test.[11] Suppose we do two separate calculations with different choices of U(k) and hence different starting approximations. If both calculations converge, then they should converge to the same answer.

We need an approximation scheme to solve the coupled cluster equations. The scheme used so far is based on the idea that nuclear matter is a low-density system. Therefore, correlations among pairs are more important than correlations among triples, etc. There are two versions of this scheme: the hole-line expansion[5,12] and the Bochum truncation.[7] For nuclear matter the difference between the two versions is less than the present numerical uncertainties in the calculations.

In the lowest-order, or 2-body, approximation, we neglect S_3 and S_4 in eq. (6) and solve for S_2. The resulting approximation to the energy is

$$E^{(2)} = T_\phi + \frac{1}{2} \sum_{p_1,p_2<k_F} (p_1p_2|V|\psi^{(2)}_{p_1p_2}) \quad . \qquad (11)$$

The kinetic energy T_ϕ ranges from 23 to 40 MeV per particle as the density ρ runs from ρ_0 to $2\rho_0$, where ρ_0 is the empirical saturation density corresponding to $k_F \approx 1.36\ \text{fm}^{-1}$. The matrix element in the second term of (11) is the Brueckner reaction matrix. This interaction term is typically -35 to -45 MeV per particle.

In Fig. 1 we plot $\psi^{(2)}$ against the relative coordinate r for the Reid potential, in the $^3S-^3D$ partial wave. The uncorrelated plane wave (first term of eq. (5)) has been chosen to have only an S-wave component $\phi_S = j_0(k_0r)$, where k_0 is the relative momentum. The tensor force couples the S and D waves and causes ψ to have both S- and D-wave components. We see that ψ_S differs appreciably from ϕ_S only at small distances, $r \leq 0.5$ fm, which is much less than the average interparticle distance of 1.5 fm. The tensor correlations are represented by ψ_D. They have a longer range but are weak in the sense that $\psi_D << \phi_S$. Thus the strong correlations have a range that is small compared to the average interparticle distance. This is the sense in which we have a low-density system.

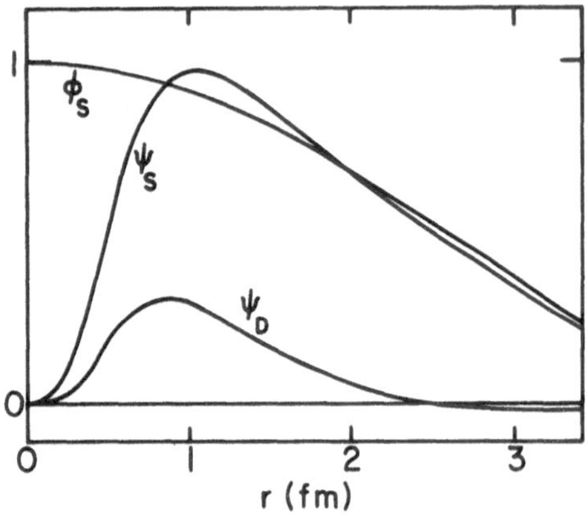

Fig. 1

This idea is made more precise by defining the dimensionless quantity κ by

$$\kappa = \rho \int |\phi-\psi|^2 \, d\tau \quad . \tag{12}$$

Using the two-body approximation $\psi \approx \psi^{(2)}$, one obtains the two-body approximation $\kappa^{(2)}$ to κ. For the Reid potential $\kappa^{(2)} = 0.14$ to 0.25 for $\rho_0 \leq \rho \lesssim 2\rho_0$. The fact that $\kappa << 1$ means we have a low-density system. We expect 3-body effects to be smaller than 2-body effects by roughly a factor of κ.

In the 3-body approximation, S_4 and S_5 are neglected in eqs. (6) and (7), which are then solved for S_2 and S_3. A new feature arises here: the possible buildup of long-range correlations. One usually looks for this in the ring diagrams of perturbation theory, the first three of which are shown in Fig. 2. The dashed lines represent V, upgoing lines represent particles above the Fermi sea, and downgoing lines represent holes in the Fermi sea. Each of these diagrams gives a contribution to $(k_1k_2|S_2|p_1p_2)$. In diagram (c), we have particles 1 and 2 excited above the Fermi sea. Then 2 falls back into the sea and excites 3; next, 3 falls back and excites 4. This chain effect can build up long-range correlations, and this is what happens, for example, in the electron gas.

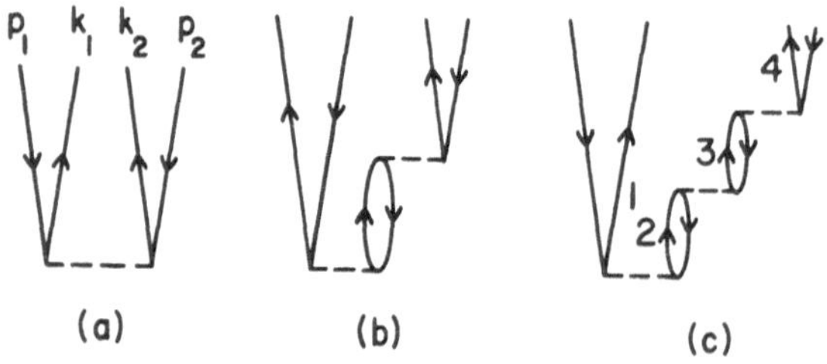

Fig. 2

This effect is contained in the coupled cluster equations at the 3-body level, and this is illustrated in Fig. 3, which shows contributions to the 3-body approximation $S_2^{(3)}$ to $(k_1k_2|S_2|p_1p_2)$. The first term, Fig. 3(a), is the 2-body approximation $S_2^{(2)}$. In the next term, Fig. 3(b), particles 1 and 2 are excited above the

Fermi sea. They are strongly correlated, the correlations being determined by $\psi^{(2)}(1,2)$. Therefore, if particle 1 tries to interact with a third particle in the Fermi sea, particle 2 will get in the way. This means that we are forced to consider a 3-body problem. The solution of the 3-body equation is represented by the box labelled $\mathcal{M}$, which eventually produces a new 2-particle-2-hole state. In Fig. 3(c), $\mathcal{M}$ acts twice, and in higher orders $\mathcal{M}$ can act any number of times.[8,13]

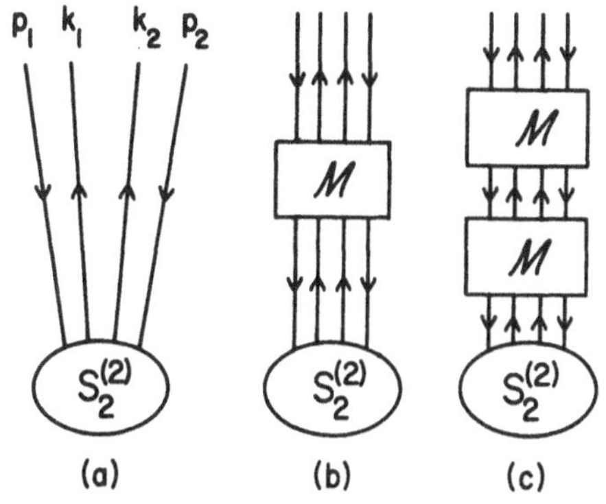

Fig. 3

The Bochum truncation at the 3-body level includes this entire series Fig. 3 a, b, c,... . These diagrams include all the ring diagrams of Fig. 2, plus a great many more. Therefore, already at the 3-body level, we are allowing the system to build up long-range correlations if it wants to.

In the 3-hole-line approximation $\mathcal{M}$ is treated only to first order, so that only diagrams 3 a, b are included. In the Bochum truncation $\mathcal{M}$ is included to all orders. The series of Fig. 3 is found to converge quite rapidly,[13] which allows the two methods to give similar numerical results.[14]

The interaction of a particle above the sea with particles in the sea is given by $\mathcal{M}$, which is more complicated than a single-particle potential. The intuitive reason for this is that, as explained above, particle 2 finds itself above the sea <u>only</u> because of correlations with particle 1. This situation is therefore completely different from exciting a <u>single</u> particle above the Fermi sea, as is done in calculating the physical excitation spectrum.

At the 3-body level we face a 3-body equation, which can be solved to an accuracy of about 10%.[13] At the 4-body level we have a 4-body equation, and accurate solutions are not available. The numerical results given below include 4-body terms. These are based on reasonable numerical estimates,[14,15] but the uncertainty is of order 100%.

III. Test Calculations

All our numerical results are calculated with the conventional single-particle spectrum of eq. (8). The first thing to look at is the convergence of the energy. This is shown in Table 1 for the Paris potential at $k_F = 1.6\ \text{fm}^{-1}$, for which $\kappa^{(2)} = 0.127$. The 2, 3, and 4-body contributions to the energy per particle are denoted ε_2, ε_3, ε_4 and are given in MeV. The hole-line results are very similar to those of the Bochum truncation. The uncertainty shown in the total is an estimate of the numerical uncertainty in computing the 3- and 4-body terms. The convergence is good. If it continues the same way in higher orders, their contribution will be small compared to the numerical uncertainty. For other 2-body potentials, the convergence is similar, except that when $\kappa^{(2)}$ is larger, ε_3, ε_4 and the numerical uncertainty are somewhat larger.

Table 1

	Hole-line	Bochum
T_ϕ	31.8	31.8
ε_2	-43.2	-43.5
ε_3	-5.0	-4.8
ε_4	-1.2	-0.6
Total	-17.6±1.3	-17.1±1.3

A second test of the method is to make calculations with two different single-particle spectra.[11] For this we compare results from the conventional spectrum with those from the conventional spectrum modified by taking $U(k>k_F) = \Delta$, where Δ is small and is independent of k. The sensitivity to the spectrum is measured by $d\varepsilon/d\Delta$, where ε is the calculated energy per nucleon. For the Paris potential at $k_F = 1.6\ \mathrm{fm}^{-1}$, we find $d\varepsilon_2/d\Delta = 0.127$, $d(\varepsilon_2+\varepsilon_3)/d\Delta = 0.030$. This is an encouraging result: improving the approximation by including 3-hole-line terms reduces the sensitivity to the single-particle spectrum.

It is important to compare Brueckner-Bethe results with variational ones. For purely central forces, Monte-Carlo variational calculations give reliable upper bounds to the ground-state energy. An important test of the Brueckner-Bethe method is whether it gives a result no higher than this upper bound. For potentials with tensor forces, Monte-Carlo upper bounds are not available. Whether the available variational results are as reliable as the Brueckner-Bethe results is not clear, but a comparison is nevertheless interesting.

For the central potential V_2,[16] the Brueckner-Bethe four-hole-line result BB(4) lies slightly below the Monte-Carlo upper bound.[17] For the V_6 (Reid) potential,[17] which has a strong tensor force, variational and Brueckner-Bethe four-hole-line results are consistent with each other.[17] For the purely central potential V_1[16] at $k_F = 1.8\ \mathrm{fm}^{-1}$, a highly reliable Fermi-hypernetted-chain (FHNC) upper bound is -141.4 MeV,[18] while BB(2) lies 21 MeV higher at -120.5 MeV. Do the BB 3- and 4-hole-line terms make up the difference? A recent evaluation[19] of the three-hole-line term gave only -7 MeV out of the required -21 MeV. However, the method of calculation involved several untested approximations. I have recalculated this term using a method[13] that avoids these untested approximations. The results are BB(3) = -140.7 MeV, BB(4) = -144.7±2.5 MeV, which is consistent with the FHNC upper bound. In summary, there is no discrepancy between Brueckner-Bethe and variational results.

Let us summarize the results of these test calculations. The numerical convergence is good. Possible long-range correlations that are described by ring diagrams are taken into account. The hole-line expansion and Bochum truncation give consistent results. The sensitivity of the calculated energy to small changes away from the conventional single-particle spectrum is satisfactory. The results are consistent with variational calculations. These results taken together strongly suggest that the Brueckner-Bethe method gives reliable ground-state energies. The present accuracy of the calculated saturation point is about ±0.1 fm^{-1} in k_F and ±2 MeV in the energy per particle.[17]

IV. Comparison with Experiment

Let us now apply the method to 2-body potentials that are fitted to NN scattering data and the deuteron. The results are summarized in Fig. 4. The solid circles are saturation points for various potentials, calculated in the BB(2) approximation. They lie on a narrow band called the Coester band that misses the empirical saturation point, which lies inside the box. The point RSC(E) means the Reid[2] potential augmented by interactions[13] in partial waves with j>2 that are consistent with empirical phase shifts. Adding 3- and 4-hole-line terms shifts the Reid and Paris saturation points to higher density and more binding, as indicated by the arrows. The same behavior is expected when 3- and 4-body terms are added for the other 2-body potentials. Therefore, to have any possibility of reaching the empirical region in the box, one should use a 2-body potential that saturates in BB(2) approximation near the HJ (Hamada-Johnston) point. Whether 3- and 4-body terms will actually bring such a potential into the box can only be learned by doing the calculation.

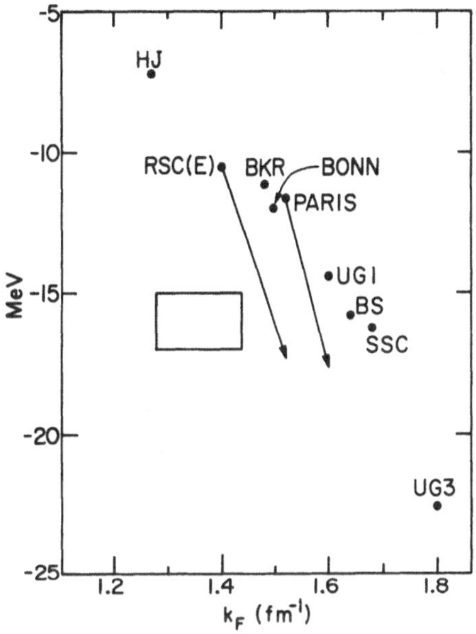

Fig. 4

Even if this happens, we will still be in trouble. That is because accurate calculations by Zabolitzky[7] using the Bochum truncation show that the HJ potential does not account for the saturation properties of the light nuclei ^{4}He, ^{16}O, and ^{40}Ca. In ^{16}O, for example, HJ predicts too small a charge radius and a binding energy of only 4.5 MeV per particle, compared with the experimental value of 8 MeV. The Reid result for ^{16}O is also interesting. It predicts too low an average density and too little binding, while in nuclear matter it predicts too high a saturation density and slightly too much binding. This is an interesting trend with mass number.

Thus the available evidence suggests that no 2-body potential that is fitted to scattering data and deuteron properties can account for nuclear saturation properties. This means that treating the nucleus as a collection of point nucleons interacting through 2-body potentials is inadequate to account for saturation. Some additional physical effect is required, and finding it will be a fascinating and challenging problem.

References

1. T. Hamada and I. D. Johnston, Nucl. Phys. 34, 382 (1962).
2. R. V. Reid, Ann. Phys. (N.Y.) 50, 411 (1968).
3. M. Lacombe, B. Loiseau, J. M. Richard, R. Vinh Mau, J. Côté, P. Pirès, and R. de Tourreil, Phys. Rev. C 21, 861 (1980).
4. K. Holinde and R. Machleidt, Nucl. Phys. A 247, 495 (1975).
5. B. D. Day, Rev. Mod. Phys. 39, 719 (1967).
6. F. Coester, Lectures in Theoretical Physics, Vol. 11, ed. K. T. Mahanthappa (Gordon and Breach, New York, 1969).
7. H. Kümmel, K. H. Lührmann, and J. G. Zabolitzky, Phys. Repts. 36C, 1 (1978).
8. B. D. Day, in Proceedings of the International School of Physics "Enrico Fermi," Varenna, 1980.
9. A. Lejeune and C. Mahaux, Nucl. Phys. A 295, 189 (1978).
10. G. A. Baker, Phys. Rev. C 17, 1253 (1978) and papers cited therein.
11. C. Mahaux, Nucl. Phys. A 163, 299 (1971).
12. B. H. Brandow, Phys. Rev. 152, 863 (1966).
13. B. D. Day, submitted to Phys. Rev. C.
14. B. D. Day and J. G. Zabolitzky, in preparation.
15. B. D. Day, Phys. Rev. 187, 1269 (1969).
16. V. R. Pandharipande, R. B. Wiringa, and B. D. Day, Phys. Lett. B57, 205 (1975).
17. B. D. Day, in The Meson Theory of Nuclear Forces and Nuclear Matter, eds. D. Schütte, K. Holinde, and K. Bleuler (Bibliographisches Institut, Zurich, 1980), p. 1.
18. J. G. Zabolitzky, Phys. Lett. B 64, 233 (1976).
19. P. Grangé and A. Lejeune, Nucl. Phys. A 327, 335 (1979).

Coupled Cluster Description of Relativistic Many Body Systems +)

H. Kümmel

Argonne National Laboratory, Argonne, Illinois, 60439 ++)

I. Introduction

Most of the many body theories we are dealing with assume that there exists a potential between particles from which the forces can be derived. Yet, nature does never provide us with potentials: actually the interaction always is due to the exchange of particles. Thus there are compelling reasons for incorporating particle exchange in many body theory, which implies particle creation and annihilation - and thus relativistic effects. This is a terribly complex problem compared to everything done in non-relativistic many body theory. Thus one would like to find out whether in principle there exists an "effective potential" acting between selected particles. For example, in the meson-nucleon system it would be desirable to have an effective potential acting only between (bare) nucleons, incorporating mesonic degrees of freedom only implicitly inside this potential. We would be willing to pay for this by living with a rather complicated (at least non-local) potential. Indeed, it is rather trivial that such effective operators exist, as long as the physical situation permits it. For instance, the meson-nucleon system below meson threshold can be replaced exactly by a system of many nucleons without mesons. Even relativistic invariance is kept intact - as befits an exact theory. Of course the mesons are hidden inside the potential, and to compute the latter requires the inclusion of mesonic degrees of freedom. Yet, this is a useful procedure for two reasons: first of all this method may suggest practicable and reasonable truncation schemes - as known for instance from the corresponding non-relativistic coupled cluster description. Second, the mere existence of such an effective potential to some extent justifies most of the work done in non-relativistic many body theory, which after all is based on the assumption that a potential exists.

What I have described so far is merely one rather evident reason for doing relativistic many body theory - i.e. quantum field theory (QFT).There is another more technical reason, which is not so evident: We have a lot of experience with approximation methods in many body theory. Some of it should be utilized in QFT. Not surprisingly, I think in terms of the coupled cluster method (CCM) /1,2/. Although there are characteristic differences - mainly related to the self energy and renormalization problem - nonrelativistic many body theory and QFT have much in common. Both use a decomposition intoan infinite set of unphysical (particle-hole or for instance - nucleon-antinucleon-meson) amplitudes. Some of the truncation schemes suggested in CCM and proven extremely successful /1,3/ should at least be tried in QFT.

+) Supported in part by the Deutsche Forschungsgemeinschaft and by a NATO grant.

++) Permanent address: Institut für Theoretische Physik, Ruhr-Universität Bochum.

This paper will describe - necessarily very briefly - both the implementation of (closed and open shell-) CCM in QFT as well as some recent results. More work is in progress.

II. Effective Operators in "Normal" Relativistic Many Body Systems

We assume that perturbation theory can be used if the coupling constant is small. This defines a normal system. Abnormal systems will be discussed briefly in the next chapter. The normal physical vacuum can be written in three forms

$$|\Psi_{vac}\rangle = \frac{U(0,-\infty)|\phi_0\rangle}{\langle\phi_0|U(0,-\infty)|\phi_0\rangle} = (1+F)|\phi_0\rangle = e^S|\phi_0\rangle, \tag{2.1}$$

where $|\phi_0\rangle$=bare vacuum. $U(0,-\infty)$ is the time-evolution operator. The first form is the well known Gell-Mann-Low /4/ wave function. Taking the meson-nucleon system as a prototype, the wave function in this description is a superposition of diagrams of Fig. 1,

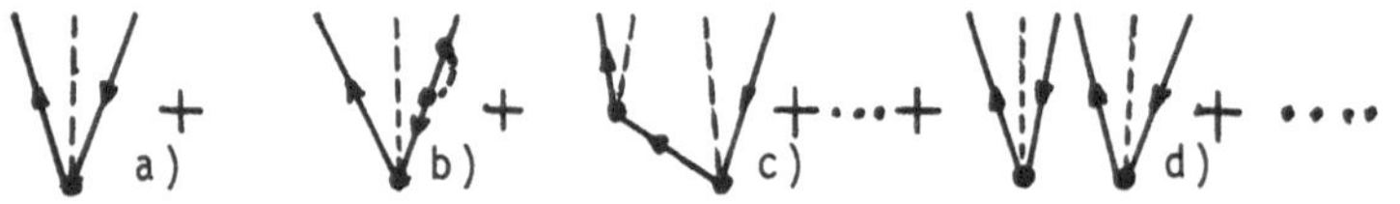

Fig.1

i.e. F contains $(N\bar{N})$-pairs and mesons. No vacuum-vacuum diagrams like in Fig. 2 occur. On the other hand, "unlinked" terms of the type of Fig.(1d) show up. It is well known that S in the last form of (1) does not contain any such unlinked terms, i.e. S is the sum of terms of Fig.1 without the unlinked ones.

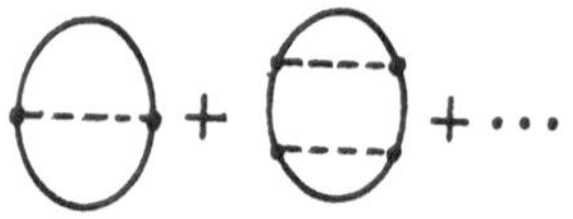

Fig.2

Thus, one could follow the standard CCM procedure by writing the Schrödinger equation for the four momentum P_μ in the form

$$\exp(-S)\, P_\mu \exp(S)|\phi_0\rangle = \Pi_\mu^{vac}|\phi_0\rangle \tag{2.2}$$

and projecting onto $|\phi_0\rangle$ and states with one-, two-...$(N\bar{N})$pairs and mesons, thereby obtaining equations for S. If there would be no renormalization problem, this would be a perfectly respectable way to calculate the vacuum. Of course $\Pi_\mu^{vac} = 0$ is required in a correct theory. Fortunately, in some QFT's one can apply a trick, the light-front dynamics /5/. In this representation instead of defining dynamical operators like P_μ as integrals over the volume for t=const., one integrates over the light front $x_\nu x_\nu$=0. One consequence is that the new creation and annihilation operators are such that the physical and bare vacuum are the same; i.e. one has <u>solved exactly the vacuum problem.</u> So, sometimes we have S=0 and life is much easier.

Going over to the one nucleon problem we closely follow the receipe of nonrelativistic CCM for open shell systems /2/: The wave function is written as

$$|\Psi_p\rangle = \exp(S)(1+F^{(1)})\, a^+(p)|\phi_0\rangle \tag{2.3}$$

(where $a^+(p)$ creates a nucleon with momentum p). S is "known" and $F^{(1)}$ is the new amplitude to be computed. It creates $(N\bar{N})$-pairs and mesons and changes the nucleon with momentum p into another one with momentum p'. If we define $P_0^{(1)}=\int d^3p\,|\phi_p\rangle\langle\phi_p|$ with $|\phi_p\rangle = a^+(p)|\phi_0\rangle$, $Q_0^{(1)}=1-P_0^{(1)}$, then $F^{(1)}=Q_0^{(1)}F^{(1)}P_0^{(1)}$. The Schrödinger equation for the energy is used in the form

$$e^{-S} H e^{S} (1+F^{(1)})|\phi_p\rangle = E_p(1+F^{(1)})|\phi_p\rangle. \tag{2.4}$$

We have similar equations for the momenta. By projecting onto $P_0^{(1)}$ and $Q_0^{(2)}$ we obtain two sets of equations. The "secular equation"

$$\langle p|H_{eff}^{(1)}|p'\rangle = \delta^3(p-p')\left(E_p - E_{vac}\right) \tag{2.5}$$

and the "equation for $F^{(1)}$"

$$0=\langle Q_0^{(1)}|\{e^{-S}H e^{S}(1+F^{(1)})a^+(p)\}_{\mathcal{L}}|\phi_0\rangle - \int d^3p'\langle Q_0^{(1)}|F^{(1)}|\phi_{p'}\rangle\langle p'|H_{eff}^{(1)}|p\rangle. \tag{2.6}$$

Here the "effective operator"

$$\langle\phi_p|\{e^{-S}H e^{S}(1+F^{(1)})a^+(p')\}_{\mathcal{L}}|\phi_0\rangle = \langle p|H_{eff}^{(1)}|p'\rangle \tag{2.7}$$

acts only in the "model space" of the bare nucleon and for the one nucleon case must be diagonal in the three momentum. $\{\ \}_{\mathcal{L}}$ means that all operators are linked via contractions. This symbol occurs, since the unlinked parts reproduce the eigenvalue equation for the vacuum and this has been subtracted out. Solving (5) and (6) by iteration, the well known linked valence folded diagram expansion is obtained. Again, without the renormalization problem the solution could be obtained using the truncation scheme of CCM, including everything up to a certain number of ($N\bar{N}$) pairs and mesons and solving the resulting coupled set of equations for the amplitudes in $F^{(1)}$. In this way the mass could be computed. The need for renormalization in most field theories to some extent spoils this program.In any case the renormalization constants must be chosen such that the experimental mass results. One example will be discussed later.

It is straight forward to go on to the two nucleon case. As in the two valence particle case /2/ we put

$$|\Psi_\alpha\rangle = \int d^3p_1\int d^3p_2\, e^{S}(1+F^{(1)}+\tfrac{1}{2}:F^{(1)2}:+F^{(2)})\,a^+(p_1)a^+(p_2)|\phi_0\rangle\, c(\alpha:p_1 p_2), \tag{2.8}$$

where now S and $F^{(1)}$ are "known". This ansatz incorporates what we know from the one particle problem and from the vacuum; i.e. the "dressing" of the individual nucleons is taken into account. $F^{(2)}$ takes the lesser burden of carrying only those corrections which are due to the interaction between the two nucleons. $F^{(2)}$ does not involve scattering boundary conditions, if all two nucleon components occuring in (2.8) are linearly independent., see /6/ . The factor $\frac{1}{2}$ takes care that we don't count the pair of dressed nucleons twice. The normal product has been introduced to remove contractions between the $F^{(1)}$, i.e. to obtain a true independent motion of this pair such that all interaction is contained in $F^{(2)}$. We note that (8) (as well as (3)) are completely general, if we assume that the exact wave function has a bare nucleon component. Defining again the projection into the "model space" by $P_0^{(2)} = \int d^3p \int d^3p' |\Phi_{pp'}\rangle\langle\Phi_{pp'}|$ with $|\Phi_{pp'}\rangle = a^+(p)a^+(p')|\Phi_0\rangle$, projecting onto $P_0^{(2)}$ we obtain a "secular equation" in the space of bare nucleons

$$\int d^3p_1' \int d^3p_2' \left[\langle p_1 p_2 | H_{eff}^{(2)} | p_1' p_2' \rangle + \langle p_1 | H_{eff}^{(1)} | p_1' \rangle \delta^3(p_2 - p_2') + \langle p_2 | H_{eff}^{(1)} | p_2' \rangle \delta^3(p_1 - p_1') \right] \times \tag{2.9}$$
$$\times\, C(\alpha : p_1' p_2') = \left(E^\alpha - E_{vac}\right) C(\alpha : p_1 p_2)$$

and an "equation for $F^{(2)}$" of a structure similar to (6), which I don't write down. Here

$$\langle p_1 p_2 | H_{eff}^{(2)} | p_1' p_2' \rangle = \langle \Phi_{p_1 p_2} | \{ e^{-S} H e^{S} (1 + F^{(1)} + \tfrac{1}{2} :F^{(1)2}: + F^{(2)}) a^+(p_1) a^+(p_2) \}_{\mathcal{L}} | \Phi_0 \rangle \tag{2.10}$$

is the "true" effective two body operator. The unlinked parts, extracted as before, produce the two one body operators on the left hand side of (9). Thus, in this equation in the "model space" of two bare nucleons the total effective Hamiltonian is $\hat{H}_{eff}^{(2)} = H_{eff}^{(2)}(1,2) + H_{eff}^{(1)}(1) + H_{eff}^{(1)}(2)$. We call this the "decomposition theorem": the two body effective operator contains a two body part + one body parts, the latter being known from the one body problem.

It is remarkable - and well known from degenerate CCM /2/ - that this feature persists to all particle numbers. Without writing down any further formulae here, we just state the general structure: for the N nucleon problem the effective four momentum operator is

$$\hat{P}_{\mu\, eff}^{(N)} = P_{\mu\, eff}^{(N)} + \sum P_{\mu\, eff}^{(N-1)} + \cdots + \sum P_{\mu\, eff}^{(1)} \tag{2.11}$$

Here $P_{\mu\, eff}^{(n)}$ for $n < N$ is determined by the n particle problem. Note that $\hat{P}_{\mu\, eff}^{(n)}$ and $P_{\mu\, eff}^{(n)}$ are not hermitian.

There exist hermitian versions, too /7/, but for them there is no decomposition corresponding to (11). For instance, the hermitian three body operator contains two body operators depending on the center of mass momentum: the third particle "feels" the two other ones due to the c.m. motion. It is no true two body operator.

Some final remarks: The Schrödinger equations can be solved for all four momenta simultaneously: since P_μ commute with the mass operator $M^2 = P_\mu P_\mu$, we have to consider the mass spectrum of e.g. the meson-nucleon system. Below meson threshold the eigenvalue of M will be the mass of N interacting nucleons. Then no states except the N nucleon states themselves need to be considered.

If in addition all N nucleon states are of the general form

$$\Psi_N = \int d^3p_1 .. \int d^3p_N \, (1 + F) | \Phi_{p_1 .. p_N} \rangle C(\alpha : p_1 .. p_N) \tag{2.12}$$

the set of equations given above is equivalent to the original Schrödinger equation and effective operators exist. If the wave functions are not of the form (12), there will be trouble. Since (12) is completely general if Ψ_N has a bare N nucleon component, the method may fail only if some or all states Ψ_N do not have such a component. Perturbational methods assume the form(12) and we expect it to be more general. But for coherent states this might be wrong.

Of course, above meson threshold states with physical mesons will show up and mix with the states without mesons. In this case there will be no effective operators for nucleons alone.

III. Coherent Vacua

In the last years solutions which never can be obtained by expansions in powers of the coupling constant have become very important. Already the physical vacua may be of this kind. I discuss here the famous Goldstone-Boson serving as a prototype /8/. The equation of motion is

$$(\partial_t^2-\partial_x^2)\chi(x,t)-m^2\chi(x,t)+\lambda\chi^3(x,t)=0 \tag{3.1}$$

It is a one time-one space dimensional field theory for the Boson field χ. The Lagrangian generating this equation of motion will be written down in two forms

$$\begin{aligned}\mathcal{L} &= \tfrac{1}{2}(\partial_t\chi)^2-\tfrac{1}{2}(\partial_x\chi)^2+\tfrac{1}{2}m^2\chi^2-\lambda/_4\,\chi^4\\ &= \tfrac{1}{2}(\partial_t\chi)^2-\tfrac{1}{2}(\partial_x\chi)^2-\tfrac{1}{2}M^2\chi^2-\tfrac{1}{2}(m^2+M^2)\chi^2-\lambda/_4\,\chi^4=\mathcal{L}_0+\mathcal{L}_{int},\end{aligned} \tag{3.2}$$

with a new artifical mass parameter M. Note that in (1) and (2) the mass term with m^2 occurs with the wrong sign.The mass is imaginary. Looking for classical constant solutions of the field equation one finds that there are three of them

$$\chi_0=0 \quad , \quad \chi_\pm=\pm\frac{m}{\sqrt{\lambda}} . \tag{3.3}$$

(One also may find stationary "soliton"-solutions: we don't discuss them here, however, although they are of great interest).

The energies going with it are $E_0=0$ and $E_\pm=\frac{m^2}{4\lambda}L$, (where L=normalization volume). Thus we may plot the energy as a function of χ symbolically as in fig. 3.

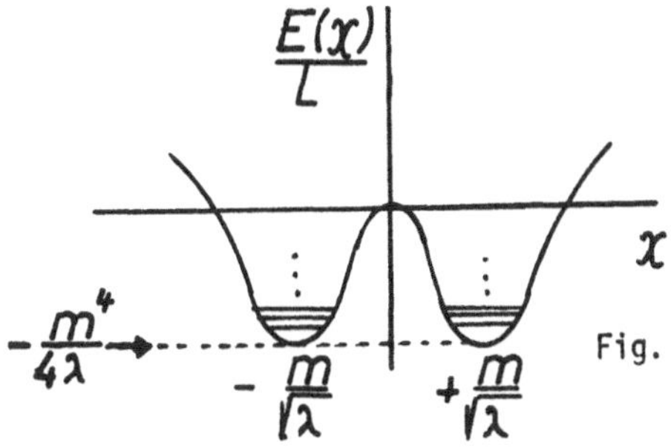

Fig. 3

Since the two minima are the solutions with lowest possible energy, they must be the vacua. They are coherent states since according to (3) they cannot be obtained by a power expansion in terms of the coupling constant λ.

Several questions immediately come into our mind. For instance, can we obtain the same result using quantum mechanics?

The answer is yes. Even CCM is helpful. However, the usual CCM using the Schrödinger equation in the form (2.2) and projecting onto a complete set of states will not work. We have some experience with such situations which tells us that a variational procedure

$$\delta\langle H\rangle = \delta\left[\frac{\langle e^S\phi_0|H|e^S\phi_0\rangle}{\langle e^S\phi_0|e^S\phi_0\rangle}\right] = 0 \tag{3.4}$$

is needed. Of course (4) and (2.2) are equivalent as long as no approximations are made. However, any truncation $S_n=0$ for $n>n_0$ will lead to different results and (4) then contains much more terms than the standard form. This is why (4) is more powerful (although much more complicated than (2.2)).
In lowest order we may try as eigenstate to zero momentum

$$|\psi\rangle = e^{S_1}|\phi_0\rangle \tag{3.5}$$

with

$$S_1 = S_1(0)\,a^+(0)\,, \tag{3.6}$$

where $a^+(o)$ creates a particle with zero momentum. But which particle? The only particle occurring in the Lagrangian has imaginary mass and therefore cannot be used as basis. Here the second form in (2) for $\mathcal{L}$ comes into play: we introduce a real mass M, which defines the basic set of creation and annihilation operators $a^+(p)$ and $a(p)$. Thus (5) and (6) are well defined by now.
Performing the variation with respect to S_1 we obtain two states

$$|\psi^\pm_{vac}\rangle \quad \text{with} \quad S_1^\pm(0) = \pm\frac{Mm^2L}{2\lambda}$$

and with

$$\langle\chi\rangle_\pm \equiv \frac{\langle\psi^\pm_{vac}|\chi|\psi^\pm_{vac}\rangle}{\langle\psi^\pm_{vac}|\psi^\pm_{vac}\rangle} = \pm\frac{m}{\sqrt{\lambda}}\,, \quad E^\pm_{vac} \equiv \langle H\rangle_\pm = -\frac{m^4}{4\lambda}L\,, \tag{3.7}$$

i.e. the two "classical" vacua. The physical quantities do not depend on the parameter M. (We remark that ψ^+_{vac} and ψ^-_{vac} are mutually orthogonal as $L\to\infty$).

The lesson to be learned from these considerations is that CCM is able to deal with coherent states as well as with normal ones. Although one has to be cautious concerning the generality of this result, there might be more realistic cases where it can be used as well.

Returning to the Goldstone Boson, we must admit that the theory in this form is not very satisfactory. First of all I have cheated by ignoring some ultraviolet divergent terms. This can be remedied by replacing all operator products in $\mathcal{L}$ by their normal products, e.g. $:\chi^2:$. But even then $E^\pm_{vac}\neq 0$, which is a very undesirable feature. In addition, the occurrence of the artificial parameter M has to be understood. Both questions will be settled in the next section, where the excited states ("mesons") around the two minima of Fig.3 are considered.

IV. One Particle States

We study once again the Goldstone-Boson to illustrate the method proposed in the second section.

We first use $\mathcal{L}$ in the standard form given in (3.2). The vacua (3.7) were at most rough approximations. We may construct approximate eigenstates with momentum $p \neq 0$

$$|\psi_p^{\pm}\rangle = a^+(p)|\psi_{vac}^{\pm}\rangle . \qquad (4.1)$$

Then we may optimize these states by chosing M such that

$$\frac{\partial}{\partial M}\left[\langle H\rangle_p^{\pm} - \langle H\rangle_{\pm}\right] = 0 . \qquad (4.2)$$

This can be done exactly and yields for small p for both $|\psi_p^+\rangle$ and $|\psi_p^-\rangle$

$$M = \sqrt{2}\, m \quad \text{with} \quad \langle H\rangle_p^{\pm} = \sqrt{p^2+M^2} \qquad (4.3)$$

as best value for M. Thus we have the desired interpretation of M as mass of one "meson": Both minima of Fig. 3 allow for "excited states" or "states of a certain number of mesons" of the form

$$\prod a_{\pm}^+(p_i)|\psi_{vac}^{\pm}\rangle . \qquad (4.4)$$

This can be only a very approximate description, especially since the objections raised against these approximations for the vacuum states still persist.

One can do much better by the following (well known) trick. Replace in $\mathcal{L}$ from (3.2) χ by

$$\chi'_{\pm} = \chi \pm \frac{m}{\sqrt{\lambda}} , \qquad (4.5)$$

as suggested by (3.7), and obtain a completely respectable new Lagrangian

$$\mathcal{L}'_{\pm} = \underbrace{\tfrac{1}{2}:\partial_t\chi'^2: - \tfrac{1}{2}:\partial_x\chi'^2: - \tfrac{1}{2}M^2:\chi'^2:}_{\mathcal{L}'_0} \underbrace{- \tfrac{\lambda}{4}:\chi'^4: \mp \sqrt{\tfrac{\lambda}{2}}M:\chi'^3:}_{\mathcal{L}'_{int}} , \qquad (4.6)$$

with $M = \sqrt{2}\, m$, if all constants are ignored. Note that the equations of motion of χ and χ' are the same, but that the physical observables from $\mathcal{L}'$ now differ by constants from the ones obtained from $\mathcal{L}$: indeed, the vacua now have classical four momenta (and therefore: mass) equal to zero, as it should be. Thus, for differences of physical quantities $\mathcal{L}'_{\pm}$ are equivalent to $\mathcal{L}$. However, technically there is a great change, since due to the subtraction of $\pm \frac{m}{\sqrt{\lambda}}$ the coherence has been removed. Furthermore, for the states built around ψ^+_{vac} or ψ^-_{vac} the corresponding $\mathcal{L}'_+$ or $\mathcal{L}'_-$ are the convenient choices (leading to two completely separated Hilbert spaces spanned by the (+) or (-) states /8/).

Studying $\mathcal{L}'$ we see that $\mathcal{L}'_0$ is the Lagrangian of free mesons with mass M and eigenstates $\prod a^+(p_i)|\phi_0\rangle$ with energy $E = \sum_i \sqrt{p_i^2+M^2}$. This is the starting point to be used in searching for exact eigenvalues and eigenstates, including now the new interaction

The great advantage of using $\mathcal{L}_{\pm}'$ instead of $\mathcal{L}$ is that now in the absence of coherence the light front dynamics (LFD) can be used. LFD is standard in QFT and of no interest to many body theorists. Thus I don't give any details. It suffices to say that we know the exact physical vacuum with S=0. Four momentum and mass vanish. Thus we may go on trying to compute the one meson state, including now the interaction terms. This will change the eigenvalue $\hat{M}^2$ of $P_\mu P_\mu$ as a function of the coupling constant ($\hat{M}(\lambda=0)=M$). The techniques in principle have been described in the second section: we have to determine $F^{(1)}=\sum F_n^{(1)}$. Remember, $F^{(1)}$ produces states different from the one meson states. It is $F_0^{(1)}=F_1^{(1)}=0$ and we have to determine $F_2^{(1)}$, $F_3^{(1)}$, ... by using the projected equation (2.6). By using LFD we have a substantially reduced number of terms and it is rather easy to keep track of the relativistic invariance consistent with the approximations made by neglecting $F_n^{(1)}$ for $n>n_0$. I have performed a SUB(2) calculation in the language of CCM ($F_2^{(1)}$ only), a SUB(3) calculation ($F_2^{(1)}$ and $F_3^{(1)}$) and a first and second order perturbation theory for comparison. The resulting masses are given in Fig. 4.

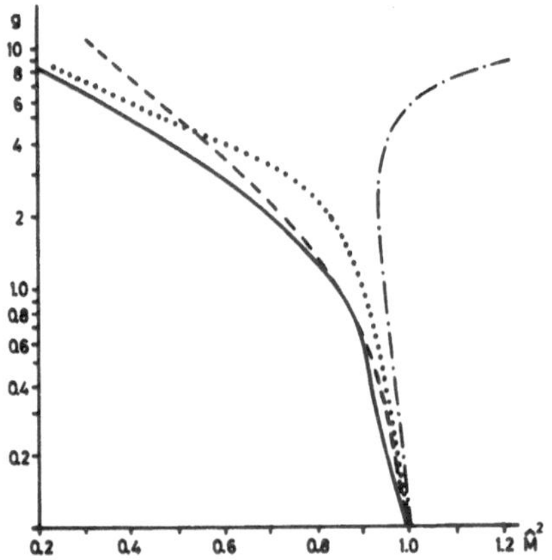

Fig. 4

Here the solid/dashed line correspond to SUB(3)/SUB(2), whereas the dotted/dashed-dotted line correspond to first/second order perturbation theory. Up to the rather large coupling constant $g=\frac{\lambda}{M^2}\approx 8$ the convergence is quite good, anyway much better than in perturbation theory. This is not surprising since a very large set of self energy diagrams has been summed up, see Fig. 5.

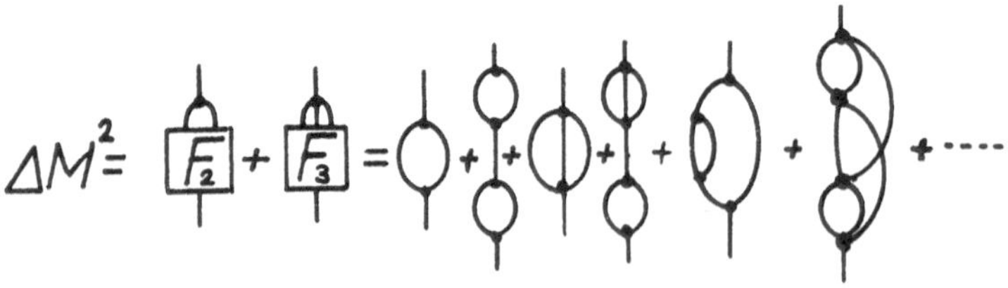

Fig. 5

This result is quite encouraging. We expect a similar quality for the two meson system: due to the "work done" on the one nucleon problem, we have already performed a lot of partial summations inside the two meson mass diagrams, see fig. 6, where some contributions to $F^{(2)}$ due to $F^{(1)}$ are shown.

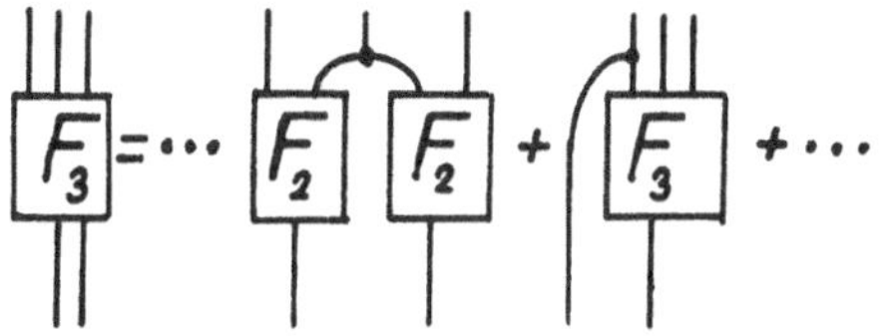

Fig. 6

V. Summary and Conclusions

We have seen that the concept of effective operators and the approximation scheme of CCM both work well in QFT's. At least in the simple model studied here neither coherence nor the renormalization posed any problems. For realistic QFT's the latter certainly is a major nuisance, however: The known renormalization procedures are tied to power series in the coupling constant, whereas CCM is not. I believe, however, that by investigating more realistic super-renormalizable QFT's one finally will be able to develop the techniques needed here.

Acknowledgment

The author wishes to thank the Argonne National Laboratory for the hospitality extended to him, H. Lee, J. Parmentola and especially F. Coester for many discussions and critical remarks. For the progress of this work discussions with the Quantum Theory Project at the University of Florida in Gainesville were very important. This was made possible by a NATO grant.

References

/1/ For a review: H. Kümmel, K.H. Lührmann, J.G. Zabolitzky Phys. Reports 36C (1978) 1
/2/ For open shell systems: R. Offermann, H. Kümmel, W. Ey, Nucl.Phys. A273 (1976) 349; W. Ey, Nucl. Phys. A296 (1978) 189
/3/ J.G. Zabolitzky, W. Ey, Nucl.Phys. A328 (1979) 5o7; K. Emrich, J.G. Zabolitzky, Nucl. Phys., in print
/4/ Gell-Mann, F. Low, Phys.Rev. 84 (1951) 35
/5/ For instance: S.J. Chang, R.G. Root, T.M. Yan, Phys.Rev. 7D (1973) 1133
/6/ H. Kümmel, submitted to Phys.Rev. D.
/7/ B. Brandow, Rev.Mod.Phys. 39 (1967) 771
/8/ R. Jackiw, Rev.Mod.Phys. 49 (1977) 681

VARIATIONAL MATRIX PADE APPROXIMANTS
APPLIED TO FEW BODY PROBLEMS

L.P. Benofy and J.L. Gammel

Department of Physics, Saint Louis University
St. Louis, Missouri 63103

1. Introduction

When contemplating the many body problems of quantum mechanics, one has in mind not only the by now classical nuclear matter problem and the problems associated with other very large fermion and boson systems, but also the problem of finite nuclei and small and large atoms and molecules. First steps in the application of the method described in the present paper to such problems, specifically to the calculation of the energy levels of the He atom and the hydrogen molecule, to the improvement of calculations including electron correlation in the Hartree-Fock method, and to the three body problem of nuclear physics have been started, principally by us (with the assistance of E. Bernardi) and by Professor G. Turchetti and others in Bologna. In the present paper, we restrict ourselves to a simple three body problem, actually two interacting particles confined to a box in one dimension. This problem is analogous to the He or H_2 problems in which the Coulomb attraction of the electrons to the nuclei serves as the box, and the interaction of the two particles in the box is the Coulomb repulsion of the two electrons.

2. Generalized Brillouin-Wigner Perturbation Theory

The Brillouin-Wigner perturbation series[1,2] for the energy E of a bound quantum mechanical system is

$$E = E_p^{(0)} + (p|V|p) + \sum_k (p|V|k) \frac{1}{E - E_k} (k|V|p) + \dots \quad , \qquad (1)$$

where $|p)$ is an energy eigenstate associated with an unperturbed Hamiltonian H_o. $H = H_o + V$, where V is the perturbation, is the Hamiltonian of the system. $E_p^{(0)}$ is the eigenvalue of H_o associated with $|p)$, that is, $H_o|p) = E_p^{(0)}|p)$. As is well known Eq. (1) is not an explicit expression for E because E appears on both sides of the equation. We call the E on the left E_{out} for "output" and the E on the right side E_{in} for "input". If one sums the right hand side with some E_{in}, he obtains E_{out}. When $E_{out} = E_{in}$, the result is said to be self-consistent, and then $E = E_{out} = E_{in}$.

We now generalize Eq. (1) by making E_{out} a matrix,

$$(a|E_{out}|b) = (a|V|b) + \sum_k (a|V|k) \frac{1}{E_{in} - E_k} (k|V|b) + \dots \quad , \qquad (2)$$

where the set of functions from which $|a)$ and $|b)$ are selected need not be eigenfunctions of H_o, but ought to include $|p)$ because our original E_{out} is

$$E_{out} = E_p^{(0)} + (p|E_{out}|p) \quad . \tag{3}$$

The $E_p^{(0)}$ has to be added because we have omitted the $E_p^{(0)}$ in Eq. (2).

It is possible to sum the right hand side of Eq. (2) exactly,

$$(a|E_{out}|b) = (a|V|b) + \sum_k (a|V|k) \frac{1}{E_{in} - E_k} (k|E_{out}|b) \quad . \tag{4}$$

If we define a Möller[6] wave matrix ψ_b,

$$E_{out}|b) = V\psi_b \quad , \tag{5}$$

we find

$$\psi_b = |b) + \sum_k |k) \frac{1}{E_{in} - E_k} (k|V|\psi_b) \quad . \tag{6}$$

3. The Schwinger Variational Principle

Equation (6) follows from a variational principle,

$$\delta_\phi I_{ab} = 0 \quad , \tag{7}$$

where

$$\begin{aligned} I_{ab} = {} & (\phi|V|b) + (a|V|\psi) - (\phi|V|\psi) \\ & + \sum_k (\phi|V|k) \frac{1}{E_{in} - E_k} (k|V|\psi) \quad . \end{aligned} \tag{8}$$

Proof:

$$\delta_\phi I_{ab} = (\delta\phi|V|b) - (\delta\phi|V|\psi) + \sum_k (\delta\phi|V|k) \frac{1}{E_{in} - E_k} (k|V|\psi) = 0 \quad , \tag{9}$$

but because $\delta\phi$ is arbitrary

$$V|b) - V|\psi) + \sum_k V|k) \frac{1}{E_{in} - E_k} (k|V|\psi) = 0 \quad , \tag{10}$$

and multiplying through by V^{-1} yields Eq. (6). Q.E.D.

The conjugate equation follows from

$$\delta_\psi I_{ab} = 0 \tag{11}$$

and yields nothing new.

4. The Cini-Fubini Ansatz

We take

$$|\phi^*) = |\psi) = c_p|p) + c_q|q) \quad , \tag{12}$$

where c_p and c_q are variational parameters and $|q)$ is an <u>arbitrary</u> function (not necessarily an eigenfunction of H_o!) which, however, we will not vary for the moment. We then find

$$\begin{aligned} I_{pp} &= c_p(p|V|p) + c_q(p|V|q) + c_p^*(p|V|p) + c_q^*(q|V|p) \\ &+ (c_p^*p + c_q^*q| - V + \sum_k V|k) \frac{1}{E_{in} - E_k} (k|V|c_pp + c_qq) \quad , \end{aligned} \tag{13}$$

so that

$$\begin{aligned} \frac{\partial I_{pp}}{\partial c_p} &= (p|V|p) - c_p^*[(p|V|p) - (p|V_2|p)] \\ &\quad - c_q^*[(q|V|p) - (q|V_2|p)] = 0 \quad , \\ \frac{\partial I_{pp}}{\partial c_q} &= (p|V|q) - c_p^*[(p|V|q) - (p|V_2|q)] \\ &\quad - c_q^*[(q|V|q) - (q|V_2|q)] = 0 \quad , \end{aligned} \tag{14}$$

where

$$(p|V_2|q) = \sum_k (p|V|k) \frac{1}{E_{in} - E_k} (k|V|q) \quad . \tag{15}$$

That is,

$$\sum_{\ell=p,q} c_\ell^*(\ell|V - V_2|m) = (p|V|m) \quad , \quad m = p \text{ or } q \quad , \tag{16}$$

the solution of which is

$$c_\ell^* = (p|V \times \frac{1}{V - V_2} |\ell) \quad , \tag{17}$$

where x means multiplication in a 2 x 2 space. The reciprocal $(V - V_2)^{-1}$ is computed in the same space.

Similarly,

$$c_\ell = (\ell| \frac{1}{V - V_2} \times V|p) \quad . \tag{18}$$

Thus

$$|\psi) = \sum_{\ell=p,q} |\ell) (\ell| \frac{1}{V - V_2} \times V|p) \quad , \tag{19}$$

and

$$(p|V|\psi) = (p|V \times \frac{1}{V - V_2} \times V|p) \quad ,$$
$$= (p|E_{out}|p) \quad . \tag{20}$$

5. Matrix Padé Approximations

Consider a matrix function F,

$$F = \begin{pmatrix} F_{11} & F_{12} \\ & \\ F_{21} & F_{22} \end{pmatrix} \quad , \tag{21}$$

such that each element of F has a power series expansion in a parameter z,

$$F(z) = zF^{(1)} + z^2 F^{(2)} + \ldots \quad . \tag{22}$$

The (J/J) matrix Padé approximant to F is

$$(J/J) = (N_1 z + N_2 z^2 + \ldots N_J z^J) \frac{1}{I + D_1 z + \ldots + D_J z^J} \quad , \tag{23}$$

where $N_1, \ldots, N_J, D_1, \ldots, D_J$ are matrices such that if the right hand side of Eq. (23) is expanded in a power series in z this expansion agrees with Eq. (22) through order 2J. As is well known, this results in a set of linear equations for the matrices N and D.[3]

Equation (20) is the (p|p) element of the (1/1) matrix Padé approximant to the series obtained by iterating Eq. (4). |a) and |b) are selected from the two functions |p) and |q).

It must be emphasized that in calculating V_2, from Eq. (15), k runs over a complete set of eigenfunctions of H_o.

These results are far more general than we prove here.[7] If one takes the next Cini-Fubini Ansatz,

$$|\psi) = c_p|p) + c_p' \sum_k |k) \frac{1}{E_{in} - E_k} (k|V|p)$$
$$+ c_q|q) + c_q' \sum_k |k) \frac{1}{E_{in} - E_k} (k|V|q) \quad , \tag{24}$$

one obtains from the Schwinger principle the (2/2) matrix Padé approximant to the series obtained by iterating Eq. (4), again selecting |a) and |b) from the two functions |p) and |q). In the iteration process, all sums over intermediate states must be performed exactly; that is, all sums over k must be over a complete set of eigenfunctions of H_o. But (to say it once more) |q) is <u>not</u> an eigenfunction of H_o. |p) is an eigenfunction of H_o, otherwise we could not calculate E_{out} from Eq. (3).

6. The Method

We come now to the heart of the method. A most extraordinary thing happens; namely, that with ψ given by Eq. (19),

$$I_{pp} = (p|E_{out}|p) \tag{25}$$

Proof:

$$\begin{aligned} I_{pp} &= \sum_{\ell=p,q} \{(p|V|\ell)c_\ell + c_\ell^*(\ell|V|p)\} \\ &\quad - \sum_{\ell=p,q}\sum_{m=p,q} c^*(\ell|V - V_2|m)c_m \quad , \end{aligned} \tag{26}$$

$$\begin{aligned} I_{pp} &= \sum_{\ell=p,q} (p|V|\ell)(\ell|\ \frac{1}{V - V_2} \times V|p) \\ &\quad + \sum_{\ell=p,q} (p|V \times \frac{1}{V - V_2}\ |\ell)(\ell|V|p) \\ &\quad - \sum_{\ell=p,q}\sum_{m=p,q} (p|V \times \frac{1}{V-V_2}\ |\ell)(\ell|V-V_2|m)(m|\ \frac{1}{V-V_2} \times V|p) \quad , \\ &= 2(p|V \times \frac{1}{V-V_2} \times V|p) - (p|V \times \frac{1}{V-V_2} \times (V-V_2) \times \frac{1}{V-V_2} \times V|p) \\ &= (p|V \times \frac{1}{V - V_2} \times V|p) \quad . \\ &= (p|E_{out}|p) \end{aligned} \tag{27}$$

This means that we may require $(p|E_{out}|p)$ to be stationary with respect to variations in any parameters occurring in the arbitrary function $|q)$.

7. Results

We have written about this method before.[4] However, we had not realized, even so short a time ago, the full power of the method. In this earlier work, we chose $|q)$ from the set of eigenfunctions of H_o.

We treat the same problem as we did on p. 351-353 of ref. 4, but we take

$$|q) = \frac{1}{2}\left(e^{-\gamma(x_1-x_2)} + e^{-\gamma(2L-x_1-x_2)}\right)e^{-\alpha|x_1-x_2|}\ |p) \quad , \tag{28}$$

where

$$|p) = \frac{2}{L} \sin \frac{\pi x_1}{L} \sin \frac{\pi x_2}{L} \quad . \tag{29}$$

$|q)$ has to be symmetrical with respect to interchange of x_1 and x_2 and with respect to reflection of $(x_1 + x_2)$ through the center of the box which is at $x = L/2$.

At first, we supposed the correlation between the two particles could be described by $\exp(-\alpha|x_1 - x_2|)$; that is, by a function of their relative coordinate only. This turned out not to be the case, and this result is of considerable interest with regard to applications to electron correlation in atomic structure.

The ordinary Padé approximant (that is, a one-dimensional matrix Padé approximant based on the state $|p)$ only) yields

$$E = -2.0544 \text{ MeV} \quad , \tag{30}$$

for a box size of 30F (the potential we used is described in ref. 4). From the calculations of ref. 4 it is certain that the correct value is $E = -2.0704$ MeV. Without γ, we find

$$E = -2.0625 \text{ MeV} \ , \quad \alpha L = 0.0125 \ , \quad \gamma L = 0 \ . \tag{31}$$

With γ, we find

$$E = -2.0698 \text{ MeV} \ , \quad \alpha L = 2.25 \quad \gamma L = 1.191 \tag{32}$$

8. Conclusion

The method is quite accurate with simple trial functions, but is it necessary that the effect on the center-of-mass motion of two-particle correlation be included.

One verifies the importance of the center-of-mass motion by trying a three (or more) state calculation with correlation functions depending on the relative coordinate, namely,

$$|q_1) = e^{-\alpha|x_1-x_2|} \sin \frac{\pi x_1}{L} \sin \frac{\pi x_2}{L} \ ,$$
$$|q_2) = e^{-\beta|x_1-x_2|} \sin \frac{2\pi x_1}{L} \sin \frac{2\pi x_2}{L} \ , \tag{33}$$

and finding that he still does not get as accurate an energy as we have shown in Eq. (32).

We close with a brief summary of the advantages of this method.

1) It is never necessary to compute what the kinetic energy operator[5] operating on our trial function is. This result is hidden in the calculation of the

second term in the Brillouin-Wigner series,

$$\sum_k (p|V|k) \frac{1}{E_{in} - E_k} (k|V|q) \quad ,$$

but never requires the calculation of a derivative of any sort of the trial function q).

2) The normalization[5] of |q) is irrelevant since c_q absorbs this factor.

3) The quantity ultimately made stationary is E_{output} itself, not the Schwinger I. In calculations not based on the Cini-Fubini Ansatz (that is, not based on Padé approximants) one has had the awkward feature that he makes I stationary when he wants a different quantity, namely, E.

9. References

1. Morse and Feshbach, Methods of theoretical physics (McGraw-Hill, New York, 1953) v. II, p. 1008 et seq.

2. R. C. Young, L. C. Biedenharn, and E. Feenberg, Phys. Rev. 106 (1957) 1151.

3. G. A. Baker, Jr., Essentials of Padé approximants (Academic Press, New York, 1975). See especially chap. 21 on matrix Padé approximants.

4. Padé and rational approximation, edited by E. B. Saff and R. S. Varga (Academic Press, New York, 1977) p. 339.

5. Compare Clark and Westhuas, Phys. Rev. 141 (1966) 833.

6. C. Möller, Kgl. Danske Videnskab. Selskab, Nat.-fys. Medd. 23 (1945) I.

7. For the general case, see J. Nuttall, The connection of Padé Approximants with stationary variational principles and the convergence of certain Padé approximants in The Padé approximant in theoretical physics, edited by G. A. Baker, Jr. and J. L. Gammel (Academic Press, New York, 1970).

NUCLEAR MATTER AND NUCLEAR HAMILTONIAN†

V. R. Pandharipande

Department of Physics
University of Illinois at Urbana-Champaign
Urbana, IL 61801

The primary objective of the many body theory of nuclei and neutron stars is to explain and predict properties of these systems from a single nuclear hamiltonian. In principle it should be possible to obtain this hamiltonian from the meson theory[1)] by eliminating all but the nucleon degrees of freedom. However our understanding of the nucleon interactions from this point of view is rather incomplete, and hence the nuclear hamiltonian must be determined from the nucleon-nucleon scattering and other bound state data.

In this talk I will suggest that it may be possible to explain the low energy nuclear properties with a hamiltonian of the form:

$$H = \sum_i -\frac{\hbar^2}{2m}\nabla_i^2 + \sum_{i<j} v_{ij} + \sum_{i<j<k} V_{ijk} \; . \tag{1}$$

The need for the three nucleon interaction (TNI) in the nuclear hamiltonian is exhibited. The TNI has been studied with the meson theory[1)] in some details, however we use a simple pedagogical model of V_{ijk} in our many-body calculations.

Over the past few years there have been significant technical developments in all methods proposed to treat many body hamiltonians that contain v_{ij} (but not V_{ijk}). The results obtained with such hamiltonians, for nuclear matter and 3 and 4 body nuclei, with the variational method,[2,3)] are now in agreement with those obtained with the Brueckner-Bethe-Goldstone method[4)] (for nuclear matter), Faddeev equations[5)] (for triton) and by the coupled cluster method[6)] (for ^{4}He nucleus). We use variational calculations of nuclear matter, ^{3}H and ^{4}He nuclei to study the nuclear hamiltonian. The limitations and approximations in these calculations are summarized below.

It is simpler to use the variational method with hamiltonians in which v_{ij} is expressed as a sum of simple operators. The v_N models of v_{ij} are defined as:

$$v_{N,ij} = \sum_{p=1,N} v^p(r_{ij})O^p_{ij} \; . \tag{2}$$

The first four operators define a spin-isospin dependent central potential:

$$O^{p=1,4}_{ij} = 1,\ \sigma_i\cdot\sigma_j,\ \tau_i\cdot\tau_j \text{ and } \sigma_i\cdot\sigma_j\tau_i\cdot\tau_j \; , \tag{3}$$

the next four describe tensor and spin-orbit potentials:

$$O^{p=5,8}_{ij} = S_{ij},\ S_{ij}\tau_i\cdot\tau_j,\ \vec{L}\cdot\vec{S} \text{ and } \vec{L}\cdot\vec{S}\tau_i\cdot\tau_j \; . \tag{4}$$

---

†Supported by NSF PHY 78-26582.

The presence of these eight operators is uniquely indicated by the scattering data. The v_8 models contain only these eight terms. They can explain the deuteron and S and P wave scattering data, however v_{14} models are needed to explain all (S,P,D and F wave) scattering data.[7] The choice of operators O_{ij}^{9-14} is not uniquely indicated by the data. To simplify many body calculations we take them as:

$$O_{ij}^{p=9,14} = L^2, L^2\sigma_i\cdot\sigma_j, L^2\tau_i\cdot\tau_j, L^2\sigma_i\cdot\sigma_j\tau_i\cdot\tau_j, (\vec{L}\cdot\vec{S})^2 \text{ and } (\vec{L}\cdot\vec{S})^2\tau_i\cdot\tau_j \; . \tag{5}$$

The Paris potential[13] has different $O_{ij}^{p=9,14}$, but the phase shifts obtained with our v_{14}[7] and the Paris models are very similar.

The variational calculations use a variational wave function of the form:

$$\Psi_v = (S \prod_{i<j} F_{ij})\,\Phi \, , \tag{6}$$

where Φ is the noninteracting Fermi gas wave function, S the symmetrizer, and F_{ij} is a pair correlation operator:

$$F_{ij} = \sum_{p=1,8} f^p(r_{ij}, d, d_t, \alpha) O_{ij}^p \; , \tag{7}$$

d, d_t and α are the variational parameters[8] that vary the range and magnitude of $f^p(r_{ij})$. The approximations are:

1) The <H> in nuclear matter is calculated using an expansion obtained by carrying out Fermi-hypernetted-chain and single-operator-chain summations.[9] The convergence of this expansion has been tested by Wiringa,[10] and the estimated error in <H> at $k_F < 1.6\ \text{fm}^{-1}$ is $\lesssim .5$ MeV.[8] The <H> in ^3H and ^4He nuclei is calculated exactly by Monte Carlo methods.

2) The variational equations $\partial\langle H\rangle/\partial f^p(r) = 0$ are not solved exactly. The <H> is minimized with respect to few parameter such as d, d_t and α[8], and thus the variational energy could be above the true energy. The error due to this approximation could be quite small, the energy does not decrease significantly when the set of variational parameters is expanded.[8] It may also be possible to eliminate this approximation by generalizing the methods developed by Lantto and Siemens[11] to solve the variational equation exactly for Jastrow correlations.

3) The Ψ_v given by eq. (6) is not general enough. It does not contain explicit back-flow and three-body correlations.[12] In ^3H and ^4He nuclei it was possible to reduce the variational energy by $\sim$ .2 and .4 MeV per nucleon respectively, by introducing three-body factors in the tensor correlation.[3] The Ψ_v also does not contain correlations associated with operators O_{ij}^{9-14}, thus the difference $v_{14}-v_8$ is treated semi-perturbatively.[2] These may be the most serious approximations in variational calculations of nuclear systems. To remove them we first must develop techniques to calculate <H> with more general wave functions. In any variational calculation Ψ_v must be restricted to the set of functions for which <H> can be reliably calculated.

By comparing the present variational results with the available exact Greens function Monte Carlo[3] calculations, and Brueckner-Bethe and coupled-cluster results at low densities, we may speculate that the present variational energies are $\lesssim$ 4%, 8% and 12% above the true energies in ^3H, ^4He nuclei and nuclear matter respectively.

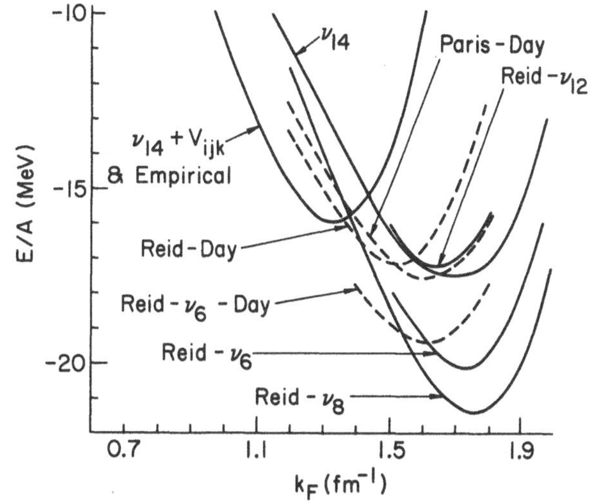

Fig. 1. $E(k_F)$ of nuclear matter. The dashed curves show Day's results with the Brueckner-Bethe method, and the full curves show results of variational calculations.

The results obtained for nuclear matter are shown in Fig. 1. The two Reid-v_6 curves indicate the extent to which variational and Brueckner methods agree, while the difference between Reid-v_8 and Reid-v_{12} curves indicates the contribution of $O_{ij}^{p\geq 9}$ terms. The v_{14} and Paris[13] models fit the scattering data very well, while Reid-Day and particularly Reid-v_{12}[7] fit the D and F waves rather crudely. The Reid-v_8 and v_6 models are pedagogical. The realistic models Reid-Day, Reid-v_{12}, v_{14} and Paris give similar results that disagree with the empirical $E(k_F)$.

The structure of the v_{14} interaction[7] is as follows:

$$v_{14} = v_s + v_I + v_\pi \ . \tag{8}$$

v_s is a purely phenomenological short range core and v_π is the one pion exchange (OPEP) potential:

$$v_\pi(r) = 3.488(Y_\pi(r)\sigma_i\cdot\sigma_j\tau_i\cdot\tau_j + T_\pi(r)S_{ij}\tau_i\cdot\tau_j)\text{MeV} \ , \tag{9}$$

$$Y_\pi(r) = \frac{e^{-\mu r}}{\mu r}\,(1-e^{-2r^2}) \ , \quad \mu = .7\ \text{fm}^{-1} \ , \tag{10}$$

$$T_\pi(r) = (1 + \frac{3}{\mu r} + \frac{3}{\mu^2 r^2})\,(1-e^{-2r^2})Y_\pi(r) \ , \tag{11}$$

and its tensor part is much stronger than the Yukawa part. The intermediate range v_I is attributed to the two pion exchange potential (TPEP) illustrated in part in Fig. 2. A detailed discussion of TPEP may be found in Holinday's talk.[14] Phenomenologically we take:

$$v_I = \sum_{p=1,14} T_\pi^2(r) I^p O^p_{ij} \ , \tag{12}$$

and adjust the strengths I^p to fit scattering data. In agreement with theory[15] it is found that $I^{p=1} >> I^{p\neq 1}$ so that v_I is mostly a central attraction. The expectation values of v_π, v_I and v_s in nuclear matter at $k_F = 1.33\ fm^{-1}$ are respectively -27.3, -148.9 and 124.1 MeV per nucleon.

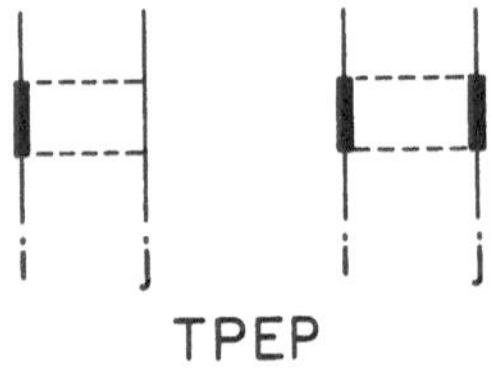

Fig. 2. Some processes that contribute to TPEP. Thick and thin lines denote Δ and nucleon propogators, and dashed lines denote exchanged pions.

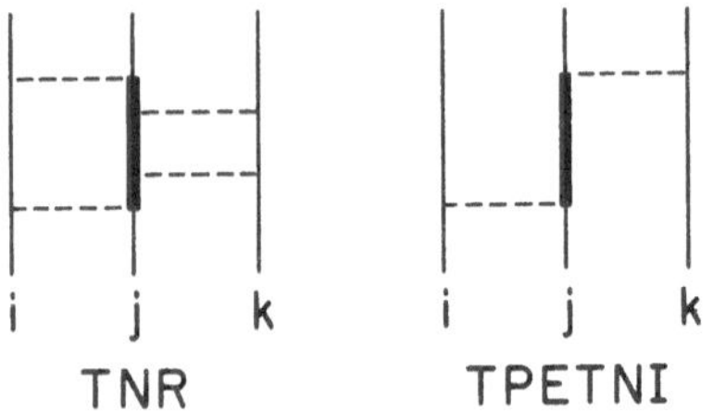

Fig. 3. The first diagram illustrates one of the many processes that can contribute to TNR, the second is the standard TPETNI diagram.

Comparing the $E(k_F)$ obtained from realistic models of v_{ij} with the empirical, we see that the V_{ijk} contribution must be attractive at small densities and repulsive at larger densities. The simplest V_{ijk} that can give such a contribution is of the form:

$$V_{ijk} = \sum_{cyc} U_o T_\pi^2(r_{ij}) T_\pi^2(r_{jk}) + U_2 T_\pi(r_{ij}) T_\pi(r_{jk}) P_2(\cos\theta_j) \sigma_i \cdot \sigma_k \tau_i \cdot \tau_k \ . \tag{13}$$

The first term called TNR is repulsive, and it is meant to simulate the quenching of v_I in matter from processes illustrated in Fig. 3. The second is a part of the conventional two pion exchange three nucleon interaction (TPETNI) shown in Fig. 3. The TPETNI has a tensor and other parts which we have not yet considered; their inclusion may not have a qualitative effect on the subsequent discussion.

A density dependent two-body interaction is obtained by integrating the TNR over $\vec{r}_k$. In this approximation we get:

$$v_{14} + V_{ijk} \sim v_s + v_I e^{-\gamma_1 \rho} + v_\pi + \text{TPETNI} \ , \tag{14}$$

$$-v_I(r_{ij}) \gamma_1 \rho \sim U_o T_\pi^2(r_{ij}) \int \rho T^2(r_{ik}) g(r_{ik}) d^3 r_{ik} \ . \tag{15}$$

Here $g(r_{ik})$ is the two-nucleon distribution function, and theoretically it is more appropriate to use $e^{-\gamma_1 \rho}$ than $(1-\gamma_1 \rho)$ in eq. (14).[2] The TPETNI cannot be treated in this way; it goes to zero if we integrate over $\vec{r}_k$. Its main contribution should be via three-nucleon correlations, and we approximate it by a function TNA(ρ):

$$TNA(\rho) = \gamma_2\, \rho^2\, e^{-\gamma_3 \rho}\, (3-2(\frac{N-Z}{A})^2)\ . \tag{16}$$

The ρ^2 comes from the three-body nature of the interaction, the $e^{-\gamma_3\rho}$ is meant to simulate the suppression of correlations at high density, and the N-Z dependence comes from the expectation value of $(\sigma_i \cdot \sigma_k)^2 (\tau_i \cdot \tau_k)^2$. The parameters γ_1, γ_2 and γ_3 are adjusted to obtain the equilibrium energy (-16 MeV), $k_F(1.33\ fm^{-1})$ and compressibility (240 MeV) of nuclear matter. The $E(k_F)$ obtained with this $v_{14} + V_{ijk}$ model is shown in Fig. 1.

Table I: Composition of Nuclear Matter Energy[2)]

$k_F(fm^{-1})$	1.13	1.23	1.33	1.43	1.53
<T> MeV	27.5	32.8	38.7	45.3	52.6
$<v_{14}>$	-37.3	-44.3	-52.1	-60.5	-69.2
TNR	1.4	2.2	3.5	5.4	8.2
TNA(ρ)	- 5.3	- 6.0	- 6.1	- 5.6	- 4.6
E(ρ)	-13.6	-15.3	-16.0	-15.3	-13.0

Contributions of various terms to nuclear matter energy are given in Table I. The <TNR> is small, but its rapid increase with k_F is necessary to obtain the empirical equilibrium density. Recently Horikawa, Thies and Lenz[16)] have estimated the real part of Δ-nucleus optical potential U_Δ to be ∿ -25 MeV, while the average value of nucleon-nucleus optical potential $\bar{U}_N$ for occupied states is ∿ -60 MeV. The Δ-percentage P_Δ in nuclear matter is estimated by Wiringa[17)] to be ∿ 6-7%. Thus the dispersion correction $P_\Delta(U_\Delta-\bar{U}_N)$ ∿ 2.3 MeV may account for most of TNR. There are also other processes that can contribute to TNR,[17)] and thus the magnitude of <TNR> in the present model appears to be reasonable.

The TNA(ρ) required to obtain the correct equilibrium energy is ∿ twice as large as the estimated contribution of TPETNI.[18)] However field theoretical calculations of TPETNI are somewhat uncertain, and the available estimates of TPETNI contribution to nuclear matter energy use first order perturbation theory which, as we will see later, may underestimate it. Also it should be noted that the present calculation of TNA(ρ) is influenced by the errors in our variational calculations of nuclear matter. It is conceivable that a better nuclear matter calculation with the v_{14} + TNR interaction will give somewhat lower energies, than the present, and consequently will need a smaller TNA(ρ) to obtain the empirical binding energy. The smallness of the contribution of V_{ijk}, as compared to that of v_{14}, suggests that the effects of four and more nucleon interactions in nuclear matter may be negligible (<< 1 MeV).

The $E(\rho,\beta = N-Z/A)$ over the range $\beta = 0-1$,[19] the equation of state of hot and dense nuclear ($\beta = 0$) and neutron ($\beta = 1$) matter at temperatures $\leq$ 20 MeV,[20] the structure of neutron stars,[20] and the real part of the optical potential $U(e < 300$ MeV) in nuclear matter[21] have been studied with this $v_{14} + V_{ijk}$ model of nuclear hamiltonian. The general agreement with experimental data is very encouraging. The calculated symmetry energy has a reasonable value of 30 MeV at $k_F = 1.33$. The $U(e,\rho_o)$ (Fig. 4) compares favorably with the depths of Woods-Saxon wells (shown by dots) needed to explain the scattering and bound state data. However at $e \lesssim e_F$ and $e \gtrsim e_F$ we seem to respectively over and under estimate the $U(e,\rho_o)$. This problem is probably due to the behavior of the effective mass at $k = k_F$[22,23] which is not correctly given by the present variational calculation. Fig. 5 shows the surface red shift $\phi = (1-2GM/RC^2)^{1/2}$ calculated as a function of neutron star mass. The vertical and horizontal lines indicate somewhat model dependent limits[20] on neutron star masses, and the surface red shift of the pulsar in Dorado constellation. The solid lines below the figure give model independent masses[24] of three neutron stars. The TNA(ρ) has little effect on the structure of the neutron stars, thus the observations on neutron stars primarily support our model for TNR. It is unlikely that neutron stars as heavy as Vela X-1 can be supported with justifiable two-nucleon interactions without any TNR.

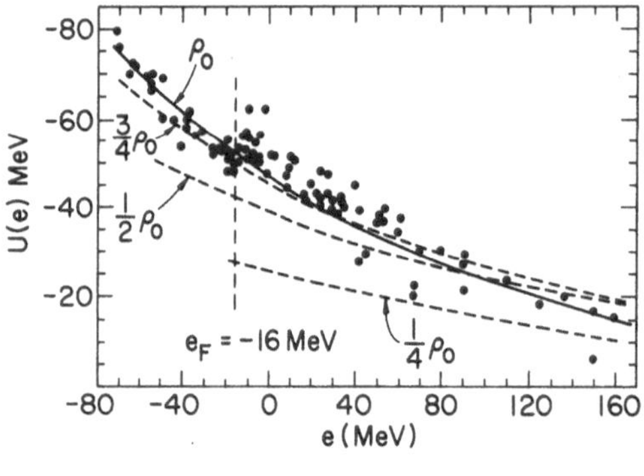

Fig. 4. The calculated $U(e,\rho)$ at various densities. The dots show empirical Woods-Saxon well depths that should be compared to $U(e,\rho_o)$.

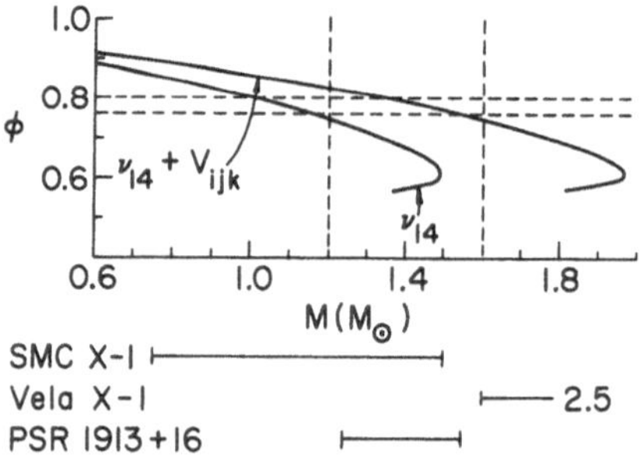

Fig. 5. The calculated red shift ϕ as a function of neutron star mass. The dashed lines show model dependent evolutionary/observational limits, while the solid lines below the figure show limits on the masses of three neutron stars.

In context of the above results, the main problem of nuclear matter appears to be the calculation of TNA(ρ) from TPETNI. It has three parts: i) determine TPETNI from experimental data; ii) calculate the three-body correlation f_{ijk} it generates, and iii) calculate $\langle H \rangle$ with Ψ_v containing f_{ijk}. All of these are simpler in three and four-body nuclei. I will discuss some methods we[25] have developed to resolve these problems, and present preliminary results. All our calculations have been done with a rather simple "homework" model of nuclear hamiltonian that consists of v_{ij} given by Reid-v_8 model, and

$$V_{ijk} = \sum_{cyc} .003\ T_\pi^2(r_{ij})T_\pi^2(r_{ik}) - 3U_2T_\pi(r_{ij})T_\pi(r_{jk})P_2(\cos\theta_j)\ . \qquad (17)$$

The strength .003 MeV of TNR is obtained from the γ_1(= .15 fm^3) in nuclear matter calculations, and the $\sigma_i\cdot\sigma_k\tau_i\cdot\tau_k$ is approximated by -3 in the S-shell nuclei ^{3}H, ^{3}He and ^{4}He. We neglect the coulomb interaction so ^{3}H ≡ ^{3}He. The strength U_2 is determined by reproducing the binding energy of ^{3}H, and we examine the effect of the TPETNI on the proton distribution in ^{3}He. Sick[26] has deduced the point proton distribution in ^{3}He from electron scattering data, by taking into account relativistic, exchange current, and proton size effects approximately. The deduced $\rho_p(r)$ has a dip at r = 0 that cannot be explained with realistic two-nucleon interactions.

Let Ψ_o be the wave function obtained with $H_o = H - \Sigma V_{ijk}$, it is taken from Monte Carlo variational calculations with Reid-v_8 potential.[3] The variational wave function Ψ_v for the hamiltonian H is taken as:

$$\Psi_v = \prod_{i<j<k} f_{ijk}(\vec{R})\ \Psi_o\ , \qquad (18)$$

where $\vec{R}$ represents $\vec{r}_i$, $\vec{r}_j$ and $\vec{r}_k$. Two choices for f_{ijk} are being studied:

$$1: \quad f_{ijk}(\vec{R}) = (1 - pV_{ijk}(\vec{R}))\ , \qquad (19)$$

$$2: \quad f_{ijk}(\vec{R}) = (1 - p\int \frac{e^{-q|\vec{R}-\vec{R}'|}}{(q|\vec{R}-\vec{R}'|)^7} V_{ijk}(\vec{R}')d\vec{R}')\ , \qquad (20)$$

The second is suggested by Greens-function theory and has two parameters p and q. In simple one-body problems the second choice works very well. Fig. 6 shows the wave function $\psi_o(r)$ for a particle in a well described by hamiltonian H_o. ψ_+ and ψ_- are exact wave functions obtained with the hamiltonians $H_o \pm V(r)$. $\psi_{V1\pm}$ and $\psi_{V2\pm}$ are obtained with variational calculations using $\psi_{V1} = (1 - p\ V(r))\psi_o$ and

$$\psi_{V2} = (1 - p\int \frac{e^{-q|\vec{r}-\vec{r}'|}}{q|\vec{r}-\vec{r}'|} V(\vec{r}')d^3r')\psi_o\ . \qquad (21)$$

The energies obtained with wave functions ψ_o, ψ_{V1}, ψ_{V2} and ψ are also shown.

We have done ^{3}H and ^{4}He with the more approximate choice 1 of $f_{ijk}(\vec{R})$, so the results should be viewed qualitatively. The variation of <H> with p is shown in Fig. 7. We obtain the experimental triton energy with U_2 = -1.5 MeV. The <TNR> is only .4 MeV in triton, and also note that the f_{ijk} reduces the triton energy from -7.5 MeV (the value one would obtain by treating the V_{ijk} in first order perturbation theory) to -8.5 MeV. The $\rho_p(r)$ is shown in Fig. 8 for two values of p near the minimum. The $\Sigma T_\pi(r_{ij})T_\pi(r_{jk})P_2(\cos\theta_j)$ term makes $f_{ijk} < 1$ when particles i, j and k are in a line (only these configurations can contribute to $\rho_p(r \sim 0)$),

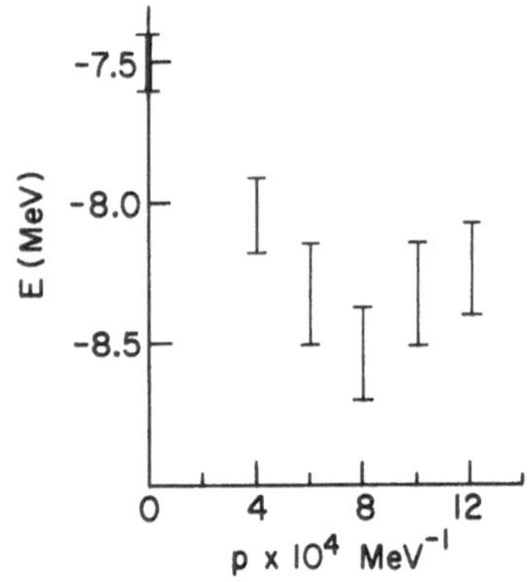

Fig. 7. The variation of ^3H energy with p (eq. 19).

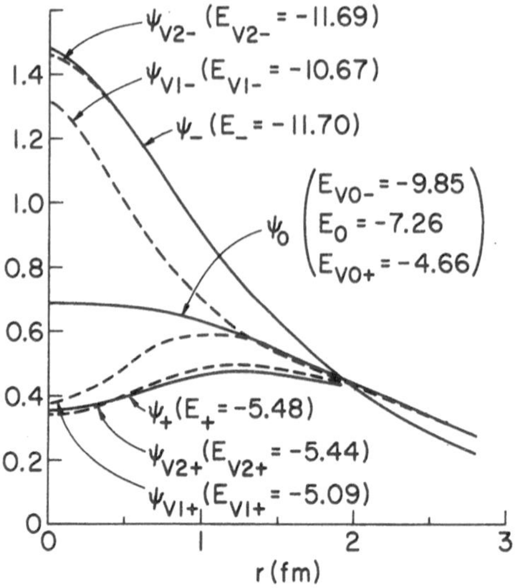

Fig. 6. Results of exact and variational calculations of a particle in a well perturbed by a strong potential V(r).

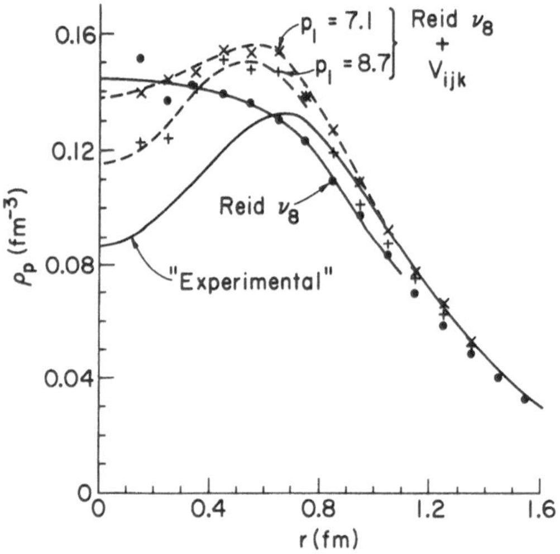

Fig. 8. The $\rho_p(r)$ in ^3He nucleus. The ·, + and x's show results of Monte Carlo calculations with Reid-v_8, and Reid-v_8 + V_{ijk} models respectively ($p_1 = p \times 10^4$).

and it also makes $f_{ijk} > 1$ when particles i, j and k form an equilateral triangle with a hole in the middle.

There are several outstanding problems in this approach to determine the nuclear hamiltonian: i) The Reid-v_8 + V_{ijk} model fitted to ^3H binding energy does not account for the magnitude of the observed dip in $\rho_p(r)$ in ^3He (Fig. 8), and secondly it overbinds the ^4He nucleus (Table II). Note that the true ^4He energy could be 1-2 MeV below the calculated upperbound. ii) The v_{14} model (without V_{ijk}) binds ^3H with almost the correct energy, but underbinds the ^4He nucleus. It appears that the v_{14} + V_{ijk} model fitted to ^3H binding energy may have ^4He underbound. iii) It may be that a new v_{14} model, in between the present v_{14} and Reid, with a V_{ijk} having $|U_2| < 1.5$, can explain ^3H and ^4He binding energies, but it may not be able to get the magnitude of the dip in $\rho_p(r)$. Improved calculations with f_{ijk} of eq. (20), and a more realistic model of V_{ijk} are needed to resolve these problems. The present model of V_{ijk} is quite pedagogical.

Table II: Calculated ^3H and ^4He Binding Energies (MeV)

	^3H	^4He
Experimental (no coulomb)	-8.54	-29.
Reid-v_8	-6.86 $\pm$ .08	-22.9 $\pm$ .5
Reid-v_8 + $V_{ijk}(U_2 = -1.5)$	-8.53 $\pm$ .15	-30.7 $\pm$ 1.0
v_{14}	-8.46 $\pm$ .06	-25.1 $\pm$ 0.4

References

1. The Meson Theory of Nucleon Forces and Nuclear Matter. Editors H. D. Schütte, K. Holinde and K. Bleuler, B. I. Wissenschaftsverlag (1980).
2. I. E. Lagaris and V. R. Pandharipande, Nuclear Physics (1981) in press.
3. J. Lomnitz Adler, V. R. Pandharipande and R. A. Smith, Nuclear Physics (1981) in press.
4. B. D. Day, in ref. 1, and in the proceedings of this meeting.
5. R. A. Brandenburg, Y. E. Kim and A. Tubis, Phys. Rev. C12 (1975) 1368.
6. H. Kümmel, K. H. Lührmann and J. G. Zabolitsky, Phys. Rep. C36 (1978) 1.
7. I. E. Lagaris and V. R. Pandharipande, Nuclear Physics (1981) in press.
8. R. B. Wiringa and V. R. Pandharipande, Phys. Lett. (1981) in press.
9. V. R. Pandharipande and R. B. Wiringa, Rev. Mod. Phys. 51 (1979) 821.
10. R. B. Wiringa, Nucl. Phys. A338 (1980) 57.
11. L. J. Lantto and P. J. Siemens, Phys. Lett. B68 (1977) 308.
12. K. E. Schmidt and V. R. Pandharipande, Nucl. Phys. A328 (1979) 240.
13. M. Lacombe, B. Loiseau, J. M. Richard, R. Vinh Mau, J. Côté, P. Pires and R. de Tourreil, preprint (1979).
14. K. Holinde, in proceedings of this meeting.
15. R. A. Smith and V. R. Pandharipande, Nucl. Phys. A256 (1976) 327.
16. Y. Horikawa, M. Thies and F. Lenz, Nucl. Phys. A345 (1980) 386.
17. R. B. Wiringa, private communication.
18. S. A. Coon, in proceedings of this meeting.
19. I. E. Lagaris and V. R. Pandharipande, to be submitted to Nucl. Phys.
20. B. Friedman and V. R. Pandharipande, Nucl. Phys. (1981) in press.
21. B. Friedman and V. R. Pandharipande, Phys. Lett. (1981) in press.
22. J. P. Jeukenne, A. Lejeune and C. Mahaux, Phys. Rept. 25C (1976) 83.
23. B. Friman, in proceedings of this meeting.
24. R. Kelley and S. Rappaport, in proceedings of IAU Symposium No. 95, Bonn 1980.
25. J. Carlson and V. R. Pandharipande, to be published.
26. I. Sick, Int. Conf. Few Body Problems and Nucl. Forces, Graz 1978.

EXCITATIONS AND TRANSPORT IN QUANTUM LIQUIDS

David Pines, Phys. Dept., UIUC, Urbana, IL 61801

Introduction

In this talk I should like to describe a new approach to the theory of excitations and transport in quantum liquids, in which the consequences of the strong interactions found in such systems as the helium liquids, nuclear matter, or metallic electrons, are described in terms of self-consistent fields whose strengths are determined by physical arguments, static measurements, and sum rule considerations. The theory makes possible a unified treatment of the elementary excitation spectra of the normal Fermi liquid, ^{3}He, and the Bose liquid, superfluid He II, and yields results in excellent agreement with experiment for both the excitations and transport properties of these systems, as well as providing a quantitative account of the normal-superfluid transition of liquid ^{3}He. It would seem to hold considerable promise as well for neutron and nuclear matter, and for the strongly coupled electron liquid. In presenting the theory I shall follow rather closely its historical development, because in this way it is easier to bring out the basic physical arguments. Wherever possible, I shall compare our theory with experiment.

The initial impetus for the theory came from neutron scattering experiments on liquid ^{4}He by David Woods,[1)] who studied the variation with temperature of the energy and lifetime of excitations of wave-vector ~ 0.38 A^{-1}; he found no appreciable change in the energy of these excitations (~ 7.4 K) and only a modest increase in their lifetime on going from temperatures (~ 1.6 K) well below the λ-point to those (~ 2.55 K) substantially above it. These results posed a serious challenge to the accepted dogma of the time, that the physical origin of the phonon-maxon-roton spectrum of liquid He II was the existence of a condensate, a macroscopically occupied single quantum state. They led me to propose that we abandon dogma, and treat the excitations as the analogue, for a neutral system, of the electron plasma oscillations: collective modes whose physical origin lies in a self-consistent field of strength sufficient to place their energies well above those characteristic of single-particle excitations.[2)] Like plasmons, then, their energies would be little affected by the changes in the single-particle-like excitations brought about by the transition from a normal to a superfluid (superconducting) phase. The proposal represented a restatement of the argument which David Bohm and I had earlier put forth (on the basis of RPA calculations) for the existence of collective modes in strongly interacting neutral systems.[3] By introducing a scalar polarization potential, proportional to the density fluctuation, to describe the polarization effects responsible for such modes, I was then able to use sum rules to prove that this early RPA-based argument was valid in the strong coupling limit [which is very nearly the case for both ^4He and ^3He]. Hence one should not have been surprised that a well-defined collective (or zero-sound) mode exists in liquid ^4He above the λ-point; it seemed natural to conclude that the zero sound modes in ^3He and ^4He possessed a common origin, and I was thus led to predict the existence of a well-defined zero sound mode in liquid ^3He at momenta, energies, and temperatures for which the Landau theory does not apply. This mode was observed some ten years later by Kurt Sköld and his collaborators in an elegant application of neutron scattering techniques to ^3He.[4]

In the meantime, Charles Aldrich and I developed the underlying theory in much more detail and applied it to superfluid ^4He and to ^3He-^4He mixtures as well as to ^3He.[5] Our approach can perhaps best be appreciated by considering what physical effects must be taken into account if one wishes to go beyond the Landau theory of a normal Fermi liquid to describe, say, the density fluctuation excitation spectrum for wavevectors comparable to the Fermi momentum, p_F. There are, in general, three possible contributions to this spectrum:

(i) a zero-sound mode
(ii) single quasiparticle-quasihole pair excitations
(iii) multipair excitations

In Landau theory, the interaction between quasiparticles on the Fermi surface gives rise to a local restoring force for a possible zero-sound mode; the contribution from single pairs is described by a low-frequency spectrum whose maximum energy is qp_F/m_o^*, where m_o^* is the quasiparticle effective mass determined by the specific heat; while the influence of multipair excitations (involving the excitation of two or more quasiparticles and quasiholes from the Fermi sea) can be neglected. In attempting to develop a theory valid for $q \sim p_F$, one must therefore:

i) allow for the possibility of a non-local restoring force for possible zero-sound modes and develop a model to calculate it.
ii) calculate the changes in the single-pair spectrum expected because pairs of net momentum $\sim p_F$ are, in general, formed from quasiparticles and quasiholes which lie far from the Fermi surface, and hence do not necessarily possess an effective mass, m_o^*.

iii) take multipair excitations into account.

Polarization Potential Theory

In our theory the non-local restoring force produced by the average self-consistent fields of the particles acting in concert is obtained from the scalar polarization potential of Ref. 3,

$$\phi^{c}(\underset{\sim}{q}\omega) = f_q^s \langle\rho(\underset{\sim}{q}\omega)\rangle \ , \tag{1}$$

where $\langle\rho(\underset{\sim}{q},\omega)\rangle$ is the average particle density fluctuation; we develop a detailed physical model for the strength of this restoring force by considering f_q^s to be the Fourier-transform of a pseudopotential, $f^s(r)$, which describes the effective particle interaction in the liquid. We argue that in the liquid the strong short-range repulsive interaction at $r \lesssim 2.7$ A between two bare helium atoms is strongly screened (the repulsion is so strong that two particles will seldom be close enough together to feel the full influence of this force), while the long range part of the interaction is unaffected (the van der Waals interaction is not screened by other particles). As a result the pseudopotentials, $f^s(r)$, and their Fourier-transforms, f_q^s, will take a form similar to that depicted in Fig. 1(b) and 1(a).

The pair excitation spectrum at finite q is calculated by using a simple model for backflow--the current fluctuations induced in the liquid by particle motion which act back on it to modify its mass. We describe the reaction of the liquid to single pair excitations in terms of an induced vector polarization potential, proportional to the particle current fluctuation, $\langle J(\underset{\sim}{q}\omega)\rangle$,

$$\underset{\sim}{A}^{c}(\underset{\sim}{q},\omega) = f_q^v \langle J(\underset{\sim}{q},\omega)\rangle \ ; \tag{3}$$

which couples to the particle current density. As a result single pair excitations behave as though they possess an effective mass,

$$m_q^* = m_o + Nf_q^v \tag{4}$$

The influence of these potentials on the spectrum is easily calculated by linear response theory; the density-density response function is given by

$$\chi^{c}(\underset{\sim}{q},\omega) = \frac{\chi_{sc}^{c}(\underset{\sim}{q},\omega)}{1-[f_q^s+(\omega^2/q^2)f_q^v]\chi_{sc}(\underset{\sim}{q}\omega)} \tag{5}$$

where $\chi_{sc}^{c}(\underset{\sim}{q}\omega)$ is the response of the density fluctuations to an external field plus the induced polarization potentials, (1) and (3). The expression (5) provides a formal basis for a unified theory of normal liquid ^{3}He and superfluid ^{4}He. The physical basis is that if it is the strong interaction between the particles rather than their quantum statistics which plays a dominant role, then the influence of statistical correlations (or temperature) should be a minor one, so that the strength of the polarization potentials, f_q^s and f_q^v should be very nearly the same for ^{3}He and ^{4}He at the same density. In other words, effects of statistics or temperature come in only through $\chi_{sc}(q,\omega)$. If one neglects the influence of the multiparticle excitations, then one has as $T \to 0$, $\chi_{sc}^c = \chi_o^*$, where χ_o^* is the single quasiparticle or single pair response; thus for

$$^4\text{He:} \qquad \chi_o^* = \frac{Nq^2/m_q^*}{\omega^2 - \varepsilon_q^2} \tag{6a}$$

$$^3\text{He:} \qquad \chi_o^* = \chi_L^* \tag{6b}$$

where the expression (6a) represents the excitation of quasiparticles of energy $\varepsilon_q = q^2/2m_q^*$ from the condensate, while in (6b), χ_L^* is the Lindhard response function for quasiparticle pairs of effective mass, m_q^*, and in both cases m_q^* is given by (4). With the choice (6b), $\chi(q,\omega)$ reduces to the usual Landau theory result in the limit of long wavelengths, provided one makes the obvious identification,

$$\lim_{q\to 0} f_q^s = f_o^s = F_o^s/[\nu/(0)] \tag{7a}$$

$$\lim_{q\to 0} f_q^v = f_o^v = (F_1^s/3)/[\nu(0)] \tag{7b}$$

where F_o^s and F_1^s are the usual dimensionless Landau parameters which describe the $\ell = 0$ and $\ell = 1$ components of the interaction between quasiparticles on the Fermi surface.

The theory thus provides a natural extension of Landau theory to finite q and ω (and T, through a corresponding modification in χ_o^*) as well as providing an equivalent theory for excitations in Bose liquids. To the extent that one is interested in excitations with $q \lesssim 2\ \text{A}^{-1}$ the exact details of the very short-range part of the pseudopotential $f^s(r)$ are unimportant; if we assume that the attractive part of the interaction is unscreened, then given the range, r_c, of the effective repulsion, its strength is determined by the requirement that $f_o^s = \int d^3r f^s(r)$, where f_o^s is given by (7a) for ^{3}He, and by ms^2/n, where s is the first sound velocity and n is the density, for ^{4}He. For ^{4}He we find that $r_c \sim 2.68$ A at all densities, so that the variation with pressure of $f^s(r)$ is uniquely determined by the observed variation of s. For ^3He we choose the effective repulsion to have a comparable range and strength. Note that the assumption that statistical correlations play a minor role in these liquids is a self-consistent one, because their effective range (~ the inter-particle spacing, r_o) is $\lesssim r_c$ for both ^3He and ^4He.

Finally, because multiparticle excitations are present with non-negligible statistical weight in the finite q density fluctuation excitation spectrum of both liquid ^4He and ^3He, it is necessary to take these into account in determining $\chi(q,\omega)$. Because their average energies (~20 K to ~40 K) are comparatively large, these multipair (for ^3He) or multiquasiparticle (for ^4He, corresponding to exciting two or more quasiparticles from the condensate) contributions act to lower collective mode energies calculated neglecting their influence. If one assumes the coupling to these modes is similar to that between the quasiparticles, i.e. is of strength $f_q^s + (\omega^2/q^2)f_q^v$, then the resulting response function may still be written in the form, (5), with, however,

$$\chi_{sc}^c(q,\omega) = \alpha_q\ \chi_o^*(q,\omega) + (1 - \alpha_q)\ \chi_m^c(q,\omega) \tag{8}$$

where α_q takes into account the reduction in the contribution made by the single particle (pair) excitations to χ_{sc}^c which necessarily accompanies the presence of multiparticle modes, while the multiparticle contribution may be written in the form,

$$\chi_m^c(q,\omega) = \frac{Nq^2}{m_q^*} \int_o^\infty dz\ \frac{\rho_m(q,z)}{\omega^2 - z^2} \tag{9}$$

where the spectral density, $\rho_m(q,z)$ is normalized to unity, $\int_o^\infty dz\ \rho_m(q,z) = 1$, in order to satisfy the f-sum rule. It is often a good approximation to treat the multiparticle excitations as possessing an average energy sufficiently large that $\chi_m^c(q,\omega)$ may be approximated by its static limit, $\chi_m^c(q,0)$, so that

$$\chi_{sc}^c(q,\omega) \cong \alpha_q\ \chi_o^*(q,\omega) - NA_q \tag{10a}$$

where

$$NA_q = [(1-\alpha_q)(Nq^2/m_q^*) \int_o^\infty dz\ [\rho_m(q,z)/z^2] \ . \tag{10b}$$

For both ^3He and ^4He the coupling to multiparticle excitations is thus manifested through a vertex correction which acts to reduce the influence of both scalar and vector polarization potentials,

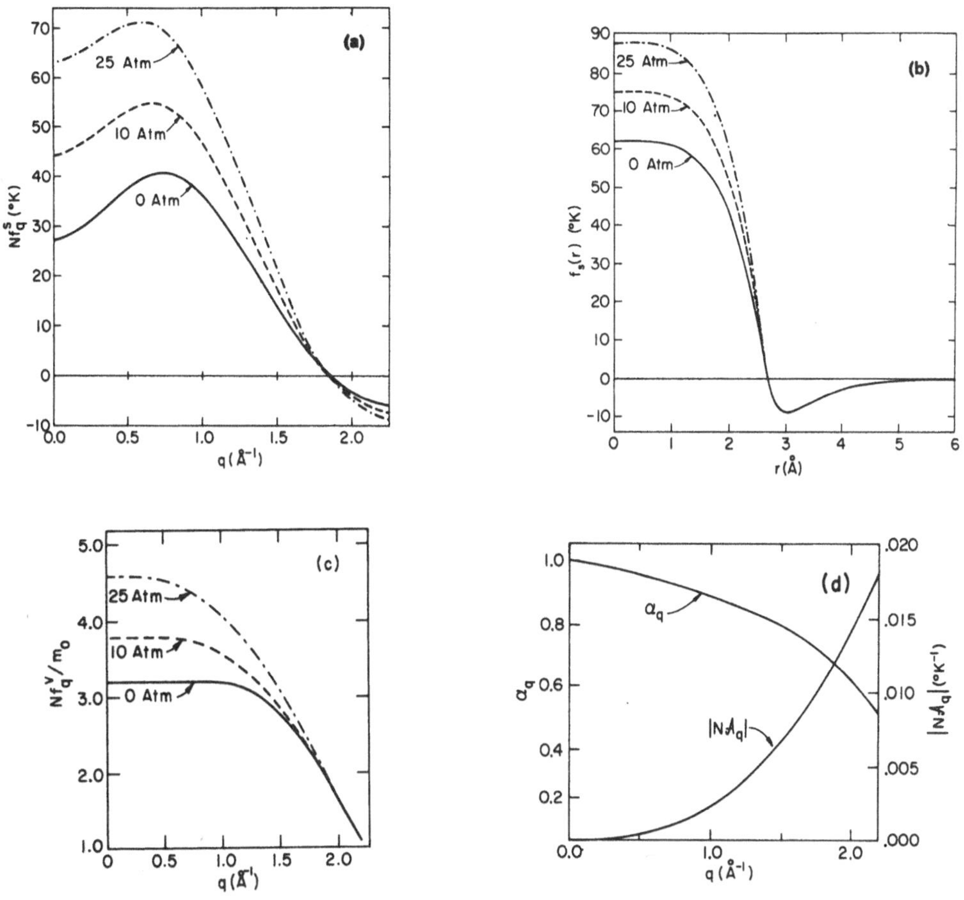

Fig. 1 Polarization potentials and mode coupling parameters.

$[f_q^s + (\omega^2/q^2)f_q^v] \rightarrow \alpha_q\ [f_q^s + (\omega^2/q^2)f_q^v]$, and hence the collective mode energies; a further reduction in these energies comes from A_q. In the long wavelength limit one finds that $\lim_{q\to 0} \alpha_q = 1 - \alpha_2 q^2$, while A_q is of order q^4.

The phonon-maxon-roton spectrum of He II at low temperatures

The polarization potentials and multiparticle mode coupling parameters which Aldrich and I used to calculate the density fluctuation excitation spectra of He II[6)] at various pressures are shown in Fig. 1, while the resulting spectra are compared with experiment in Fig. 2. The long wavelength limit of f_q^v, f_o^v, is assumed to be identical to that of ^{3}He at the same density, while its q dependence is fixed by a fit to excitations in the vicinity of 1.85 A^{-1}, where $f_q^s \cong 0$. As I have noted, once the range of repulsive part of the pseudopotential

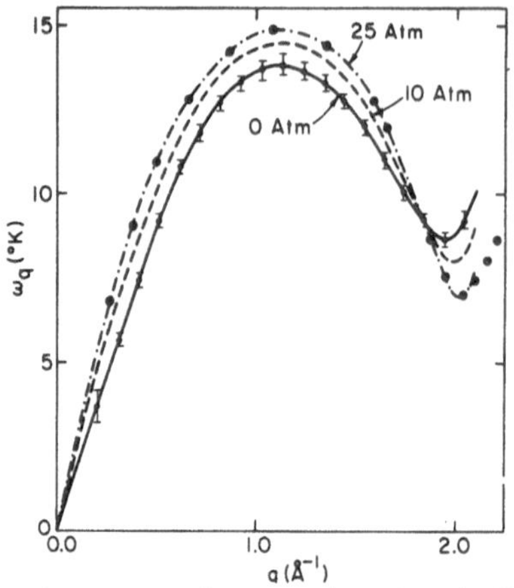

Fig. 2 Comparison of theory and experiment at three pressures.

$f^s(r)$ is fixed, f^s_q is essentially determined; hence its behavior at higher pressures is a prediction of the theory. We chose α_q and A_q to obtain the best fit to the spectrum at svp, and assumed that to a first approximation these did not vary with pressure. Hence there are no free parameters available in our calculation of the excitation spectrum at 10 and 25 atm; the agreement between theory and experiment thus provides a critical test of our theory, which it is gratifying to note that it passes.

Phonon Dispersion of He II

Further confirmation of the correctness of our model for f^s_q comes from calculation of phonon dispersion at long wavelengths ($q \lesssim 0.6$ Å^{-1}) and low temperatures.[7] For examples, Dynes and Narayanamurti,[8] in an investigation of the propagation of high frequency phonons in He II as a function of pressure, find that a well-defined threshold wave-vector q_c and energy, E_c exist, beyond which the so-called "anomalous" three phonon processes (in which one phonon may decay into two) can no longer occur. Our pseudopotential, $f^s(r)$, provides an immediate explanation for the existence of an anomalous region; at svp, for example, because f^s_q initially increases with increasing q, phonon dispersion (ω_q/sq) will likewise tend to increase. This initial increase in f^s_q may be traced to the strong screening in the liquid of the repulsive part of the bare atom-atom potential, which makes it possible for the moment, $\int_0^\infty dr\, f^s(r)r^4$, to be attractive; its influence on the phonon dispersion is, however, opposed by mode-mode coupling, which through α_q acts to reduce the phonon energy; the result of the interplay between these two physical effects is shown in Fig. 3a, while our theoretical results are compared with experiment in Fig. 3b.

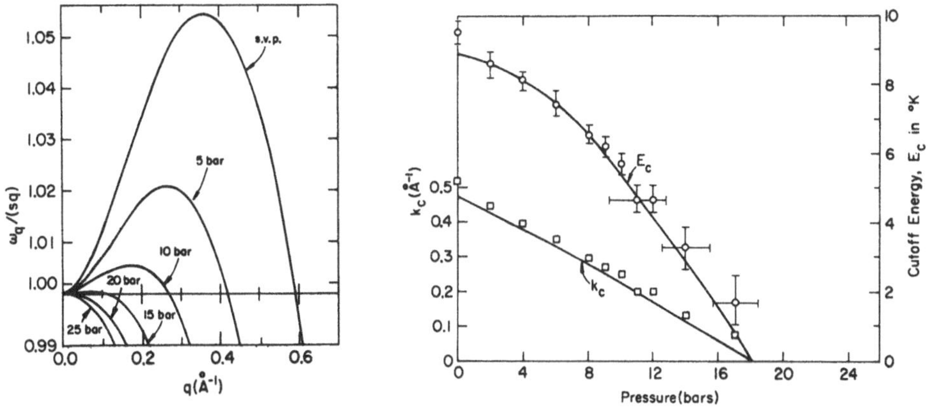

Fig. 3 (a) Theoretical results for phonon dispersion;[7] (b) comparison of calculated values of q_c and E_c (solid lines)[7] with experiment.[8]

Note that the reduction with pressure of both the magnitude of the anomalous dispersion and the wavevector regime over which it exists find their explanation in the fact that as the pressure increases the strength of the repulsive part of the pseudopotential likewise must increase (since according to our theory the attractive part is unchanged) in just such a way as to yield the experimentally observed increase in the first (or zero) sound velocity. This reduction in the screening of the bare atom-atom repulsive interactions in turn leads to a less rapid initial rise in f^s_q (the moment, $[\int_0^\infty dr\, f^s(r) r^4]$, is less attractive), so that mode coupling is more effective in opposing anomalous dispersion, to the point that at some 18 atm, the upward dispersion ends.

Aldrich, Pethick and I also showed that at very long wavelengths the phonon dispersion relation takes the form,

$$\lim_{q\to 0} \omega_q = s_o q\, [1 + \omega_2 q^2 + \omega_3 q^3 + \dots] \qquad q \lesssim 0.2\ A^{-1} \tag{11}$$

where the ω_3 term originates in the suggestion of Feenberg[9] and, independently, Kemoklidze and Pitaevskii,[9] that the power series expansion of any potential which, like ours, possesses a van der Waals tail,

$$f^s_q = \frac{4\pi}{V} \int_o^\infty dr\, f_s(r) r^2 (\sin qr)/(qr) \tag{12}$$

takes the form

$$\lim_{q\to 0} N f^s_q = ms^2 (1 + f^s_2 q^2 + f^s_3 q^3 + \dots) \qquad q \lesssim 0.2\ A^{-1} \tag{13}$$

since for a van der Waals interaction the expansion of (12) in even powers of q breaks down at order q^4. Quite recently Aldrich and I (unpublished) have been able to determine ω_2 from a fit of (11) to the superb low temperature specific heat measurements of Greywall;[10] we find at svp, $\omega_2 = (1.95\pm0.05)\ A^2$, while $\omega_3 = -3.36\ A^3$ on using the result,[9] $f_3^s = (\pi^2\rho/12Nf_o^s) \lim_{r\to\infty} [f_s(r)r^6]$.

Polarization Potentials and Elementary Excitations in Liquid ^3He

According to the physical picture which forms a basis for our unified theory of excitations in ^3He and ^4He, the polarization potentials responsible for the density fluctuation excitation should, at the same density, be little changed as one goes from ^4He to ^3He. For the effective mass field, f_q^v, we have made a virtue out of necessity by taking the long wavelength values, f_o^v, to be identical; we are thus able to infer these values for ^4He from those measured for ^3He, and can likewise take the values of the momentum dependent effective mass to be comparable. Aldrich and I have proposed that this same procedure be followed for the range of the pseudopotential, $f^s(r)$; thus we take its range to be ~2.68 A for ^3He at 20 atm [since it is only at this pressure that the density of ^3He is equal to that of ^4He at svp] and keep it fixed at higher pressures. What, however, is the range of $f^s(r)$ for ^3He at pressures lower than 20 atm? Initially, we took it to be 2.68 A; we then obtained a zero sound mode which became strongly Landau damped at $q_c \sim 1.7\ A^{-1}$, at which wave vector its energy was ~ 19K.[5] When the pioneering neutron scattering experiment by Scherm et al.[11] at ILL showed no evidence for an undamped zero sound mode at wavevectors $\gtrsim 1.4\ A^{-1}$, we decided the most likely explanation was that the range of the effective repulsion was somewhat larger; for example, with $r_c \sim 3.1$ A, and minor adjustments to f_q^v to fit the Scherm data at large q, we found a zero sound spectrum which was essentially flat (at ~ 12 K) from 0.8 A^{-1} to ~ 1.4 A, where it became Landau damped. That led us to the theoretical dispersion relation for zero sound shown in Fig. 4, where it is compared with the experimental results obtained subsequently by Sköld et al.[4] As noted in reference 12, given the fact that the theoretical dispersion relation was calculated in advance of the experimental measurements, the agreement between theory and experiment is highly satisfactory. Polarization potential theory was not only capable of explaining the excitation spectrum of liquid ^4He as a function of pressure, but made possible a successful quantitative prediction of the density fluctuation excitation spectrum of liquid ^3He.

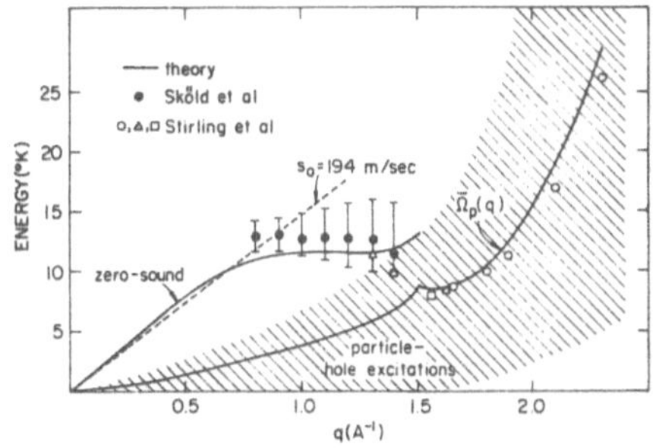

Fig. 4 Comparison between theory and experiment for the zero sound energy and the mean excitation energy of a single density pair.[12)]

Note that although the polarization potentials are very similar the excitations do not possess the maxon-roton character found in ^{4}He; the reason is that the single pair excitations present in ^{3}He act to push the spectrum up significantly for $q \gtrsim 1\ A^{-1}$ in just such a way as to eliminate a roton-like dip in the spectrum. The experiments of Scherm et al.[11)] and of Stirling and his collaborators[13)] provide a useful experimental check on the assumed momentum dependence of the single pair effective mass, m_q^*; the calculated and experimental mean single pair excitation energies are also shown in Fig. 4. The physical origin of the increase in the range of $f^s(r)$ between 20 atm and svp is likely the zero point motion of the ^{3}He atoms in the liquid, since, as Aldrich and I have pointed out,[14)] the mean square vibration amplitude of the ^{3}He atoms in the liquid at 20 atm is comparable to that of ^{4}He at svp, while at svp it is twice that of ^{4}He at the same pressure.

What is measured experimentally in an inelastic neutron scattering experiment is a combination of dynamic structure factors, $\sigma_c S^c(q\omega)$ + $\sigma_I S^I(q\omega)$, where σ_c and σ_I are the coherent and incoherent cross sections, respectively, for the scattering of neutrons from a ^{3}He nucleus, while S^c and S^I are dynamic structure factors, which are related to response functions by $S^{c,I}(q\omega) = -\,\mathrm{Im}\,\chi^{c,I}(q,\omega)/\pi$ in the low temperature limit. $\chi^c(q\omega)$ is the usual density-density response function we have been considering, while $\chi^I(q\omega)$ is the corresponding spin-spin response function. Initially Aldrich, Pethick, and I used Landau theory to calculate $\chi^I(q\omega)$ in the long wavelength limit and then extended our results to finite wavevectors with the aid of sum rule arguments;[12)] we were able to obtain good qualitative agreement with the experimental results of Sköld et al,[4)] which showed a strong "paramagnon" enhancement of the spin fluctuation excitation spectrum.

Subsequently Aldrich and I calculated the spin fluctuation exctation spectrum using an expression which is directly analogous to (5),

$$\chi^{I}(q\omega) = \frac{\chi_{sc}^{I}(q\omega)}{1-\left(f_q^a + (\omega^2/q^2)\, f_q^{v,a}\right)\chi_{sc}^{I}(q\omega)} \tag{14}$$

while taking for $\chi_{sc}^{I}(q\omega)$ the spin analogue of the expression, (8). Because spin current is not conserved, the multipair response functions $\chi_m^I(q\omega)$ and $\chi_m^c(q\omega)$ are not equal, while one can show using sum rule arguments that the mode coupling parameter, α_q^I, is not equal to α_q^c. From the expression, (14), one finds readily the low frequency enhancement of the spin pair excitations [by a factor of $(1 + f_q^a\, \nu(0))^{-2}$] which may be expected to persist as long as $f_q^a \nu(0) \approx -1$.

In the course of our calculations we had to develop a physical model for f_q^a, as well as examining more closely our earlier model for f_q^s. We treated these effective field parameters as the Fourier transforms of the spin symmetric and anti-symmetric combinations of pseudopotentials $f^{\uparrow\uparrow}(r)$ and $f^{\uparrow\downarrow}(r)$ which describe the effective interaction between quasiparticles and quasiholes of parallel and antiparallel spin respectively; thus

$$f_q^{s,a} = \frac{4\pi}{V} \int_0^{\infty} dr \left[\frac{f^{\uparrow\uparrow}(r) +,- f^{\uparrow\downarrow}(r)}{2} \right] r^2 \frac{\sin qr}{qr} \tag{15}$$

The success of our calculation of the zero sound spectrum for ^3He tells us that the pseudopotentials $f^{\uparrow\uparrow}(r)$ and $f^{\uparrow\downarrow}(r)$ must resemble $f^s(r)$ for ^4He. Qualitatively it is clear that $f^{\uparrow\uparrow}(r)$ and $f^{\uparrow\downarrow}(r)$ cannot differ substantially from each other, or from $f^s(r)$, since at all pressures the spin antisymmetric combination possesses a quite small (negative) zeroth moment, $Nf_0^a = -0.67$ K, as compared to the spin-symmetric moment, $Nf_0^s = 11.3$ K, at svp, and the latter quantity increases with pressure. We found it possible to place a limit on the relative differences in the range of the repulsive interaction for parallel spins, $r_{\uparrow\uparrow}$, and that for antiparallel spins, $r_{\uparrow\downarrow}$,

$$\delta = (1 - r_{\uparrow\downarrow}/r_{\uparrow\uparrow}) . \tag{16}$$

δ furnishes a direct measure of the relative effectiveness of the Pauli principle [compared to that of the strong repulsive interaction] in keeping particles of parallel spin apart. We showed that if $\delta \gtrsim 2.5\%$, then an antiferromagnetic instability would appear at finite q; because Sköld et al. do not observe such an instability in their neutron scattering experiments, δ must be less than 2.5% at svp. However, difficulties in experimental resolution at long wavelengths $(q \lesssim 0.5\ \text{Å}^{-1})$ have precluded our making a definitive deter-

Following the initial experiments in which they discovered zero sound and established the existence of enhanced low frequency spin pair excitations, Sköld and Pelizzari[15] extended their neutron scattering results down to momentum transfers ~ 0.4 A, and carried out experiments both at 40 mK and at 1.2 K in order to study the temperature variation of the spin pair and the zero sound excitations. A preliminary report on the comparison between our theory and these experiments was given in Reference 14. Two aspects of that comparison merit special mention: first, in accord with the general physical arguments which form the basis for our unified theory, no change in the peak position of zero sound or any high frequency broadening of that mode was observed for $q \lesssim 1.2\ A^{-1}$ even though one has gone from a temperature region ($\sim$ 40 mK) in which the Landau theory is surely valid to one (1.2^{o} K) in which it is surely not. Second, at moderate wave-vectors ($q \sim p_F \sim 0.8\ A^{-1}$) there continues to be structure in the spin fluctuation portion of the excitation spectrum despite the broadening which necessarily accompanies a temperature $\sim E_F$. The likely explanation for this is that, in accord with our prediction, f_q^a has not changed, and the $(1 + f_q^a)^{-2}$ enhancement is not completely wiped out by the temperature broadening of χ_{sc}.

In Fig. 5 our recent (and as yet unpublished) calculations of the experimentally determined quantity, $S(\theta,\omega)$, are compared with the experimental results of Sköld and Pelizzari for a scattering angle of 31.2^{o}, which corresponds to momentum transfers between $0.8\ A^{-1}$ and

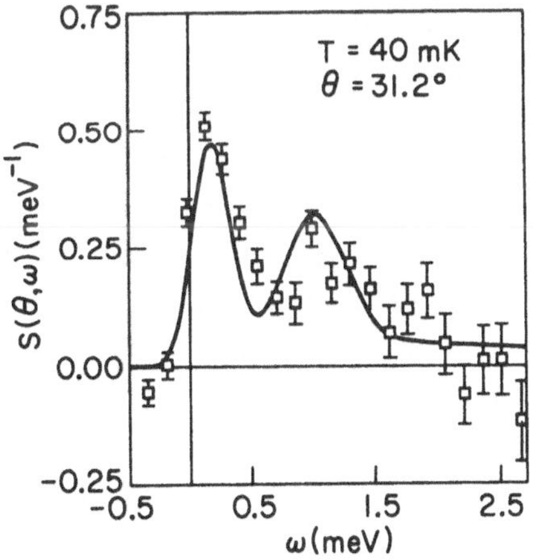

Fig. 5 Comparison with experiment of calculated spin pair excitation (the low frequency peak) and zero sound excitation (the high frequency peak) contributions to $S(\theta,\omega)$.

$0.85\ A^{-1}$. In these calculations we have taken into account instrumental broadening and the effects of both particle and spin multipair excitations; as may be seen, a good fit to the experimental results for the peak position, height, and width of both the low frequency spin pair excitations and the zero sound mode is found. Our fit to the data is equally satisfactory at larger scattering angles; at smaller angles it is necessary to introduce phenomenologically an additional broadening of the zero sound mode in order to fit the peak height and width satisfactorily. Hilton et al.[16] have independently reached the same conclusion--that the zero sound mode for $q \lesssim 0.8$ A is broader than can be explained by the effects of intrumental resolution. The likely physical explanation for this is that because of the substantial anomalous dispersion found in the ^3He zero sound excitation, decay of one phonon into two is sufficiently rapid to be detectable in neutron scattering experiments.

Transport Properties of ^3He

On a somewhat optimistic note, Aldrich, Pethick and I suggested in 1976, following our initial success in explaining the pioneering experiments of Sköld et al, that given better neutron scattering experiments and the development of physical models for $f^{\uparrow\uparrow}(r)$ and $f^{\uparrow\downarrow}(r)$, it might prove possible to develop a unified treatment of the Landau parameters and the transport properties of both normal and superfluid ^3He.[12] During the past year Kevin Bedell and I have been engaged in carrying out such a program, and I am pleased to report that it has met with success.[17] What we have done is to use the previously determined Aldrich-Pines polarization potentials, with δ regarded as a free parameter, to construct scattering amplitudes for quasiparticles in ^3He. The AP polarization potentials represent effective quasiparticle-quasihole interactions; to obtain the quasiparticle scattering amplitudes we obtain an analytic solution of the generalized Bethe-Salpeter equation which connects this particle-hole interaction to the scattering amplitude, and then use an algorithm which corresponds to a generalization of the s-p apprximation of Dy and Pethick[18] to construct triplet and singlet scattering amplitudes. The four measurable transport properties of normal ^3He, the thermal conductivity, K, viscosity, η, spin diffusion, D, and quasiparticle lifetime, τ, can then be calculated by taking various angular averages of the squared triplet and singlet scattering amplitudes; our results are copared with experiment and with the s-p approximation in Fig. 6a; the agreement with experiment is good at all pressures up to 34 atm.

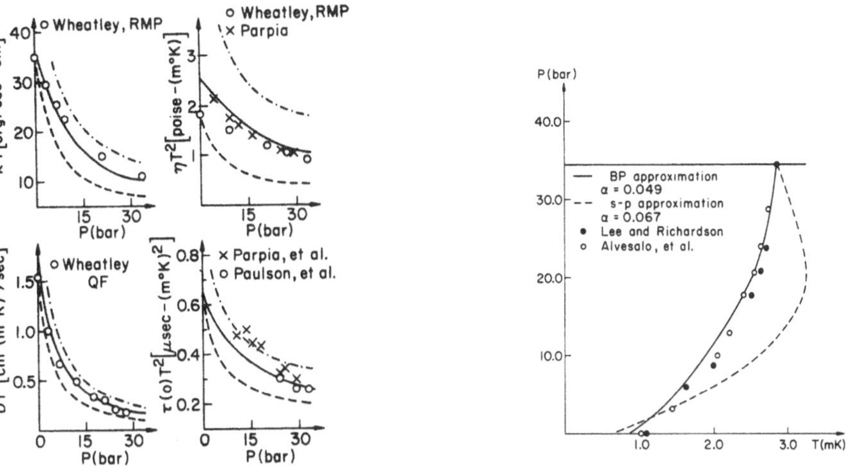

(a) (b)

Fig. 6a Comparison of theory with experiment for transport properties of ^{3}He.[17] The solid lines represent the best fit Bedell and Pines obtain by varying δ; the dot-dashed line corresponds to $\delta=0$; the dashed line gives results found in the s-p approximation. 6b. Comparison of theory with experiment for the superfluid transition temperature.[19] The solid line is the BP calculation; the dashed line the s-p approximation result.

The Pauli-principle parameter δ which yields the best fit to experiment is 0.11 at svp and decreases gradually to 0.0037 at ~20 atm, after which point it is approximately constant for pressures up to 35 atm. This pressure dependence of δ may be interpreted as zero-point motion enhancement of the influence of the Pauli principle at pressures $\lesssim$ 20 atm. Since, as we have noted earlier, such zero point motions do not appear to play a substantial physical role for pressures $\gtrsim$ 20 atm, it is appealing to interpret the high pressure value, δ = .0037, present from 20 atm to 34 atm, as describing the effect of the "pure" Pauli principle.[19] It should be emphasized that while δ is small, it plays an important role in determining the transport properties, as may be seen in Fig. 6a. The sensitivity of the transport properties to δ may be traced to their comparatively great sensitivity to f_q^a, which is itself quite sensitive to changes in δ.

The Superfluid Transition Temperature of Liquid ^{3}He

Since both the temperature, T_c, at which the normal Fermi liquid makes a transition to the superfluid phase, and the nature of the pairing in that phase, depend upon these same quasiparticle scattering amplitudes, it was to be expected that Bedell and I would investigate the superfluid transition. For ℓ-state pairing, the transition temperature is given by[20]

$$T_c = 1.13\ \gamma_\ell\ T_F \exp 1/g_\ell \qquad (17)$$

where g_ℓ, which must be attractive for the transition to occur, is an angular average of the scattering amplitudes, and γ_L is a renormalization constant which provides a measure (relative to E_F) of the energies at which the actual frequency-dependent interaction shifts from being attractive to repulsive. In carrying out our calculations, we used our previously determined quasiparticle scattering amplitudes to calculate g_ℓ; we found that at all pressures the pairing interaction is most attractive in the ^3P state, in accord with experiment.[19] The physical origin of the ^3P pairing may be traced to the difference in the range of the effective repulsion between particles of parallel and anti-parallel spin, which in turn determines the momentum dependence of f_q^a. We then determined $\alpha_L [\cong 0.05]$ by fitting our calculated T_c to experiment at a single pressure, the melting pressure, and keeping this value for all lower pressures. Our results are compared with experiment, and with the s-p approximation, in Fig. 6b. Given the fact that the transition temperature depends exponentially upon the averaged scattering amplitudes, the agreement between theory and experiment is not only satisfying--it is surprisingly good.

Concluding Remarks

Some problems to which polarization potential theory is currently being applied, and for which promising results have already been obtained, include:

(i) Roton-roton interactions in liquid ^{4}He (DP with Bedell and Zawadowski).

(ii) Excitations and transport in ^{3}He-^{4}He mixtures (DP with Aldrich, Bedell and W-C. Hsu).

(iii) Strong-coupling corrections and specific heat discontinuities for the three superfluid phases of ^{3}He (K. Bedell).

(iv) Transport properties of ^{3}He-B (Bedell and Hsu).

Finally, there are a host of other problems for which it may prove useful to apply the polarization potential formulation presented here: neutron and nuclear matter, the strongly coupled electron liquid, electron-electron and electron-ion interactions in solids, etc, by developing the appropriate model pseudopotentials for the particular problem at hand.

Acknowledgement

I should like to take this opportunity to thank my collaborators in this research, C. H. Aldrich, K. Bedell and C. J. Pethick for numerous helpful and stimulating discussions, and to acknowledge the support of the National Science Foundation (NSF Grant DMR 78-21068).

List of References

1. A.D.B. Woods, Phys. Rev. Letters 14, (1965) 355.
2. D. Pines, in Quantum Fluids, D. Brewer, ed. (North-Holland, Amsterdam, 1966), p. 257.
3. D. Pines and D. Bohm, Phys. Rev. 85, 338 (1952).
4. K. Sköld, C. A. Pelizzari, R. Kleb, and C. E. Ostrowski, Phys. Rev. Lett. 37, 842 (1976).
5. C. H. Aldrich III, Ph.D. Thesis, Univ. of Ill., 1974 (unpublished).
6. C. H. Aldrich III and D. Pines, J. Low Temp. Phys. 25, 677 (1976).
7. C. H. Aldrich III, C. J. Pethick, and D. Pines, J. Low Temp. Phys. 25, 691 (1976).
8. R. C. Dynes and V. Narayanamurti, Phys. Rev. B 12, 1720 (1975).
9. E. Feenberg, Phys. Rev. Lett. 26, 301 (1971); M. P. Kemoklidze and L. P. Pitaevskii, Sov. Phys. JETP 32, 1183 (1971).
10. D. Greywall, Phys. Rev. B18, 2127 (1978).
11. R. Scherm, W. G. Stirling, A.D.B. Woods, R. A. Cowley, and G. J. Coombs, J. Phys. C 7, L341 (1974).
12. C. H. Aldrich, C. J. Pethick and D. Pines, Phys. Rev. Lett. 37, 845 (1976).
13. W. G. Stirling, R. Scherm, P. A. Hilton, and R. Cowley, J. Phys. C 9, 1643 (1976).
14. C. H. Aldrich and D. Pines, J. Low Temp Phys. 32, 689 (1978).
15. K. Sköld and C. A. Pelizzari, J. Phys. C 11, L589 (1978).
16. P. A. Hilton, R. A. Cowley, R. Scherm, and W. G. Stirling, J. Phys. C 13, L295 (1980).
17. K. Bedell and D. Pines, Phys. Rev. Lett. 45, 39 (1980).
18. K. Dy and C. J. Pethick, Phys. Rev. 185, 373 (1969).
19. K. Bedell and D. Pines, Phys. Lett. 78A, 281 (1980).
20. See, for example, B. Patton and A. Zaringhalam, Phys. Lett. 55A, 95 (1975).

Appendix

In this Appendix, I should like to assess the current status of polarization potential theory in Socratic fashion, calling attention to some of the questions (Q) which have been or can be raised concerning the present formulation and then giving my current, occasionally tentative, answers (A).

Q: To what extent is one justified in treating the polarization potentials, f_q^s, f_q^v, etc, as frequency independent? A: Insofar as these represent physically either the effect of the single pair (or quasiparticle) fields or, as is the case with the short-range screening, describe in a quasistatic limit what is basically a comparatively high frequency phenomenon associated with virtual multipair excitations, it should be a good approximation to neglect any frequency dependence, as we have done. On the other hand, when one begins to look closely at the role played by the real multipair or multiparticle excitations under circumstances that these are not widely separated in energy from the collective modes whose energy we seek to calculate, it is clear that the static appraoch to this coupling begins to break down. Fred Zawadowski, Aldrich, and I have been studying this problem; we have found that we can reformulate the Ruvalds-Zawadowski hybridization approach (involving the coupling of a possible two-roton bound state or resonance) in polarization potential language, and that what results is a frequency dependent coupling of the multiparticle excitations to the collective modes (calculated in the absence of mode-mode coupling) which may in turn be described by a new set of polarization potentials which are frequency-dependent. Our goal here is an amalgam of hybridization and polarization potential theory.

Q: Does one run into problems of "double-counting?" A: Not in the static limit, because the influence of the virtual multiparticle excitations on the polarization potentials is treated implicitly, through the short-range screening effects, while the influence of the real multiparticle excitations is dealt with explicitly; in the dynamic model mentioned above, one can also avoid problems of double-counting by subtracting off any contributions to the restoring forces which have already been taken into account.

Q: How accurate are the present parameters which describe the strengths of the polarization potentials for the Helium liquids? A: For ^{3}He the excellent fit to experiment which Bedell and I obtain with the angular averages of the scattering amplitudes calculated from the AP polarization potentials leads me to suspect that the latter might be known to something like the 10% level of accuracy. f_q^s is likely known to some 10% for ^{4}He at all pressures; if one tries radically different values, based, say on physical models for $f_s(r)$ in which there is considerable screening of the attractive part of the interaction or one uses a quite different range, r_c, for the repulsive part, one finds that a detailed quantitative fit to the neutron scattering experiments is no longer possible. On the other hand, for ^4He f_q^v is known somewhat less well, (say to 20-25%), in part because if one comes up with a somewhat different value for f_o^v for ^4He or its momentum dependence, Aldrich and I find that it is possible to compensate for this to a considerable extent by choosing a different set of mode-mode coupling parameters. We lack, at present, a clear physical basis for choosing either to the same accuracy that we can pin down f_q^s for that system.

QUASIPARTICLE PROPERTIES IN NUCLEAR MATTER

B. L. Friman*

Department of Physics
University of Illinois at Urbana-Champaign
Urbana, IL 61801

1. Introduction

I would like to discuss here work I have been doing, in collaboration with J.-P. Blaizot, on understanding the nucleon effective mass m^* in nuclear matter.

The analysis of empirical data on the density of single-particle levels around the Fermi surface in nuclei[1,2] is consistent with $m^* \simeq m$, the free nucleon mass. On the other hand the energy dependence of the optical potential[3] and the energy of deeply bound states[4] suggest that far away from the Fermi surface the effective mass is more like 0.7 m. Thus, as noted already by Brown et al.[1] the effective mass is apparently enhanced near the Fermi surface.

Bertsch and Kuo[5] pointed out that the coupling of the particle to two-particle one-hole (and two-hole one-particle) configurations, like the process shown in Fig. 1a enhances the effective mass near the Fermi surface.

FIG. 1

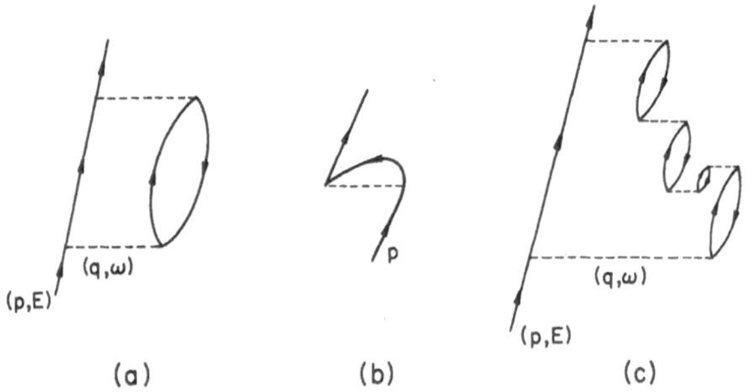

Later Jeukenne et al.[6] showed that the enhancement is present also in infinite nuclear matter.

The effect can be understood as follows. For momenta p close to the Fermi surface, most of the more complicated states lie further away from the Fermi surface than the single particle (or single hole) states. This is due to the Pauli exclusion principle restricting the phase space of the two-particle one-hole states. Thus the energy denominators corresponding to the intermediate state in diagram 1a are predominantly negative (positive for hole states), which pushes the quasiparticle states towards the Fermi surface, thereby enhancing the effective mass.

---

* Research Supported in part by NSF grants DMR 78-21068, PHY 78-26582 and DMR 78-21069.

The process shown in Fig. 1b on the other hand gives a contribution to the effective mass which varies slowly with energy.

In this work we evaluate the nucleon self-energy in nuclear matter in a simple model which enables us to study the effect of certain higher order diagrams, not previously included in this context. The model includes π- and ρ-exchange, the effect of shortrange correlations and nucleon form factors. Our main aim is to study the effect of screening of the two-particle one-hole contribution to the self-energy (Fig. 1c). We also examine the dependence of m^*/m on the range of the interaction.

2. The Nucleon Self-Energy

The quasiparticle energy is given by

$$\varepsilon_p = p^2/2m + \Sigma(p, \varepsilon_p) , \tag{1}$$

where $\Sigma(p,\varepsilon)$ is the nucleon self-energy. The effective mass is related to the density of single-particle levels by

$$m^* = p \frac{dp}{d\varepsilon_p} . \tag{2}$$

From eqs. (1) and (2) it follows that

$$\frac{m^*}{m} = \frac{\left[1 - \frac{\partial \Sigma(p,\varepsilon)}{\partial \varepsilon}\right]_{\varepsilon=\varepsilon_p}}{\left[1 + \frac{m}{p} \frac{\partial \Sigma(p,\varepsilon)}{\partial p}\right]_{\varepsilon=\varepsilon_p}} \tag{3}$$

The denominator

$$\frac{\tilde{m}}{m} = \left[1 + \frac{m}{p} \frac{\partial \Sigma(p,\varepsilon)}{\partial p}\right]^{-1}_{\varepsilon = \varepsilon_p} , \tag{4}$$

which gets most of its contribution from diagrams like 1b, is slowly varying. Therefore all the structure in m^*/m comes from the numerator

$$\frac{\bar{m}}{m} = \left[1 - \frac{\partial \Sigma(p,\varepsilon)}{\partial \varepsilon}\right]_{\varepsilon = \varepsilon_p} . \tag{5}$$

Here we use the notation of Jeukenne et al.[6]

Let us now assume that the particle-hole interaction $V^m_{ST}(q)$ depends only on the momentum q, spin S, spin projection m and isospin T of the particle-hole pair. Then the contribution to the nucleon self-energy from diagrams 1a and c is of the form[7]

$$\Sigma(p,\varepsilon) = i \sum_{S,m,T} (2T+1) \int \frac{d^3q\,d\omega}{(2\pi)^4} G(p-q, \varepsilon-\omega) \left[V^m_{ST}(q)\right]^2 \chi^m_{ST}(q,\omega) , \tag{6}$$

where $G(p,\varepsilon)$ is the nucleon Green's function and $\chi^m_{ST}(q,\omega)$ is the response function in the channel S,m,T. To lowest order (Fig.1a) $\chi^m_{ST}(q,\omega)$ is just the Lindhard function[7] $U(q,\omega)$, which is independent of spin and isospin.

Let us first take a look at the imaginary part of $\Sigma(p,\varepsilon)$. Close to the Fermi surface[8,9]

$$|\mathrm{Im}\Sigma(p,\varepsilon)| = \alpha\,(\varepsilon-\varepsilon_F)^2 - \beta|\varepsilon-\varepsilon_F|^3 \quad , \tag{7}$$

where ε_F is the Fermi energy (including the potential part). This form is obtained by analyzing the phase-space available for the particle (hole) to decay into two particles and a hole (two holes and a particle) (see Fig. 2). This process dominates the imaginary part close to the Fermi surface. Off the energy shell ($\varepsilon \neq \varepsilon_p$) the imaginary part is actually a bit more complicated,[10] but for the present discussion (7) is adequate.

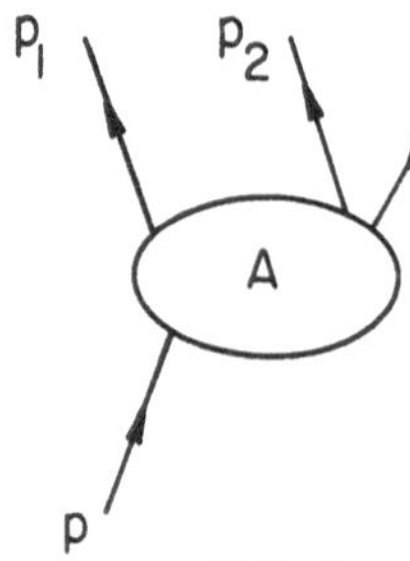

FIG. 2

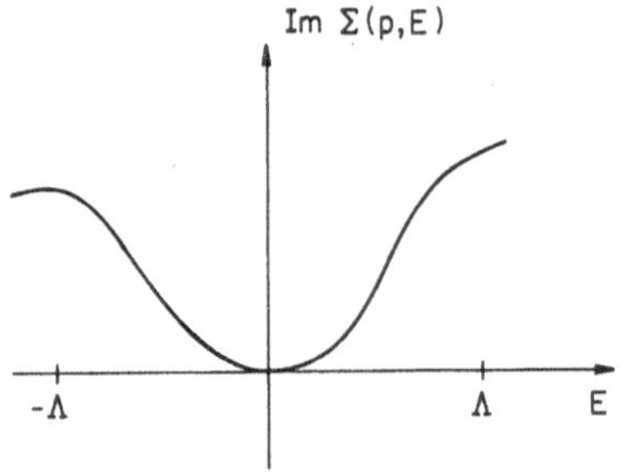

FIG. 3

In Fig. 3 we show the typical form of $\mathrm{Im}\Sigma(p,\varepsilon)$. For a well behaved interaction $\mathrm{Im}\Sigma(p,\varepsilon)$ vanishes as $\varepsilon \to \infty$, so that the real and imaginary parts of $\Sigma(p,\varepsilon)$ are related by an <u>un</u>subtracted dispersion relation[8,11]

$$\mathrm{Re}\Sigma(p,\varepsilon) = V_{HF}(p) + (P/\pi)\int_{-\infty}^{\infty} \frac{|\mathrm{Im}\Sigma(p,\varepsilon')|}{\varepsilon-\varepsilon'}\, d\varepsilon' \quad . \tag{8}$$

Here $V_{HF}(p)$ is the Hartree-Fock potential and P denotes principal value.

Inserting eq. (7) into (8) we find

$$\mathrm{Re}\Sigma(p,\varepsilon) - V_{HF}(p) = -2\alpha(\varepsilon-\varepsilon_F)\Lambda - 2\beta(\varepsilon-\varepsilon_F)^3 \ln\left|\frac{\varepsilon-\varepsilon_F}{\Lambda}\right| , \tag{9}$$

where we have kept only the leading terms. The integral (8) is cut off at $|\varepsilon'-\varepsilon_F| = \Lambda$. The cutoff parameter Λ should satisfy $\Lambda \ll \alpha/\beta$, so that the expansion (7) of $\mathrm{Im}\Sigma(p,\varepsilon)$ is valid over the range of integration. The effective mass is then of the form

$$\frac{\bar{m}}{m} = 1 + 2\alpha\Lambda + 6\beta(\varepsilon-\varepsilon_F)^2 \ln\left|\frac{\varepsilon-\varepsilon_F}{\Lambda}\right| , \tag{10}$$

where we again have retained only leading terms. By comparing (7) and (10) we see that the quadratic term in $\mathrm{Im}\Sigma(p,\varepsilon)$ gives rise to a constant enhancement of $\bar{m}/m$. The cubic term on the other hand, gives a state dependent contribution, which cuts down the effective mass as the particle moves away from the Fermi surface. Thus we see that the state dependence of the effective mass is due to the general form of $\mathrm{Im}\Sigma(p,\varepsilon)$ near ε_F. Therefore an enhancement of m^*/m near the Fermi surface is a general property of normal Fermi liquids, as pointed out by Jeukenne et al.[6] The magnitude of the effect is of course dependent on the details of the interaction.

If we assume that the interaction A in Fig. 2 is of the Yukawa form

$$A = V(q) = V_o \frac{\mu^2}{q^2+\mu^2} \tag{11}$$

(and neglect exchange terms), we find that the coefficient α is given by

$$\alpha = \frac{V_o^2 N_o}{16\pi v_F^2} q_c \ , \tag{12}$$

where

$$q_c = \int_o^{2p_F} (V(q)/V_o)^2 \, dq = p_F \left(\frac{\mu^2}{\mu^2+4p_F^2} + \frac{\mu}{2p_F} \tan^{-1} \frac{2p_F}{\mu} \right) , \tag{13}$$

$N_o = 2p_F m^*/\pi^2$ is the density of states at the Fermi surface and v_F is the Fermi velocity. For $\mu = m_\pi$ $q_c = 0.4\ p_F$, for $\mu = 5.5\ m_\pi$ $q_c = 1.5 p_F$ ($p_F = 1.35\ \mathrm{fm}^{-1}$) and in the limit $\mu \to \infty$ $q_c = 2p_F$. Thus we would expect the short-range part of the interaction to be most effective at producing an enhancement of the effective mass.

The coefficient β on the other hand is much less dependent on μ. The leading contribution

$$\beta = \frac{V_o^2 N_o}{24\pi v_F^3} \tag{14}$$

comes from the small q region of the integral in (6), and is therefore independent of μ (for $\mu >> \Lambda/v_F$). In this approximation there are additional terms[10] coming from $q \sim 2p_F$. (It is unclear whether these terms survive when repeated scattering of pairs of quasiparticles is taken into account.[12]) For $\mu = m_\pi$ this contribution is negligible, while for $\mu = 5.5\ m_\pi$, it increases β by 40%.

Let us now discuss the effect of screening, i.e., Fig. 1c. In our model V(q) is then replaced by[10]

$$V_{eff}(q) = \frac{V(q)}{1 - V(q)U(q,0)} \ . \tag{15}$$

The effect of screening is twofold; an overall change in the strength of the interaction and a change in the range. This is illustrated in Fig. 4, where we show V(q) (dashed line) and $V_{eff}(q)$ (full line) for $\mu = m_\pi$ (left figure) and $\mu = 5.5\ m_\pi$ (right figure). We choose $V_o = f_\pi^2/m_\pi^2$, so that

$$V(0)\ U(0,0) = -V_o N_o = -0.9 \ . \tag{16}$$

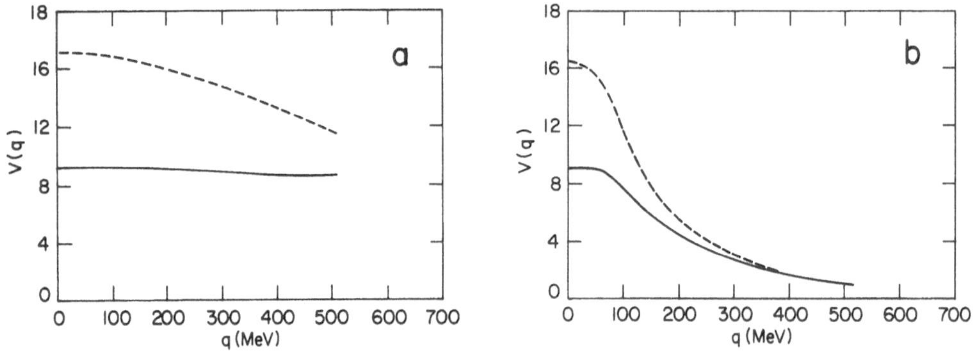

FIG. 4

At $q = 0$ the strength of $V_{eff}(q)$ is thus reduced by roughly 50% compared to $V(q)$. The q-dependence of $V_{eff}(q)$ is much smoother, reflecting a shorter range in r-space. As a consequence of the smoother q-dependence, the parameter q_c is increased, when we include screening; $q_c = 0.6p_F$ for $\mu = m_\pi$ and $1.9p_F$ for $\mu = 5.5\ m_\pi$. Thus, since the coefficients α and β are strongly reduced by screening, we expect a similar reduction of the effective mass enhancement.

3. Results and Discussion

In this section we discuss results of numerical calculations of the nucleon self-energy.[10]

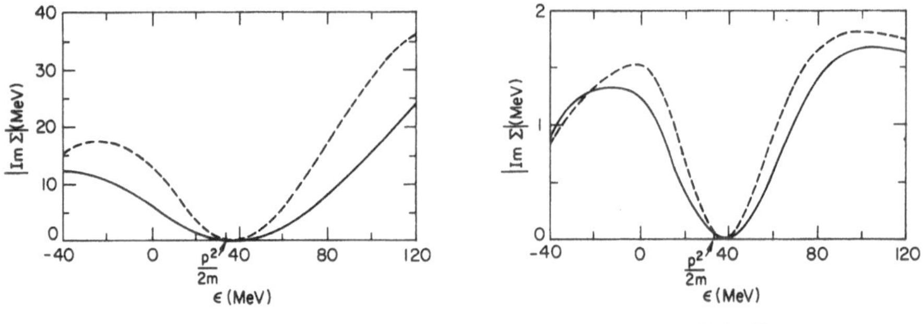

FIG. 5a FIG. 5b

In Fig. 5 we show the imaginary part of $\Sigma(p,\varepsilon)$ in lowest order (dashed line) and with screening (full line) for $p = 250$ MeV/c ($p_F = 266$ MeV/c). The interaction is the same as above (eq. 11) with $V_o = f_\pi^2/m_\pi^2$. Figure 5a is for $\mu = 5.5\ m_\pi$ and 5b for $\mu = m_\pi$. As expected $\mathrm{Im}\Sigma(p,\varepsilon)$ is much smaller for the long range interaction (5b). Note also that $\mathrm{Im}\Sigma(p,\varepsilon)$ peaks closer to the Fermi surface for μ smaller. This is, at least partly, due to the μ-dependence of q_c since the energy-scale in (7) is set by $\alpha/\beta \simeq v_F q_c$.

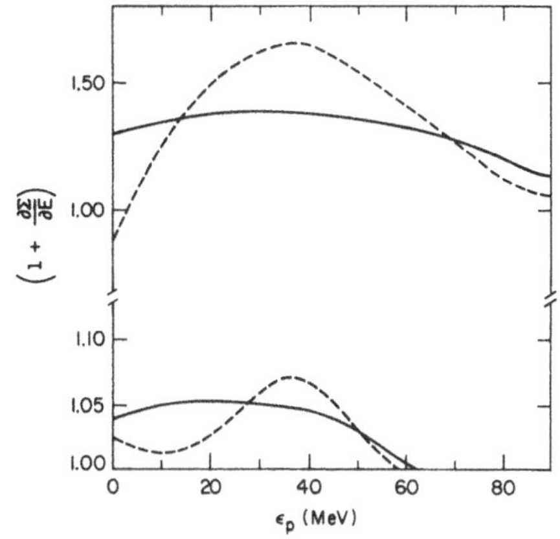

FIG. 6

In Fig. 6 we show $\bar{m}/m$ for the same interaction. The notation is the same as in Fig. 5, with the $\mu = m_\pi$ results in the lower part of the figure. The quasiparticle energy was approximated by the kinetic energy. Here we clearly see how the short range interaction produces a much larger and wider enhancement than the one pion exchange. The increase in the width of the enhancement with increasing μ reflects the widening of the imaginary part discussed in connection with Fig. 5. Further, the inclusion of screening dramatically reduces the magnitude and increases the width of the enhancement. This is consistent with our expectations, since screening reduces both the strength and the range of the interaction.

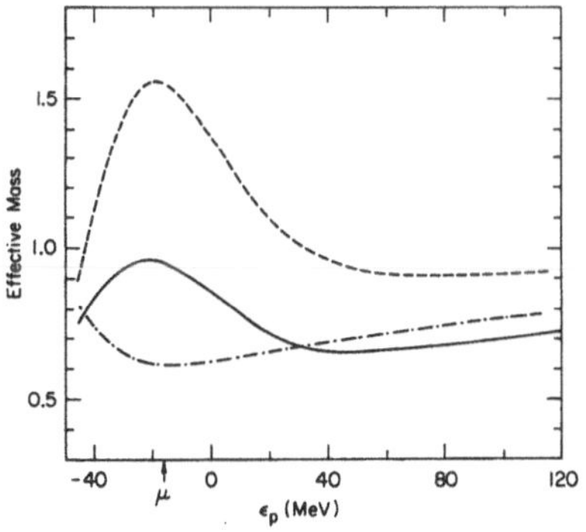

FIG. 7

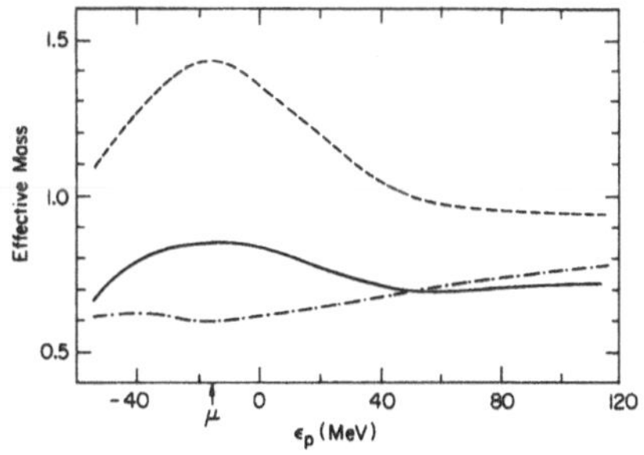

FIG. 8

In Figs. 7, 8 and 9 we show the results of a more ambitious calculation. Here we take the particle-hole interaction to include π- and ρ-exchange, form factors and the effect of short-range correlations. The $q = 0$ limit is adjusted so as to give a reasonable value for the Landau-Migdal parameter G_o'. We take $G_o' = N_o(f_\pi^2/m_\pi^2)g'$ = 1.33 i.e., $g' = 0.5$ (see e.g. ref. 13). This gives a reasonable model for the particle-hole interaction in the S = T = 1 channel.[14] Since this channel carries the largest statistical weight (see eq. (6)) it is the most important one for determining the quasiparticle properties.

The same interaction is used in evaluating the Hartree-Fock potential. However, in order to reproduce roughly the overall energy dependence of the single particle potential[15] we increase the magnitude of $V_{HF}(p)$ by 25%. This decreases the Hartree-Fock effective mass from 0.72 m to 0.65 m. The latter value is consistent with both variational[15] and Brueckner type[13] calculations of m^*/m. Further details may be found in ref. 10. The scales of Figs. 7, 8 and 9 are adjusted so that the Fermi energy ε_F = -16 MeV and the single-particle potential $U(\varepsilon_p=\varepsilon_F)$ = -54 Mev.

In Fig. 7 we show $\bar{m}/m$ (dashed line), $\tilde{m}/m$ (dash-dot line) and m^*/m (full line) as functions of the quasiparticle energy ε_p. The self-energy includes the Hartree-Fock potential and the particle-hole contribution to lowest order (Fig. 1a). Fig. 8 is the same, except that the particle-hole bubble was summed to all orders.

By comparing the two figures we see how the inclusion of screening strongly smoothens the state dependence of m^*/m. The effective mass varies from 0.95 at ε = -16 MeV to 0.65 at ε = 40 MeV in the lowest order results, while the corresponding values are 0.85 at the Fermi surface and 0.69 at ε = 60 MeV, when screening is included. Thus, we conclude that a realistic calculation of the quasiparticle effective mass must include screening.

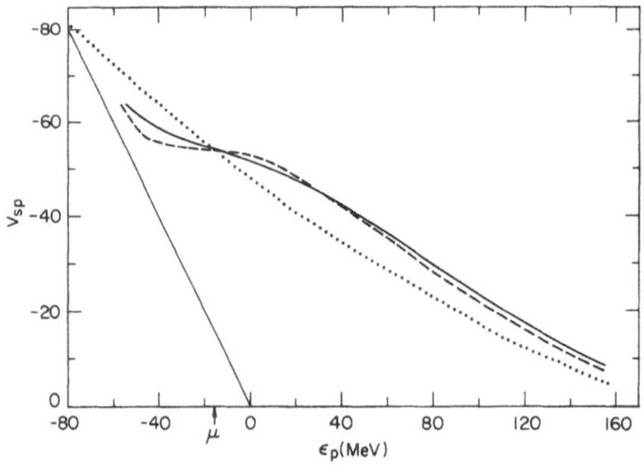

FIG. 9

In Fig. 9 we show the single-particle potential $U(\varepsilon_p)$ as a function of energy. The three curves represent different approximations to the self-energy:

i) dotted line; the Hartree-Fock potential.

ii) dashed line; the H-F potential plus the particle-hole contribution to lowest order.

iii) full line; the H-F potential plus the particle-hole contribution with screening.

The straight line in the left part of the Figure is $U(\varepsilon_p) = \varepsilon_p$. Since $\varepsilon_p - U(\varepsilon_p)$ is the kinetic energy, which is positive, $U(\varepsilon_p)$ must always lie to the right of this line.

The two curves that include particle-hole effects show a clear plateau around the Fermi surface. Both the magnitude and range (in energy) of this effect are in rough agreement with the data.[15] A more detailed comparison is not possible, since the data is too scattered.

References

1. G. E. Brown, J. H. Gunn and P. Gould, Nucl. Phys. 46, 598 (1963).
2. J. P. Blaizot, Phys. Reports 64C, 171 (1980).
3. A. Bohr and B. R. Mottelson, Nulear Structure, Vol. 1, Benjamin, New York, 1969.
4. J. Mougey et al. Nucl. Phys. A262, 461 (1976).
5. G. F. Bertsch and T. T. S. Kuo, Nucl. Phys. A112, 204 (1968).
6. J. P. Jeukenne, A. Leujenne and C. Mahaux, Phys. Reports 25C, 83 (1976).
7. A. L. Fetter and J. D. Walecka, Quantum Theory of many-particle systems, McGraw-Hill, New York, 1971.
8. J. M. Luttinger, Phys. Rev. 121, 942 (1961).
9. D. J. Amit, J. W. Kane and H. Wagner, Phys. Rev. 175, 313 (1968).
10. J. P. Blaizot and B. L. Friman, preprint 1981.
11. R. Sartor and C. Mahaux, Phys. Rev. C21, 1546 (1980).
12. G. Baym and C. J. Pethick, in Physics of liquid and solid helium, Vol. 2, eds. K. H. Bennemann and J. B. Ketterson, Wiley-Interscience, New York, 1978.
13. S.-O. Bäckman, O. Sjöberg and A. D. Jackson, Nucl. Phys. A321, 10 (1979).
14. G. E. Brown, S.-O. Bäckman, E. Oset and W. Weise, Nucl. Phys. A286, 191 (1977).
15. B. Friedman and V. R. Pandharipande, preprint 1980.

NEW APPROACHES TO THE STUDY OF COLLECTIVE EXCITATIONS IN STRONGLY-INTERACTING FERMI SYSTEMS

D.G. Sandler[†] and N.-H. Kwong
W.K. Kellogg Radiation Laboratory
California Institute of Technology
Pasadena, California 91125/USA

and

J.W. Clark[‡]
McDonnell Center for the Space Sciences and Department of Physics
Washington University
St. Louis, Missouri 63130/USA

and

E. Krotscheck[††]
Department of Physics
State University of New York
Stony Brook, New York 11794/USA

Abstract: We discuss an extension of the random-phase approximation (RPA) which permits use of the strong two-body forces present in nuclear matter, finite nuclei and liquid ^{3}He. A method is outlined for solving the RPA equations at finite momentum transfer for infinite Fermi systems when renormalized single-particle energies and exchange matrix elements of the interaction are included explicitly. Results are given for nuclear matter using a schematic nucleon-nucleon potential. We conclude with a discussion of current applications and possibilities for gaining further insight into the low-lying excitations of dense Fermi systems.

The fact that enormous advances have occurred during the past few years in our quantitative description and understanding of the ground-state properties of many-fermion systems is strikingly displayed by many of the contributions to this conference. In addition, since the Trieste meeting[1] we have seen that the variational method[2] can be applied to the microscopic study of excitations in extended Fermi systems which are essentially "single-particle" in nature.[3,4] Certainly, it is time to extend the existing achievements by attempting a microscopic description of small-amplitude collective excitations of these uniform systems as well as the low-lying collective states of finite closed-shell nuclei. Of course, by "microscopic" we mean that we take as fundamental input the free-space two-body potential between constituent particles. This goal seems particularly relevant in light of the fact that a large portion of the empirical information we have at our disposal results from (or relates to) probing these excited states.

Past success, on a less microscopic level, suggests that we may accomplish our goal through a linearized description of the small-amplitude excitations of a strongly-correlated Fermi sea, by means of a semi-classical analysis of small oscillations about an equilibrium configuration into which strong correlations are directly built. The method of correlated basis functions[2,5] (CBF) presents an ideal framework from which to proceed. Briefly summarized, the CBF method constructs a non-orthogonal basis $\{|\psi_m\rangle\}$, each member of the form $|\psi_m\rangle = F|\Phi_m\rangle\, I_{mm}^{-1/2}$, where

---

[†]Supported in part by the National Science Foundation [PHY79-23638] at Caltech.
[‡]Supported in part by the National Science Foundation [DMR80-08229] at Washington U.
[††]Supported in part by the Duetsche Forschunsgemeinschaft and the U.S. Department of Energy [DE-AC02-76ER13001].

$I_{mm} = \langle \Phi_m | F^\dagger F | \Phi_m \rangle$. The set $\{|\Phi_m\rangle\}$ is a basis of model states which provides a zeroth-order description in that they include Fermi statistics and the essential symmetries of the A-body system. We will take $|\Phi_m\rangle$ to be a Slater determinant of orbitals $m = (m_1\, m_2 \ldots m_A)$, so that $|\Phi_o\rangle$ is the HF ground state. The operator $F(1\ldots A)$ incorporates into the basis certain strong correlations induced by the bare two-body potential $v(ij)$. This symmetric correlation operator is required to obey the cluster property[2]; a common choice is the state-independent Jastrow form $F = \Pi_{i<j} f(r_{ij})$.

The basic ingredients appearing in the implementation of CBF theory are the matrix elements of the Hamiltonian and the identity in the correlated basis , respectively denoted $H_{mn} = \langle \psi_m | H | \psi_n \rangle$ and $N_{mn} = \langle \psi_m | \psi_n \rangle$. (Note that in general $N_{mn} \neq \delta_{mn}$.) For example, the exact energy eigenvalues are solutions of the secular equation

$$\det\,(H_{mn} - E\, N_{mn}) = 0. \tag{1}$$

As an alternative to solving (1) by brute-force diagonalization, one usually adopts a systematic or approximate method of solution; for example, perturbation theory performed in the correlated basis results in an expansion for the ground state energy similar in form to the Rayleigh-Schrödinger series.[2,5] The first term in the expansion is $H_{oo} = E_o = \langle \psi_o | H | \psi_o \rangle$, the expectation value of H in the trial ground state $|\psi_o\rangle = F|\Phi_o\rangle\, I_{oo}^{-1/2}$. Typically, the specific nature of the correlations to be included in F is determined by minimizing E_o, evaluated by cluster expansion to some given order. Further CBF-based schemes resulting in equations strikingly similar to those arising within standard many-body formalisms will be discussed in another talk.[6] Suffice it to say that the emergence of new, more "highly-correlated" versions of conventional theories is not accidental — it follows from and is illustrative of the conceptual simplicity and richness of the CBF method.

We will now briefly outline a route leading to a correlated theory of small-amplitude collective excitations. (Details of this particular "correlated time-dependent-Hartree-Fock (TDHF)" approach will be given elsewhere[7]; a similar derivation leading to the same essential results can be found in Ref. 8.) Consider the time-dependent trial state

$$|\psi(t)\rangle = F|\Phi(t)\rangle / \langle \Phi(t)|F^\dagger F|\Phi(t)\rangle^{1/2}, \tag{2a}$$

where $|\Phi(t)\rangle$ is of the Thouless form[9,10]

$$|\Phi(t)\rangle = e^{-\frac{i}{\hbar} E_o t} \exp\{\sum_{ph} c_{ph}(t)\, a_p^\dagger a_h\}\, |\Phi_o\rangle\,, \tag{2b}$$

and the amplitudes $c_{ph}(t)$ of particle (p) - hole (h) excitations are small in magnitude. Equations (2a),(2b) may be alternatively written

$$|\psi(t)\rangle = |\chi(t)\rangle / \langle \chi(t)|\chi(t)\rangle^{1/2}\,, \tag{3a}$$

$$|\chi(t)\rangle = e^{-\frac{i}{\hbar} E_o t} \exp\{\sum_{ph} c_{ph}(t)\, \alpha_p^\dagger \alpha_h\}\, F|\Phi_o\rangle\,, \tag{3b}$$

where the "correlated creation and destruction operators" are defined by the relation $\alpha_p^\dagger \alpha_h F|\Phi_o\rangle I_{oo}^{-1/2} = F a_p^\dagger a_h |\Phi_o\rangle I_{ph,o}^{-1/2}$. We have adopted a notation convenient to the situation where 0p0h, 1p1h and 2p2h model states $|\Phi_m\rangle$ are pertinent: i.e., matrix elements are labelled as $\mathcal{O}_{oo} = \mathcal{O}_{o,o}$; $\mathcal{O}_{ph,o}$; $\mathcal{O}_{php'h',o}$; $\mathcal{O}_{ph,p'h'}$. Now define the correlated-TDHF functional

$$\mathcal{J}(t) = \langle \psi(t) | H - i\hbar \frac{\partial}{\partial t} | \psi(t) \rangle\,, \tag{4}$$

and require $\delta\mathcal{J}(t) = 0$, subject to the condition that F remains fixed. Explicitly we have

$$\delta\mathcal{J}(t) = \sum_{ph} \left[R_{ph}(t)\, \delta c^*_{ph}(t) + S_{ph}(t)\, \delta c_{ph}(t) \right] , \tag{5}$$

so that our constrained variation becomes equivalent to the statement

$$R_{ph}(t) = S_{ph}(t) = 0 , \tag{6}$$

for all ph and all t. The functions R_{ph} and S_{ph} have a rather complicated structure, involving sums of products of $\{c_{p'h'};\ c^*_{p'h'};\ \dot{c}_{p'h'};\ \dot{c}^*_{p'h'}\}$ and $\{H_{o,php'h'};\ H_{ph,p'h'};\ N_{o,p'h'};\ N_{o,ph};\ N_{ph,p'h'}\}$. Simplifications occur upon invoking the "correlated Brillouin condition"[2,8]

$$H_{ph,o} - E_o\, N_{ph,o} = 0, \qquad \text{all ph.} \tag{7}$$

Solution of Eq. (7) defines a correlated equilibrium configuration: it is satisfied trivially due to translational invariance if $|\Phi_o\rangle$ is the Fermi-gas ground state corresponding to the uniform extended system. It is now apparent that to investigate semi-classical oscillations about the correlated equilibrium configuration, we adopt the canonical decomposition

$$c_{ph}(t) = x_{ph}\, e^{-i\omega t} + y_{ph}\, e^{i\omega t} . \tag{8}$$

When (8) is inserted into Eq. (6), we arrive at the supermatrix equation

$$\begin{pmatrix} A & B \\ B^* & A^* \end{pmatrix} \begin{pmatrix} X \\ Y \end{pmatrix} = \hbar\omega \begin{pmatrix} M & 0 \\ 0 & -M^* \end{pmatrix} \begin{pmatrix} X \\ Y \end{pmatrix} , \tag{9}$$

where the elements of the component matrices are given by

$$\begin{aligned} A_{ph;p'h'} &= H_{ph,p'h'} - H_{oo}\, N_{ph,p'h'} , \\ B_{ph;p'h'} &= H_{php'h',o} - H_{oo}\, N_{php'h',o} \\ M_{ph;p'h'} &= N_{ph,p'h'} - N_{ph,o}\, N_{o,p'h'} . \end{aligned} \tag{10}$$

Equation (9) is identical to that of the celebrated random-phase approximation[9,10] (RPA) with the exception of the appearance of a non-trivial metric matrix. Actually, RPA-like equations with even more complicated metric matrices have arisen before in another context[11], although their origin can always be traced to the non-orthogonality of the basis employed.

It is obvious that solution of the "correlated RPA" (CRPA) equations requires evaluation of the 2p2h matrix elements of (10). For a finite system, it is at present only practical to do so in low-cluster order[5]; for sufficiently light nuclei (e.g., ^{16}O) this may suffice. However, for a uniform extended system at moderate densities and beyond, it is expected that a more highly-summed evaluation of these quantities will be necessary to achieve convergence of the CRPA frequencies and amplitudes. It is instructive to display the explicit integral-equation version of (9), (10) for the infinite system:

$$(e_p - e_h)x^{(n)}_{ph} + \sum_{p'h'} (\mathcal{V}_{h'p,p'h} - \mathcal{V}_{h'p,hp'})x^{(n)}_{p'h'} + \sum_{p'h'} (\mathcal{V}_{pp',hh'} - \mathcal{V}_{pp',h'h})\, y^{(n)}_{p'h'}$$

$$= \hbar\omega_n \sum_{p'h'} (\mathcal{N}_{h'p,p'h} - \mathcal{N}_{h'p,hp'})x^{(n)}_{p'h'}$$

$$(e_p - e_h)y^{(n)}_{ph} + \sum_{p'h'} (\mathcal{V}_{hh',pp'} - \mathcal{V}_{hh',p'p})x^{(n)}_{p'h'} + \sum_{p'h'} (\mathcal{V}_{hp',ph'} - \mathcal{V}_{hp',h'p})y^{(n)}_{p'h'}$$

$$= -\hbar\omega_n \sum_{p'h'} (\mathcal{N}_{hp',ph'} - \mathcal{N}_{hp',h'p})y^{(n)}_{p'h'} \qquad (11)$$

In (11), e_k, $\mathcal{N}$ and $\mathcal{V}$ are, respectively, the single-particle energies, compact correlation operator and (non-local) effective interaction of CBF theory[5,6]; for the Jastrow choice of F, e_k and the relevant matrix elements of $\mathcal{N}$ and $\mathcal{V}$ can be evaluated to FHNC level of accuracy.[6,12,13]

A few general remarks on CRPA are in order. First, when F = 1, the CRPA equations (11) (or, more generally, (9), (10)) collapse to the equations of the "generalized RPA"[14] for a HF potential; that is, HF single-particle energies and exchange matrix elements of the potential are explicitly included. Second, the presence of the non-trivial metric matrix does not destroy the conjugate relation between CRPA solutions, nor does it alter the theorem which guarantees real eigenfrequencies when the CRPA matrix is positive definite. In fact, one can derive a local stability criterion for the correlated equilibrium state $F|\Phi_0\rangle$ which assumes exactly the same form as the analogous HF condition.[8]

We now address the problem of solving the generalized-RPA equations for a uniform extended system. More specifically, we wish to solve (11) for the case F = 1 and the two-body potential v(12) is well-behaved at the origin. This is an important task for two reasons: i) we must have a viable and accurate method of solution for the HF problem to have any hope of solving the CRPA equations, and ii) we may be able to discern interesting features of the spectrum of low-lying excitations for nuclear matter. Suppressing spin and isospin, a reduction of the basis achieved by defining $\mathbf{q} = \mathbf{p} - \mathbf{h} = \mathbf{h}' - \mathbf{p}'$ results in the coupled equations

$$[\epsilon(\mathbf{k}+\mathbf{q}) - \epsilon(\mathbf{k})]x(\mathbf{k},\mathbf{q}) + \int_{Q_x} d\mathbf{k}'[\tilde{v}(\mathbf{q}) - \tilde{v}(\mathbf{k}-\mathbf{k}')]x(\mathbf{k}',\mathbf{q})$$
$$+ \int_{Q_y} d\mathbf{k}'[\tilde{v}(-\mathbf{q}) - \tilde{v}(\mathbf{k}'-\mathbf{k}-\mathbf{q})]y(\mathbf{k}',-\mathbf{q}) = \hbar\omega\, x(\mathbf{k},\mathbf{q})$$

$$[\epsilon(\mathbf{k}+\mathbf{q}) - \epsilon(\mathbf{k})]\, y(\mathbf{k},-\mathbf{q}) + \int_{Q_y} d\mathbf{k}'[\tilde{v}(\mathbf{q}) - \tilde{v}(\mathbf{k}-\mathbf{k}')]\, y(\mathbf{k}',-\mathbf{q}) \qquad (12)$$
$$+ \int_{Q_x} d\mathbf{k}'[\tilde{v}(-\mathbf{q}) - \tilde{v}(\mathbf{k}-\mathbf{k}'-\mathbf{q})\, x(\mathbf{k}',\mathbf{q}) = -\hbar\omega\, y(\mathbf{k},-\mathbf{q}).$$

The regions of integration $Q_{x,y}(\mathbf{k},\mathbf{q},k_F)$ are the appropriate Pauli-restricted subsets of the Fermi sphere corresponding to $x(\mathbf{k},\mathbf{q})$ and $y(\mathbf{k},-\mathbf{q})$, respectively. Upon expanding $x(\mathbf{k},\mathbf{q}) = \sum_m x_m(k,\theta_k^{\pm})e^{im\varphi}$, and similarly for $y(\mathbf{k},-\mathbf{q})$, Q_x and Q_y are defined in terms of $\cos\theta_k^{\pm}$. Solution of (12) then proceeds by partial-waving the various Fourier transforms $\tilde{v}(\mathbf{q},\mathbf{k},\mathbf{k}')$ of v(r), and imposing Gaussian quadrature on the sets $\{k\}$,$\{\cos\theta_k^{\pm}\}$. At a given $k_F = [(3\pi^2\rho)/2]^{1/3}$ for each q considered, diagonalization of a 2N x 2N matrix yields N distinct eigenvalues ω_n, n = 1. . .N, and their associated amplitudes $x^{(n)}$ and $y^{(n)}$. From these, one constructs the dynamic structure function[2,14]

$$S(q,\ \omega) = \frac{1}{A}\sum_{n} |\langle n|\rho_{\underset{\sim}{q}}|0\rangle|^2\ \delta(\omega - \omega_n) \quad , \tag{13}$$

where $\rho_{\underset{\sim}{q}} = \sum_{\underset{\sim}{k}} a^{+}_{\underset{\sim}{k}+\underset{\sim}{q}}\ a_{\underset{\sim}{k}}$ is the density-fluctuation operator. The spectrum $\omega(q)$ follows from the investigation of $S(q,\ \omega)$ over the entire range of q being considered.

We have carried through this procedure for nuclear matter using the schematic interaction of ref. 15. It consists of a repulsive barrier of height U_o = 1700 MeV and width c = 0.488 fm in all states, with an even-state attractive well of depth V_o = -52 MeV and range R_o = 2.10 fm. This simple potential leads to saturation (in HF) at k_F = 1.5 fm^{-1}, with a binding energy per nucleon of 11.3 MeV. (A new parameterization of this potential[16], yielding very accurate saturation properties, is now being treated within this framework. Further details of the method outlined here will be given in ref. 17). Of course, the spin-isospin dependence of the force necessitates use of the Pandya relations [15]in generalizing (12) to dependence on the spin-isospin ($\tilde{S}$, $\tilde{T}$) in the uncoupled p-h channel.

Some illustrative results of our generalized-RPA calculations at k_F = 1.5fm^{-1} shown in fig. 1. In (a) - (e), results of a smoothed histogram of $\rho(E) \equiv S(q,\ \omega)$ (in arbitrary units) for various values of q are sketched. All results shown are for the ($\tilde{S}$, $\tilde{T}$) = (1, 1) channel, associated with the giant dipole resonance. (Note that due to its SU(4)-invariance, the only other distinct channel is ($\tilde{S}$, $\tilde{T}$) = (0, 0). No collective modes appear here). Figures 1(a) and 1(b) compare $\rho(E)$ at q/k_F = 1/15, respectively corresponding to generalized RPA and "lowest-order" RPA, i.e., using a free spectrum for single-particle energies and only the direct matrix elements of the potential. Note the drastic raising of the resonance frequency and its relative suppression in strength due to the effective Pauli repulsion in the p-h interaction. Similarly, we find for the speed of sound in the (1, 1) channel $C_s^{1,1} \approx$ 0.6c, whereas the analysis of ref. 15, in which a free-spectrum was employed, gives $C_s^{1,1}$ = 0.33 c. Since the Fermi velocity at k_F = 1.5 fm^{-1} is v_F = 0.32 c, one sees that the ability of the interaction to support zero sound depends critically on inclusion of HF self-energies.[17] (Results of extended treatment, which parallels that of ref. 15 is in accord with this statement [16]). Figures 1(c) - 1(e) show $\rho(E)$ for q/k_F = 2/5, 2/3, 1, respectively. Landau damping gradually sets in, and ultimately the resonance peak completely merges with the continuum. In fig. 1(f), the spectrum $\omega(q)$ is displayed, illustrating the characteristic turn-off in the spectrum at this point of merging.

In conclusion, we briefly mention possible applications for the new CRPA formalism (some of which are under way), and we point out some aspects of the theory which deserve further attention or require clarification.

1) Calculations of $S(q,\ \omega)$ and $\omega(q)$ for nuclear matter, neutron matter and liquid ^3He are being carried out using singular coordinate-space potentials (e.g., Reid soft-core, Lennard-Jones). Also being performed are CRPA stability analyses for the p-h degree of freedom in nuclear and neutron matter, of relevance to pion condensation in these systems.[18]

2) We plan to calculate strength functions and sum rules for excitation of nuclear matter by various one-body operators $\mathcal{O} = \sum \mathcal{O}_i$. Within the framework of the CRPA, the amplitude for excitation from the ground state to state n is given by

$$P^{CRPA}_{0\to n} = (X^{(n)+}\ Y^{(n)+}) \begin{pmatrix} M & 0 \\ 0 & M^* \end{pmatrix} \begin{pmatrix} \mathcal{O}_{ph} \\ \mathcal{O}_{hp} \end{pmatrix} \quad . \tag{14}$$

3) The CRPA could be a vehicle toward determination of the quasiparticle

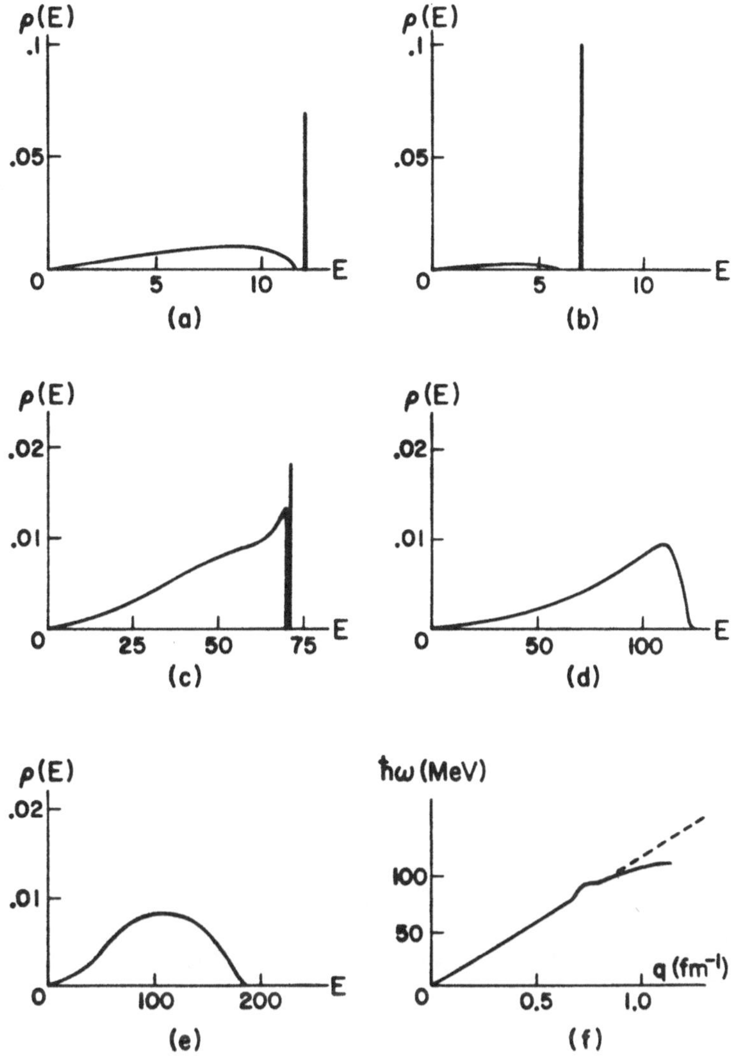

Fig. 1. Results of generalized-RPA calculations for nuclear matter using schematic interaction of Ref. 15.

interaction at the Fermi surface in nuclear matter and liquid ^3He. In the low-q limit, Eq. (9) reduces to a transport-like equation. If we make the usual correspondence with the Landau transport equation[14], extraction of the quasiparticle interaction should be straightforward. It has been shown[6] that the lowest-order quasiparticle-interaction should be given by the matrix element of the CBF effective interaction V on the Fermi surface. However, it is clear that what arises from the CRPA low-q limit is the matrix element of V "dressed" in some way by the M-matrix. We clearly need more insight into how (and why) this occurs.

4) The "Green function" of the CRPA is

$$G^{CRPA} = \sum_n \frac{1}{\omega_n - \omega - i\delta} \begin{pmatrix} \tilde{X}^{(n)} \\ \tilde{Y}^{(n)} \end{pmatrix} (\tilde{X}^{(n)+}\ \tilde{Y}^{(n)+}), \quad S^{+}MS = I, \quad S^{-1}X^{(n)} = \tilde{X}^{(n)} \quad (15)$$

It is of fundamental interest to ascertain the relationship of this "kernel" to the conventional RPA Green function, obtained from the ladder approximation to G. In other words, what is the diagrammatic function of M?

5) Finally, we mention that we are currently investigating the effects of varying the correlated-TDHF functional with respect to the correlation operator. Whether the short-range correlations are appreciably different in the ground state and a correlated collective state is an outstanding question. In addition, this study may shed some light on the related issue concerning the "orthogonality" of the high-momentum excitations included in F and the low-momentum excitations forthcoming from the CRPA.

D.G.S. and N.-H.K. acknowledge stimulating conversations with S. E. Koonin and K. R. Sandhya-Devi.

References

1. See the proceedings of the First International Conference on Recent Progress in Many-Body Theories, Trieste, Italy, 2-7 October 1978, in Nucl. Phys. A328 (1979).
2. J. W. Clark, in Progress in Particle and Nuclear Physics, ed. D. H. Wilkinson (Pergamon, Oxford 1979), vol. 2, p. 89; E. Feenberg, Theory of Quantum Fluids (Academic, N. Y., 1969).
3. V. R. Pandharipande, contribution to these proceedings; B. Friedman and V. R. Pandharipande, preprint.
4. J. C. Owen, contribution to these proceedings.
5. J. W. Clark, L. R. Mead, E. Krotscheck, K. E. Kürten and M. L. Ristig, Nucl. Phys. A328, 45 (1979).
6. E. Krotscheck, contribution to these proceedings.
7. D. G. Sandler, J. W. Clark and E. Krotscheck, to be published.
8. J. W. Clark, in "The Many-Body Problem, Jastrow Correlations versus Brueckner Theory" (Springer Verlag, Berlin), to be published.
9. D. J. Thouless, The Quantum Mechanics of Many-Body Systems (Academic, N. Y., 1972).
10. G. E. Brown, Unified Theory of Nuclear Models and Forces (North Holland, Amsterdam, 1971).
11. D. J. Rowe, Nuclear Collective Motion (Methuen, London, 1970).
12. E. Krotscheck and J. W. Clark, Nucl. Phys. A328, 73 (1979).
13. E. Krotscheck and R. A. Smith, to be published.
14. D. Pines and P. Nozières, The Theory of Quantum Liquids (Benjamin, N. Y., 1966).
15. W. M. Alberico, R. Cenni and A. Molinari, Riv. del Nuovo Cim. 1, 1 (1978).
16. W. M. Alberico, private communication.
17. N.-H. Kwong and D. G. Sandler, to be published.
18. G. Bertsch and M. B. Johnson, Phys. Rev. D 12, 2230 (1975).

IMPURITY QUASIPARTICLE AS A WEAK INHOMOGENEITY

A.Kallio, M.Puoskari+ and P.Pietiläinen

University of Oulu, Oulu, Finland

Moving impurity particle causing back flow in a background Bose- or Fermi liquid at T = 0 is treated as a weak inhomogeneity applying the inhomogenious HNC-theory. The Euler-Lagrange equation for the density fluctuation is simple but non-linear. In the case of charged impurity in electron gas the equation is solved in closed form using uniform limit approximation. This already goes beyond the conventional linear responce theory of Friedel oscillations. For the impurity in He-liquids the present theory can be applied to calculate the quasiparticle effective mass and correction terms to the effective mass approximation.

1. Introduction

In the conventional field theoretical approach [1] the impurity problem is usually treated with linear responce theory [2]. Hence in the Coulomb system the responce to the impurity is determined by the dielectric function. For the low-density Coulomb system or for systems such as ^{4}He, ^{3}He, H etc, where the interactions are very strong the perturbation theory is inapplicable or becomes very complicated. Nevertheless the original idea of Landau-Pomeranchuck quasiparticle [3] works well in practice [4,5] even for strongly interacting case and one should attempt to derive the quasiparticle properties microscopically. Here we will only consider a single quasiparticle but the generalization to two quasiparticles may also be attempted. This would offer the possibility to study microscopically also the interaction of quasiparticles.

In a previous paper [7] the HNC-theory for inhomogenious system (GHNC) was applied to calculations of properties of quantum solids and the liquid solid phase transition. The purpose of present paper is to show that with GHNC many other problems can be formulated. Here we will consider the screening of a foreign charge in electron gas and the theory of impurity quasiparticle in a uniform system in general.

In this respect the mixture of ^{3}He in liquid ^{4}He is interesting. Since the pioneering work of Feynman [8] several calculations [5,9,10,11] have been made to understand this system. The interesting experimental numbers are [5] the binding of ^{3}He $E_0 = -2.785$ K and the effective mass $M_3^* = 2.34\ m_3$. In general the calculations agree rather well with experiment, but the situation is not completely clear [11]. In the present work we have no extensive numerical calculation for this particular problem and we do not expect that the present theory to work in its simlified form where we have used s.c. global density approximation for the radial distribution function to avoid solving the GHNC-equation.

2. Formulation of the problem

We formulate the problem first for general case and then indicate the appropriate changes for the Coulomb system. We start with Hamiltonian

$$H = H_0 - \frac{\hbar^2}{2M} \nabla_A^2 + \sum_{i=1}^{N} V(\vec{r}_i - \vec{r}_A)$$

$$H_0 = - \frac{\hbar^2}{2m} \sum_{i=1}^{N} \nabla_i^2 + \sum_{i<j} v(\vec{r}_i - \vec{r}_j) . \tag{1}$$

Here H_0 is the Hamiltonian for the N-body Bose or Fermi system and the subindex A refers to the impurity. We assume that the ground state density is n and the radial distribution function is $g(r_{12})$. We perform the calculations assuming that the ground state wave function is

$$\Psi_0 = \phi \cdot \mathrm{Exp}\left[\frac{1}{2} \sum_{i<j} u(ij)\right] , \tag{2}$$

where

$$|\phi|^2 = \begin{cases} 1, & \text{for bosons} \\ \exp\left[\sum_{i<j} u_0(ij)\right], & \text{for fermions.} \end{cases} \tag{3}$$

For fermions eq(3) contains the Lado approximation [12]. This approximation has been shown to work rather well both for ^{3}He-system [13,14] and the electron gas [15]. In what follows we calculate everything for Fermi-system, whence the Bose system is obtained by setting $u_0(r_{ij}) = 0$. With the HNC-connection the radial distribution functions of the free and interacting Fermi gas $g_F(r)$ and $g(r)$ and the corresponding liquid structure functions $S_F(k)$ and $S(k)$ are tied together with the corresponding correlation factors with the equations

$$u_0(r) = \log g_F(r) - \frac{1}{(2\pi)^3 n}\int e^{i\vec{k}\vec{r}}\,\frac{[S_F(k)-1]^2}{S_F(k)}\,d\underset{\sim}{k}$$

$$u_0(r) + u(r) = \log g(r) - \frac{1}{(2\pi)^3 n}\int e^{i\vec{k}\vec{r}}\,\frac{[S(k)-1]^2}{S(k)}\,d\underset{\sim}{k}\,. \tag{4}$$

Here we have for simplicity overlooked the elementary diagrams which can be added f.e. with Padé technique [16]. All these quantities are assumed to be known for the ground state N-body system.

In order to take into account the impurity we use the Feynman wave function

$$\Psi_k = e^{i\vec{k}\vec{r}_A}\,\Pi\,\varphi(r_i - r_A)\,\Psi_0$$

$$\varphi(\vec{r}) = R(\vec{r})e^{i\delta(\vec{r})}\,, \tag{5}$$

where the real part R takes care of the correlations of the impurity with the back-ground particles and $\delta(\vec{r})$ describes the backflow. In writing the wave function (5) we have made the approximation that $u(r_{12})$ remains unchanged even near the impurity. Correcting for this would bring us the same degree of complication as one would have in the case of quantum surface or a quantum solid namely the solution of the GHNC-equation. The main difficulty there is the lost translational invariance. Due to this approximation we may not be able to treat the case of ion in He-liquids, but u and g_{12} should also depend upon r_A.

Since the total energy of the impurity system does not depend upon $\vec{r}_A$ we may leave that coordinate unintegrated while calculating the expectation values. The expectation values are calculated with the density functionals of which we need only the two lowest ones

$$\rho_A(\vec{r}_1) = \rho(\vec{r}_1 - \vec{r}_A) = N\,\frac{\int |\Psi_0|^2\,\Pi R_i^2\,d\tau_1}{\int |\Psi_0|^2\,\Pi R_i^2\,d\tau}$$

$$d\tau = d\underset{\sim}{r}_1\,d\tau_1 = d\underset{\sim}{r}_1\,d\underset{\sim}{r}_2\ldots d\underset{\sim}{r}_N$$

$$\rho_A^{(2)}(\vec{r}_1, \vec{r}_2) = \rho_A(\vec{r}_1)\rho_A(\vec{r}_2)\,g_{12} \tag{6}$$

$$g_{12} = f_{12}^2\,e^{N_{12}+E_{12}} = e^{u_0+u+N_{12}+E_{12}}\,.$$

In what follows the impurity is looked upon as producing an external field where all the other particles move. Therefore from here on the GHNC can be applied. The energy can be expressed in terms of ρ_A, $R(\vec{r})$, $\mathcal{S}(\vec{r})$ and quantities connected with ground state. The function $R(\vec{r})$ can be eliminated from the energy expressions by the BGY-equation and the Euler-Lagrange equations connect the functions $\rho_A(\vec{r})$ and $\mathcal{S}(\vec{r})$. The BGY-equation is obtained from eqs (6) and it reads

$$\vec{\nabla}_1 \log \frac{\rho_A(r_1)}{R^2(r)} = \int d\underset{\sim}{r}_2 \rho_A(\vec{r}_2)\ g_{12} \vec{\nabla}_1 [u(1,2)+u_0(1,2)]. \tag{7}$$

In what follows we will approximate g_{12} by the ground state radial distribution function. Correcting for this means solving also the GHNC-equation

$$N_{12} = \int d\underset{\sim}{r}_3 \rho_A\ (r_3)[g_{13}-1-N_{13}][g_{23}-1]. \tag{8}$$

In the non-Coulombic case the function $\rho_A(\underset{\sim}{r}) = n[1+\gamma(\vec{r})]$ is normalized by

$$n \int \gamma(\vec{r})\, d\underset{\sim}{r} = 0\ .$$

In the Coulombic case the cancellation of infinite terms by the background requires the normalization to be

$$n \int \gamma(\vec{r})\, d\underset{\sim}{r} = \frac{Q}{Z}\ .$$

Here $-Ze$ is the charge of the particle in the N-body system and Q that of the impurity. Due to our approximation for g_{12} the sequential condition

$$\int \rho_A(r_2)\,(g_{12}-1)\,dr_2 = -1$$

is not satisfied if $\vec{r}_2$ is near $\vec{r}_A$. The experience from the uniform case tells us that unless this condition is satisfied optimization may become unstable.

Denoting by $N\ \epsilon_0$ the ground state energy of the N-body system the energy expression becomes

$$\begin{aligned} E_K = N\epsilon_0 &+ \frac{\hbar^2}{2\mu}\int(\nabla\sqrt{\rho_A})^2 d\underset{\sim}{r} + 2n\tau_0\int\gamma(\vec{r})d\underset{\sim}{r} \\ &+ \rho_A(\vec{r})V(\vec{r})d\underset{\sim}{r} + \frac{1}{2}n^2\iint\gamma(\vec{r}_1)\gamma(\vec{r}_2)x_{12}d\underset{\sim}{r}_1 d\underset{\sim}{r}_2 \\ &+ \frac{\hbar^2}{2M}\left[\vec{k}-\int\rho_A\nabla S d\underset{\sim}{r}\right]^2 + \frac{\hbar^2}{2\mu}\int\rho_A(\nabla S)^2 d\underset{\sim}{r} \\ &+ \frac{\hbar^2}{2M}\iint\rho_A(\vec{r}_1)\rho_A(\vec{r}_2)\vec{\nabla}_1 S(\vec{r}_1)\vec{\nabla}_2 S(\vec{r}_2)(g_{12}-1)d\underset{\sim}{r}_1 d\underset{\sim}{r}_2 \ . \end{aligned} \tag{9}$$

Here the function $x(r_{12}) = x_{12}$ contains the information of the ground state

$$\begin{aligned} x_{12} = &- \frac{\hbar^2}{4m} g_{12}\left[\nabla_{12}^2(u-u_0)-(\nabla_{12}u_0)^2\right] \\ &+ \frac{\hbar^2}{4\mu}\nabla_1\left[g_{12}\nabla_1(u+u_0)\right] + g_{12}v_{12} \ . \end{aligned} \tag{10}$$

We have further used the notations

$$\mu = \frac{mM}{m+M}$$

$$\begin{aligned} \tau_0 = &- \frac{\hbar^2}{8m} n\int g_{12}\left\{\nabla^2[u-u_0]-(\nabla u_0)^2\right\} d\underset{\sim}{r} \\ &+ \frac{1}{2} n\int g(r)v(r)d\underset{\sim}{r} \ . \end{aligned}$$

For Bose system one has to set $u_0(r) = 0$. For Fermi system there is additional three-body term which is known to be small [13,15] both for electron liquid and ^{3}He and hence we have neglected it here.

3. Euler-Lagrange equations

To obtain the ground state of the impurity system we set $k = S = 0$. The variation with respect to $\sqrt{\rho_A}$ of the quantity

$$E_0 - \lambda n\int\gamma(\vec{r})d\underset{\sim}{r}$$

gives the characteristic Schrödinger equation with $\lambda = 2\tau_0$:

$$- \frac{\hbar^2}{2\mu}\nabla^2\sqrt{\rho_A} + [V(r)+W(r)]\sqrt{\rho_A} = 0 \ . \tag{11}$$

The induced potential now has the form

$$W(r_1) = n\int \gamma(\vec{r}_2)\mathcal{X}_{12}\,d\underset{\sim}{r}_2 \ . \tag{12}$$

Since $\rho_A = n(1+\gamma)$ this equation is non-linear but otherwise is very simple. One may attempt to solve it in the Coulombic case using the uniform limit approximation

$$\sqrt{\rho_A} - \sqrt{n} \cong \frac{1}{2}\sqrt{n}\,\gamma(\vec{r}) \ . \tag{13}$$

For the Coulombic case the following changes in equations (9) and (10) have to be made

$$\int \rho_A V(r)\,d\underset{\sim}{r} \longrightarrow -QZe^2 n\int \frac{\gamma(\vec{r})}{r}\,d\underset{\sim}{r} \tag{14}$$

$$g_{12}v_{12} \longrightarrow g_{12}\,\frac{Z^2e^2}{r_{12}} \ . \tag{15}$$

Using the Fourier-transforms

$$\gamma(\vec{r}) = \frac{1}{(2\pi)^3 n}\int e^{i\vec{q}\vec{r}}\gamma(\vec{q})\,d\underset{\sim}{q}$$
$$\mathcal{X}_{12} = \frac{1}{(2\pi)^3 n}\int e^{i\vec{q}\vec{r}}\mathcal{X}(\vec{q})\,d\underset{\sim}{q} \tag{16}$$

one obtains the solution

$$\gamma(\vec{k}) = \frac{4\pi nQZe^2}{\frac{\hbar^2}{4\mu}\,k^4 + k^2\mathcal{X}(\vec{k})} \ . \tag{17}$$

The normalization is $\gamma(k=0) = Q/Z$ since the most singular term in $\mathcal{X}(k)$ near $k = 0$ comes from the potential term of eq(15)

$$\lim_{k\to 0} k^2\mathcal{X}(k) = 4\pi Z^2e^2 n \ . \tag{18}$$

Furthermore $\gamma(k)$ has finite volume integral so that $\gamma(r=0)$ is finite. The Friedel oscillations [17] are now produced by the function $\mathcal{X}(k)$ having discontinuity at $k = 2k_F$ due to the function $u_0(r)$ appearing in $\mathcal{X}_{12}$. Unfortunately the connection here given by eqs (4) is more complicated than in the ref. 17 and we are unable to deduce the amplitude of Friedel oscillations mathematically.

For the case of ^{3}He in ^{4}He-liquid one deduces the following asymptotic behaviour for small k

$$\gamma(k) \sim ak^2 + bk^3$$

which corresponds to finite correlation length in r-space for $\gamma(\vec{r})$. One notices however that eq. 11 cannot have a solution without some external constraint therefore its application demands the simultaneous solution of the GHNC equation (8). This would bring in an other piece to the induced potential W(r) from the variation

$$\frac{\delta N_{12}}{\delta \sqrt{\rho_A}}$$

which would require heavy numerical computing. In order to get stability we have added a term

$$c \int \frac{[S(k)-1]^2}{S(k)} \gamma^2(k)\ d\underset{\sim}{k} \tag{19}$$

in the energy expressions of eq(9) with parameter c. In the lowest order GHNC-equation produces terms of this type yet the effect of this term in the energy is very small.

4. Generalized backflow equations

The case of the moving impurity can now be handled by simultaneous variation of energy expressions (9) with respect to $\sqrt{\rho_A}$ and $\mathcal{S}(\vec{r})$. Hence we obtain a coupled system of which the second equation simply expresses the current concervation since we are calculating stationary states. The equation are the following

$$\begin{gathered} -\frac{\hbar^2}{2\mu} \nabla^2 \sqrt{\rho_A} + [V(r)+W(\vec{r})+W_s(\vec{r})]\sqrt{\rho_A} = 0 \\ \vec{\nabla}\left\{ \rho_A(\vec{r})\left[\frac{M}{\mu} \vec{\nabla}\mathcal{S} + \vec{k}[p(r)+\beta-1]\right]\right\} = 0 . \end{gathered} \tag{20}$$

Here we have defined

$$\begin{gathered} \int \rho_A \nabla\mathcal{S}\ d\underset{\sim}{r} = \beta\underset{\sim}{\vec{k}} \\ \int \rho_A(\vec{r}_2)\nabla_2\mathcal{S}(\vec{r}_2)(g_{12}-1)dr_2 = p(\vec{r}_1)\vec{k} \\ W_s(r) = \frac{\hbar^2}{2\mu}(\nabla\mathcal{S})^2 + \frac{\hbar^2}{M}\vec{k}\cdot[p(r)+\beta-1]\vec{\nabla}\mathcal{S} . \end{gathered} \tag{21}$$

The usual effective mass approximation is obtained by neglecting the coupling term $W_s(r) = 0$

$$E_K = E_0 + \frac{\hbar^2 k^2}{2M^*} , \tag{22}$$

where E_0 is the $k = 0$ ground state of the impurity system and the effective mass is $M^* = (1-\beta)^{-1}M$. Equation (22) is correct also if we keep the coupling term W_s but now E_0 may depend upon k hence giving correction terms to the effective mass approximation [6].

For the backflow our equation is very similar to the one in ref. 8 generalized to have the foreign mass dependence and given more precise meaning for the other function ρ_A which cannot be the ground state radial distribution function but is the variable density of the background system. Unfortunately one cannot deduce much more without explicite numerical work.

5. Numerical results

We have calculated the density fluctuation $\gamma(r)$ from equation (17) in the case of a positron in uniform electron gas at the density $r_s = 1.5$. The result is compared with the two approximations of ref. 18 in Fig. 1. It is seen that for large distances we have qualitative agreement and hence the Friedel oscillations are produced. At $r = 0$ our value is too small by factor of three [19] yet it is large enough to invalidate the approximation of eq(13). For $r_s = 6$ $x(k)$ produces a pole at a finite value of k in eq(17) for $\gamma(k)$ which may be an indication of a bound state formation [20].

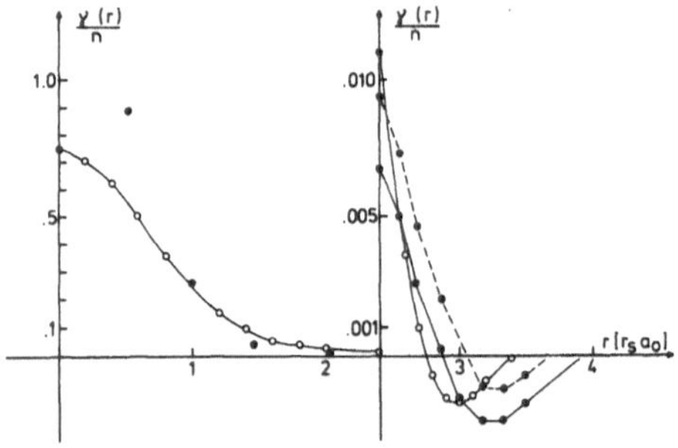

Fig. 1. Friedel oscillations at $r_s = 1.5$. Legend:
—o— present calc.
--●-- Δn_H of ref. 18
—●— Δn_B of ref. 18

We have also calculated the effective mass of ^3He in ^4He by stabilizing the Euler-equation with the additional term of eq(9). The resulting density function $\rho_A(r)$ is identical with $g_{44}(r)$ up to $r = 3Å$ but has higher and broader maximum due to the different normalization. With all the terms included we get $M_3^* = 1.87$ and $\mu_3 = -3.1$ ^oK. The backflow function is different from the classical function $A(\hat{k}\cdot\hat{r})r^{-2}$,hence there is some hope to break away from the dead lock of the conventional approach [11], where the backflow function remains classical.

Clearly in both cases as a next step one should also solve the GHNC-equation (8) even if u(r) of the ground state is kept. Since such a calculation is numerically rather formidable the solution $\rho_A(r)$ proposed here with eq(19) should serve as a good starting point. In the positron case this method gives $\gamma(r)$ which shows Friedel oscillations and central densities become higher than in ref. 18 but in agreement with ref. 19.

References

[1] Gell-Mann M. and Brueckner K., Phys. Rev. 106 (1957) 364.

[2] Fetter A.L. and Walecka J.D., "Quantum Theory of Many-particle Systems" McGraw-Hill 1971.

[3] Landau L.D. and Pomeranchuck I., Dokl. Akad. Nauk, SSSR (1948) 669.

[4] Bardeen J., Baym G. and Pines D., Phys. Rev. 156 (1967) 207.

[5] Baym G. and Pethick C., in "The Physics of Liquid and Solid Helium", Ed. Benneman K. & Ketterson I., Wiley, 1978.

[6] Disatnik Y. in "Liquid and Solid Helium", Ed. Kuper C.G., Lipson S.G. and Revzen M., Israel Univ. Press, 1974.

[7] Kallio A., Pietiläinen P., Puoskari M. and Toropainen P., Physica Scripta 22 (1980) 91.

[8] Feynman R.P. and Cohen M., Phys. Rev. 102 (1956) 1189.

[9] Davison T.B. and Feenberg E., Phys. Rev. 178 (1969) 306.

[10] Woo C.W. in Ref. [5].

[11] Owen J.C., Preprint 1980 (Manchester M 139 PL, UK).

[12] Lado F., J. Chem. Phys. 47 (1967) 5369.

[13] Puoskari M. and Kallio A., Preprint 1980 (Univ. of Oulu, Dept. Th. Phys., to be publ. in Physica Scripta).

[14] Ristig M.L., Lam P.M. and Nollert D., J. de Physique Colloq. 41 (1980) C7-213.

[15] Zabolitsky J.G., Phys. Rev. B 22 (1980) 2353

[16] Smith R.A., Kallio A., Puoskari M. and Toropainen P., in Proceedings of Trieste Conf., Nucl. Phys. A 328 (1979) 186.

[17] Friedel J., Phil. Mag. 43 (1952) 43
Nuovo Cim. 7 Supl. 2 (1958) 287.

[18] Langer J.S. and Vosko S.H., J. Phys. Chem. Solids 12 (1960) 196.
Kohn W. and Vosko S.H., Phys. Rev. 119 (1960) 912.

[19] Arponen J. and Pajanne E., Ann. Phys. 121 (1979) 343.

[20] Lowy D.N. and Jackson A.D., Phys. Rev. B 12 (1975) 1689.

+ present address: NORDITA, Blegdamsvej 17, DK-2100 København Ø, Denmark

EFFECTIVE FREE ENERGY FOR NONLINEAR DYNAMICS

Alexander L. Fetter
Institute of Theoretical Physics, Department of Physics
Stanford University, Stanford, CA 94305, USA

Abstract

Nonlinear dissipative systems often become dynamically unstable at critical values of a control parameter h. For $h<h_c$, all normal modes of the linearized theory decay, whereas, when h exceeds h_c, linearized modes with wavenumbers near a nonzero critical value $\underset{\sim}{q}_c$ grow exponentially with time. A nonlinear analysis of the original free energy yields an effective renormalized free energy for the slowly varying envelope function that modulates the plane wave $\exp(i\underset{\sim}{q}_c\cdot\underset{\sim}{x})$. For several cases of interest in connection with superfluid ^{3}He-A, the sign of the resulting quartic coupling produces catastrophic growth, indicating a failure of the small amplitude perturbation theory.

Dynamical instabilities of nonlinear systems typically involve a restricted set of normal modes. Familiar examples are the Bénard convection rolls or hexagonal cells in a fluid layer heated from below and the Taylor vortices in Couette flow between rotating cylinders.[1] In both cases, the original uniform state becomes unstable at a critical value of the applied stress, leading to a stationary but spatially periodic distortion with a characteristic wavenumber q_c, implying a corresponding loss of translational symmetry. When the applied stress exceeds the critical value, the amplitude of the spatially periodic normal mode generally increases with a square-root behavior.[2] For still larger stress, this mode itself becomes unstable with respect to more complicated time-dependent motions.[3-7]

Although many of these questions first arose in the problem of classical viscous fluids, the resulting techniques have much wider applicability. One interesting and general example[8] is a dynamical system specified by a set of n coupled fields $u_i(\underset{\sim}{x},t)$ $(i=1,\cdots,n)$ and a free-energy density $f[u_i,u_{i\lambda},h]$, where the Greek subscript indicates a spatial derivative $u_{i\lambda} = \partial u_i/\partial x_\lambda$, and h is a parameter subject to external control. These fields are assumed to obey coupled dynamical equations

$$\dot{u}_i \equiv \frac{\partial u_i}{\partial t} = -\frac{\delta f}{\delta u_i} \equiv \frac{\partial}{\partial x_\lambda}\frac{\partial f}{\partial u_{i\lambda}} - \frac{\partial f}{\partial u_i}, \tag{1}$$

where a sum is implied over repeated indices. Evidently, f acts as a Lagrangian density, with the dynamics governed by a dissipation function[9] $D = \frac{1}{2}\int d^3x\, \dot{u}_i\dot{u}_i$. As a result, u_i can be interpreted as a generalized coordinate that can be chosen to simplify the particular problem in question.

For definiteness, assume an extended sample of volume V subject to periodic boundary conditions. The fields can then be expanded in plane waves

$$u_i(\underset{\sim}{x}) = V^{-\frac{1}{2}} \Sigma_{\underset{\sim}{q}} e^{i\underset{\sim}{q}\cdot\underset{\sim}{x}} u_i(\underset{\sim}{q}), \tag{2}$$

with $u_i(\underset{\sim}{q})^* = u_i(-\underset{\sim}{q})$. Correspondingly, the dissipation function has the form

$$D = \frac{1}{2} \Sigma_{\underset{\sim}{q}_1\underset{\sim}{q}_2} \delta_{\underset{\sim}{q}_1+\underset{\sim}{q}_2,0} \; \dot{u}_i(\underset{\sim}{q}_1)\dot{u}_i(\underset{\sim}{q}_2). \tag{3}$$

The Fourier components themselves thus become the generalized coordinates, subject to the dynamical equation

$$u_i(\underset{\sim}{q}) = \frac{\partial D}{\partial u_i(-\underset{\sim}{q})} = - \frac{\partial F}{\partial u_i(-\underset{\sim}{q})}, \tag{4}$$

where $F = \int d^3x\, f$ is the total free energy. Assume that the equilibrium configuration has a spatially uniform free-energy density $f^{(0)}$, and let u_i denote the deviation from this equilibrium. Hence the free energy has an expansion of the form

$$F = F^{(0)} + F^{(1)} + F^{(2)} + \cdots \tag{5}$$

where the superscripts denote the powers of u. The zero-order term is just $F^{(0)} = Vf^{(0)}$. The first-order term follows directly by expanding the free-energy density

$$f^{(1)} = f_i^{(1)} u_i(\underset{\sim}{x}) + f_{i\lambda}^{(1)} u_{i\lambda}(\underset{\sim}{x}). \tag{6a}$$

where

$$f_i^{(1)} \equiv (\partial f/\partial u_i)_0, \quad f_{i\lambda}^{(1)} \equiv (\partial f/\partial u_{i\lambda})_0, \tag{6b}$$

with the derivatives evaluated in the equilibrium configuration. A simple calculation yields

$$F^{(1)} = V^{\frac{1}{2}} f_i^{(1)} u_i(\underset{\sim}{q}=0). \tag{6c}$$

In a similar way, the second-order term becomes

$$F^{(2)} = \frac{1}{2} \Sigma_{\underset{\sim}{q}_1\underset{\sim}{q}_2} N_{ij}(\underset{\sim}{q}_1,\underset{\sim}{q}_2) u_i(\underset{\sim}{q}_1) u_j(\underset{\sim}{q}_2) \delta_{\underset{\sim}{q}_1+\underset{\sim}{q}_2,0} \tag{7a}$$

where

$$N_{ij}(\underset{\sim}{q}_1,\underset{\sim}{q}_2) = f_{i,j}^{(2)} + iq_{1\lambda} f_{i\lambda,j}^{(2)} + iq_{2\lambda} f_{i,j\lambda}^{(2)} - q_{1\lambda} q_{2\mu} f_{i\lambda,j\mu}^{(2)}. \tag{7b}$$

Here the coefficients are second-order partial derivatives of f, again evaluated in equilibrium. The third and higher-orders are direct generalizations, with coefficients that are symmetric under the interchange of any pair of variables

$$N_{ijk\cdots}(\underset{\sim}{q}_1,\underset{\sim}{q}_2,\underset{\sim}{q}_3,\cdots) = N_{jik\cdots}(\underset{\sim}{q}_2,\underset{\sim}{q}_1,\underset{\sim}{q}_3,\cdots) \tag{8}$$

and obey the complex-conjugate relations

$$N_{ijk\cdots}(\underset{\sim}{q}_1,\underset{\sim}{q}_2,\underset{\sim}{q}_3,\cdots)^* = N_{ijk\cdots}(-\underset{\sim}{q}_1,-\underset{\sim}{q}_2,-\underset{\sim}{q}_3,\cdots) \tag{9}$$

The zero-order term in the dynamical equation (4) gives the condition $f_i^{(1)} = 0$ that determines the equilibrium. Deviations from this static configuration then satisfy the coupled equations

$$-\dot{u}_i(\underset{\sim}{q}) = N_{ij}(-\underset{\sim}{q},\underset{\sim}{q}) u_j(\underset{\sim}{q}) + \cdots, \tag{10}$$

where the terms omitted are second- and higher-order in the variables u_i. It is convenient first to study the linearized equations of motion, which constitute an eigenvalue problem for the hermitian matrix $N_{ij}(-\underset{\sim}{q},\underset{\sim}{q})$. In general, there are n real eigenvalues, which will be denoted $-\lambda^{(s)}(\underset{\sim}{q})$ $(s = 1,\cdots,n)$, and the corresponding eigenvectors $e_i^{(s)}(\underset{\sim}{q})$ satisfy the relations

$$N_{ij}(-\underset{\sim}{q},\underset{\sim}{q})e_j^{(s)}(\underset{\sim}{q}) = -\lambda^{(s)}(\underset{\sim}{q})e_i^{(s)}(\underset{\sim}{q}) \tag{11a}$$

$$e_i^{(s)}(\underset{\sim}{q})^* = e_i^{(s)}(-\underset{\sim}{q}),\quad e_i^{(s)}(\underset{\sim}{q})^* e_i^{(s')}(\underset{\sim}{q}) = \delta_{ss'}. \tag{11b}$$

The associated modal matrix[10] $A_{is}(\underset{\sim}{q}) \equiv e_i^{(s)}(\underset{\sim}{q})$ is unitary and diagonalizes the matrix N_{ij}

$$[A(\underset{\sim}{q})^\dagger N(-\underset{\sim}{q},\underset{\sim}{q})A(\underset{\sim}{q})]_{ss'} = -\lambda^{(s)}(\underset{\sim}{q})\delta_{ss'}. \tag{12}$$

It is convenient to introduce normal coordinates $\zeta^{(s)}(\underset{\sim}{q},t)$ through the relation

$$u_i(\underset{\sim}{q},t) = A_{is}(\underset{\sim}{q})\zeta^{(s)}(\underset{\sim}{q},t), \tag{13}$$

and the linearized equations obtained from Eq. (10) then assume the transparent form (no summation over s)

$$\dot{\zeta}^{(s)}(\underset{\sim}{q},t) = \lambda^{(s)}(\underset{\sim}{q})\zeta^{(s)}(\underset{\sim}{q},t). \tag{14}$$

For small values of the control parameter h, assume that all the eigenvalues are negative, implying that the equilibrium is locally stable with respect to arbitrary small-amplitude perturbations. When h reaches a critical value h_c, however, one of the eigenvalues (with s = 1, say) becomes zero at some nonzero critical wavevector $\pm \underset{\sim}{q}_c$. The function $\lambda^{(1)}(\underset{\sim}{q})$ has a local maximum near $\underset{\sim}{q}_c$ that is negative for $h<h_c$ but becomes positive for $h>h_c$. For $|h-h_c| \ll h_c$ and $|\underset{\sim}{q}-\underset{\sim}{q}_c| \ll q_c$, the function $\lambda^{(1)}(\underset{\sim}{q})$ has the form

$$\lambda^{(1)}(\underset{\sim}{q}) \approx \lambda^{(1)}(\underset{\sim}{q}_c) - (q-q_c)_\mu(q-q_c)_\nu m_{\mu\nu}, \tag{15}$$

where $m_{\mu\nu}$ is a real symmetric matrix with nonnegative eigenvalues, and $\lambda^{(1)}(\underset{\sim}{q}_c)$ has the sign of $h-h_c$. Given the matrix $N_{ij}(-\underset{\sim}{q},\underset{\sim}{q})$, the determination of $h_c, \underset{\sim}{q}_c, \lambda^{(1)}(\underset{\sim}{q}_c)$, and $m_{\mu\nu}$ is merely a numerical problem. Just beyond threshold $(h \gtrsim h_c)$, the condition $\lambda^{(1)}(\underset{\sim}{q}) = 0$ defines some small ellipsoidal region in wavevector space; let the semi-major axis (say) of this region be $\varepsilon\, q_c$, where $\varepsilon \ll 1$. More generally, it is natural to express wavevectors in this vicinity as $\underset{\sim}{q} = \underset{\sim}{q}_c + \varepsilon\, \underset{\sim}{w}$, where $\underset{\sim}{w}$ is of order unity; the function $\lambda^{(1)}(\underset{\sim}{q})$ may then be written as

$$\lambda^{(1)}(\underset{\sim}{q}) \approx \varepsilon^2(s\lambda - w_\mu w_\nu m_{\mu\nu}). \tag{16}$$

Here, λ is a positive constant of order unity, $s \equiv \mathrm{sgn}\,(h-h_c)$, and $(h-h_c)/h_c$ is of order $s\,\varepsilon^2$.

For most purposes, the normal coordinates themselves provide the most convenient set of dynamical variables. Below threshold $(h < h_c)$, all infinitesimal-amplitude perturbations decay, but when h exceeds h_c, normal modes with s = 1 and $\underset{\sim}{q}$ near $\pm\, \underset{\sim}{q}_c$ exhibit slow exponential growth at a rate proportional to ε^2, at least in the linearized theory.

To study the evolution of the new configuration that appears beyond threshold, it is essential to include the nonlinear terms. The free energy F may be rewritten in terms of the normal coordinates:

$$F^{(2)} = -\frac{1}{2}\Sigma_{\underset{\sim}{q}}\Sigma_i\ \lambda^{(i)}(\underset{\sim}{q})\zeta^{(i)}(\underset{\sim}{q})\zeta^{(i)}(\underset{\sim}{q})^* \tag{17a}$$

$$F^{(3)} = \frac{1}{6}\Sigma_{\underset{\sim}{q}_1\underset{\sim}{q}_2\underset{\sim}{q}_3}\ \delta_{\underset{\sim}{q}_1+\underset{\sim}{q}_2+\underset{\sim}{q}_3,0}\ \bar{N}_{ijk}(\underset{\sim}{q}_1,\underset{\sim}{q}_2,\underset{\sim}{q}_3)\zeta^{(i)}(\underset{\sim}{q}_1)\zeta^{(j)}(\underset{\sim}{q}_2)\zeta^{(k)}(\underset{\sim}{q}_3) \tag{17b}$$

where

$$\bar{N}_{ijk}(\underset{\sim}{q}_1,\underset{\sim}{q}_2,\underset{\sim}{q}_3) \equiv N_{i'j'k'}(\underset{\sim}{q}_1,\underset{\sim}{q}_2,\underset{\sim}{q}_3)A_{i'i}(\underset{\sim}{q}_1)A_{j'j}(\underset{\sim}{q}_2)A_{k'k}(\underset{\sim}{q}_3) \tag{17c}$$

is the projection of N onto the appropriate eigenvectors. As a result, the normal coordinates obey the nonlinear dynamical equation

$$\dot{\zeta}^{(i)}(\underset{\sim}{q}) = \lambda^{(i)}(\underset{\sim}{q})\zeta^{(i)}(\underset{\sim}{q}) - (2V^{\frac{1}{2}})^{-1}\ \Sigma_{\underset{\sim}{p}}\ \bar{N}_{ijk}(-\underset{\sim}{q},\underset{\sim}{p},\underset{\sim}{q}-\underset{\sim}{p})\zeta^{(j)}(\underset{\sim}{p})\zeta^{(k)}(\underset{\sim}{q}-\underset{\sim}{p})$$
$$- (6V)^{-1}\ \Sigma_{\underset{\sim}{p}\underset{\sim}{p}'}\ \bar{N}_{ijk\ell}(-\underset{\sim}{q},\underset{\sim}{p},\underset{\sim}{p}',\underset{\sim}{q}-\underset{\sim}{p}-\underset{\sim}{p}')\zeta^{(j)}(\underset{\sim}{p})\zeta^{(k)}(\underset{\sim}{p}')\zeta^{(\ell)}(\underset{\sim}{q}-\underset{\sim}{p}-\underset{\sim}{p}')+\ldots \tag{18}$$

We are particularly interested in the normal modes $\zeta^{(1)}(\underset{\sim}{q})$ that become unstable beyond threshold; it is helpful to give them a special symbol

$$\zeta^{(1)}(\underset{\sim}{q}_c+\varepsilon\underset{\sim}{w}) \equiv \varepsilon\ \psi(\underset{\sim}{w}), \tag{19}$$

where the factor ε makes explicit an amplitude proportional to $(h - h_c)^{\frac{1}{2}}$. The factor ε^2 in Eq. (16) suggests the introduction[11-13] of a "slow" time variable $\tilde{t}=\varepsilon^2 t$, in which case, the unstable normal mode obeys the approximate equation

$$\varepsilon^3\ [\frac{\partial\psi(\underset{\sim}{w})}{\partial\tilde{t}} - (s\lambda - w_\mu w_\nu m_{\mu\nu})\psi(\underset{\sim}{w})]$$
$$= -(2V^{\frac{1}{2}})^{-1}\Sigma_{\underset{\sim}{p}}\bar{N}_{ijk}(-\underset{\sim}{q}_c-\varepsilon\underset{\sim}{w},\underset{\sim}{p},\underset{\sim}{q}_c+\varepsilon\underset{\sim}{w}-\underset{\sim}{p})\zeta^{(j)}(\underset{\sim}{p})\zeta^{(k)}(\underset{\sim}{q}_c+\varepsilon\underset{\sim}{w}-\underset{\sim}{p})$$
$$- (6V)^{-1}\Sigma_{\underset{\sim}{p}\underset{\sim}{p}'}\bar{N}_{ijk\ell}(-\underset{\sim}{q}_c-\varepsilon\underset{\sim}{w},\underset{\sim}{p},\underset{\sim}{p}',\underset{\sim}{q}_c+\varepsilon\underset{\sim}{w}-\underset{\sim}{p}-\underset{\sim}{p}')\zeta^{(j)}(\underset{\sim}{p})\zeta^{(k)}(\underset{\sim}{p}')\zeta^{(\ell)}(\underset{\sim}{q}_c+\varepsilon\underset{\sim}{w}-\underset{\sim}{p}-\underset{\sim}{p}')$$

All other normal modes ($i \neq 1$ or $\underset{\sim}{q}$ far from $\pm\ \underset{\sim}{q}_c$) have negative growth rates of order 1; thus they can arise only through the nonlinear terms in Eq. (18) and will therefore be <u>at least</u> of order ε^2. Consequently, the only contribution to the last term of Eq. (20) occurs when <u>all three</u> normal modes are critical ones implying that $\underset{\sim}{p}$, $\underset{\sim}{p}'$, and $\underset{\sim}{q}_c-\underset{\sim}{p}-\underset{\sim}{p}'$ must simultaneously be near $\pm\ \underset{\sim}{q}_c$. Only three such possibilities exist, and the permutation symmetry allows us to combine them as

$$-(\varepsilon^3/2V)\Sigma_{\underset{\sim}{w}'\underset{\sim}{w}''}\bar{N}_{1111}(-\underset{\sim}{q}_c,-\underset{\sim}{q}_c,\underset{\sim}{q}_c,\underset{\sim}{q}_c)\psi(\underset{\sim}{w}')\psi(\underset{\sim}{w}'')\psi^*(\underset{\sim}{w}'+\underset{\sim}{w}''-\underset{\sim}{w}),$$

apart from higher-order corrections in ε.

The quadratic term in Eq. (20) is more complicated, for there are important contributions with $\underset{\sim}{p}$ near $\pm\ \underset{\sim}{q}_c$, $\underset{\sim}{0}$, and $2\underset{\sim}{q}_c$. Symmetry reduces these terms to

$$-\varepsilon V^{-1}\Sigma_{\underset{\sim}{w}'}\Sigma_k[\bar{N}_{11k}(-\underset{\sim}{q}_c-\varepsilon\underset{\sim}{w},\underset{\sim}{q}_c+\varepsilon\underset{\sim}{w}',\varepsilon\underset{\sim}{w}-\varepsilon\underset{\sim}{w}')\psi(\underset{\sim}{w}')\zeta^{(k)}(\varepsilon\underset{\sim}{w}-\varepsilon\underset{\sim}{w}')$$
$$+\bar{N}_{11k}(-\underset{\sim}{q}_c-\varepsilon\underset{\sim}{w},-\underset{\sim}{q}_c+\varepsilon\underset{\sim}{w}',2\underset{\sim}{q}_c+\varepsilon\underset{\sim}{w}-\varepsilon\underset{\sim}{w}')\psi^*(-\underset{\sim}{w}')\zeta^{(k)}(2\underset{\sim}{q}_c+\varepsilon\underset{\sim}{w}-\varepsilon\underset{\sim}{w}')],$$

and it is therefore necessary to examine the normal modes with $\underset{\sim}{q}$ near $\underset{\sim}{0}$ and $2\underset{\sim}{q}_c$. To order ε^2, Eq. (18) shows that these modes have rapidly decaying transients plus a slowly growing term driven by the quadratic coupling to the critical modes:

$$\zeta^{(k)}(\varepsilon\underset{\sim}{w}-\varepsilon\underset{\sim}{w}') \approx [\varepsilon^2/V^{\frac{1}{2}}\lambda^{(k)}(\varepsilon\underset{\sim}{w}-\varepsilon\underset{\sim}{w}')]\ \Sigma_{\underset{\sim}{w}''}\ \bar{N}_{11k}(-\varepsilon\underset{\sim}{w}+\varepsilon\underset{\sim}{w}',q_c+\varepsilon\underset{\sim}{w}',-\underset{\sim}{q}_c+\varepsilon\underset{\sim}{w}-\varepsilon\underset{\sim}{w}'-\varepsilon\underset{\sim}{w}'')$$
$$\times\ \psi(\underset{\sim}{w}'')\psi^*(\underset{\sim}{w}'+\underset{\sim}{w}''-\underset{\sim}{w})$$

with a similar expression for $\tilde{\zeta}^{(k)}(2\underset{\sim}{q}_c+\varepsilon\underset{\sim}{w}-\varepsilon\underset{\sim}{w}')$. A combination of these various contributions leads to the final approximate dynamical equation

$$\frac{\partial}{\partial\tilde{t}}\psi(\underset{\sim}{w}) = (s\lambda-w_\mu w_\nu m_{\mu\nu})\psi(\underset{\sim}{w})$$
$$+\ (\Gamma/V)\Sigma_{\underset{\sim}{w}'\underset{\sim}{w}''}\psi(\underset{\sim}{w}')\psi(\underset{\sim}{w}'')\psi^*(\underset{\sim}{w}'+\underset{\sim}{w}''-\underset{\sim}{w}), \qquad (21)$$

where

$$\Gamma \equiv \frac{1}{2}\Sigma_k\left[\frac{2|\bar{N}_{11k}(-\underset{\sim}{q}_c,\underset{\sim}{q}_c,0)|^2}{-\lambda^{(k)}(0)} + \frac{|\bar{N}_{11k}(-\underset{\sim}{q}_c,-\underset{\sim}{q}_c,2\underset{\sim}{q}_c)|^2}{-\lambda^{(k)}(2\underset{\sim}{q}_c)}\right]$$
$$-\ \frac{1}{2}\bar{N}_{1111}(-\underset{\sim}{q}_c,-\underset{\sim}{q}_c,\underset{\sim}{q}_c,\underset{\sim}{q}_c). \qquad (22)$$

This dynamical equation has the familiar Landau form for the amplitude of an unstable normal mode near threshold.[2] Note, however, that the first two terms of Eq. (22) are positive definite because the growth rates $\lambda^{(k)}(0)$ and $\lambda^{(k)}(2\underset{\sim}{q}_c)$ are negative definite; on the other hand, the last term can be either positive or negative. Thus, the overall sign of Γ can be determined only by a detailed study of the nonlinear contributions to the original free-energy density f.

Consider first the behavior very close to threshold, when only the single mode $\psi(\underset{\sim}{w}=0)\equiv\psi$ is critical. In that case, the dynamical equation becomes

$$\partial\psi/\partial\tilde{t} = s\lambda\psi + (\Gamma/V)|\psi|^2\psi. \qquad (23)$$

If Γ is negative, the amplitude $\psi = 0$ is stable below threshold ($s = -1$), and the amplitudes $|\psi| = \pm(\lambda V/|\Gamma|)^{\frac{1}{2}}$ are stable above ($s = 1$). This is the familiar normal bifurcation[4] to a new static equilibrium configuration characterized by the normal mode $\zeta^{(1)}(\underset{\sim}{q}_c)$ with amplitude of order ε. If Γ is positive, in contrast, then small-amplitude deviations from zero with $|\psi|^2 < \lambda V/\Gamma$ remain stable below threshold, but the critical normal mode grows catastrophically once h exceeds h_c; this latter situation represents an inverted bifurcation.[4] Inspection of Eq. (22) shows that a theory based on a free-energy density f can undergo a normal bifurcation only if the quartic terms in f predominate over the cubic ones. Such could be the case, for example, if f contains only even powers of the original fields u_i. Conversely, dynamical equations based on a free-energy density f will always exhibit an inverted bifurcation if the nonlinear couplings are solely quadratic.

A second situation of interest arises further beyond threshold, when a group of normal modes has become unstable. It is then natural to construct a "wave packet" of such modes, writing a general disturbance as

$$u_i(\underset{\sim}{x},t) = V^{-\frac{1}{2}}\Sigma_{\underset{\sim}{q}}\, e^{i\underset{\sim}{q}\cdot\underset{\sim}{x}}\, u_i(\underset{\sim}{q},t)$$

$$\approx V^{-\frac{1}{2}}\, \Sigma_{\underset{\sim}{w}}[\, e_i^{(1)}(\underset{\sim}{q}_c+\varepsilon\underset{\sim}{w})\zeta^{(1)}(\underset{\sim}{q}_c+\varepsilon\underset{\sim}{w})\, e^{i(\underset{\sim}{q}_c+\varepsilon\underset{\sim}{w})\cdot\underset{\sim}{x}} + \text{c.c.}]$$

$$\approx 2\varepsilon \mathrm{Re}[\, e_i^{(1)}(\underset{\sim}{q}_c)\, e^{i\underset{\sim}{q}_c\cdot\underset{\sim}{x}}\psi(\underset{\sim}{r},\tilde{t})], \tag{24}$$

where $\underset{\sim}{r} = \varepsilon\, \underset{\sim}{x}$ is a rescaled spatial variable and [see Eq. (19)]

$$\psi(\underset{\sim}{r},\tilde{t}) = V^{-\frac{1}{2}}\, \Sigma_{\underset{\sim}{w}}\psi(\underset{\sim}{w},\tilde{t})\, e^{i\underset{\sim}{w}\cdot\underset{\sim}{r}}\, . \tag{25}$$

Equation (24) shows that the disturbance has a characteristic wavevector $\underset{\sim}{q}_c$ but is modulated with an envelope $\psi(\underset{\sim}{r},\tilde{t})$ that varies slowly in space and time. This envelope function obeys an equation obtained as the spatial Fourier transform of Eq. (21)

$$\frac{\partial\psi(\underset{\sim}{r},\tilde{t})}{\partial\tilde{t}} = s\lambda\psi(\underset{\sim}{r},\tilde{t}) + \Gamma|\psi(\underset{\sim}{r},\tilde{t})|^2\psi(\underset{\sim}{r},\tilde{t}) + m_{\mu\nu}\frac{\partial^2}{\partial r_\mu \partial r_\nu}\psi(\underset{\sim}{r},\tilde{t})\, . \tag{26}$$

Alternatively, if we introduce a renormalized free-energy density

$$\tilde{f} = -s\lambda|\psi(\underset{\sim}{r})|^2 + m_{\mu\nu}\frac{\partial\psi(\underset{\sim}{r})}{\partial r_\mu}\frac{\partial\psi^*(\underset{\sim}{r})}{\partial r_\nu} - \frac{1}{2}\Gamma|\psi(\underset{\sim}{r})|^4, \tag{27}$$

then the dynamical equation (26) is equivalent to

$$\frac{\partial\psi(\underset{\sim}{r},\tilde{t})}{\partial\tilde{t}} = -\frac{\delta\tilde{f}}{\delta\psi^*(\underset{\sim}{r},\tilde{t})}\, . \tag{28}$$

Note that $\tilde{f}$ depends only on the rescaled variables $\underset{\sim}{r}$ and $\tilde{t}$; moreover, its parameters involve only the eigenvalue of the unstable normal mode (through λ and $m_{\mu\nu}$) and the coupling to modes with $\underset{\sim}{q} = 0$ and $2\underset{\sim}{q}_c$ (through Γ). If Γ is negative, this amplitude equation can determine the stable structures beyond threshold,[11-13] but if Γ is positive, the resulting inverted bifurcation again implies a catastrophic instability that cannot saturate at some nearly configuration.

One interesting application of this formalism is to superfluid ^{3}He-A, where the p-wave Cooper pairs have an internal unit angular momentum vector $\hat{\ell}$. In a uniform external hydrodynamic flow, $\hat{\ell}$ aligns itself along the flow ($\hat{z}$, say). Application of a parallel magnetic field H produces a sequence of effects.[14,15] (1) When H exceeds a first critical field H_0, the system undergoes a <u>normal</u> bifurcation to a helical configuration, in which $\hat{\ell}$ deviates from $\hat{z}$ by an angle proportional to $(H-H_0)^{\frac{1}{2}}$ and displays a spatial periodicity whose scale depends on the external flow. (2) When H reaches a second critical value H_1, the helix itself becomes unstable with respect to long-wavelength perturbations, but this second transition is an <u>inverted</u> bifurcation because the corresponding Γ is positive definite. Hence the subsequent dynamical motion is complicated, depending on the detailed experimental configuration. A similar situation occurs when ^{3}He-A is initially placed in a uniform magnetic field and then subjected to an increasing hydrodynamic current.[16] At a field-dependent critical current, the uniform configuration undergoes a dynamical instability through an inverted bifurcation to a spatially periodic structure characterized by a nonzero wavevector $\underset{\sim}{q}_c$. This transition presumably signals the onset of dissipation.

The preceding analysis has concentrated on systems whose dynamics are derivable from a free energy through Eq. (4). It would be interesting to consider more general cases, in which the right-hand side of Eq. (4) is replaced by an arbitrary function of u and its spatial gradients. (After completing this work, I learned that Cross[17] has developed similar techniques to derive an amplitude equation for the Rayleigh-Bénard problem which exemplifies this type of problem. In this case, the sign of the nonlinear coupling term turns out to ensure a normal bifurcation.) In particular, is there a simple criterion for determining when the nonlinear dynamics can lead to an amplitude equation of the form (21), and can the character of the bifurcation be inferred directly from the structure of the original dynamical equations? These intriguing questions await further study.

Acknowledgments

I am grateful to B. A. Huberman and M. R. Williams for valuable discussions. This research was supported in part by the National Science Foundation through Grant No. DMR 78-25258.

References

1. S. Chandrasekhar, Hydrodynamic and Hydromagnetic Stability (Oxford University Press, Oxford, 1961), Chaps. II and VII.
2. L. D. Landau and E. M. Lifshitz, Fluid Mechanics (Pergamon, London, 1959), Chap. III.
3. D. D. Joseph, Stability of Fluid Motion (Springer-Verlag, Berlin, 1976), Chaps. V and XI.
4. C. Normand, Y. Pomeau, and M. G. Velarde, Rev. Mod. Phys. **49**, 581 (1977).
5. F. H. Busse, Rep. Prog. Phys. **41**, 1929 (1978).
6. G. Ahlers, in Fluctuations, Instabilities, and Phase Transitions, edited by T. Riste (Plenum, New York, 1975), p. 181.
7. D. Coles, J. Fluid Mech. **21**, 385 (1965).
8. P. C. Hohenberg and B. I. Halperin, Rev. Mod. Phys. **49**, 435 (1977).
9. H. Goldstein, Classical Mechanics (Addison-Wesley, Reading, MA, 1950), pp. 21-22.
10. A. L. Fetter and J. D. Walecka, Theoretical Mechanics of Particles and Continua (McGraw-Hill, New York, 1980), Sec. 22.
11. A. C. Newell and J. A. Whitehead, J. Fluid Mech. **38**, 279 (1969).
12. L. A. Segel, J. Fluid Mech. **38**, 203 (1969).
13. E. Coutsias and B. A. Huberman, to be published.
14. A. L. Fetter and M. R. Williams, Phys. Rev. Lett. **43**, 1601 (1979), and Phys. Rev. B, to be published.
15. Y. R. Lin-Liu, D. Vollhardt, and K. Maki, Phys. Rev. B**20**, 159 (1979).
16. A. L. Fetter, to be published.
17. M. C. Cross, Phys. Fluids **23**, 1727 (1980).

GREEN'S FUNCTION MONTE CARLO AND THE MANY-FERMION PROBLEM*

M. H. Kalos
Courant Institute of Mathematical Sciences
New York University
New York, N.Y. 10012/USA

Abstract. The application of Green's function Monte Carlo to many body problems is outlined. For boson problems, the method is well developed and practical. An "efficiency principle", importance sampling, can be used to reduce variance. Fermion problems are more difficult because spatially antisymmetric functions must be represented as a difference of two density functions. Naively treated, this leads to a rapid growth of Monte Carlo error. Methods for overcoming the difficulty are discussed. Satisfactory algorithms exist for few-body problems; for many-body problems more work is needed, but it is likely that adequate methods will soon be available.

1. Boson Problems.

The application of Green's function Monte Carlo methods has been reviewed in published[1] and unpublished material[2] but it is useful to restate briefly the basic ideas. Consider the Schrödinger equation in coordinate space, $R \equiv \{x_1, x_2, \ldots x_N\}$, for an N body problem

$$H\psi(R) = E\psi(R) \quad . \tag{1}$$

we may shift the energy scale by V_0 so as to obtain

$$(H+V_0)\psi(R) = (E+V_0)\psi(R) . \tag{2}$$

If E_0 is the lowest eigenvalue and

$$V_0 + E_0 > 0 , \tag{3}$$

then Green's function for $H + V_0$ which obeys

*Supported by the U.S. Department of Energy under Grant AC 02-79ER 10353.

$$(H+V_0)G(R,R') = \delta(R-R') \tag{4}$$

is non-negative. Then, the Schrödinger equation may be written in integral form

$$\psi(R) = (E_0+V_0) \int G(R,R')\psi(R')dR' \ . \tag{5}$$

We may iterate this equation to define a sequence of functions

$$\phi_{n+1}(R) = (E_0+V_0) \int G(R,R')\phi_n(R')dR' \tag{6}$$

starting from some assumed initial $\phi_0(R)$.

Basic principles of probability theory shows that if

a) a set of values of R^n, viz., R_1^n, R_2^n,... be sampled at random from $\phi_n(R^n)$; and if

b) for each R_k^n draw values of R^{n+1} at random using $(E_0+V_0)G(R^{n+1},R_k^n)$ as a transition density from R_k^n to R^{n+1}; then the density of the configurations R^{n+1} is $\phi_{n+1}(R)$.

Step b) is possible because $G(R,R') \geq 0$. In this way a sequence of populations may be drawn successively from the functions ϕ_n. It is important to recognize that knowledge of these functions is derived entirely from the population $\{R^n\}$. Ratios of integrals with respect to these functions may be evaluated numerically as

$$\frac{\int f(R)\phi_n(R)dR}{\int g(R)\phi_n(R)dR} \cong \frac{\sum_m f(R_m^n)}{\sum_m g(R_m^n)} \tag{7}$$

Note that the crucial step is sampling R^{n+1} from $G(R^{n+1},R_k^n)$ conditional on R_k^n. This is important since, although $G(R,R')$ is in general not known, it is possible to specify random walk algorithms that generate the required populations.

It is interesting to analyze the sequence ϕ_n when $\phi_0(R) = \delta(R-R_0)$. We use the eigenfunction expansion of $G(R,R')$, viz

$$G(R,R') = \sum_\ell \frac{\psi_\ell(R)\psi_\ell(R')}{(E_\ell + V_0)} \ . \tag{8}$$

Then, iteration of Eq. (6) shows that for large n,

$$\phi_n(R) = \sum_\ell \left(\frac{E_0+V_0}{E_\ell+V_0}\right)^n \psi_\ell(R)\,\psi_\ell(R_0) \to \psi_0(R)\psi_0(R_0) \tag{9}$$

The last equation illustrates three important points:

a). The density of configurations R is asymptotically $\psi_0(R)$, independent of the starting R_0.

b). This asymptotic behavior does not require the use of E_0, the ground state eigenvalue in the iteration.

c). A configuration at R_0 contributes toward the asymptotic population with a coefficient proportional to $\psi_0(R_0)$.

It is a general result in Monte Carlo that if a configuration at some R contributes $\chi(R)$ toward a desired result, then an integral equation for $\chi(R)\phi(R)$ will give lower statistical error than an equation for the original density. In this case $\psi_0(R)$ plays the role of $\chi(R)$. Accordingly, we modify Eq. (6) to

$$\tilde{\phi}_{n+1}(R) \equiv \psi_0(R)\phi_{n+1}(R) = (E_0+V_0)\int \left[\psi_0(R)G(R,R')/\psi_0(R')\right]\times\psi_0(R')\phi_n(R')dR' \tag{10}$$

The analysis that gave Eq. (9), then yields

$$\tilde{\phi}_n(R) = \psi_0(R)\sum_{\ell}\left(\frac{E_0+V_0}{E+V_0}\right)^n \psi_\ell(R)\psi_\ell(R_0)/\psi_0(R_0) \rightarrow \psi_0^2(R) \tag{11}$$

so that for every n the size of the population, $\int \tilde{\phi}_n(R)dR$, is unity, independent of R_0 or any other details of the random walk. This biassing or "importance sampling" leads in principle to the estimation of the eigenvalue with zero sampling error. Of course it requires use of $\psi_0(R)$, which, if known, would make the calculation unnecessary. Theoretical arguments and much experience indicate that the use of a trial function ψ_T instead of ψ_0 in the importance sampling transformation reduces the Monte Carlo error, and for a many-body problem is essential to a practical calculation.

This procedure has been applied successfully to ^{4}He in liquid and crystal states. Figure 1 shows recent results[3] for the liquid

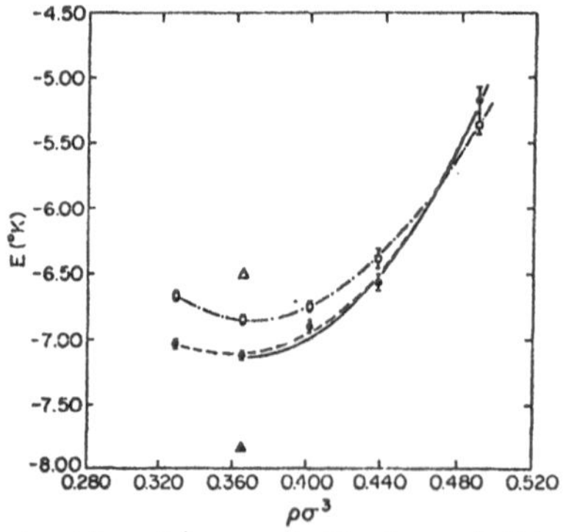

Fig. 1 Equation of State of Liquid ^{4}He
—— Expt; -.-.-, LJ; --- HFDHE2 Potentials

calculated with potentials[4] that fit gas phase data of He better than the Lennard-Jones potential generally used. It shows that the HFDHE2 potential of reference 4 agrees well with the experimental equation of state. Similar agreement is obtained for the crystal phase and the liquid structure function. Inclusion of perturbative estimates of the Axilrod-Teller three-body force weakens significantly the quality of the agreement.

2. Fermion Problems

The necessity of using Monte Carlo methods to perform the many-dimensional quadratures leads to special difficulties in the treatment of fermion systems. This may not be apparent at first glance.

Formally, if ϕ_0 is antisymmetric, and thus orthogonal to all but antisymmetric ψ_ℓ, then the Green's function expansion (Eq.(8)) shows that all ϕ_n are antisymmetric. Where, then, lies the problem?

It is that Monte Carlo methods can deal only with positive density functions. Clearly an antisymmetric spatial function cannot be positive everywhere, so we must write an antisymmetric function ϕ_{An} as a difference:

$$\phi_{An}(R) = \phi_n^+(R) - \phi_n^-(R); \quad \phi^\pm \geq 0 \tag{12}$$

Straightforward Monte Carlo treatment of Eq. (6) leads to separate iterations for $\phi^\pm$, that is, to

$$\phi_{n+1}^\pm(R) = (E_0+V_0)\int G(R,R')\phi_n^\pm(R')dR' \tag{13}$$

Now, if ϕ_{A0} is given, it is natural to set

$$\phi_0^\pm(R) = \frac{1}{2}\phi_{A0}(R) \pm \frac{1}{2}|\phi_{A0}(R)| \quad . \tag{14}$$

Unfortunately $|\phi_{A0}|$ is a symmetric function, not othogonal to the Bose ground state. The leading boson and fermion components contributing to $\phi_n^\pm$ are the respective ground states:

$$\phi_n^\pm \sim \psi_B(R) + \left(\frac{E_B+V_0}{E_F+V_0}\right)^n \alpha_0\psi_F(R) \tag{15}$$

Since $E_B < E_F$, the second term decays geometrically.

In the following sections, we discuss various attempts to deal with this basic obstacle.

2.1 The Fixed Node Approximation

A spatially antisymmetric continuous function in 3N dimensions must vanish on one or more manifolds of 3N-1 dimensions. The latter, which are called "nodal surfaces", separate configuration space into domains in which the wave function has one sign. It is important to note that the condition that $\psi_F(R) = 0$ when $\vec{r}_i = \vec{r}_j$ gives a manifold of 3N-3 dimensions; therefore it defines a submanifold of the nodal surface, not the whole.

Furthermore, the nodal surfaces, except in special cases, are not specified in advance by symmetries of the problem, but are determined in the course of finding ψ_F. Nevertheless much attention has been given to them, because if the nodal surfaces were known, the boundary condition that ψ_F must vanish there would reduce the problem to that already solved for bosons. It is therefor tempting to use an approximate nodal surface, such as that of an approximate trial function ψ_{FT}. Typically the nodal surface would be determined by zeros of a Slater determinant. One can show that the energy obtained with an approximate nodal surface has an error which is of second order in small displacements from the exact surface.

Calculations of this kind were first carried out by Anderson(5) for a few-electron systems. Ceperley has applied the method to a number of many-body systems including the electron gas(6) and liquid ^3He(7) for which he quotes an estimate (probably an upper bound) of -2.06K/atom using Lennard-Jones forces.

2.2 Use of Antisymmetrized Green's Functions.

Formally, one may introduce an antisymmetrized Green's function by a sum over permutations P

$$G_A(R,R') = \sum_P (-1)^P \, G(R,PR') \tag{16}$$

But, if in using this, positive and negative functions are treated separately, it leads to the same problems as before. Specifically, the sum over even permutations is a symmetric Green's function which preferentially propagates the symmetric part of any $\phi^{\pm}$ (c.f. Eq.(15)).

2.3 Transient Estimates

Although the separate iteration of ϕ_n^+ and ϕ_n^- yields a decaying antisymmetric component, integration with any antisymmetric factor projects it out. Thus if ψ_{FT} is an antisymmetric trial function, Eq. (15) indicates that

$$\int \phi_n^+(R)\psi_{FT}(R)dR \to \alpha_0 \left(\frac{E_B+V_0}{E_F+V_0}\right)^n \langle\psi_F|\psi_{FT}\rangle$$

$$\int \phi_n^+(R)\ H\ \psi_{FT}(R)dR \to \alpha_0 \left(\frac{E_B+V_0}{E_F+V_0}\right) \quad \langle H\psi_F|\psi_{FT}\rangle \tag{17}$$

Then the ratios

$$E_n \equiv \frac{\int \phi_n^+\ H\psi_{FT}\ dR}{\int \phi_n^+\ \psi_{FT}\ dR} \to \frac{\langle H\psi_F|\psi_{FT}\rangle}{\langle\psi_F|\psi_{FT}\rangle} = E_F \tag{18}$$

converge toward the fermion energy. Note that if ψ_{FT} were exact then $E_n = E_F$ for every n. But when ψ_{FT} is not exact, it is useful to examine the variance of the Monte Carlo estimates of the integrals of Eq. (17). For the first,

$$\text{Variance}\{\int \phi_n^+\ \psi_{FT}dR\} = \int \phi_n^+\ \psi_{FT}^2dR - \{\phi_n^+\ \psi_{FT}dR\}^2 \tag{19}$$

$$\sim \int \psi_B\ \psi_{FT}^2dR$$

That is, because ψ_{FT}^2 is symmetric, the boson function is projected out. This has the effect that the "signal" for either integral decays while the noise is constant. For fixed computing effort for each n, the relative error is defined as

$$\frac{[\text{Variance}]^{\frac{1}{2}}}{\text{Value}} \propto \left[\frac{E_F+V_0}{E_B+V_0}\right]^n .$$

Ceperley and Alder[6] have pointed out that if the population is made to grow as $(E_F+V_0)^n/(E_B+V_0)^n$ then the variance at each iterate will be constant. It seems very impractical to have the work grow exponentially. But before the computing time becomes excessive it is possible to derive useful estimates of E_F . This was done for the electron gas in reference 6.

An improvement upon this was applied to ^{3}He by Lee et al.[8] who noticed that if $\phi_0 = \psi_{FT}$ in Eq. (18) then a rigorous bound on E_F is obtained, i.e.

$$E_n \geq E_F \tag{20}$$

for every n . Thus these "transient estimates" form a sequence of upper bounds. Such a sequence is shown in Figure 2 for a system of

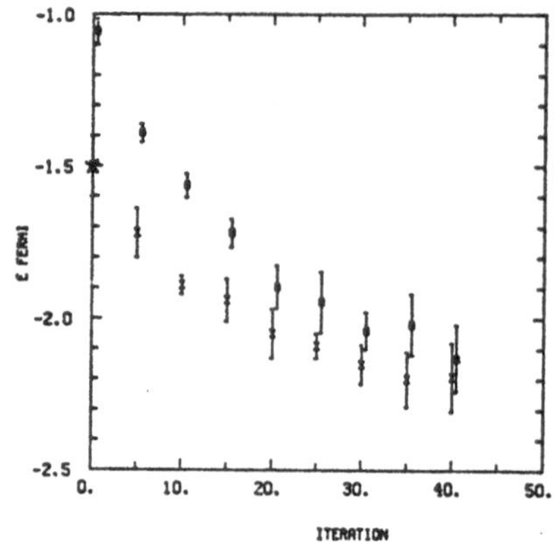

Fig. 2. Energy per atom of ^{3}He at experimental equilibrium density. I ψ_{FT} = Jastrow × det(plane waves); ▪ψ_{FT} has triplet correlations and backflow orbitals.

38 ^{3}He atoms interacting by the HFDHE2 potential at the experimental equilibrium density. It is not clear whether the sequence has converged, but the last three bounds can be combined to give the relationship

$$E_F \leq -2.20 \pm 0.05 \text{ K/atom.}$$

For the same system, variational bounds[9] have been found as follows

$$E_F < -1.03 \pm 0.03 \;;\; \psi_{FT} = \text{Jastrow} \times \det\{\text{plane waves}\}$$

$$< -1.52 \pm 0.03 \;;\; \text{triplet correlations added to } \psi_{FT}$$

$$< -1.91 \pm 0.02 \;;\; \text{"backflow" added to orbitals of } \psi_{FT}$$

In Figure 2 the crosses show E_n using the last type of ψ_{FT} (but not optimal parameters); circles use the first and simplest function. The better ψ_{FT} always gives smaller upper bounds.

2.4. Green's Function Filtering

In the analysis of the growth of the symmetric component ψ_B relative to the required antisymmetric component ψ_F, a key element lay in the assumption that ϕ^+ and ϕ^- are iterated independently. That is, configurations drawn from ϕ^+ have no effect upon those drawn from ϕ^-. If our integral equations were solved by finite difference methods, then cancellation between positive and negative domains of ϕ_A would be a natural and necessary aspect of convergence to the ground state. A formal analog in Green's function Monte Carlo is the following. We use the decomposition of ϕ_A into $\phi^{\pm}$ as prescribed by Eq. (12) to write the basic iteration (Eq. (6)) as

$$\phi_{n+1}(R) = (E_0+V_0)\left\{\int G(R,R')\phi_n^+(R')dR' - \int G(R,R')\phi_n^-(R')dR'\right\} \tag{21}$$

Because

$$\int \phi_A(R)dR = 0 \Rightarrow \phi^+(R)dR = \int \phi^-(R)dR$$

we may - and should - demand equal populations in the functions $\phi^{\pm}$. Now write Eq. (21) as the sum over the K configurations in the populations $\phi_n^{\pm}$ respectively:

$$\phi_{n+1}(R) = \frac{E_0+V_0}{K} \sum_k \left[G(R,R_k^+) - G(R,R_k^-)\right] \tag{22}$$

Since Eqs. (21) and (22) are linear, the expected value of the right side is unaltered by any particular correspondence that we choose to relate configurations R_k^+ and R_k^- ; statistical independence is not required. We will, in fact, try to pair R_k^+ to a nearby R_k^- so as to generate as much cancellation between the corresponding terms in Eq. (22). Of course each such pair of terms is positive in some domain and negative in another. These determine respectively $\phi_{n+1}^{\pm}$. Define

$$G^{\pm}(R;R_k^+,R_k^-) = \begin{matrix}\max\\ -\min\end{matrix}\left[G(R,R_k^+) - G(R,R_k^-),\ 0\right] \tag{23}$$

and then set

$$\phi_{n+1}^{\pm}(R) = \frac{E_0+V_0}{K} \sum G^{\pm}(R;R_k^+,R_k^-) \tag{24}$$

In Eq. (21), boson components of $\phi_{\pm}^{(n)}$, which are equal on the average, cancel on the average. In the Monte Carlo realization, Eqs. (22)-(24), the cancellation is incomplete, but if the population is large enough, the boson components will be stabilized relative to the fermion components. This has proved true for several few-body problems. The most interesting is a class of three "neutron" bound states[10]. Neutral particles of spin one half were assumed to interact by forces with square well attractions and repulsions. The ground state is required to be spatially antisymmetric with respect to the exchange of coordinates of the two particles of like spin. In one set of problems, the potential energy of the remaining particle was assumed to be a function of its distance from the center of mass of the other two. This interaction leads to analytically soluble three-particle wavefunctions, but does not otherwise affect the character of the Monte Carlo. It proved possible to find stable densities $\phi^{\pm}$ and accurate determination of the known energies. When the potentials where made pair-wise additive, the populations were again stable, giving energies close to but deeper than simple variational treatments of the same systems.

It seems unlikely that the method proposed here will, without extensive refinement, be capable of treating many-body problems. Zabolitzky[11] has suggested one possibility, namely that pairing like that of Eq. (22) be carried out on the integral equation in momentum space.

2.5 Approximate Filtering

Except for the "fixed node" method, no approximations have been made in the Monte Carlo quadratures. If carried out long enough - regrettably the time required may grow rapidly with n and with the number of particles - they are capable of giving exact numerical estimates. Perhaps one should give up this requirement in favor of approximations whose effect can be assessed or bounded and which more easily yield stable populations $\phi^{\pm}$.

One possibility is to use trial functions ψ_{FT} and ψ_{BT} to partition the population at stage n into a primarily boson set and a primarily fermion set. Technically, suppose that the set R_k^+, $k = 1,2,...K$ is ordered so that

$$\sum_{k=1}^{K_1} \psi_{BT}(R_k^+) \cong 0$$
$$\sum_{k=K_1+1}^{K} \psi_{FT}(R_k^+) \cong 0 \tag{25}$$

The first K_1 are the "fermion subset", the last $K-K_1$ the "boson subset". The quality of the partition could be tested by performing separate GFMC calculations on the two subsets. Each such calculation should exhibit smooth convergence to the appropriate energy and other observables. The filtering can be done repeatedly and each sub-population should converge separately. Departures from smooth convergence provide a measure of the error introduced and possibly a means of correction.

3. Prospects.

As we have seen, there exist algorithms for boson systems which are exact, stable, practical, and efficient. We have shown how much more difficult is the corresponding treatment of fermion systems. Exact, stable and reasonably efficient methods for few fermion problems have been developed. As yet there is none for the many-body problem, although with arbitrarily large computing efforts, increasingly better upper bounds can be obtained in a systematic way. It may be necessary to introduce various numerical approximations, whose effects can be estimated, to obtain efficient computational procedures for large systems.

References

1. M.H. Kalos, D. Levesque, and L. Verlet, Phys. Rev. A 9 2178 (1974); D. Ceperley and M.H. Kalos, Chapter 4 of "Monte Carlo Methods in Statistical Physics", K. Binder, Ed. Springer-Verlag (1979); P. Whitlock et al. Phys. Rev B 19, 5598 (1979).

2. P.A. Whitlock and M.H. Kalos, Lecture Notes on Monte Carlo, Chapters 15-18 (unpublished).

3. M.H. Kalos, M.A. Lee, P.A. Whitlock, and G.V. Chester, submitted to Phys. Rev. B.

4. R.A. Aziz et al., J. Chem. Phys. 70 4330 (1979).

5. J. Anderson, J. Chem. Phys. 73 3897 (1980) and references therein.

6. D. Ceperley and B. Alder, Phys. Rev. Letts, 45, 566 (1980).

7. D. Ceperley, private communication.

8. M. A. Lee, K.E. Schmidt, M. H. Kalos, and G.V. Chester, submitted to Phys. Rev. Letts.

9. M. H. Kalos, K.E. Schmidt, M.A. Lee, and G.V. Chester, to be published.

10. D. M. Arnow, M.H. Kalos, K.E. Schmidt, and M.A. Lee, Bull. Amer. Phys. Soc. 26, 35 (January 1981).

11. J. Zabolitzky, private communication.

THE STOCHASTIC SOLUTION OF THE MANY-BODY SCHROEDINGER EQUATION FOR FERMIONS

D. M. Ceperley
National Resource for Computation in Chemistry
Lawrence Berkeley Laboratory
University of California
Berkeley, California 94720 USA

This paper describes the basis of calculations we have made to compute the ground state properties of many-fermion systems. Elsewhere in this volume Kalos will discuss the Green's Function Monte Carlo (GFMC) approach to this problem. Earlier quantum calculations have been reviewed in ref. [1]. These GFMC methods of which the present method is a variant, are not to be confused with variational Monte Carlo, as first utilized by McMillan [2]. The latter give properties of model trial functions, whereas GFMC gives properties of the exact ground state.

We start by considering the Schroedinger equation for N particles written in imaginary time:

$$-\frac{\partial\phi(R,t)}{\partial t} = H\phi(R,t) = (-\frac{\hbar^2}{2m}\nabla^2 + V(R) - V_0)\phi(R,t) \tag{1}$$

Here R is the 3N set of coordinates, V(R) is the potential energy function and V_0 an arbitrary constant. In order to make a many-body calculation feasible, importance sampling is introduced by multiplying this equation by a known trial function $\Psi_T(R)$ and making the substitution: $f(R,t) = \Psi_T(R)\phi(R,t)$. With some algebra we arrive at the expression:

$$-\frac{\partial f}{\partial t} = \frac{\hbar^2}{2m}[-\nabla^2 f + \nabla(fF_Q(R)] + [E_L(R) - V_0]f \tag{2}$$

where $F_Q(R)$ and $E_L(R)$ are defined as:

$$F_Q(R) = \nabla \log \Psi_T{}^2 = \text{quantum trial "force"} \tag{3}$$

$$E_L(R) \equiv \Psi_T{}^{-1} H \Psi_T(R) = \text{local energy} \tag{4}$$

Equation (2) can be interpreted as a simple Markov process. Suppose we construct an ensemble of systems (points in 3N dimensional space) with the probability density f(R,0). Usually the initial condition taken is $f(R,0) = \Psi_T(R)^2$. These systems then diffuse and branch such that their probability distribution is given by eq. (2). The three terms on the right hand side are interpreted respectively as random diffusion, drift and branching. By branching, it is meant that a particular system is either eliminated from the ensemble, if the local energy is less than V_0, or duplicated in the ensemble, otherwise.

A steady state population requires that V_0 be equal the lowest eigenvalue, E_0. It is easily shown that for large time $f(R,t) = \Psi_T(R)\phi_0(R)$ where ϕ_0 is the exact ground state eigenfunction. The eigenvalue then can be estimated as:

$$E_0 = \frac{\int \Psi_T H \phi_0}{\int \Psi_T \phi_0} = \frac{\int \phi_0 H \Psi_T}{\int \Psi_T \phi_0} = \langle E_L(R) \rangle \qquad (5)$$

where <...> means the average over the distribution f(R,t) for large enough t. By the usual statistical formulas the variance of the energy estimate is given by:

$$\mathrm{var}(E_0) = \langle (E_L(R) - E_0)^2 \rangle / M \qquad (6)$$

where M is the number of independent sample points. As usual with Monte Carlo methods the error bar on E_0 is proportional to the inverse square root of computer time. However by choosing a good trial function Ψ_T, we can reduce the coefficient dramatically. In the limit as Ψ_T approaches ϕ_0 the variance vanishes. In practice by using Slater-Jastrow trial functions we can eliminate many of the singularities in $E_L(R)$. The importance sampling by Ψ_T is also essential in controlling fluctuations and instabilities in the branching process.

Our algorithm for this process essentially uses a short time approximation to the exact Green's function. This approximation has been described for a classical polymer system in ref. [3]. Equation (2) is identical to the classical Smoluckowski equation except for the presence of the local energy term. In this approximation if a system has diffused from R_0 to R within a time t, then the expected number of copies at time t will be:

$$\exp\left[-t(E_L(R) + E_L(R_0) - 2V_0)/2\right]$$

This method is much simpler than the exact GFMC method of Kalos (0), but only exact in the limit as $t \to 0$.

The above method is easily implemented to calculate the ground state properties of boson systems. But for fermion systems, there are serious and, as yet, not resolved difficulties. The crux of the problem is that unless ϕ_F (the exact fermion eigenfunction) and Ψ_T have the same sign everywhere, then $f = \phi_F \Psi_T$ cannot be interpreted as a probability density. Hence the nodes of ϕ_F are required. Except for one dimensional or few particle problems exact specification of the nodal locations is an extremely tough problem. However important progress has been made in circumventing this difficulty for several systems.

The Fixed-Node Method

A simple, though approximate, method of accounting for antisymmetry is simply to let the nodes of a trial function act as an absorbing barrier to the diffusion.

Suppose $\Psi_T(R)$ is an antisymmetric trial function. The nodes of Ψ_T divide the configuration space into connected volumes. Using the above Monte Carlo method we can so obtain the eigenfunctions (ϕ_k) and eigenvalues (e_k) inside each nodal volume (V_k) and which vanish outside that volume:

$$\begin{aligned} H\phi_k &= e_k\phi_k \\ \phi_k(R) &= 0 \quad R \notin V_k \end{aligned} \tag{7}$$

Each of the eigenvalues e_k is an upper bound to the fermion energy, E_F, since the antisymmetric function:

$$\hat{\phi}_k(R) = \sum_P (-)^P \phi_k(PR) \tag{8}$$

has a variational energy e_k, where P is a permutation.

$$e_k = \frac{\int \hat{\phi}_k H \hat{\phi}_k}{\int \hat{\phi}_k^2} \geq E_F \tag{9}$$

It is easily shown that $\hat{\phi}_k$ is non-zero, since at each point, only permutations of the same sign can contribute to the sum in eq. (8). Otherwise the volume V_k would contain both positive and negative regions of $\Psi_T(R)$. This variational argument can be easily generalized to include spin.

In practice the fixed-node method is carried out in the following way. The initial ensemble is chosen as before, to be $f(R,0) = \Psi_T(R)$. If the population is large enough, all the different nodal regions will be populated. The diffusion and branching process precedes as in the boson case, except that whenever a system crosses a node of Ψ_T that system is eliminated from the ensemble. It is easily seen that the V_0 necessary to maintain a stationary population in the ensemble is given by $e_{FN} \equiv \min_k \{e_k\}$. Thus if e_k depends on the nodal volume, the diffusion process will select out those with the lowest e_k.

In practice this method achieves a good upper bound to E_F because all of the correct many-particle boson correlations are in $\hat{\phi}_k$. Because the shape of the volume V_k is in general incorrect, $\hat{\phi}_k$ is a solution of the Schroedinger equation everywhere except at the nodes of Ψ_T where it has a discontinuous gradient. (The discontinuity will not contribute to the integral in eq. (9) since $\hat{\phi}_k$ is zero there.) By the usual arguments ($e_k - E_F$) must vanish quadratically as the nodes of Ψ_T approach those of ϕ_F. In principle one could vary the nodal locations to obtain the best upper bound but the highly dimensional nodal surfaces are difficult to parameterize in a systematic fashion.

Nodal Relaxation

If the nodes of Ψ_T are sufficiently close to those of ϕ_F, an improvement in the fixed-node procedure will give the exact fermion eigenfunction. The basic idea is that, if the diffusion process begins in an antisymmetric state (i.e., with configurations carrying $\pm$ signs depending on the sign of Ψ_T) the diffusion process, including allowing diffusion across the nodes, will maintain the antisymmetry and must converge to the antisymmetric ground state. This procedure is however unstable since a fluctuation of the boson ground state will grow and dominate at large times.

We can represent the diffusion process, with the importance function Ψ_T, in terms of its Greens function as:

$$f\ (R,t) = \int dR_0\ \Psi_T(R) e^{-(H - V_0)t}\ \Psi_T^{\ -1}(R_0)\ f_A(R_0,0) \tag{10}$$

where $e^{-(H-V_0)t}$ is the Green's function for eq. (1). If $f_A(R,0)$ is symmetric and $\Psi_T(R)$ is antisymmetric then:

$$\lim_{t\to\infty} f_A\ (R,t) \propto \Psi_T(R)\phi_F(R) e^{-(E_F - V_0)t} \tag{11}$$

But the diffusion process can only have positive importance functions--otherwise one is lead to negative probabilities; the sign of Ψ_T must be taken out as a weight. Let $\sigma(R) = \text{sign}\ (\Psi_T(R)) = \pm 1$. Then we can rewrite eq. (10) as

$$f_A(R,t) = \int dR_0 \sigma(R)\sigma(R_0) |\ \Psi_T(R)\ |\ e^{-(H - V_0)t} |\Psi_T(R_0)|^{-1} f_A(R_0,0) \tag{12}$$

Since the time for which this algorithm is stable is short, it is desirous to take the initial distribution as close as possible to the limiting distribution in eq. (11). A convenient choice is the fixnode distribution, $f_A(R,0) = \Psi_T(R)\phi(R)$.

The integral in eq. (12) can be performed by a very simple extension of the fixed-node diffusion process. Suppose we wish to calculate the fermion eigenvalue Define:

$$E_A(t) = \frac{\int \Psi_T H\ \Psi_T^{\ -1} f_A(R,t)}{\int f_A\ (R,t)} = \frac{\langle \sigma(R)\sigma\ (R_0) E_L(R) \rangle}{\langle \sigma(R)\ \sigma\ (R_0) \rangle} \tag{13}$$

Now from our initial conditions $E_A(0) = e_{FN}$ and from eq. (11) $E_A(\infty) = E_F$. $E_A(t)$ will relax from the fixed-node energy to the fermion energy. Each system is now assigned a new clock, denoted s. Until a system diffuses across a node for the first time, s = 0, but as soon as a system crosses a node of Ψ_T then s begins counting. If a system branches the clock is reproduced in all copies. When s reaches some maximum value s_M, the system is eliminated from the ensemble. The collection of all systems at all times for which s = 0 is simply the fixed-node distribution, $f_A(R,0)$. The collection of all systems at all times for which $s \leq t$ with the weight $\sigma(s)\sigma(0)$ has the distribution $f_A(R,t)$, since this collection represents all the systems which evolved from the fixed node distribution in time t. Then to calculate $E_A(t)$ one merely performs the averages in eq. (13) over all configurations for which $s \leq t$. This will give $E_A(t)$ for $0 \leq t \leq s_M$. Since $|\Psi_T|$ contains nodes, care must be taken in constructing the short time Green's function so that systems can tunnel through the nodes properly.

Because of the the instability referred to earlier, the number of configurations needed in order to determine $E_A(s_M)$ grows with s_M like $\exp(s_M(E_F - E_0))$. Hence an upper limit to s_M, assuming a reasonable computer run, is determined by the difference between the boson and fermion energies, in other words, the node crossing frequency. Thus we are limited, by this algorithm, to problems for which the nodes of Ψ_T can be repaired in a time small compared with $(E_F - E_0)^{-1}$. The available evidence on jellium suggests that we have met this criterion; the potential is soft enough so that the Hartree-Fock nodes are rather good. The situation on ^{3}He, at this moment, does not appear as promising; more sophisticated algorithms or methods may be needed.

Applications

The applications, to date, have been to jellium in two and three dimensions, liquid ^{3}He and hydrogen (as a two component system of protons and electrons). I will not discuss the latter two systems as our calculations are still incomplete. The electron gas has been discussed in refs. [4,5]. The following phases were studied: the boson fluid, the normal paramagnetic fluid, the spin-polarized or ferromagnetic fluid and the Wigner crystal. The ground state energies as obtained by the nodal relaxation method are given in Table I with the error bars in parentheses. Because of importance sampling, the errors are much smaller than usual with Monte Carlo. Table II contains the energy differences in going from variational Monte Carlo with a Jastrow-Slater trial function to the fixed-node energy and then to the exact energy. At low density, we find, that the normal electron gas undergoes a phase transition at r_s = 75 to a ferromagnetic fluid and at r_s = 100 to a Wigner crystal. The boson system undergoes Wigner crystallization at r_s = 160 (r_s is the Wigner sphere radius in units of Bohr radii).

Table I

r_s	E_{PMF}	E_{FMF}	E_{BF}	E_{BCC}
1.0	1.174(1)	---------	--------	---------
2.0	0.0041(4)	0.2517(6)	-0.4531(1)	---------
5.0	-0.1512(1)	-0.1214(2)	-0.21663(6)	---------
10.0	-0.10675(5)	-0.1013(1)	-0.12150(3)	---------
20.0	-0.06329(3)	-0.06251(3)	-0.06666(2)	---------
50.0	-0.02884(1)	-0.02878(2)	-0.02927(1)	-0.02876(1)
100.0	-0.015321(5)	-0.015340(5)	-0.015427(4)	-0.015339(3)
130.0	------------	-----------	-0.012072(4)	-0.012037(2)
200.0	------------	-----------	-0.008007(3)	-0.008035(1)

The ground state energy of the charged Fermi and Bose systems. The density parameter, r_s, is the Wigner sphere radius in units of Bohr radii. The energies are Rydbergs and the digits in parenthesis represent the error bar in the last decimal place. The four phases are: paramagnetic or unpolarzed Fermi fluid (PMF); the ferromagnetic or polarized Fermi fluid (FMF); the Bose fluid (BF); and the Bose crystal with a BCC lattice.

Table II

r_s	δ_{PMF}	γ_{PMF}	δ_{FMF}	γ_{FMF}	δ_{BF}	δ_{BCC}
2	40	9	11.0	---	12.0	-----
5	17	2	7.2	---	6.8	-----
10	11	1	6.5	1.8	5.1	-----
20	6.7	0.7	3.0	1.0	3.3	-----
50	2.9	0.31	1.6	0.25	1.7	2.0
100	1.7	----	1.2	----	1.2	0.41
130	----	----	---	----	1.1	0.30

The error in the variational approximation in 10^{-4} Rydbergs for four different phases. $\delta = E_v - E_o$ (the difference between the Jastrow trial function and the exact ground state energy).
$\gamma = E_{FN} - E_o$ (the difference between the 'fixed-node' energy with plane wave nodes and the exact ground state energy).

Errors

Finally, I would like to close with a discussion of the type of errors which limit the accuracy of a GFMC calculation. They are ranked in order of increasing importance in the one system we have studied most extensively, the three dimensional electron gas.

1) Numerical errors. (Truncation errors or the use of a short time Green's function, round-off errors, use of pseudo-random numbers, etc.) These errors with a sufficient amount of programming effort can be made very small. For the diffusion model the exact Kalos algorithm [1] can be used.
2) Convergence of f(R,t) to $\Psi_T\ \phi_0$. With homogeneous quantum systems and good trial functions the convergence is very rapid, indicating only local diffusion processes are needed to convert Ψ_T into ϕ_0.
3) Statistical errors. As demonstrated above these errors depend on the fluctuations in the local trial energy and on the number of independent systems that can be generated. For the electron gas, the Jastrow-Slater trial function is good enough so that this error is quite small, compared with the accuracy of other types of calculations, and small enough to determine the phase transitions. But for more complicated systems the trial functions will be more difficult to construct.
4) The fermion problem. Our experience with the electron gas at many different densities suggests that the Hartree-Fock nodes are good enough for the present method to converge to the exact ground state. But clearly this error is not under control. We need both better algorithms and more experience.
5) Finite system effects. With present supercomputers we have been able to simulate up to 250 fermions. For the electron gas, even with periodic boundary conditions, this is far from the thermodynanic limit. Our simulations show that the correction to the kinetic energy behaves like $N^{-2/3}$ and to the potential energy as N^{-1}. To extrapolate to the limit of large N, we have taken some simple model, for example Hartree-Fock, with some adjustable parameters and have analytically calculated the finite system effects. Then using simulations at a variety of values of N, we have both fixed the adjustable parameters, and tested the model. The model is satisfactory for the electron gas although the range of N is rather limited. Faster computers and algorithms will help, both of which are in the offing.

Acknowledgments

This work was supported by the Director, Office of Energy Research, Office of Basic Energy Sciences, Chemical Sciences Division of the U. S. Department of Energy under Contract No. W-7405-ENG-48 and under a grant from the National Science Foundation (Grant No. CHE-7721305.)

References

1. D. M. Ceperley and M. H. Kalos on "Monte Carlo Methods in Statistical Physics," ed. K. Binder, Springer-Verlag (1979).
2. W. L. McMillan, Phys. Rev. A 138, 442 (1965).
3. D. Ceperley, M. H. Kalos, and J. L. Lebowitz, "Computer Simulation of the Static and Dynamic Properties of Polymer Chain," submitted to J. Chem. Phys. (1980).
4. D. M. Ceperley and B. J. Alder, Phys. Rev. Letts. 45, 566 (1980).
5. D. M. Ceperley and B. J. Alder, Journal de Physique C7,41, 295 (1980).

RECENT DEVELOPMENTS AND FUTURE PROSPECTS IN CBF THEORY*

E. Krotscheck and R. A. Smith
Department of Physics, State University of New York
Stony Brook, New York 11794, U.S.A.

and

J. W. Clark
McDonnell Center for the Space Sciences
and Department of Physics, Washington University
St. Louis, Missouri 63130, U.S.A.

I. INTRODUCTION

This series of conferences on many-body theories and its predecessors, the Urbana Workshops of 1973 and 1977, have witnessed the remarkable success of variational methods in the microscopic calculation of ground-state properties of dense, strongly interacting systems. To many it may seem time to rewrite the classic textbooks on many-body theory [1-3]. However, we shall demonstrate in this contribution that prominent examples of conventional many-body methods, namely perturbative correction of the ground-state energy and the quasiparticle interaction, the BCS approach to pairing phenomena, and Brueckner-Bethe-Goldstone theory, may be transparently reformulated in terms of nonorthogonal, correlated wave functions. (The same holds for the RPA, which will be discussed in a separate talk [5].) These methods will be applied herein to the CBF theory of liquid ^{3}He. Details are relegated to Refs. [6,7].

The key idea of the CBF scheme is to use a correlation operator F to construct a basis of correlated states

$$|\psi_m^{(A)}\rangle = I_{mm}^{-1/2} F_A|\Phi_m^{(A)}\rangle \ , \qquad I_{mm} = \langle\Phi_m^{(A)}|F_A^\dagger F_A|\Phi_m^{(A)}\rangle \ , \tag{1.1}$$

$|\Phi_m^{(A)}\rangle$ being an A-particle Slater determinant built from plane-wave orbitals specified by the subscript $m = (m_1,\ldots m_A)$. The filled Fermi sea will carry the subscript o.

The definition (1.1) sacrifices the convenient orthogonality of the usual basis states. On the other hand, with a suitable F_A one can guarantee the finiteness of all matrix elements of the Hamiltonian, and build other essential physical properties of the system into the basis. The loss of orthogonality is not very severe, in that the required modifications of expressions for physical quantities are well understood. In essence, the CBF method provides a tool for the systematic construction of weak, effective interactions from strong, bare two-body potentials.

*Research supported in part by the Deutsche Forschungsgemeinschaft, the U.S. Department of Energy under Contract No. DE-AC02-76ER13001 and the U.S. National Science Foundation under Grant No. DMR80-08229.

It is important to note that one can implement a second-quantized formulation by introducing creation and annihilation operators $\alpha_k^\dagger$, α_k for correlated states, according to

$$\alpha_k^\dagger|\psi_m^{(A)}\rangle = I_{nn}^{-\frac{1}{2}} F_{A+1} a_k^\dagger|\Phi_m^{(A)}\rangle , \qquad I_{nn} = \langle\Phi_m^{(A)} a_k|F_{A+1}^\dagger F_{A+1}|a_k^\dagger\Phi_m^{(A)}\rangle ,$$
$$\alpha_k|\psi_m^{(A)}\rangle = I_{\ell\ell}^{-\frac{1}{2}} F_{A-1} a_k|\Phi_m^{(A)}\rangle , \qquad I_{\ell\ell} = \langle\Phi_m^{(A)} a_k^\dagger|F_{A-1}^\dagger F_{A-1}|a_k\Phi_m^{(A)}\rangle . \tag{1.2}$$

These "correlated" creation and annihilation operators fulfill the same commutator rules as the corresponding field operators $a_k^\dagger$, a_k, the only caveat being that we have to define their adjoints separately and to distinguish carefully between operators acting to the left and to the right.

II. OPTIMIZED JASTROW FUNCTIONS

The general CBF theory carries an intrinsic redundancy: We may handle a physical effect either by a suitable choice of the correlation operator F (as in the case of hard cores) or through the proper selection of perturbative corrections. A Jastrow F,

$$F_A = \prod_{i<j}^{A} f(r_{ij}) , \qquad f(r) = \exp \frac{1}{2}u(r) , \tag{2.1}$$

presently offers the best tradeoff between formal and computational effort. With optimal choice of f(r) one may already obtain good low-order results. Optimization reduces the size of perturbation corrections and leads naturally to backflow in the effective interaction.

Methods for optimizing the Jastrow function have been previously discussed [7-11]. In the end, one solves the Euler equation, which in momentum space reads

$$(\hbar^2k^2/4m)[S(k) - 1] + S'(k) = \omega(k) = 0 \quad . \tag{2.2}$$

Here, S(k) is the familiar static structure function. The generalized structure function [4] S'(k) may be expressed as

$$S'(k) = \int d^3r[\delta S(k)/\delta u(r)]v_{JF}(r) + \text{kinetic-energy terms} \quad , \tag{2.3}$$

where

$$v_{JF}(r) = v(r) - (\hbar^2/4m)\nabla^2 u(r) \tag{2.4}$$

is the Jackson-Feenberg (JF) effective interaction, and the "kinetic-energy terms" include additional contributions to S'(k) involving the usual derivatives of the Slater function in the JF kinetic energy. The "prime" equations for S'(k) and its ingredients--the FHNC' equations [7]--are generated from the FHNC equations by applying the linearization (2.3). By-products of this method are the quantities $\Gamma'_{dd}(r)$ and $X'_{cc}(k)$, the analogs of the FHNC quantities $\Gamma_{dd}(r)$ and $X_{cc}(k)$ [12]; $\Gamma'_{dd}(r)$ will in fact supply an essential local piece of the effective interation, while $X'_{cc}(k)$ is a key constituent of the single-particle excitation spectrum.

A simplified Euler equation is obtained by dropping from (2.2) all de and non-trivial ee diagrams (those not expressible in terms of the non-interacting structure function $S_F(k)$) and linearizing in the dd quantities $\tilde{\Gamma}_{dd}(k)$ and $\tilde{\Gamma}'_{dd}(k)$:

$$\tilde{\Gamma}'_{dd}(k) \approx (\hbar^2 k^2/4m)[1 - 2/S_F(k)]\tilde{\Gamma}_{dd}(k) \quad . \tag{2.5}$$

It is found [6] that the optimization condition (2.2) is typically 80% exhausted by this approximation.

All optimizations were done using the JF energy, the Lennard-Jones 6-12 interaction, and the FHNC/C approximation for the FHNC and the FHNC' equations [7]. The Jastrow energy expectation values for normal and fully-spin-polarized systems (denoted ^{3}He and ^{3}He$\ddagger$, respectively) are given in Table 1. Expectedly, the optimization does not produce a substantial lowering of $\langle H\rangle = H_{oo}$ with respect to the result for Schiff-Verlet trial functions. A conservative estimate of the elementary diagrams may be obtained by comparison of the JF and PB forms of the kinetic energy [12]. Details of the calculations leading to Table 1 may be found in Ref. [7]. Table 1 reveals a rather uncomfortable feature of the Jastrow wave function, namely instability of the unpolarized state relative to the fully polarized one. This is patently a deficiency of the state-independent Jastrow ansatz and not due to the neglect of elementary diagrams [13]. It is therefore necessary to improve upon the Jastrow description of the ground state of normal liquid ^{3}He.

Table 1: Variational ground-state energy of ^{3}He and ^{3}He$\ddagger$. All energies are given in °K per particle.

ρ (Å^{-3})	$\langle H\rangle$ (^{3}He)	$\langle H\rangle$ (^{3}He$\ddagger$)
0.0076	-0.72	-0.54
0.0112	-1.00	-1.07
0.0130	-0.98	-1.29
0.0142	-0.91	-1.34
0.0148	-0.84	-1.36
0.0166	-0.52	-1.34
0.0180	-0.13	-
0.0200	0.66	-

III. EFFECTIVE INTERACTION

The variational ground-state energy is just one matrix element in the correlated basis. Indeed we may calculate any (on- or off-) diagonal element of H or 1,

$$H_{mn} = \langle\psi_m|H|\psi_n\rangle \quad , \quad N_{mn} = \langle\psi_m|\psi_n\rangle \quad , \tag{3.1}$$

and especially

$$H'_{mn} = H_{mn} - H_{oo}N_{mn} \quad , \tag{3.2}$$

the latter being the combination in which the generic matrix elements of H naturally occur. As a special case, the diagonal matrix elements of H in 1p1h states

$|\psi_m\rangle = \alpha_p^\dagger \alpha_h |\psi_o\rangle$ generate the particle-hole excitation spectrum

$$H'_{mm} = e(p) - e(h) \quad . \tag{3.3}$$

The most important off-diagonal matrix elements are those in which the states $|\psi_m\rangle$ and $|\psi_o\rangle$ differ by exactly two orbital labels. The ph nonorthogonality operator and effective interaction are defined respectively through

$$\langle pp'|N_{ph}(12)|hh'\rangle_a = \langle \psi_o^{(A)} \alpha_h^\dagger \alpha_{h'}^\dagger \alpha_{p'} \alpha_p |\psi_o^{(A)}\rangle \quad ,$$

$$\langle pp'|V_{ph}(12)|hh'\rangle_a = \langle \psi_o^{(A)} \alpha_h^\dagger \alpha_{h'}^\dagger \alpha_{p'} \alpha_p |H - H_{oo}^{(A)}|\psi_o^{(A)}\rangle \quad . \tag{3.4}$$

The subscript a indicates antisymmetrization. The corresponding pp and hh expressions are constructed similarly.

The effective interactions $V_{pp}(12)$, $V_{ph}(12)$, and $V_{hh}(12)$ and the corresponding nonorthogonality operators are genuine many-body quantities; they depend on the underlying state $|\psi_o\rangle$. In particular, they are not Galilean invariant. Their analytic structure also depends on the form of the correlation operator F. With the Jastrow choice (2.1) for F, the three nonorthogonality-correction operators turn out to have the same $\underset{\sim}{r}$-space representation to leading order in the particle number. We accordingly drop the subscripts pp, ph, and hh. Moreover, the effective interactions à la (3.4) have the simple form

$$\langle ij|V(12)|k\ell\rangle = \langle ij|W(12)|k\ell\rangle + \tfrac{1}{2}[\pm e(i) \pm e(j) \pm e(k) \pm e(\ell)]\langle ij|N(12)|k\ell\rangle = V_{ij,k\ell} , \tag{3.5}$$

in which the plus sign applies for particle labels and the minus sign for hole labels. Again, the $\underset{\sim}{r}$-space representations of the (non-local) two-body operator $W(12)$, for pp, ph, and hh cases, coincide to leading order in the particle number.

Rather than giving more details on the analytic structure of the operators $N(12)$ and $W(12)$ [14,9], we shall concentrate here on the physical meaning of the effective ph interaction, by reporting its most important properties:

(i) The diagonal limit of the particle-hole interaction (note that $V_{ph}(12)$ has been defined only for off-diagonal channels) is identical with the Jastrow piece of the quasiparticle interaction,

$$f_{oo}(\underset{\sim}{k}\sigma,\underset{\sim}{k}'\sigma') = \delta^2 H_{oo}/\delta n(\underset{\sim}{k},\sigma)\delta n(\underset{\sim}{k}',\sigma') = \lim_{q\to 0} \langle \underset{\sim}{k}+\underset{\sim}{q}\sigma, \underset{\sim}{k}'-\underset{\sim}{q}\sigma'|W(12)|\underset{\sim}{k}\sigma,\underset{\sim}{k}'\sigma'\rangle_a$$
$$\equiv \langle \underset{\sim}{k}\sigma\underset{\sim}{k}'\sigma'|W(12)|\underset{\sim}{k}\sigma\underset{\sim}{k}'\sigma'\rangle_a \quad . \tag{3.6}$$

Some caution is, however, required due to the non-commutivity of the diagonal limit and the limit of infinite box size when long-range correlations are assumed. Also, it is much simpler to sum additional diagrams for the quasiparticle interaction than for the general ph interaction.

(ii) For optimized correlation functions the weighted average

$$\sum_{hh'} (I_{mm}/I_{oo})^{\frac{1}{2}} H'_{mo} = A\omega(q) \tag{3.7}$$

of the ph interaction, with $|\Phi_m\rangle = a^\dagger_{\underset{\sim}{h}+\underset{\sim}{q}}\, a^\dagger_{\underset{\sim}{h}'-\underset{\sim}{q}}\, a_{\underset{\sim}{h}'} a_{\underset{\sim}{h}} |\Phi_o\rangle$, vanishes identically in the momentum transfer q. This leads to substantial cancellations between central and non-central components of the effective interaction, since the ratio (I_{mm}/I_{oo}) is a positive and rather slowly varying function of its arguments. One consequence of such cancellations is considerably improved convergence of the CBF perturbation expansion [7].

(iii) The (dominant) local contributions to $\mathcal{W}(12)$ and $\mathcal{N}(12)$ are

$$\begin{aligned} \mathcal{W}_{loc}(r) &= \Gamma'_{dd}(r) + (\hbar^2/4m)[\nabla,[\nabla,\Gamma_{dd}(r)]] \quad , \\ \mathcal{N}_{loc}(r) &= \Gamma_{dd}(r) \quad . \end{aligned} \tag{3.8}$$

Using the approximate relation (2.5) and bare single-particle energies in the definition (3.4) of $\mathcal{V}(12)$, we obtain

$$\mathcal{V}(r_{12}) \cong (\hbar^2/2m)[V_c(r_{12}) + \Gamma_{dd}(r_{12})\underset{\sim}{r}_{12}\cdot(\nabla_1 - \nabla_2)] \quad , \tag{3.9}$$

where the Fourier transform of $V_c(r)$ is $\tilde{V}_c(q) = q^2\, \tilde{\Gamma}_{dd}(q)[1 - S_F(q)]^{-1}$. We note that the average (3.7) of (3.9) over the hole states vanishes identically. The result (3.9) shows clearly that the "backflow" character [15] of the effective ph interaction arises naturally in the Jastrow-correlated basis.

IV. PERTURBATION CORRECTIONS

The effective interaction and its multiparticle generalizations are the keys to formulating perturbative corrections to the ground-state energy and the quasiparticle interaction. The leading terms of the expansion for the ground-state energy E may be found in Ref. [16]. The two-body effective interaction of (3.4) contributes to the second-order energy correction in the correlated basis [12,17] an amount

$$\Delta E^{(2,2)} = -\frac{1}{4} \sum_{pp'hh'} \left|\mathcal{V}_{pp'(hh')_a}\right|^2 \Big/ [e(p) + e(p') - e(h) - e(h')] \quad . \tag{4.1}$$

Straightforward variation of the $E-H_{oo}$ expansion with respect to occupation numbers gives the perturbation corrections to the quasiparticle interaction. The standard formulas for

$$\Delta f^{(2,2)}(\underset{\sim}{k}\sigma,\underset{\sim}{k}'\sigma') = \delta^2\left(\Delta E^{(2,2)}\right)/\delta n(\underset{\sim}{k}\sigma)\delta n(\underset{\sim}{k}'\sigma') \tag{4.2}$$

found in textbooks (e.g., Eq. (21.13) of [3]) apply. We employed Monte Carlo integration to calculate the perturbation corrections to E, m*/m, and the magnetic susceptibility which are shown in Table 2.

The variational estimates for m*/m and χ_F/χ were not obtained from the diagonal limit of the effective 2p2h interaction (3.4). Instead, we calculated m*/m by numerical differentiation of the single-particle energies (3.3) at k_F, and evaluated χ_F/χ by performing the functional variation (3.6) of the FHNC energy expectation value. This resummation of larger classes of diagrams is feasible in the Landau limit, since in each partial wave only one linear integral equation arises. More details of the procedures used will be given elsewhere [6,7].

Table 2: Perturbation corrections to the ground-state energy, effective-mass ratio, and magnetic susceptibility. Column 3 gives the sum of the variational energy (Table 1) and the CBF correction. Columns 4 and 6 give the variational estimates for m*/m and χ_F/χ; Columns 5 and 7 the same quantities when CBF corrections are included.

ρ ($\mathring{A}^{-3}$)	$\Delta E^{(2,2)}$	E_2	$(m^*/m)_V$	$(m^*/m)_2$	$(\chi_F/\chi)_V$	$(\chi_F/\chi)_2$
0.0076	-0.22	-0.94	1.13	1.79	0.23	0.38
0.0112	-0.35	-1.35	0.94	1.65	-0.09	0.06
0.0130	-0.44	-1.42	0.87	1.66	-0.22	-0.08
0.0142	-0.51	-1.42	0.83	1.69	-0.30	-0.15
0.0148	-0.55	-1.39	0.81	1.70	-0.35	-0.19
0.0166	-0.68	-1.20	0.75	1.78	-0.45	-0.31
0.0180	-0.81	-0.94	0.71	1.86	-0.53	-0.39

Additional contributions from some third-order diagrams arising from variations of the effective interaction with respect to $n(\underset{\sim}{k}\sigma)$ turned out to be of relative order 10^{-2}. The results of Table 2 are encouraging, particularly the corrected values for the effective mass. That the improvement in m*/m is the most significant effect in implementing the CBF scheme is understandable from the discussion of Sec. III. However, we still feel a need to go farther in the perturbation expansion of E: differences between the experimental and theoretical binding energies may not be completely attributed to the omission of elementary diagrams, and the instability toward spin-alignment, though weaker, persists. Valuable insights will be provided by the calculation, in progress, of 3p3h and third-order perturbation corrections. More sweeping improvements of the CBF treatment may be pursued by means of the integral-equation techniques of the correlated coupled-cluster theory described in the next section.

V. CORRELATED COUPLED-CLUSTER THEORY [18]

In the CBF scheme, how can we account for higher-order effects analogous say to those summed by the Bethe-Goldstone equation? This could be accomplished by systematically calculating selected portions of higher-order CBF corrections [16]. More efficiently, we may generalize conventional many-body methods to the correlated basis. The coupled-cluster (CC) theory of Coester and Kümmel [19] is most suitable for this purpose. The exact ground state is written in the form

$$|\Psi\rangle = \exp(S)|\Phi_o\rangle \quad , \tag{5.1}$$

where S is a sum of npnh operators ($n \geq 2$). The operator S is determined by projecting the Schrödinger equation on a complete set $\{|\Phi_m\rangle\}$ of Slater determinants:

$$<\Phi_m\, e^{-S}|H|e^S\, \Phi_o> = E_o \delta_{mo} \quad . \tag{5.2}$$

This theory contains the Bethe-Goldstone equation as a special case. In the correlated analog of CC theory, S is defined by the correlated creation and annihilation operators (1.2) acting on the correlated basis; in addition the non-orthogonality of the basis states must be treated properly. Thus one begins with an energy expression

$$E_o = <\psi_o\, e^{-S}|H|e^S\psi_o>/<\psi_o\, e^{-S}|e^S\psi_o> \tag{5.3}$$

and the correlated coupled-cluster equations

$$<\psi_m\, e^{-S}|H - E_o|e^S\psi_o>/<\psi_o\, e^{-S}|e^S\, \psi_o> = 0 \quad . \tag{5.4}$$

Further elaboration of the theory follows a standard pattern outlined below. We restrict S to its two-body (2p2h) component $S^{(2)}$ and expand the ground-state energy (5.3) in powers of S. This expansion is very similar to the cluster expansions of the energy expectation value for state-dependent correlation operators, though the graphical representation, designed to take proper account of the particle-hole operator structure of S, is closer to the Goldstone-like diagrammatic scheme of conventional CC theory. There are "operator chains" (ring diagrams) and graphs with "parallel connections" (ladder diagrams). However, no "commutator diagrams" arise since all S operators commute due to their particle-hole structure. (The counterparts of the latter diagrams make their way into the formalism via the higher S amplitudes determined by higher-subsystem equations.) In a parallel analysis of the Schrödinger equation (5.4), identification of sets of (sub-)diagrams in common with the energy expansion suggests the definition of a renormalized 2p2h operator $\$$ in terms of an infinite series of diagrams containing S and the nonorthogonality correction N. The new $\$$ operator supercedes S, and infinite series of diagrams such as the rings, the ladders, etc. are eliminated in favor of an appropriate generating equation. The complete 2p2h approximation to the ground-state energy becomes

$$E_o^{(2)} = H_{oo} + \frac{1}{4} \sum_{pp'hh'} V_{hh'(pp')_a}\, \$_{pp'(hh')_a} \quad . \tag{5.5}$$

Subsequent approximations enter through the equation employed to determine $\$$ for insertion into (5.5). In nuclear problems, we expect that the most important prescription for determining $\$$ will be the correlated Bethe-Goldstone equation. This approximation, corresponding to the retention of only the particle-particle ladder diagrams in the full CCC equation for $\$$, is

$$0 = [e(p) + e(p') - e(h) - e(h')]\$_{pp'(hh')_a} + V_{pp'(hh')_a} + \frac{1}{2} \sum_{p_1p_2} \Big\{ V_{pp'(p_1p_2)_a} - [e(h) + e(h')] N_{pp'(p_1p_2)_a} \Big\} \$_{p_1p_2(hh')_a} \quad . \tag{5.6}$$

VI. FUNCTIONAL DERIVATIVE METHOD: CORRELATED BCS THEORY [20]

Our final application of the ingredients of CBF theory to the description of many-body phenomena, the correlated BCS theory, is a special case of a much more general method. The derivation of the correlated RPA equations as reported by Sandler [5] proceeds along similar lines, and further exemplifications, such as a quantitatively reliable investigation of pion-condensation instabilities, can be foreseen.

The correlated BCS state is written in the form

$$|CBCS> = \sum_{m,N} |\psi_m^{(N)}><\Phi_m^{(N)}|BCS> \quad , \tag{6.1}$$

$$|BCS> = \prod_{\underset{\sim}{k}} (u_k + v_k\, a^\dagger_{\underset{\sim}{k}\uparrow}\, a^\dagger_{-\underset{\sim}{k}\downarrow})\,|0> \tag{6.2}$$

(The extension to more general forms of pairing [21] is straightforward [20].) One could use standard cluster-expansion techniques [22] and FHNC-like resummations of planar diagrams (cf. [23]) to evaluate

$$<\hat{H} - \mu\hat{N}>_s = <CBCS|\hat{H} - \mu\hat{N}|CBCS>/<CBCS|CBCS> \quad , \tag{6.3}$$

but this approach requires the early specification of the correlation operator and obscures the relation to conventional weak-coupling BCS theory. We choose, rather, to express the expectation value (6.3) in terms of the deviations of u_k and v_k from their normal-state values. Accordingly, we begin with

$$<\hat{H} - \mu\hat{N}>_s = H_{oo} - \mu A + 2\sum_{\underset{\sim}{k}}^{k<k_F} v_k^2 \left\{e_k[u_k,v_k] - \mu\right\} - 2\sum_{\underset{\sim}{k}}^{k>k_F} u_k^2 \left\{e_k[u_k,v_k] - \mu\right\}$$
$$+ \sum_{\underset{\sim}{k},\underset{\sim}{\ell}} u_k v_k u_\ell v_\ell P_{k\ell}[u_k,v_k] \quad , \tag{6.4}$$

which has the structure seen in conventional BCS theory, except that the pairing interaction now depends on the BCS amplitudes u_k, v_k. However, this dependence does not contribute to the first two variational derivatives of (6.4) with respect to the BCS amplitudes. Experience gained in ordinary BCS theory [24] together with the estimate that the deviations of u_k and v_k from their normal-state values are of order $m^*\Delta/\hbar^2 k_F^2$ suggest that retaining the full dependence of the pairing interaction on the BCS amplitudes produces at most a 1% correction to the condensation energy. The stability of the normal phase against pairing is (rigorously) unaffected.

In the indicated approximation, we recover the CBF single-particle energies and pairing interaction of Eqs. (3.3), (3.5):

$$e_k[u_k,v_k] = e(k) \tag{6.5}$$

$$P_{k\ell}[u_k,v_k] = <\underset{\sim}{k}\uparrow,-\underset{\sim}{k}\downarrow|\mathcal{W}(12)|\underset{\sim}{\ell}\uparrow,-\underset{\sim}{\ell}\downarrow>_a$$
$$+ [|e(k)-\mu| + |e(\ell) - \mu|]<\underset{\sim}{k}\uparrow,-\underset{\sim}{k}\downarrow|\mathcal{N}(12)|\underset{\sim}{\ell}\uparrow,-\underset{\sim}{\ell}\downarrow>_a \quad . \tag{6.6}$$

For zero center-of-mass momentum, the summation of all planar diagrams is feasible. In Table 3, we display the dimensionless pairing matrix elements $\delta(\lambda) = k_F^2 P_{k_F k_F} m^*/(2\pi^2/\hbar^2)$ for the $\lambda = {}^1S_0$, ${}^3P_0$, and ${}^1D_2$ partial waves, obtained by summing planar diagrams for optimized Jastrow correlations. That the present description does not show a ${}^3P_0$ pairing instability is presumably due to the absence of spin-density fluctuations from the correlation-operator ansatz. (Such a correlation component is expected to lead to more attraction in triplet and more repulsion in singlet states [25,26].) This failure of the Jastrow model is in concert with its qualitatively poor predictions of Fermi liquid parameters, reported above. However, the CBF perturbative improvements to the quasiparticle interaction encourage similar correction of the pairing interaction. Work involving suitable generalizations of coupled-cluster or related theories [27,28] is in progress.

Table 3: Dimensionless pairing matrix elements

ρ (Å^{-3})	δ (${}^1S_0$)	δ (${}^3P_0$)	δ (${}^1D_2$)
0.0076	0.74	0.03	-0.64
0.0112	1.08	0.55	0.09
0.0130	1.20	0.76	0.36
0.0142	1.25	0.87	0.51
0.0148	1.28	0.92	0.58
0.0166	1.35	1.06	0.77
0.0180	1.39	1.17	0.90

VII. OUTLOOK

The formal development of CBF methods is currently in a transient, hence exciting state. A number of important examples indicate that CBF methods do allow explicit construction of a universal, weak effective interaction from a strong, bare interaction. Conventional many-body models are still recognizable when formulated in terms of correlated basis states.

Though strong evidence suggests that the correspondence may be still deeper, a general rule for translating a given formalism into the correlated-basis language awaits discovery. The next big step is obviously an attempt to recreate time-dependent perturbation theory and the one- and two-body Green functions for correlated wave functions. The ideas are still vague, though more insight may be derived soon from studies of the correlated RPA and BCS theories.

Considerably more numerical work remains to be done to exploit the power of the existing formalism. We stress, however, that liquid ^{3}He is arguably the hardest problem on the scene. For this system many of the established approximations of nuclear physics (e.g., effective-mass approximation of the single-particle spectrum; angle-averaging of the Pauli operator) are poor. The reader is also reminded that the actual difficulties center on energy calculations where an accuracy of a few tenths of a degree is required. For applications to nuclear systems [29], minor modifications of standard methods should suffice.

References

[1] D. J. Thouless, The Quantum Mechanics of Many-Body Systems (Academic Press, New York, 1972).

[2] D. Pines and P. Nozières, The Theory of Quantum Liquids (Benjamin, New York, 1966).

[3] G. E. Brown, Many Body Problems (North Holland, Amsterdam, 1971).

[4] E. Feenberg, Theory of Quantum Fluids (Academic Press, New York, 1969).

[5] D. G. Sandler, J. W. Clark, and E. Krotscheck, these proceedings; and to be published.

[6] E. Krotscheck and R. A. Smith, to be published.

[7] E. Krotscheck, R. A. Smith, J. W. Clark, and R. M. Panoff, Phys. Rev. B, to be published.

[8] E. Krotscheck, Phys. Letters 54A, 123 (1975), J. Low Temp. Phys. 27, 199 (1977).

[9] E. Krotscheck, Phys. Rev. A 15, 397 (1977).

[10] J. C. Owen, Phys. Letters 89B, 303 (1980), Phys. Rev. B, in press.

[11] E. Krotscheck, Nucl. Phys. A317, 149 (1979).

[12] J. W. Clark, in Progress in Particle and Nuclear Physics, ed. D. H. Wilkinson (Pergamon Press, Oxford, 1979), vol. 2.

[13] D. Levesque, Phys. Rev. B 21, 5159 (1980).

[14] E. Krotscheck and J. W. Clark, Nucl. Phys. A328, 73 (1979).

[15] R. P. Feynman and M. Cohen, Phys. Rev. 102, 1189 (1956).

[16] J. W. Clark, L. R. Mead, E. Krotscheck, K. E. Kürten, and M. L. Ristig, Nucl. Phys. A328, 45 (1979).

[17] C. W. Woo, Phys. Rev. 151, 138 (1966).

[18] E. Krotscheck, H. Kümmel, and J. G. Zabolitzky, Phys. Rev. A 22, 1243 (1980); E. Krotscheck and J. W. Clark, in The Many Body Problem, Jastrow Correlations versus Brueckner Theory, ed. R. Guardiola and J. Ros (Springer-Verlag, Berlin, 1981).

[19] F. Coester and H. Kümmel, Nucl. Phys. 17, 477 (1960); H. Kümmel, K. H. Lührmann, and J. G. Zabolitzky, Phys. Rep. C36, 1 (1978).

[20] E. Krotscheck and J. W. Clark, Nucl. Phys. A333, 77 (1980).

[21] R. Tamagaki, Progr. Theor. Phys. 44, 905 (1970).

[22] K. Nakamura, Progr. Theor. Phys. 21, 713 (1959); 24, 1195 (1960); C.-H. Yang and J. W. Clark, Nucl. Phys. A174, 49 (1971).

[23] S. Fantoni, these proceedings; and Nucl. Phys. A, in press.

[24] R. C. Kennedy, Nucl. Phys. A118, 189 (1968).

[25] A. Layzer and D. Fay, Int. J. Magn. 1, 135 (1971).

[26] P. W. Anderson and W. F. Brinkman, Phys. Rev. Lett. 30, 1108 (1973).

[27] Y. Gerstenmaier and D. Schütte, Z. Naturf. 35a, 796 (1980).

[28] K. Emrich, these proceedings; and to be published.

[29] E. Krotscheck and R. A. Smith, Phys. Lett. B, in press.

OPTIMAL JASTROW CORRELATIONS FOR FERMI LIQUIDS

J.C. Owen
Department of Theoretical Physics,
The University,
Manchester M13 9PL, U.K.

1. Introduction

For the ground-state of particles obeying Bose statistics the Jastrow ansatz for the trial ground-state wave function has been thoroughly studied[1,2,3]. The variational problem is well understood and there is a clear sequence of trial functions (Feenberg functions) which will lead, in an apparently convergent way, to the true ground-state wave function. If the particles obey Fermi statistics however, the problem is more complicated. The simplest extension for fermions is the Slater-Jastrow ansatz

$$\psi = \prod_{i<j} f(r_{ij})\,|\phi\rangle \quad , \tag{1}$$

where $|\phi\rangle$ is a Slater determinant of single particle orbitals (for example, plane waves) and $f(r_{ij})$ is a pair correlation function. Assuming such a form, there remains the mathematical problem of evaluating the ground-state energy in this trial wave function and of varying the correlation function in order to obtain that function which actually minimises the energy.

Even with this simple trial wave function the energy cannot usually be evaluated exactly as a functional of $f(r)$. One must therefore ensure that the approximations which are introduced are 'reasonable' in the sense that they lead to values for the energy and functional forms for the optimal correlation functions which are not qualitatively different from the true optimal values. This turns out to be a significantly more difficult problem for fermions than for bosons. For example, a straightforward implementation of the hypernetted chain approximation[4,5] is reasonable for bosons, but not for fermions[6]. In this talk I will describe the formulation of the variational equations for fermions and describe some reasonable approximations which may be introduced. I will describe the power series solution of the equations for the low density hard-sphere Fermi gas and the numerical solution for liquid ^{3}He throughout the entire density range where solutions exist.

Although the simple Slater-Jastrow ansatz with the optimal correlation function does give a rather good description of the gross short-range and long-range structure of liquid ^{3}He it does not give a good description of the spin-dependent structure nor of the low-lying single particle excited states. In the final part of this talk I will discuss some of the possible ways of overcoming these problems.

2. Variational Equations

Using the trial wave function (1) the ground-state energy for a uniformly extended Fermi system interacting through two body forces may be written as

$$E/N = \frac{\hbar^2}{2m} \{\frac{3}{5} k_F^2 + \rho \int V^*(r) g(r) d\vec{r} + T_{JF} \}$$

$$\text{where} \quad V^*(r) = -\frac{1}{4} \nabla^2 \ell n f^2(r) + \frac{m}{\hbar^2} V(r) \tag{2}$$

$$\text{and} \quad T_{JF} = \frac{1}{4} \int \prod_{i<j} f^2(r_{ij}) \nabla_1^2 |\phi|^2$$

where $g(r)$ is the two-particle distribution function. Varying $\ell n f^2(r)$ in this expression for the energy leads to the Euler-Lagrange equation in the following form

$$\frac{1}{4} \rho \nabla^2 g(r) = \rho \int V^*(r') \frac{\delta g(r')}{\delta \ell n f^2(r)} d\vec{r}' + \frac{\delta T_{JF}}{\delta \ell n f^2(r)} \tag{3}$$

Both $g(r)$ and the right hand side of this equation may be expressed as functionals of $f(r)$ and so (3) does indeed represent an equation for $f(r)$. The detailed analysis of this equation is best performed using diagrammatic cluster expansions[6,7] and will not be discussed in detail here. The general development can nevertheless be illustrated by Fourier transforming equation (3) and writing it as

$$-\frac{1}{4} k^2 (S(k) - 1) = S'(k) \tag{4}$$

where $S(k)$ is the liquid structure factor and $S'(k)$ is simply defined as the Fourier transform of the right hand side of equation (3). The diagrammatic analysis of $S'(k)$ shows that it can be factored in momentum space as

$$S'(k) = S^2(k) X'(\{S\}, \{V^*\}, k) \tag{5}$$

Roughly speaking the functional $X'(k)$ contains only terms which can be expressed as non-nodal (or two-point irreducible) diagrams involving $f^2(r) - 1$, $V^*(r)$ and exchanges from the Slater determinants. It is therefore "short-range" as a function of r and in particular it has a finite Fourier transform so that $X'(k)$ is finite even for $k \to 0$. The decomposition (5) represents the summation of all ring (or chain) diagrams contributing to $S'(k)$.

Inserting the decomposition (5) into the variational equation (4) gives a quadratic equation for the explicit occurrence of $S(k)$ which is easily solved to give

$$S(k) = [-1 + (1 + 16X'(k)k^{-2})^{\frac{1}{2}}][8X'(k)]^{-1}k^2 \tag{6}$$

and for $k \to 0$, $S(k) \backsim \frac{1}{2}k\, X(0)^{-1}$. This shows immediately that, for the optimal correlation function $S(k)$ is linear in k for small k. In addition it gives the following as a necessary condition for the existence of solutions to the variational equation

$$1 + 16X'(k)k^{-2} \geq 0 \text{ for all } k. \tag{7}$$

The preceding discussion has emphasised the long-range or small-k properties of the variational equation. An alternative way to view it is as an effective two-body scattering equation in the medium. Pulling out terms which are linear in the "wave function" f(r), the variational equation may be written in the form

$$\frac{\hbar^2}{m} \nabla^2 f(r) + \int d\vec{r}' \; g_F(|\vec{r}-\vec{r}'|)V(r')f(r') = D(r) \tag{8}$$

where $g_F(r)$ is the two-particle distribution function for a non-interacting Fermi gas and, with the exception discussed below, the term D(r) contains many-body terms which are at least quadratic in f(r).

Equation (8) has the form of a state-averaged Bethe-Goldstone equation for a pair of particles scattering in the Fermi sea where the function g_F corresponds to a state-averaged Pauli operator. It is instructive to Fourier transform equation (8)

$$\hbar^2/m(\tilde{f}-1)k^2 + S_F(k)V\tilde{f} = D(k) \tag{9}$$

Now $S_F(k)$ goes to zero for small k and so even in the absence of D(k) we have

$$\tilde{f}-1 \sim \alpha k^{-1} \text{ for } k \to 0 \text{ or } f(r) \sim 1 + cr^{-2} \text{ for } r \to \infty.$$

Thus the average Pauli operator ensures the healing of the correlation function as r^{-2} for large r, rather than r^{-1} as would be the case for free scattering.

The equation (9) differs from the Bethe-Goldstone equation in that it contains the driving term D(k) on the right hand side. D(k) contains non-linear terms in f(r) which represent many-body effects due to correlations. It also contains those terms linear in f(r) which are due to the scattering of a pair of particles in exchanged states. In perturbation theory the scattering of particles in direct and exchanged states leads to the same Bethe-Goldstone equation. This is not the case in Jastrow theory (due to the state averaging in the wave function) and so exchange scattering must be included in the driving term. In the numerical calculations for liquid ^{3}He, however, it turns out that the exchange terms are small compared with the non-linear terms.

3. Hard Sphere Fermi Gas

Let us consider a low density gas of fermions with spin degeneracy ν and interacting through an infinitely repulsive potential of diameter a. From perturbation theory[9)] the energy per particle is given as a series in $k_F a$ as follows

$$E/N = k_F^2(0.3 + (\nu-1)0.106(k_F a) + (\nu-1)0.055(k_F a)^2 + \ldots)$$

The first term is the Fermi energy, the second is the energy due to the total volume which is excluded by the hard spheres and the third term is a correction due to the fact that part of the phase-space for a pair of particles scattering is excluded by the filled Fermi sea.

The corresponding power series using the Slater-Jastrow trial wave function is obtained by setting $D \equiv 0$ in eq. (9). The result is[8)]

$$E/N = k_F^2(0.3 + (\nu-1)0.106(k_Fa) + (\nu-O(1))0.08(k_Fa)^2 + \ldots)$$

The coefficient of $(k_Fa)^2$ is a power series in ν^{-1} (where ν is the spin degeneracy). The coefficient of ν comes only from scattering in "direct" states, but the remaining terms cannot be evaluated without including the linear terms in D due to scattering in exchange states. Although not difficult in principle, this would be tedious in practice. The variational theory gives the first two terms correctly and gives the correct power law behaviour for the third term. For large ν, the coefficient of this term is about 30% too large. This may be understood from the fact that f(r) is the correlation function for an average pair of particles and so is not very efficient at excluding those final scattering states which are already occupied in the Fermi sea.

The effective mass of a single particle excitation is also given by perturbation theory[9)] as

$$m^*/m = 1 + 0.2(k_Fa)^2 + \ldots$$

The variational result, obtained by simply exciting one particle in the Slater determinant, is

$$m^*/m = 1 + 0.02(k_Fa)^2 + \ldots$$

Although the power law behaviour is still correct the coefficient is wrong by a factor of ten. The Jastrow correlations do not respond to the excitation, there is no 'back-flow' or rearrangement term and so the pure Jastrow ansatz gives a poor description of the single particle excitation.

4. Numerical Results for Liquid ^{3}He

Although of theoretical interest, the low density hard sphere gas is not a good model for actual liquids as observed in the laboratory. In this section I will describe the results of the numerical solution of the variational equation for liquid ^{3}He atoms interacting with the Lennard-Jones potential. In the numerical calculation the basic quantity X'(k) in eq. (6) is evaluated in a hypernetted chain-like approximation augmented with those elementary diagrams which are necessary to ensure that X'(k) remains finite for all k. The details may be found in ref. 7).

Figure 1 shows the energy per particle evaluated for all densities where the variational equations have solutions. The minimum in the curve represents the empirical saturation density. If the density is lowered below this saturation point the solutions eventually disappear. This occurs in the region where the incompressibility becomes negative. The uniformly extended phase is unstable against small density fluctuations and this is reflected in the lack of solutions to the variational equations. If the density is increased above the saturation point the solutions again disappear. A similar phenomenon has been observed for bose particles where it has been associated with an instability against strong short-range ordering[10].

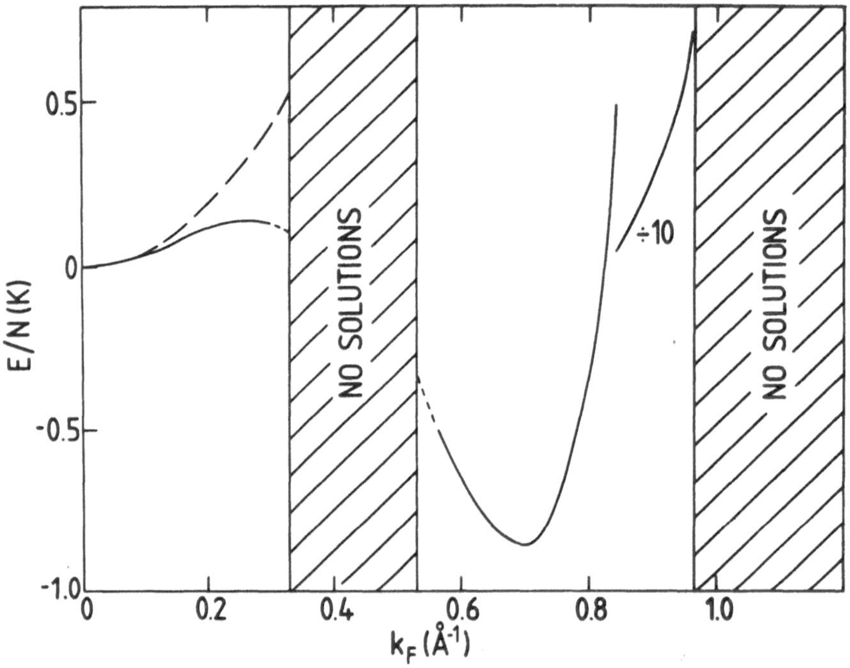

Figure 1 The energy per particle as a function of the Fermi momentum throughout the entire density range where solutions exist. The solid line shows solutions which were actually obtained and the dotted continuation lines show an estimate of probable solutions. The long dashed line shows the energy per particle of a non-interacting Fermi gas.

For the Fermi system there are also solutions for very low densities. In this region the energy is dominated by the Fermi statistics which tends to stabilise the uniform phase. The dotted line in fig. 1 shows the energy of a non-interacting Fermi gas. This region of the energy-density curve may be of significance for low concentrations of ^{3}He dissolved in ^{4}He.

Figure 2 shows the liquid structure function compared with the experimental measurements (dots) of X-ray scattering[11]. The agreement, especially for small-k is rather good.

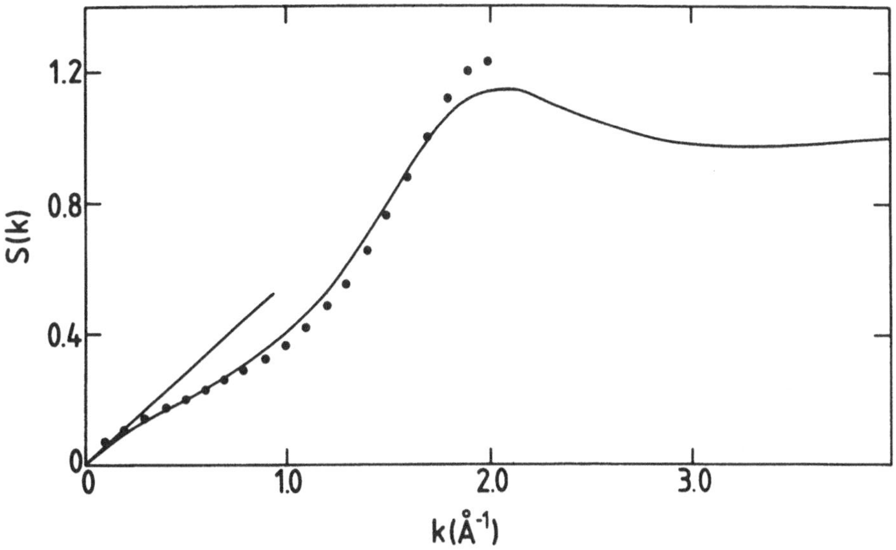

Figure 2 The calculated structure function S(k) (solid line) at a density of 0.0164Å^{-3} compared with the experimental results of X-ray scattering at a temperature of 0.4K. The straight line shows the slope of the theoretical curve at the origin.

5. Discussion

The Slater-Jastrow trial wave function seems to be able to give a rather good description of the bulk properties of liquid ^{3}He. In this simple form, however, it does not give a good description of the spin dependent properties nor of the single particle excited states. An improved ansatz which would probably correct both of these deficiencies is the state-dependent[8] ansatz

$$\psi = \prod_{i<j} f_{\alpha(i)\alpha(j)}(r_{ij}) \, |\phi>$$

where the correlations are allowed to depend on the single particle states which the non-interacting particles occupy in the Slater determinant. Accurate calculations with this wave function would be extremely difficult, however. If the state-dependence is 'weak' then it is sensible to write

$$\psi = \prod_{i<j} \bar{f}(r_{ij}) \prod_{i<j} (1 + \eta_{\alpha(i)\alpha(j)}(r_{ij})) \, |\phi>,$$

and to expand in powers of η. This is conveniently done in the context of two-particle, two-hole Correlated Basis Functions (CBF) theory and is the subject of the following talk by Krotscheck. If the η is not 'weak' it may be better to use a rather less general ansatz which will allow a more accurate evaluation of the energy. For example, in order to calculate spin-dependent quantities, it should be sufficient to generalise the correlation function[12,13] as follows

$$f(i,j) = f^C(r_{ij})\ (1 + \eta(r_{ij})\ \sigma_i \cdot \sigma_j).$$

For the single particle excited state it may be sufficient to allow backflow correlations only around the excited particles in the following form

$$\psi_q = \prod_{i<j} f(r_{ij}) \prod_j e^{i\alpha(\vec{q},\vec{r}_{Aj})} |q\rangle$$

where α is an operator which picks out the coordinate $\vec{r}_A$ of the particle which is excited in the Slater determinant. This trial wave function will give an effective mass of about 1.6 bare masses. When coupled to the spin dependent correlations it is possible that a wave function of this nature could give values for the effective mass which are considerably larger.

References

1) E. Feenberg, Theory of Quantum Fluids (Academic Press, N.Y., 1968)
2) C.C. Chang and C.E. Campbell, Phys. Rev. B15 4238 (1977)
3) A.D. Jackson, A. Lande and L.J. Lantto, Nucl. Phys. A317 70 (1979)
4) S. Fantoni and S. Rosati, Il Nuovo Cimento 25A 593 (1975)
5) L.J. Lantto and P.J. Siemens, Nucl. Phys. A317 55 (1979)
6) E. Krotscheck, Phys. Rev. A15 397 (1977)
7) J.C. Owen, Phys. Rev. B (in Press)
8) J.C. Owen, Phys. Lett. 82B 23 (1979)
9) A.L. Fetter and J.D. Walecka, Quantum Theory of Many-Particle Systems (McGraw-Hill, New York, 1971)
10) L. Castillejo, A.D. Jackson, B.K. Jennings and R.A. Smith, Phys. Rev. B20 3631 (1979)
11) R.B. Hallock, Jo. Low Temp. Phys. 9 109 (1972)
12) R.B. Wiringa and V.R. Pandharipande, Rev. Mod. Phys. 51 821 (1979)
13) J.C. Owen, Annals of Physics (N.Y.) 118 373 (1979)

CBF PERTURBATION CORRECTIONS TO THE JASTROW GROUND-STATE OF THE ELECTRON GAS

L. J. Lantto[†], E. Krotscheck, and R. A. Smith
Dept. of Physics, SUNY
Stony Brook, NY 11794, U S A

I. Introduction

The homogeneous electron gas neutralized by a uniform positive background is perhaps the most extensively studied quantum many-body system. The importance of the role of correlations was realized about 50 years ago. In the 1930s, Wigner [1] calculated the energy of the electron system in the low density limit where the Coulomb interaction dominates the kinetic energy. Twenty years later, exact results for the correlation energy were obtained in the high density limit [2] from infinite partial summations of the perturbation expansion of the correlation energy. During the past 20 years much effort has been spent on calculating the correlation energy at the intermediate densities most important for applications. However, only recently were "benchmark" calculations provided by exact Monte-Carlo evaluation of the correlation energy for wide density ranges for both para- and ferromagnetic electron fluids.

Among other methods, the Jastrow variational approach has been applied in the last two years to the problem of static correlations in the electron gas. In addition to the correlation energy, variational estimates were obtained for the pair-correlation function, the static structure function, and the single-particle momentum distribution [4-6].

The obvious step following the Jastrow estimate of the correlation energy is the inclusion of state-dependent and three-, four-, ... body correlations. For this purpose, essentially two different routes can be taken:
(i) We can improve on the variational description by explicitely inventing more complicated correlation operators describing "backflow", spin-depence, etc. [7], or
(ii) we can improve systematically on the Jastrow description of the system by doing perturbation theory with correlated wave functions (CBF) [8].

As a first step in a more detailed exploration of the CBF method, this paper reports on the CBF perturbation corrections to the correlation energy of the electron gas. The first result is a comparison of the 2-particle, 2-hole contribution to the second-order CBF correction with the discrepancy between the variational Monte-Carlo and the (presumably) exact GFMC ground-state energies over a range of densities. In addition, we present some analytic results which apply in the high-density limit.

II. Variational Correlation Energy

The essentials of CBF theory have been described in a separate contribution [9]. We therefore concentrate here on some details of the implementation of the variational/CBF- procedure for the electron gas problem. The correlated basis is generated by the standard choice of the Jastrow correlation operator

$$|\Psi_m\rangle = I_{mm}^{-1/2} F|\Phi_m\rangle, \; I_{mm} = \langle\Phi_m|F^+F|\Phi_m\rangle, \; F = \prod_{i<j} f(r_{ij}) \tag{1}$$

(For matters of notation, see the accompanying paper [9]).
The variational energy expectation value

$$H_{00} = \langle\Psi_0|H|\Psi_0\rangle \tag{2}$$

is calculated by means of the Fermi-Hypernetted-Chain (FHNC)-method [10]; the pair correlation function f(r) is calculated which minimizes the expectation (2).

In the present calculation, we used the Lantto-Siemens Lagrange-multiplier technique [11], which expresses the optimization condition for f(r) in the form

$$(-\hbar^2/m\nabla^2 + v(r) + w_B(r) + w_X(r))g^{1/2}(r) = 0, \tag{3}$$

where g(r) is the radial distribution function, v(r) the Coulomb interaction, and $w_B(r)$ and $w_X(r)$ are induced potentials which arise from dynamical (direct) and exchange correlations. The numerical method for solving eq. (3) has been described earlier [11], details on the implementation of the method for the electron-gas problem may be found in ref. [5]. In previous applications, however, all so-called "elementary" diagrams were neglected. In the present case, some elementary diagrams are approximated by their FHNC/C estimates; this approximation is motivated by the momentum-space properties of these functions as described in ref. [12]. In the present work, we adhere to the "minimal" estimate of elementary diagrams by using the /C-estimate only for the "de"-diagrams:

$$\tilde{X}_{de}^{/C}(k) = \tilde{X}_{de}^{/0}(k)\,S_F(k) \tag{4}$$

A corresponding estimate (which may also be called a "quasi-boson-approximation" [7] for the elementary exchange diagrams) for the "ee"-diagrams is also possible; however for the electron gas, a solution to the Euler-Lagrange equation of eq. (2) automatically has the correct long-wavelength behaviour for the sum of the ee-diagrams

$$1 + \tilde{X}_{ee}(0+) = 0 \tag{5}$$

In passing, we note also that the "Fermi- cancellation" - correction (4) implies a minimal loss of consistency between the momentum- and coordinate-space representations of the diagrammatical elements of FHNC-theory.

r_s	FHNC/0	FHNC/C	VMC
1	-140.	-118.	-113.
2	-98.	-86.5	-85.1
3	-79.	-70.9	-70.7
4	-67.	-61.0	-61.3
5	-58.	-54.0	-54.4
10	-37.	-35.5	-35.9
20	-23.	-22.1	-22.3
50	-11.	-10.9	-11.1

Table 1. Variational correlation energies in Ry. r_s in the average particle distance in units of the Bohr radius.

Table 1 is a compilation of the energies obtained with the original optimized FHNC/0 equations [5], the present optimized FHNC/C equations, and the variational Monte-Carlo method [3]. There is now good agreement at all densities; the FHNC/C modifications are significant at high density.

III. Perturbation Corrections

The calculation of the Jastrow-FHNC CBF- perturbation corrections requires the solution of an additional set of four linear integral equations, the so-called FHNC'- equations [13,9]. These are linearized FHNC equations, with the Jackson-Feenberg effective interaction and corresponding kinetic energy terms acting as the driving functions. Some formulations of the optimization algorithm for the Jastrow correlation function supply the solution of the FHNC'-equations as natural by-products [13]. There is, of course, a close relation between the Lagrangian multipliers of the optimization algorithm employed here and the solution of the FHNC'-equations. Rather than exploiting such a connection, we have chosen to solve the FHNC'-equations independently. One reason for this is that in order to generate a well-behaved effective interaction for the CBF perturbation corrections, one must preserve the correct long-wavelength behaviour of the exchange diagrams.

The 2-particle, 2-hole contribution to the CBF perturbation expansion has the form of an ordinary second-order perturbation correction with an effective interaction and effective single-particle energies:

$$\Delta E^{(2,2)} = -(1/4) \sum_{pp'hh'} |\langle pp'|\mathcal{V}(12)|hh'\rangle_a|^2 / (e_p + e_{p'} - e_h - e_{h'}) \qquad (6)$$

The CBF- single particle energies e_k may be constructed algebraically from the solutions of the FHNC'- equations [13]. The effective interaction $\mathcal{V}(12)$ has the structural decomposition

$$<pp'|\mathcal{V}(12)|hh'> =$$
$$D^{-1}(<pp'|\mathcal{W}^B(12)|hh'> + (1/2)(e_p + e_{p'} - e_h - e_{h'})<pp'|\mathcal{N}^B(12)|hh'>) \quad (7)$$

$$D = ((1-\tilde{X}_{cc}(p))(1-\tilde{X}_{cc}(p'))(1-\tilde{X}_{cc}(h))(1-\tilde{X}_{cc}(h')))^{1/2}. \quad (8)$$

The (dominant) local pieces of the 2-body operators $\mathcal{W}^B(12)$ and $\mathcal{N}^B(12)$ are again algebraically constructed from the FHNC and FHNC'- quantities:

$$\mathcal{N}_{loc}(r) = \Gamma_{dd}(r), \quad \mathcal{W}_{loc}(r) = \Gamma'_{dd}(r) + (\hbar^2/4m)\nabla^2\Gamma_{dd}(r) \quad (9)$$

Details on the more general structure of the two-body operators $\mathcal{W}^B(12)$ and $\mathcal{N}^B(12)$ may be found in refs. [13,14]. It suffices to mention here that additional, non-local, contributions turn out to be negligible.

r_s	$\Delta E^{(2,2)}$	$E_{GFMC}-E_{VMC}$
2	-7.0	-4.9
5	-2.0	-1.9
10	-0.84	-1.2
20	-0.40	-0.74
50	-0.14	-0.32
100	-0.06	-0.17

Table 2: Comparison between CBF correction with GFMC correlation energies. Col. 3 is due to Ceperley [15]. All energies are given in Ry.

Results for the CBF energy corrections (6) are given in Table 2. At low and intermediate densities, we find expectedly the CBF- corrections to be somewhat smaller than the difference of GFMC and VMC energies. This may be attributed partly to the fact that we have used optimized rather than parametrized Jastrow functions. Some additional correction may come from the 3p-3h contributions to the CBF perturbation expansion.

At higher densities, we find an excess of the 2p-2h CBF-correction over the the difference between GFMC and VMC correlation energies. This is somewhat disturbing, particularly in view of the considerations of the next section. The reader is, however, reminded that the smallness of the CBF-energy correction arises from rather delicate cancellations between central and non-central contributions to the effective interaction (6). One of the reasons of the excess of the 2p-2h corrections may be the somewhat inconsistent treatment of the EL-FHNC and the FHNC'- equations ("minimal" /C-corrections in the first ones, and full FHNC/C treatment of the second set), introducing inconsistencies primarily in the long-ranged correlations, which are the important ones at high-densities. It is, of course, also conceivable, but less likely, that third-, fourth, ... order diagrams of the CBF perturbation expansion become important.

IV. High Density Limit

Since the exact correlation energy of the electron gas is known in the high density limit, it is of special interest to study the same limit in the variational FHNC/CBF theory. Only long-ranged quantities are of importance in this limit; this results in some considerable simplification of the FHNC and FHNC'- equations. Most importantly, we can drop all "de-" and "ee-" diagrams but the trivial ones, which may be expressed in terms of the structure function $S_F(k)$ of the non-interacting Fermi gas. The optimization condition (3) is rewritten in momentum space [8,9] as

$$\omega(k) = (\hbar^2k^2/4m)(S(k)-1) + S'(k) = 0 \tag{10}$$

A discussion of all components of eq. (10) and the FHNC and FHNC'- equations will be given in ref. 13. In the case of interest here, the FHNC/FHNC' equations simplify to

$$\begin{aligned} S(k) &= S_F(k)(1+S_F(k)\tilde{\Gamma}_{dd}(k)) \\ \Gamma_{dd}(r) &= f^2(r)\exp(N_{dd}(r)) - 1 \\ N_{dd}(k) &= \tilde{\Gamma}_{dd}(k)(1-S_F(k)/S(k)) \end{aligned} \tag{11}$$

$$\begin{aligned} S'(k) &= S^2(k)(X'_{dd}(k) - (\hbar^2k^2/4mS_F^2(k))(S_F(k)-1)) \\ X'_{dd}(r) &= (1+\Gamma_{dd}(r))v_{JF}(r) + \Gamma_{dd}(r)N'_{dd}(r) \\ \tilde{N}'_{dd}(k) &= (S_F^2(k)/S^2(k)-1)\tilde{X}'_{dd}(k)-(\hbar^2k^2/4m)\ \tilde{\Gamma}_{dd}(k)(S_F(k)-1). \end{aligned} \tag{12}$$

$v_{JF}(r)$ is the Jackson-Feenberg effective interaction

$$\begin{aligned} v_{JF}(r) &= v(r) - (\hbar^2/4m)\nabla^2\ \ln(f^2(r)) \\ &= v(r) - (\hbar^2/4m)\nabla^2(\Gamma_{dd}(r) - N_{dd}(r)), \end{aligned} \tag{13}$$

where we have used in the last line that $\Gamma_{dd}(r)$ is small. $v(r)$ is the bare Coulomb interaction. (The Fourier- transform is defined with a density factor to be dimensionless.)

Keeping only components in $X'_{dd}(r)$ which fall off no faster than r^{-4}, we can set

$$X'_{dd}(r) = v_{JF}(r), \tag{14}$$

which leads, after some algebraic manipulations, to the high-density approximation for the static structure function

$$S_F^2(k)/S^2(k) = 1+ 4mS_F^2(k)\tilde{v}(k)/(\hbar^2k^2) \tag{15}$$

which is identical with the result of ref. [7] obtained in the quasi-Boson approximation. Consequently, the high density limit of the variational ground-state energy is

$$E_c = (9/16\pi^2)\ell n\, r_s \quad \text{Ry}. \tag{16}$$

For the computation of the high-density limit of the CBF-correction (6), we follow the same lines. We find

$$\begin{aligned} W_{loc}(k) &= \tilde{\Gamma}'_{dd}(k)-(\hbar^2k^2/4m)\tilde{\Gamma}_{dd}(k) = X'_{dd}(k)+N'_{dd}(k)-(\hbar^2k^2/4m)\tilde{\Gamma}_{dd}(k) \\ &= -(\hbar^2k^2/2mS_F(k))\tilde{\Gamma}_{dd}(k), \end{aligned} \tag{17}$$

omit the exchange term and the D^{-1}-factor, and use non-interacting single-particle energies $e_k = t_k = \hbar^2k^2/2m$. Some simplifications are gained by

$$\int d^3hd^3h'(W_{loc}(k)+(1/2)(t_{\underset{\sim}{h}+\underset{\sim}{k}} + t_{\underset{\sim}{h}'-\underset{\sim}{k}} - t_{\underset{\sim}{h}} - t_{\underset{\sim}{h}'})\tilde{\Gamma}_{dd}(k))n(h)n(h')*$$

$$(1-n(|\underset{\sim}{h}+\underset{\sim}{k}|))(1-n(|\underset{\sim}{h}'-\underset{\sim}{k}|)) = 0 \tag{18}$$

in which n(k) is the Fermi distribution $n(k)=\theta(k_F-k)$. We note that eq. (18) is a special realization of a more general average property of the effective interaction (cf. eq.(3.7) of ref. [9]).

The remaining evaluation of the high-density limit of the second-order energy correction is essentially a textbook procedure [16]. Taking the limit of small momentum transfer for the energy denominator integration, we obtain

$$\begin{aligned} \Delta E_c &= \Delta E^{(2,2)}/A = \\ &= -(3\hbar^2/16m)(32(1-\ell n2)/9-1)\int_0^\infty dkk^4(S(k)/S_F(k)-1)^2/S_F(k) \end{aligned} \tag{19}$$

which yields

$$\Delta E_c = (2(1-\ell n\ 2)/\pi^2 - 9/16\pi^2)\ell n\, r_s \quad \text{Ry} \tag{20}$$

in the limit $r_s \to 0$. We find that in the high-density limit, the second-order CBF correction gives precisely the difference between the Jastrow-variational and the exact correlation energy [16]. One can also easily demonstrate that higher order CBF-ring diagrams do not contribute in the high-density limit. They lead to integrals of the same structure as eq. (19), but with higher powers of the factor $(S(k)/S_F(k)-1)$; these higher powers effectively remove the logarithmic singularity as $r_s \to 0$.

V. Summary

We have reported in this paper the first calculations for the correlation energy within the combined FHNC variational-CBF scheme. The energy corrections turn out to be rather small and-with the exception of a somewhat uncomfortable feature at very high densities- of the expected size. In particular the exact result for the high density limit gives another indication for the efficiency of our approach.

Let us conclude with some remarks on the future prospects of applications of the CBF approach in the electron gas problem. Clearly, it is feasible to include higher-order perturbation corrections and three-body correlations. It should be kept in mind, however, that the CBF corrections to the correlation energy are already smaller than the elementary diagram contributions to the variational correlation energy, and the uncertainty arising by different choices of the Jastrow function appears to be of the same order of magnitude. If a more accurate computation of the correlation energy should be desired, it should be accompanied by a renewed attack on those aspects of the variational problem mentioned above.

The accompanying paper [9] has also explained why the 2p-2h CBF corrections are so small for state-independent interactions. This is due to the fact that the Fermi-sea average of the 2p-2h interactions is made to vanish by the optimization of the Jastrow function.

Significant contributions are to be expected, however, for quantities which do not involve Fermi-sea averages of the effective interactions. The Fermi-liquid parameter calculations of ref. [9] are an obvious starting point for future investigations.

This research was supported, in part, by the U. S. Dept. of Energy under Contract No. DE-AC02-76ER13001 and by the Deutsche Forschungsgemeinschaft.

†Present address: Dept. of Theoretical Physics, University of Oulu, SF-90570 Oulu 57, Finland.

References:

[1] E. P. Wigner, Trans. Faraday Soc. 34, 678 (1938).

[2] M. Gell-Mann and K. Brueckner, Phys. Rev. 106, 364 (1957).

[3] D. M. Ceperley, Phys. Rev. B18, 3126 (1978).
D. M. Ceperley and B. J. Adler, Phys. Rev. Lett. 45, 566 (1980).

[4] S. Chakravarty and C.-W. Woo, Phys. Rev. B13, 4815 (1976).

[5] L. J. Lantto, Phys. Rev. B22, 1380 (1980).

[6] J. G. Zabolitzky, Phys. Rev. B22, 2353 (1980).

[7] V. R. Pandharipande and R. B. Wiringa, Rev. Mod. Phys. 51, 821 (1979).
K. E. Schmidt and V. R. Pandharipande, Phys. Rev. B19, 2504 (1979).
J. C. Owen, Ann. Phys. 118 (1979) 373., Nucl. Phys. A328, 143 (1979).
R. A. Smith, Nucl. Phys. A328, 169(1979).

[8] E. Feenberg, Theory of Quantum Fluids, (Academic, NY 1969)

[9] E. Krotscheck, R. A. Smith, and J. W. Clark, these proceedings.

[10] J. W. Clark, in Progress in Particle and Nuclear Physics, Vol. 2, ed. D. H. Wilkinson (Pergamon, Oxford 1979).

[11] L. J. Lantto and P. J. Siemens, Nucl. Phys. A317, 55 (1979).

[12] E. Krotscheck, Nucl. Phys. A317, 149 (1979).

[13] E. Krotscheck, R. A. Smith, J. W. Clark and R. M. Panoff, Phys. Rev. B., to be published.

[14] E. Krotscheck and J. W. Clark, Nucl. Phys. A328, 73 (1979).

[15] D. M. Ceperley, private communication.

[16] A. L. Fetter, J. D. Walecka, Quantum Theory of Many-Particle Systems (McGraw Hill, NY 1971).

CORRELATIONS IN BOSE FLUIDS

L. REATTO

Istituto di Fisica, G.N.S.M., Università di Parma, Parma, Italy.

Introduction

The static correlations between the particles of a many body system represent a fundamental information on a quantum liquid. The simplest correlation function is the radial distribution function g(r) that is simply related to the probability of finding two particles at distance r. In my talk I shall discuss two aspects of correlations in Bose liquids that show how a careful study of g(r) gives information on the microscopic structure of quantum fluids. In the first place I shall discuss how it is possible to deduce information on the structure of the ground state wavefunction from the knowledge of the exact correlations obtained from a Green function Monte Carlo (GFMC) calculation.[1] A practical method to obtain the Jastrow wavefunction having the maximum overlap integral with the exact ground state ψ_o will be presented. In the second part of my talk I shall discuss the temperature dependence of g(r) and the two proposals that have been put forward to interpret the anomalous temperature dependence of g(r) in liquid ^{4}He will be presented. The first one attributes this effect to the temperature dependence of the Bose Einstein condensate fraction[2] and the second to the thermal population of roton states.[3]

"Best" Jastrow wavefunction

A Green function Monte Carlo computation provides us only with a set of configurations drawn from the exact ψ_o. In this way one can compute exact averages, like g(r), but we would like to have also information on the structure of ψ_o. For instance we would like to know the role played in ψ_o by Jastrow correlations, i.e. between couples of particles, by three particle correlations and so on. In the first place one has to define a suitable criterion that gives the Jastrow component of the exact ψ_o, since this function is not known in an analytic way. A convenient criterion, but not the unique, is to ask for the Jastrow function having the maximum overlap integral with ψ_o. If we call $\psi_J(u)$ the normalized Jastrow function characterized by a pseudopotential u(r):

$$\psi_J(\{r_i\};u) = Q_N^{-1/2}(u) \prod_{i<j} \exp[-\tfrac{1}{2}u(r_i-r_j)] \qquad (1)$$

where $Q_N(u)$ is the normalization: $Q_N(u)=\int dr_1..dr_N \Pi \exp(- u(r_i-r_j))$, this maximum overlap criterion asks for the extremum of the superposition integral $<\psi_J(u)|\psi_o>$:

$$\delta<\psi_J(u)|\psi_o>/\delta u(r) = 0. \tag{2}$$

This criterion is a natural one because if one expands ψ_o on a complete set of functions that contains a Jastrow function as one of its elements, the choice of this particular Jastrow function as the one having maximum overlap guarantees that there is a minimum admisture of other wavefunctions in the expansion of ψ_o. It should be noticed that in general the Jastrow function having maximum overlap does not coincide with the one that gives the minimum expectation value of the energy. We may expect, however, that the two criteria should give similar answer if a Jastrow function is a reasonable approximation to ψ_o.

The functional derivative (2) picks up two contributions, one from the explicit u dependence of ψ_J and one from $Q_N(u)$. The computation is very easy and one finds that the condition (2) is satisfied if:[4]

$$g_{OJ}(r,u^*) = g_J(r,u^*) \tag{3}$$

where we have denoted by u^* that particular pseudopotential giving maximum overlap. $g_J(r,u^*)$ represents the radial distribution function corresponding to the Jastrow function whereas g_{OJ} represents a mixed average, i.e. a density-density correlation function constructed with $\psi_o\psi_J$ in place of ψ_J^2:

$$g_J(r,u) = (V/N)^2<\psi_J(u)|\rho(r+r')\rho(r')|\psi_J(u)> - \frac{V}{N}\delta(r) \tag{4}$$

$$g_{OJ}(r,u)= (V/N)^2<\psi_J(u)|\rho(r+r')\rho(r')|\psi_o>/<\psi_J(u)|\psi_o>- \frac{V}{N}\delta(r) \tag{5}$$

In (5) one has to divide the average by $<\psi_J|\psi_o>$ since the weight $\psi_J\psi_o$ is not a normalized function. $\rho(r)$ is the local density operator.

It is known[1] that a good approximation for the exact radial distribution $g_o(r)$ is the expression obtained by considering the difference $\psi_o-\psi_J$ as a perturbation and this gives $g_o(r)=2g_{OJ}(r,u)-g_J(r,u)$. Under this approximation the condition (3) of maximum overlap implies also that

$$g_o(r) = g_{OJ}(r,u^*) = g_J(r,u^*) \tag{6}$$

These three radial distribution functions are directly available from a GFMC computation that uses the Jastrow function $\psi_J(u^*)$ as importance sampling. However, we do not know u^* in advance and a GFMC computation is in general performed with a Jastrow function with $u(r) \neq u^*(r)$ so that the three correlation functions are not equal: $g_o(r) \neq g_{OJ}(r,u) \neq g_J(r,u)$. One would like to obtain directly the difference $\delta(r)=u^*(r)-u(r)$ without having to repeat the GFMC computation with a different importance sampling function at the search of u^*. Practical schemes to obtain $\delta(r)$ can be devised and I discuss now such a scheme based on Eq.(6).

We start with the following exact relation for the radial distribution function for any Jastrow function (see, for instance, ref.5)

$$\ln g_J(r,u) = g_J(r,u)-1-C_J(r,u)-u(r)+b(r;u) \qquad (7)$$

where C_J is the direct correlation function corresponding to g_J and $b(r;u)$ is the so called bridge function, i.e. the sum of all the elementary diagrams changed of sign. The study of $b(r;u)$ in a variety of systems has shown[5] that this function has essentially the same r dependence irrespective of the precise form of $u(r)$. Therefore we make the assumption

$$b(r;u) = b(r;u^*) \qquad (8)$$

and by taking the difference of (7) from the analogous relation for $g_J\ (r,u^*)$ we obtain

$$u^*_{(o)}(r)=u(r)+g_o(r)-g_J(r,u)-(C_o(r)-C_J(r,u))+\ln(g_J(r,u)/g_o(r)) \qquad (9)$$

where we have taken into account that $g_J(r,u^*)=g_o(r)$. All the quantities of the r.h.s. of (9) can be obtained from a GFMC computation and $u^*_{(o)}$ represents a first estimate of u^*. This is approximate because of the assumption (8) but one can improve on this by performing a Monte Carlo computation with $u^*_{(o)}(r)$ and the resulting $g_J(r,u^*_{(o)})$ can now be used to obtain an improved estimate $u^*_{(1)}(r)$ following the previous procedure but with the weaker assumption $b(r;u^*_{(o)})=b(r;u^*)$. This appears as a practical method to obtain sistematically u^* from GFMC computation and the method is currently being implemented.[6]

Temperature dependence of spatial correlations

The temperature dependence of the short range order in liquid ^{4}He below the λ temperature is unique: when the fluid is heated at constant density the short range order increases. This is opposite to the behaviour found above T_λ in ^{4}He and in all other liquids. Two distinct proposals have been put forward to explain the anomalous behaviour of ^{4}He and I shall discuss them in the following.

The first proposals[2] explains the anomaly with the following physical picture: particles in the Bose Einstein condensate are delocalized and do not contribute to short range order at least for not too small distances where there is the direct influence of the hard core of the interatomic interaction. Therefore $g(r,T)-1$, that is a measure of the short range order, is expected to be proportional to the square of the number of particles outside the condensate, i.e. $(1-n_o(T))^2$, so that one writes

$$g(r,T)-1 = f(T)F(r), \quad r>\ell_o \approx 4 \text{ Å}, \tag{10}$$

$$f(T) = (1-n_o(T))^2, \tag{11}$$

where $F(r) = g(r,T_\lambda)-1$ because $n_o=0$ at $T=T_\lambda$. Since $n_o(T)$ is a decreasing function of T from (10) the anomalous behaviour of $g(r,T)$ follows for $T<T_\lambda$.

Cummings et al.[2] attempted to give a microscopic basis to (10,11) but recently their argument has been strongly criticized on different grounds.[7] Therefore let us consider relation (10) only on an empirical ground. At first sight relation (10) appear to be a rather stringent statement on $g(r,T)$. In the first place it implies that the zeros of $g(r,T)-1$ do not change position with temperature and, secondly that the amplitudes of the different loops of $g(r,T)-1$ change by a constant factor. Indeed, from experimental data[8,9] it is found that the zeros of $g(r,T)-1$ are essentially temperature independent. The data are not inconsistent with the second prediction but the accuracy of the data does not permit a real test of this second prediction. However, if we examine $g(r,T)$ of classical liquids we find exactly the same behaviour. For instance from the results of molecular dynamics computation for the classical Lennard Jones system[10] it is found that the zeros of $g(r,T)-1$ do not change position when the temperature is changed with the density kept constant. More in general the ratio $[g(r,T_1)-1]/[g(r,T_2)-1]$ is roughly constant if the two temperatures T_1 and T_2 are not too dissimilar and for distances r outside the region of the first maximum of $g(r)$. The conclusion is that (10) represents roughly the temperature dependence at constant density of any liquid provided we consider a not too large range of T and a not too large range of r. The only difference between a classical liquid and ^{4}He is that in the first case $f(T)$ is a decreasing function of temperature whereas in ^{4}He $f(T)$ increases with T up to $T \simeq T_\lambda$. Given the generality of the representation (10) of $g(r,T)$ in order to give any attendibility to relation (11) between $f(T)$ and $n_o(T)$ it is quite essential to have a microscopic justification of if and this we do not have yet if, as we said, the argument of Cummings et al. is not valid. Finally, I notice that if we accept relation (11) then it is predicted a strong decrease of the temperature variation of $g(r,T)$ between low temperature and T_λ in ^4He under pressure. In fact theory predicts a strong decrease of the ground state Bose Einstein condensate n_o when the density is increased and near solidification a reduction of n_o for a factor of three compared to the equilibrium value has been computed.[1] No data are available yet to test this prediction on $g(r,T)$.

Another explanation[3] of the anomalous T dependence of correlations in ^4He is in term of the coupling of static to dynamic properties in a quantum system. Thermal properties of ^4He are well described in terms of elementary excitations, phonon and roton excitations essentially, and bove 1^oK rotons are the dominant excitations. A roton is in the first place a density fluctuation so that the static structure factor $S(k,T) = \frac{1}{N}\langle\rho_k\rho_{-k}\rangle_T$, that is a measure of density fluctuations of wavevector k, is an increasing function of temperature for $k \simeq k_R$ due to the contribution of thermally excited rotons. k_R is the wavevector of a roton. On the other hand $k_R \approx k_o$, where k_o is the wavevector of the main maximum of $S(k)$, due to the coupling of dynamic to static properties (remember Feynman's form for the excitation spectrum $\varepsilon_k = \hbar^2k^2/2mS(k)$). The result is that the height of the main peak of $S(k)$ increases when the system is heated and this corresponds to an increasing short range order since $g(r)-1$ is the Fourier transform of $S(k)-1$. This point was made explicit and quantitative for the first time by De Michelis et al.[3] but it was implicit in an earlier work of Feenberg.[11] The Landau picture of ^4He in term of a gas of elementary excitations can be used to explicitly construct[12] the density matrix in the r representation, $\langle r_1',..r_N'|\rho_T|r_1,..r_N\rangle$, if we assume the Feynman form for the excited states, $\psi_k = \rho_k\psi_o/\sqrt{NS_o(k)}$. The diagonal part of the density matrix, from which we can compute $g(r,T)$, has the structure of a Jastrow function:

$$\langle r_1,..r_N|\rho_T|r_1,..r_N\rangle = |\psi_o(r_1,..r_N)|^2 \exp\{-\sum_{i<j}\nu_T(r_i-r_j)\}/Q_T, \quad (12)$$

where Q_T is a normalization constant and the thermal Jastrow correlation reads

$$\nu_T(r) = \frac{1}{N}\sum_k \frac{1}{S_o(k)}\{\tanh(\varepsilon_k(T)/2k_BT)-1\}e^{i\vec{k}.\vec{r}} \quad (13)$$

with $S_o(k)$ being the ground state structure factor and $\varepsilon_k(T)$ the empirical excitation spectrum at temperature T. The justification for using the empirical spectrum has already been discussed[11] and is based essentially on a mean field argument.

The computation of $g(r,T)$ from (12) is similar to the problem of computing $g(r)$ from a Jastrow function so that similar techniques can be used. De Michelis et al.[3] used the Monte Carlo method, having approximated ψ_0 with a variational Jastrow function. Since the thermal correlation ν_T is weak compared with the correlations contained in ψ_0, g(r,T) has been later[4,13] computed adapting to the present case a suitable formulation of perturbation theory developed for classical liquids, the "exponential" approximation.[15] The two methods give results in good agreement[13] and this is important because within the perturbation approximation of the exact ground state ψ_0 one needs knowing only g(r). This function can be deduced from experiment performed at low temperature and in this way no assumption on ψ_0 or on the Hamiltonian is needed and no free parameter is present in the computation of g(r,T).

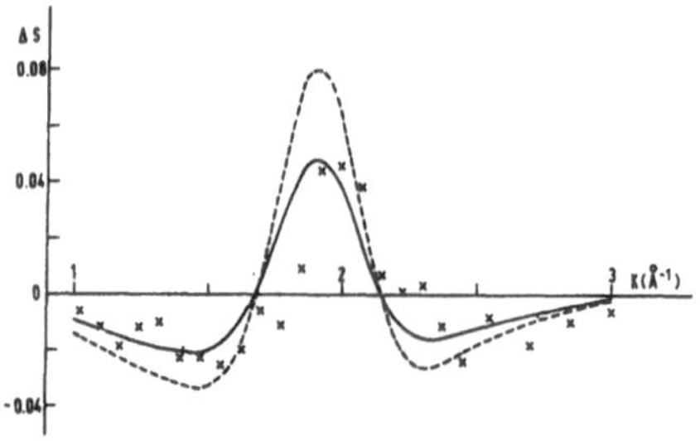

Fig.1 - $\Delta S(k,T) \equiv S(k,T) - S_0(k)$ at ρ_{eq} and $T = 2.15°$. Experimental data[14] (crosses) and theory[13] for finite linewidth (solid line) and for sharp excitation ($\delta_k = 0$) (dashed line).

As expected on qualitative ground it is found that the short range order increases at finite temperature as more rotons are thermally excited. In S(k,T) this manifests itself in an increasing and sharpening of its main peak and as an example in Fig.1 the difference $\Delta S(k,T) \equiv S(k,T) - S_0(k)$ is shown at $T = 2.15°$. Above $T \approx 2.1°$ rotons acquire an appreciable linewidth [16] and this indicates that the elementary excitation picture of ^4He is loosing validity. In order to take into account qualitatively the effect of the finite lifetime of rotons on correlations Gaglione et al.[13] have computed S(k,T) using the density matrix (12) with ν_T averaged over a lorentzian distribution of energy of rotons with a width $\delta_k(T)$ taken from experiment. The shape of $\Delta S(k,T)$ does not change (see Fig.1) but the intensity of the temperature effect is reduced and when rotons become overdamped at $T \simeq T_\lambda$ an additional heating of the system causes a decrease of the height of S(k,T) as shown in Fig.2 where the maximum of $\Delta S(k,T)$ is plotted as function of T. The overall agreement between theory and experiment is very good as it can be seen in Fig.1 and 2. At close inspection the only significant deviation is a small displacement ($\Delta k \approx 0.1$ Å^{-1}) of the theoretical S(k,T) compared with the experimental one. It is believed[13] that this is due to the effect of backflow associated to rotons that introduces three particle correlations and a computation of this effect is being performed.

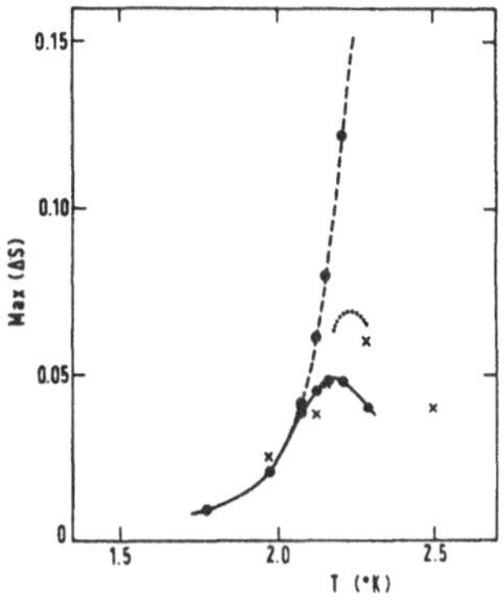

Fig.2 - Maximum value of $\Delta S(k,T)$ as function of T at ρ_{eq}. Experimental data[14] (crosses) and theory[13] for finite linewidth (dots connected by a solid line) and for sharp excitations (dots connected by a dashed line).

In conclusion the density matrix (12,13) has a strong microscopic fundation up to temperatures of order 2° and good agreement with experiment is found in the temperature dependence of S(k,T). The only limitation of the density matrix ρ_T is that it does not include the effects of backflow but ρ_T can be improved on this respect by including three particle correlations. At higher temperatures the density matrix is based on uncontrolled approximations but we believe that it includes phenomenologically in a sound way the effect of roton--roton interaction. An interesting prediction[13] obtained with the density matrix (12,13) is that the increase of the short range order at T_λ, for instance, with respect to that of the ground state strongly <u>increases</u> if the density is higher than the equilibrium one . This is shown in Fig.3 where the temperature variation of the maximum of $\Delta S(k,T)$ is given along different isocores. This behaviour is due to the fact that the energy and the effective mass of rotons decrease at high density. When we remember that exactly the opposite behaviour is predicted on the basis of the alternative description of the temperature variation of correlations in term of Bose Einstein condensation we conclude that a measurement of S(k,T) along different isocores will be a crucial experiment to discriminate between the two theories.

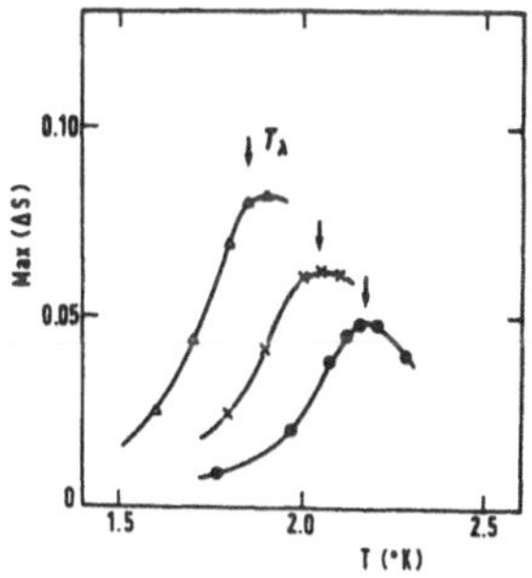

Fig.3 - Maximum value of $\Delta S(k,T)$ as function of T at three densities: ρ_{eq} (dots) $1.1 x \rho_{eq}$ (crosses) and $1.2 x \rho_{eq}$ (triangles).

References

1. M.H. Kalos, this conference; D.M. Ceperley and M.H. Kalos, in Monte Carlo Method in Statistical Physics, K. Binder ed. (Springer, Berlin, 1979).
2. F.W. Cummings, G.J. Hyland and G. Rowlands, Phys. Konden. Mater. 12, 90 (1970).
3. C. De Michelis, G.L. Masserini and L. Reatto, Phys. Lett. A 66, 484 (1978).
4. G. Gaglione, G.L. Masserini and L. Reatto, Phys. Rev. B 22, 1237 (1980).
5. Y. Rosenfeld and N.W. Ashcroft, Phys. Rev. A 20, 1208 (1979).
6. P.A. Whitlock, M. Lee and M.H. Kalos, unpublished.
7. G.V. Chester and L. Reatto, Phys. Rev. B 22, 5199 (1980); A. Griffin, Phys. Rev. B 22, 5193 (1980); A.L. Fetter, Phys. Rev. B (in press).
8. V.F. Sears and E.C. Svensson, Phys. Rev. Lett. 43, 2009 (1979).
9. H.N. Robkoff, D.A. Ewen and R.B. Hallock, Phys. Rev. Lett. 43, 2006 (1979).
10. See, for example, L. Verlet, Phys. Rev. 165, 201 (1968).
11. E. Feenberg, Ann. Phys. (N.Y.) 70, 133 (1972).
12. O. Penrose, in Proc. Int. Conf. on Low Temperature Physics, J.R. Dillinger (ed.), University of Wisconsin, Madison (1958) p.117.
13. G. Gaglione, G.L. Masserini and L. Reatto, Phys. Rev. B 23, (1981).
14. E.C. Svensson, V.F. Sears, A.D.B. Woods and P. Martel, Phys. Rev. A 21, 3638 (1980).
15. H.C. Andersen and D. Chandler, J. Chem. Phys. 57, 1918 (1972).
16. O.W. Dietrich, E.H. Graf, C.H. Huang and L. Passell, Phys. Rev. A 5, 1377 (1972).

MODEL HAMILTONIAN FOR SUPERFLUIDITY

Moorad Alexanian
Department of Physics
Montana State University
Bozeman, Montana 59717, USA
and
Departamento de Física*
Centro de Investigación y de
Estudios Avanzados
Apartado Postal 14-740
México 14, D.F., México

I. INTRODUCTION

The classic work of Bogoliubov[1] on a microscopic theory of superfluidity is successful in obtaining a phonon like low-momentum spectrum from a repulsive interaction. The crucial assumption for this weakly interacting Bose system is the macroscopic occupation of a single quantum state. However, the results[2-4] for a dilute Bose gas are given in terms of an expansion parameter--the ratio of the scattering length to the correlation length--which is large when evaluated for ^{4}He. Also, the elementary excitation spectrum[3] does not give rise to the roton minimum observed with the neutron-scattering experiments.[5,6] In a recent work,[7] a model for the description of a dense strongly interacting Bose gas has been proposed which features macroscopic occupation of infinitely many single-particle momentum states in an arbitrarily small neighborhood of the origin $k^2=0$; and so possesses a nonuniform condensate. Note that macroscopic occupation of a single quantum state need not be true for sufficiently strong interaction. This new model is a generalization of Bogoliubov's model and gives rise to not only a phonon spectrum, superfluidity and a phase transition, but also to a roton spectrum, a condensate fraction of 10% and the correct temperature and density dependence of the roton parameters and speed of first sound.[7]

II. MANY-CONDENSATE MODEL

The theory of Ref. 7 for He II in the bulk is based on the model Hamiltonian

$$\hat{H} = \frac{gN^2}{2V} \sum_{\vec q} A^2_{\vec q} + N_0 \sum_{\vec k} \xi^2_{\vec k} \varepsilon_{\vec k} + \sum_{\vec k} a^\dagger_{\vec k} a_{\vec k} \left(\varepsilon_{\vec k} - \frac{gN_0}{V} \sum_{\vec q} A^2_{\vec q}\right)$$

$$+ \frac{gN_0}{2V} \sum_{\vec k_1 \vec k_2} \left(A_{\vec k_1 + \vec k_2} a^\dagger_{\vec k_1} a^\dagger_{\vec k_2} + 4 A_{\vec k_1 - \vec k_2} a^\dagger_{\vec k_2} a_{\vec k_1}\right.$$

$$\left. + A_{\vec k_1 + \vec k_2} a_{\vec k_1} a_{\vec k_2}\right) , \qquad (1)$$

---

*Permanent Address

where $\varepsilon_{\vec{k}} = \hbar^2k^2/2\,m_{He}$ with m_{He} the mass of the helium atom. The condensate wave function

$$\psi(\vec{r}) = (N_0/V)^{1/2} \sum_{\vec{k}} \xi_{\vec{k}}\, e^{i\,\vec{k}\cdot\vec{r}} , \tag{2}$$

where $-1 \leq \xi_{\vec{k}} \leq 1$ and $\xi_{\vec{k}} = \xi_{-\vec{k}}$ so that the average linear momentum of the condensate with N_0 particles is zero. The constant matrix element g for the potential energy is related to the s-wave scattering length a in vacuum by $g = 4\pi\hbar^2 a/m_{He}$ and

$$A_{\vec{q}} \equiv \sum_{\vec{k}} \xi_{\vec{k}}\, \xi_{\vec{k}-\vec{q}} = A_{-\vec{q}} \tag{3}$$

is the Fourier component of $|\psi(\vec{r})|^2$ with $A_0 = 1$. The Hamiltonian (1) reduces to Bogoliubov's case for $\xi_{\vec{k}} = \delta_{\vec{k},0}$, that is, $A_{\vec{q}} = \delta_{\vec{q},0}$. The thermodynamic state of the system is determined by minimizing the Helmholtz free energy with respect to the condensate wave function $\psi(\vec{r})$ for given fixed density and temperature. The bilinear Hamiltonian (1) is diagonalized by means of a canonical transformation[7], and so

$$\hat{H} = E + \sum_{\vec{k}} E_{\vec{k}}\, b^{\dagger}_{\vec{k}} b_{\vec{k}} , \tag{4}$$

where

$$E = \frac{gN^2C}{2V} + N_0 \sum_{\vec{k}} \xi^2_{\vec{k}}\, \varepsilon_{\vec{k}}$$

$$+ \tfrac{1}{2} \sum_{\vec{k}} \left(E_{\vec{k}} - \varepsilon_{\vec{k}} + \frac{gN_0C}{V} - \frac{2gN_0}{V} + \frac{g^2N_0{}^2}{2\,V^2}\, \frac{C}{\varepsilon_{\vec{k}}}\right) \tag{5}$$

with

$$C \equiv \sum_{\vec{k}} A^2_{\vec{k}} \geq 1 . \tag{6}$$

The single-particle excitation spectrum $E_{\vec{k}}$ is phononlike in the long-wavelength limit and is given by[7]

$$E_{\vec{k}} = \left(2g\,\frac{N_0}{V}\, C\, \varepsilon_{\vec{k}}\right)^{1/2} \equiv \hbar\, c\, k , \tag{7}$$

while for intermediate and large values of the wavevector $\vec{k}$

$$E^2_{\vec{k}} = \varepsilon^2_{\vec{k}} + \varepsilon_{\vec{k}} \left(\frac{4gN_0}{V} - \frac{2gN_0}{V}\, C\right) + \frac{g^2N_0{}^2}{V^2}\, C(C-1) . \tag{8}$$

The excitation spectrum (8) has a roton structure near its minimum value at $\varepsilon_{\vec{k}} = (gN_0/V)(C-2)$ for $C \geq 2$. The roton parameters are given by[7]

$$\Delta = (gN_0/V)(3C-4)^{1/2} , \tag{9}$$

$$\mu_r = \frac{m_{He}(3C-4)^{1/2}}{2(C-2)} , \tag{10}$$

and

$$p_0^2 = 2m_{He}(gN_0/V)(C-2) . \tag{11}$$

In particular, it follows that

$$\mu_r = m_{He}^2 \Delta/p_0{}^2 \tag{12}$$

for any d-dimensional system. Also, with the aid of (7), (10), and (11), the speed of (first) sound becomes

$$c = \frac{1}{\sqrt{2}} \frac{p_0}{m_{He}} \frac{[1+4\mu_r^2/m^2{}_{He}]^{1/2}}{[1/4 + (3/4)(1+(32/9)(\mu_r{}^2/m^2{}_{He}))^{1/2}]^{1/2}} \tag{13}$$

valid for any d-dimensional system. Finally, the condensate fraction follows directly from (10) and (11) and is

$$\frac{N_0}{N} = \frac{3}{8}\frac{p_0{}^2}{m_{He}}\frac{V}{gN}[(1+\frac{32}{9}(\mu_r^2/m_{He}^2))^{1/2}-1] . \tag{14}$$

Note that with the help of the replacement scheme[8] $g/V \to g^{[d]}/V^{[d]} = 3\hbar^2 \pi^{d/2} a^{d-2}/m_{He}V^{[d]} \Gamma(d/2+1)$, where $V^{[d]}$ is the d-dimensional volume of the system, all the above results may be obtained in the appropriate d-dimensional system.

Since the Bogoliubov result is contained in our model Hamiltonian (1), we expect to recover the asymptotically exact perturbative result of Bogoliubov in the limit of a dilute gas with very weak potential, i.e., $C \to 1$ for $N/V \to 0$ and $g \to 0$ at low temperatures. However, this has not been shown rigorously since we do not have an expression for $E_{\vec{k}}$ for all values of $\vec{k}$.

III. CONDENSATE NATURE IN ONE AND TWO DIMENSIONS

The macroscopic occupation of infinitely many single-particle quantum states with a finite limit-point gives rise[7] to a (gauge invariance) symmetry-breaking term in the Hamiltonian (1)

$$\hat{H}_{symm} = \int_V d\vec{r} \quad \{\psi(\vec{r})\, \hat{\psi}^\dagger(\vec{r}) \quad + \psi^*(\vec{r})\, \hat{\psi}(\vec{r})\} \int_V d\vec{r}'\ V(\vec{r}-\vec{r}')|\psi(\vec{r}')|^2 . \tag{15}$$

The symmetry-breaking term $\hat{H}_{symm}$ gives rise to further condensation in the states with momenta which are integral multiples of the momenta of the original states in the condensate. Therefore, $\hat{H}_{symm}$ is dropped[7] from (1) but with the consistency proviso that the condensate consists of a macroscopic occupation of a denumerably infinite number of single-particle quantum states--states of the form $\sum_{i=1}^{M} n_i k_i$, $n_i = 0, \pm 1, \pm 2, \ldots$; M finite with fixed, arbitrarily small vectors $\vec{k}_1, \vec{k}_2 \ldots \vec{k}_M$. In proofs[8] of the absence of Bose-Einstein condensation in one- and two-dimensional interacting Bose systems, which introduce a symmetry-breaking field into the Hamiltonian in order to get a non-vanishing order parameter at finite volume, one must have that $\langle \hat{H}_{symm} \rangle \leqslant 0$. Notice that if the pair potential $V(\vec{r})$ is positive definite, then by (15) $\langle \hat{H}_{symm} \rangle \geqslant 0$. Therefore, for spatially inhomogeneous condensates, the interaction between the particles generates its own symmetry-breaking field and, thus, such a term cannot be added with arbitrary properties without regard for the dynamics of the system. Similarly, for proofs[8] which use no symmetry-breaking fields; the $1/k^2$-singularity needed for the proof based on the Bogoliubov's inequality is removed since in any arbitrary neighborhood of $\vec{k} = 0$, an unlimited number of nonvanishing $\xi_{\vec{k}}$ can be found--since the sequence $\{\vec{k}\}$ of condensates has a limit-point at $\vec{k} = 0$. Consequently, rigorous proofs[8] demand that a Bose gas in spatial dimensionality $d \leqslant 2$ must possess a nonuniform condensate as introduced in Ref. 7 for the description of superfluidity in the bulk. Therefore, the model Hamiltonian (1) is likewise applicable to one- and two-dimensional superfluids. Hence phonon and rotons exist[8] as elementary excitations at low temperatures in one- and two-dimensional Bose systems with repulsive interactions just as it occurs for similar three-dimensional systems. Recent neutron scattering work[9] further confirms the existence of a two-dimensional roton with the properties given by the roton of Ref. 8.

IV. DISCUSSION

The quantity C defined by (6) plays a fundamental role in our work. In terms of the condensate wave function (2),

$$C^{[d]} = V^{[d]} \int |\psi(\vec{r})|^4 d\vec{r} / [\int |\psi(\vec{r})|^2 d\vec{r}]^2 \, . \tag{16}$$

Suppose that for fixed T such that $T < T_c^{[2]}$--recall that $T_c^{[2]} < T_c^{[3]}$, where $T_c^{[d]}$ is the critical temperature of the d-dimensional system--the number densities $N^{[3]}/V$ and $N^{[2]}/A$ are adjusted so that $g^{[3]} N_0^{[3]}/V = g^{[2]} N_0^{[2]}/A$. It is quite plausible[8] then that $C^{[d]}$ is actually independent of d. Therefore, our approximate forms for the excitation spectrum--eqs. (7) and (8)--are actually independent of d for fixed $T < T_c^{[d-1]}$ and given $g^{[d]} \rho_0^{[d]} = g^{[d-1]} \rho_0^{[d-1]}$. Or, equivalently, the densities $\rho^{[d]}$ and $\rho^{[d-1]}$ can be so chosen for a given temperature that the excitation spectra for the (d-1)- and d-dimensional Bose systems are identical. For instance, for d=3, one has that $\frac{4}{3} a \rho_0 = \sigma_0$, where $\rho_0(\sigma_0)$ is the condensate density for d=3 (d=2). Now for $\rho = 0.143$ g/cm^3 and $T < 1.26^o$K, $N_0/N = 0.105$ and so $\sigma_0 = 4.4 \times 10^{-10}$ g/cm^2. If one supposes[9] that $\sigma/m_{He} = 0.077$ atom/Å^2, then $\sigma_0/\sigma = 0.084$, a reduction[8] of 17% from that for the bulk superfluid. Hence, for $\sigma_0 = 4.4 \times 10^{-10}$ g/cm^2 and $T \lesssim 1.26^o$K, the superfluid film has the same excitation spectrum as the bulk superfluid at the same temperature and $\rho_0 = 0.015$ g/cm^3. What density σ does σ_0 correspond to? We believe, for instance, that in a macroscopic rectangular slab of superfluid He II the two dimensional faces of the slab contain two-dimensional elementary excitations--phonons and rotons--with precisely the same properties as the excitations in the bulk of the superfluid. Now the relation between condensate fractions in the bulk and on the faces of the slab is $\frac{4}{3}\, a\, \rho_0 = \sigma_0$. However, our nonuniform condensate model does not give us a relationship between ρ and σ. Nevertheless, if the slab is placed in contact with, say a solid surface, then a self-determined equilibrium density will be established on the surface of contact which varies in a continuous fashion as one moves perpendicularly away from the surface of contact into the bulk superfluid. Thus, the elementary excitation parameters will depend[8] on the density (or pressure) and, hence, on distance from the surface of contact in accordance to the value of the self-determined density. But, otherwise, the excitations are of the precise same nature as in the bulk. This is indeed what was observed by the Brookhaven group[9]. Of course, we are supposing layer completion; otherwise, the analysis must be modified[10]. Finally, it is quite interesting that the condensate fraction as given by (14) is in very good agreement[11] with values of N_0/N obtained directly from the momentum distributions for the ^4He atoms deduced from neutron-inelastic-scattering measurements of the dynamic structure factor $S(Q,\omega)$ for large values of Q[12]. Moreover, our result (14) lends further support[11] to the prescription[13] which forms the basis of the measurements of the condensate fraction by neutron-inelastic-scattering[14] and x-ray-diffraction.[15]

REFERENCES

1. N.N. Bogoliubov, J. Phys. Moscow USSR 11, 23 (1947).
2. T.D. Lee, K. Huang, and C.N. Yang, Phys. Rev. 106, 1135 (1957).
3. S.T. Beliaev, Zh. Eksp. Teor. Fiz. 34, 433 (1958) [Sov. Phys.-Jetp 7, 299 (1958)].
4. N.M. Hugenholtz and D. Pines, Phys. Rev. 116, 489 (1959).
5. D.G. Henshaw and A.D.B. Woods, Phys. Rev. 121, 1266 (1961).
6. For a review of the structure and elementary excitations of liquid helium see A.D.B. Woods and R.A. Cowley, Rep. Prog. Phys., 36, 1135 (1973).
7. M. Alexanian and R.A. Brito, Phys. Rev. B 17, 3547 (1978).
8. M. Alexanian, Physica (Utrecht) 100A, 45 (1980).
9. W. Thomlinson, J.A. Tarvin, and L. Passell, Phys. Rev. Lett. 44, 266 (1980).
10. R.D. Puff and J.G. Dash, Phys. Rev. B 21, 2815 (1980).
11. M. Alexanian, Phys. Rev. Lett. 46, 199 (1981).
12. V.F. Sears, E.C. Svensson, P. Martel and A.D.B. Woods (to be published).
13. G.J. Hyland, G. Rowlands, and F.W. Cummings, Phys. Lett. 31A, 465 (1970); see also F.W. Cummings, G.J. Hyland and G. Rowlands, Phys. Konden. Mater. 12, 90 (1970).
14. V.F. Sears and E.C. Svensson, Phys. Rev. Lett. 43, 2009 (1979).
15. H.N. Robkoff, D.A. Ewen, and R.B. Hallock, Phys. Rev. Lett. 43, 2006 (1979).

CONSEQUENCES OF DEFECTS ON LATTICE VIBRATIONS

SEMICLASSICAL THEORY

L. Andrade

Departamento de Física, Facultad de Ciencias, Universidad Nacional Autónoma de México, México 20, D.F.

1. INTRODUCTION.

A considerable part of today's research activity in solid state physics is devoted to the investigation of the influence of defects on the behaviour of solids (1). One is no longer surprised to discover that small numbers of imperfections contribute to striking macroscopic consequences. It is very well known (2) that the effects of the surface or of a finite concentration of defects in the crystal or on its surface on the vibrational properties of the lattice can be determined rather well from a knowledge of the effects of the surface or of only one or two isolated imperfections inside the crystal or on its surface area. If this were not the case the study of most defect problems of interest would become completely out of the question.

In a point of time the earliest studies of the conseauences of defects on lattice vibrations were carried out by Lifshitz, Montroll, Hori, Wallis and others (3) and the theoretical and experimental methods have been reviewed by Lifshitz (4), Montroll et all (5), Maradudin (6) and Wallis (7). In those studies the principal trend were toward the determination of the effect of the imperfections on the individual mode frequency levels and over the determination of the frequencies of the localized defect modes when the crystals include short-range forces because the extension to lattices with long-range interactions becomes nearly insuperable due to computational problems.

Recently we have been interested in the influence of defects on the behaviour of solids when we include nearest, next-nearest and more complex interactions, we have developed a theory based in the solution of the difference equations and we have obtained some interesting findings. The purpose of this lecture is to point out that in the study of the vibrations in crystal lattices with nearest, next-nearest and more complex interactions we must consider the more complex structure of the solution of the difference equations in order to understand the behaviour of the systems and we will illustrate this here with a brief review of our main results. We restrict ourselves in this discussion to defect problems for one-dimensional lattices because in many cases the solutions can be given in closed form (analytically) and their qualitative features carry over more physical interesting real problems. The method that we present here has been applied to one-dimensional cases but can be extended toward more dimensions.

2. MONATOMIC LATTICES.

The simplest monatomic problem with nearest and next-nearest interactions is that of a single defect particle in a linear chain of particles each of mass m in which each particle interacts with its two nearest and its two next- nearest neighbours. We assume that the particle at the origen is replaced by a defect atom whose mass is m' and which is linked to its nearest neighbours and its next-nearest neighbours by bonds like the atoms of mass m whose spring constants are K and $\varkappa$ respectively. The time independent equations of motion of the lattice can be written as

$$-m\omega^2 X_n = K(X_{n+1}+X_{n-1}-2X_n)+\varkappa(X_{n+2}+X_{n-2}-2X_n) \qquad (1)$$

where X is the maximum atomic displacement from its equilibrium position and ω the frequency of the normal modes. The equation of motion of the impurity atom

$$-m'\omega^2 X_0 = K(X_1+X_{-1}-2X_0)+\varkappa(X_2+X_{-2}-2X_0) \qquad (2)$$

and the localization of the solution around defect prescribe the boundary conditions and we pose the boundary value problem. The solution of this problem using the method of the difference equations has been obtained and the discussion for general changes of masses and force constants has been given by the author (8) and the principal results will be shown here. The eigenfrequencies of the localized modes are given by

$$(1-2\varepsilon)^2\gamma^2 y^3 + 8(\gamma+1)\gamma\,\varepsilon^2(1-2\varepsilon)y^2 + \{16(\gamma+1)^2\varepsilon^2 + 8\gamma(\gamma+4)\varepsilon - 4\gamma(\gamma+4)\}\varepsilon^2 y - 16(\gamma+2)^2\varepsilon^4 = 0 \qquad (3)$$

where $\gamma = K/\varkappa$, $\varepsilon = m/m'$ and $y/4$ is the square frequency of the localized mode normalized with respect to the maximum frequency of the normal modes of the infinite crystal with nearest neighbour interactions, when $y < (\gamma+4)^2/4\gamma$ and by

$$\gamma^2\varepsilon(\varepsilon-2)(1-\varepsilon)^2y^4 + 4\gamma(\gamma+1)(1-\varepsilon)^2y^3 - 2\gamma(\gamma+4)(1-\varepsilon)^2y^2 - 4\gamma y + (\gamma+4)^2 = 0 \qquad (4)$$

where γ and y were defined before and $\varepsilon = m'/m$. In this case the frequencies of the localized modes are determined by the eq. (4) when $y > (\gamma+4)^2/4\gamma$

In Fig. (1) we have ploted the maximum atomic displacements as a function of position in the lattice for the impurity mode in the case $m' = 2m$ and $K = 4\varkappa$. It is to be emphasized that this figure is a graph and not a physical diagram. The actual displacements are parallel to the axis of the lattice. These maximum atomic displacements have the same form like in the case of the maximum atomic displacements of the localized modes in the crystal with imperfections and nearest neighbour interactions and die out rapidly with distance from the defect but they diminish their maximum atomic displacements in a very rapid drop, more rapidly than in the case of the localized modes in crystals with nearest neighbour interactions. They are symmetrical about the position of the impurity

Fig. (1). The maximum atomic displacement as a function of position in the lattice for the impurity mode when $m' = 2m$ and $K = 4\varkappa$.

atom (particles to the left being in phase with those to the right) like in the case of localized modes in crystals with nearest neighbour interactions. The rate of localization of this type mode is bigger when the mass of the impurity atom is bigger. The existence condition for this type mode is $m' > m$ and the frequency was calculated using eq. (3). Only one impurity mode exist in this case with frequency $.6\,Y_0$, where Y_0 is the maximum frequency for the normal modes in the perfect crystal with nearest neighbour interactions. This frequency lies in the range of the normal mode frequencies allowed to the perfect crystal. The frequency of this type mode is diminished when the mass of the impurity atom is larger and the range of allowed frequencies for the localized mode is inside the allowed branch for the normal modes of the infinite crystal when the mass of the impurity atom range from the mass of the regular atoms of the chain to infinite mass. The range of allowed frequencies for the impurity mode is determined by the strenght of the next nearest neighbour interaction and in this case is $0 < \gamma < .866\gamma_0$. If the strenght of the next-nearest neighbour interaction is diminished the frequency of the impurity mode is smaller and the rate of the localization of the maximum atomic displacements of the mode is greater than in the case of smaller strenght interaction.

When the mass of the impurity atom is the lighter one i.e. $m' < m$ other type of localized mode exist and its frequency is calculated using eq. (4). The tables (1) and (2) show the maximum atomic displacement as a function of position in the lattice for this impurity mode when $4m'=3m$ and $2m'=m$, respectively and $\gamma=4\chi$. These maximum atomic displacements die out rapidly with distance from the defect and are symmetrical about the position of the imperfection, but they diminish like the product of a trigonometric function

atomic position	maximum displacemen t
7	-.000592542
6	-.005977810
5	.022226526
4	-.045848669
3	.050070508
2	.042354531
1	-.363797327
0	1

Table (1). Values of the maximum atomic displacement as a function of atomic position in the lattice for the impurity mode when $4m'=3m$ and $\gamma=4\chi$.

atomic position	maximum displacement
7	-.000152265
6	.006635858
5	-.000596453
4	-.005083004
3	.025524324
2	-.035720532
1	-.160713217
0	1

Table (2). Values of the maximum atomic displacement as a function of atomic position in the lattice for the impurity mode when $2m'=m$ and $\gamma=4\chi$.

and a sequence decaying one. The atomic displacements in this impurity mode die out more rapidly with distance from the defect so it's more localized than former one and the phase of particles is changing in a more complex way than in the case of localized modes in crystals with nearest neighbour interactions. The rate of localization is bigger when the mass of impurity atom is lighter. Only one localized mode exist in each case with frequencies $1.038\gamma_0$ and $1.19\gamma_0$ respectively. These frequencies lie above the branch of allowed frequencies for the normal modes in the perfect crystal. From our simple results we can observe that as $m' \to m$, the impurity frequencies return to the top of the branch of allowed frequencies, while as $m' \to 0$, the impurity frequency becomes infinite, like in the case of localized modes in lattices with nearest neighbour interactions. If the strenght of the next-nearest neighbour interaction is weaker the frequency of the impurity mode is higher, the rate of localization of the mode is bigger and the range of allowed frequencies is different.

Since Hori and Asahi (9) introduced their transfer matrix method attemped to solve the problem of the isotopic impurity in a linear chain with nearest and next-nearest neighbour interactions but concluded that the manner of the influence of the impurity on the eigenfrequencies is similar to that in the nearest neighbour approximation if the next-nearest neighbour interaction is sufficiently weak compared with the nearest neighbour interaction because their method is inable to give the complete solution of the problem. Recently L. Andrade (10) includes nearest and next nearest neighbour interactions in a monatomic linear lattice and using the scattering matrix method finds the first type of localized mode with frequency in the allowed branch of the normal modes of the infinite crystal but with this method was incapable in order to obtain the complete solution of the problem. The comparison of this incomplete result is in agreement with the corresponding part of our solution.

Another particulary simple problem in a linear chain of particles of mass m and constants of force K and $\varkappa$ of its nearest and next-nearest neighbour interactions respectively, is the creation of a surface. One isotopic impurity on this surface introduces a little more complication. Let's consider the position of the impurity atom in the origen and we will assume that the particle at one end is replaced by a defect atom whose mass is m' and which is linked to its nearest neighbour and its next-nearest neighbours by bonds like the atoms of the perfect crystal. The time independent equations of motion of the particles of the lattice are given by eq. (1). The equations of motion of the two end atoms

$$-m'\omega^2 X_0 = K(X_1 - X_0) + \varkappa(X_2 - X_0) \tag{5}$$

$$-m\omega^2 X_1 = K(X_2 + X_0 - 2X_1) + \varkappa(X_3 - X_1) \tag{6}$$

joined with the localization of the solution in the surface region define the boundary conditions of the problem so we have a characteristic value problem. The solution using the method of the difference equations has been obtained by the author (11). We will give here the principal results. The eigenfrequencies of the localized modes are the solutions of the eigenvalue equation and are given by

$$\begin{aligned}
&\varepsilon^4\gamma^4 X^5 + \varepsilon^2\gamma^3[2\gamma(\gamma-2)\varepsilon^2 - (3\gamma^2+4\gamma+4)\varepsilon + \gamma^2]X^4 + [-4\varepsilon^4\gamma^5 \\
&+ 4\varepsilon^3\gamma^3(\gamma^2+3\gamma+4) + \varepsilon^2\gamma^2(3\gamma^3+11\gamma^2+12\gamma+6) - 2\varepsilon\gamma^4(\gamma+1)]X^3 + \\
&[2\gamma^4(\gamma+4)\varepsilon^3 - \gamma^2\varepsilon^2(7\gamma^3+32\gamma^2+46\gamma+24) + \gamma\varepsilon(\gamma^4-2\gamma^3-13\gamma^2-12\gamma-4) \\
&+ \gamma^3(\gamma+1)^2]X^2 + [-\gamma^3(\gamma+4)\varepsilon^2 + \gamma\varepsilon(2\gamma^4+17\gamma^3+48\gamma^2+48\gamma+16) \\
&- (\gamma^5+4\gamma^4+\gamma^3-5\gamma^2-4\gamma-1)]X - (\gamma^4+8\gamma^3+20\gamma^2+16\gamma+4) = 0
\end{aligned} \tag{7}$$

where $\gamma = K/\varkappa$, $\varepsilon = m'/m$ and $X = m\omega^2/K$. These frequencies must satisfy the next relationship $X < (\gamma+4)^2/4\gamma$.

In Figs. (2) and (3) we have ploted the maximum atomic displacements as a function of position in the lattice for the localized modes in the case of a clean surface when $K = 4\varkappa$. We have in this case two different type of the surface modes. In both localized modes the maximum atomic displacements decrease from the end into the interior of the crystal in a decaying sequence form.

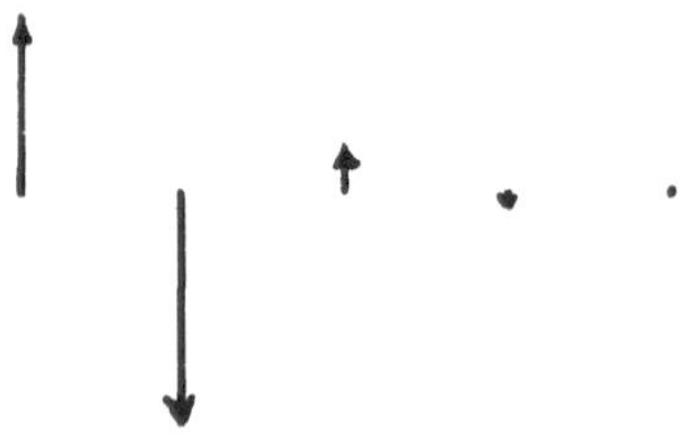

Fig. (2). The maximum atomic displacement as a function of the position in the lattice for the surface mode when $K = 4\varkappa$.

Fig. (3). The maximum atomic displacement as a function of the position in the lattice for the surface mode when $K = 4\varkappa$.

In one type mode Fig. (2), the maximum atomic displacements are different from the maximum atomic displacements in the surfaces modes of the lattices with nearest neighbour interactions because the largest maximum atomic displacement is on the second atom into the interior of the crystal. This type mode is a " subsurface " mode. The other type mode Fig. (3) has its largest maximum atomic displacement on the surface atom, but the first (surface) atom and the second (subsurface) atom are in phase. The rate of localization of the maximum atomic displacements in the subsurface mode is smaller than in the surface mode. The surface mode just now is different respect of the nearest neighbour interactions mode. In this case the frequencies are $.87\gamma_0$ for the subsurface mode and $.42\gamma_0$ for the surface mode. These frequencies lie in the range of the allowed normal mode frequencies of the perfect crystal. The mode (subsurface mode) with high frequency is less localized and the mode (surface mode) with low frequency is more localized. If the strenght of the next nearest neighbour interactions is diminished the frequencies of both type of localized modes are smaller, the form of the maximum atomic displacements are conserved and the rate of localization are larger for both type of localized modes. For different values of γ we found two localized modes with frequencies in the allowed branch of the normal modes of the infinite crystal. These modes have been found even in the case γ very large i.e. the next nearest neighbour interaction is very little in comparison with the nearest neighbour interaction. In all cases the localized modes have different forms.

The effect of the impurity atom on the surface in the frequencies and forms of the displacements of the localized modes of the clean surface is very simple. If the mass m' of the adsorbed impurity atom is less than the mass of the atoms of the perfect crystal, the frequencies of the two localized modes are larger and the forms of the atomic displacements are insensitive to this change except for the fact that the modes change from surface mode to subsurface mode. The rate of localization of the localized modes is diminished if the mass of the impurity adsorbed atom is diminished, of course, this imply that the modes are more delocalized. One of the localized mode exist in all range of the impurity masses. When the mass of the adsorbed atom is small in comparison with the mass of the atoms of the perfect crystal one of the localized mode disappears.

The Fig.(4) shows the case when $\gamma = 4$ and $2m' = m$ when one localized mode

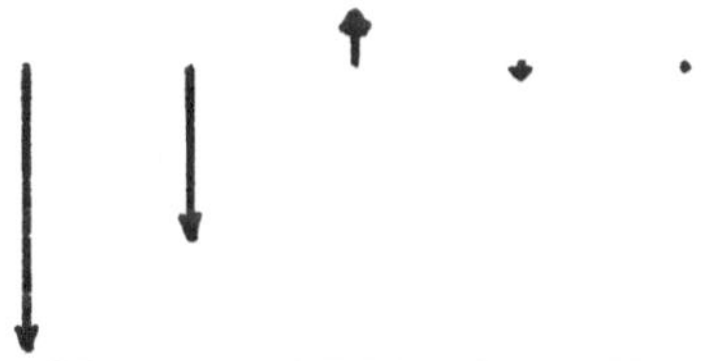

Fig.(4). The maximum atomic displacement as a function of the position in the lattice for the adsorbed mode when $m = 2m'$ and $\gamma_2 = 4\gamma$.

exist. The frequency of the adsorbed mode in this case is $.52\gamma_0$.

The clean surface problem was studied by Fukushima(12) using the Mikusinski operational calculus he concluded that if there are not modification in the atomic mass and force constant at the surface then no true surface mode appear, but only the so-called pseudo-surface mode, which is a damped oscillatory solution. This kind of mode has been mentioned by Slater (13) and Synge (14) and they asserted that not purely localized mode exist and it is probable that semilocalized modes are much more prevalent than surface modes. Fukushima can not solve the problem because he can't calculate the frequencies of the modes. The reason is that his method is inable to provides the complete solution. The assumption of the Slater and Synge is wrong because we have demonstrated (15) the existence of the purely localized modes. Recently the author (15) obtained a part of the solution in the case of the clean surface. using the scattering matrix method.

3. DIATOMIC LATTICES

The study of the consequences of defects on vibrations of alternating diatomic lattices with nearest and next-nearest neighbour interactions is more difficult than the corresponding problem for monatomic lattices even in the case of nearest neighbour interactions. As a result, different methematical techniques have to be employed in solving defect problems for the diatomic lattice. However it has been shown that in the simple cases which include nearest neighbour interactions only, the method of the difference equations gives the complete solution of the problem. For example when Wallis (16) studied the surface oscillations of the crystal lattices, he demonstrated that a general criterion can be given for the existence of surface modes, namely, the total mass of the light atoms must be less than the total mass of the heavy atoms. If surface mode exist in a diatomic linear chain with nearest neighbour interactions, the number of surface modes equals the number of end of the lattice which have light atoms. When two end atoms are of different types, the single surface mode has a frequency which lies in the middle of the forbidden gap. In general the criterion for the existence of the surface mode is to have a lighter atom at the end of the chain. Quite recently we(17) have demonstrated the existence of another surface mode when the end atom is a heavy atom, with the same frequency in the middle of the forbidden gap like the mode of the Wallis, but with different maximum atomic displacements. The case of one adsorbed atom in a diatomic linear chain has been solved (18) and some interesting findings have been found.

We now proceed in the study of the consequences of the introduction of an isotopic impurity of mass m in a diatomic linear lattice of alternating masses m_B and m_A, with nearest and next-nearest neighbour interactions. K, k and χ are the force constants associated with central force interactions of nearest neighbour, next nearest neighbour of mass m_A and next nearest neighbour of mass m_B respectively. The time independent equations of motion of the lattice can be written as

$$-m_A\omega^2 X_n = K(y_{n+1}+y_n-2X_n)+\chi(X_{n+1}+X_{n-1}-2X_n) \tag{8}$$

and

$$-m_B\,\omega^2 y_{n+1} = K(X_{n+1}+X_n-2y_{n+1})+k(y_{n+2}+y_n-2y_{n+1}) \tag{9}$$

where X is the maximum atomic displacements from its equilibrium position of the atoms with masses m_A, y is the maximum atomic displacements from its equilibrium position of the atoms with masses m_B and ω the frequencies of the normal modes. Eqs. (8) and (9) are a simultaneous equations and can be solved by the method of the difference equations. The details of the calculation and a more complete analysis will be given elsewhere. We are going to give here a brief review of some interesting findings. The frequencies of the normal modes of the perfect crystal have been shown in a numerical form by Gazis and Wallis (19) and by the author analytically. These frequencies are dependent of the next nearest neighbour interactions. On the contrary the maximum frequency of the optical branch is independent of the next-nearest neighbour interactions, this is as it should be, since there is no relative motion of the next nearest neighbour in this case. The other edge branch frequency of the optical branch and the edge branch frequency of the accoustical branch are functions of the next-nearest neighbour interaction namely

$$\omega_+^2 = 2K\left[\frac{(1+\ell_1)}{m_A}+\frac{\ell_2}{m_B}\right] \tag{10}$$

$$\omega_-^2 = 2K\left[\frac{(1+\ell_2)}{m_B}+\frac{\ell_1}{m_A}\right] \tag{11}$$

where $\rho_1 = \chi/K$ and $\rho_2 = k/K$. In the Fig. (5) we have ploted the wide of the frequency wave vector dispersion relation when $m_B = 2m_A$, $\chi = 0$ and $K = 4k$. The ordinate of the figure is the square freauency normalized with respect to the square of the maximum accoustical frequency of the normal modes of the perfect crystal with nearest neighbour interactions. The branch wide of the optical normal modes in this case is smaller than in the case of the nearest neighbour interactions. The Fig. (6) shows the wide of the frequency wave vector dispersion relation when $m_B = 2m_A$ for the crystal with nearest neighbour interactions. The branch wide of the accoustical modes is more broad now in the nearest and next-nearest neighbour interactions that in the case of the nearest neighbour interactions. The equation of motion

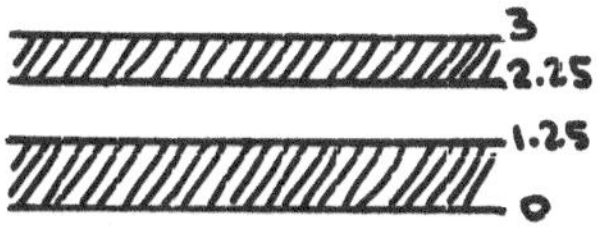

Fig. (5). Wide of the frecuency wave vector dispersion relation when $m_B = 2m_A$ $\chi = 0$ and $K = 4k$.

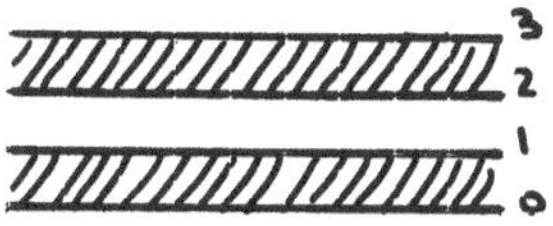

Fig. (6). Wide of the frequency wave vector dispersion relation when $m_B = 2m_A$

of the impurity atom

$$-m\omega^2 X_0 = K(Y_1 + Y_0 - 2X_0) + \chi(X_2 + X_{-2} - 2X_0) \qquad (12)$$

and the localization of the solution around defect prescribe the boundary conditions and we pose the boundary value problem. The details of the calculation and a more complete analysis will be given elsewhere. Nów we proceed in order to give a brief review of some interesting findings.

In Fig. (7) we have ploted the maximum atomic displacement as a function of position in the lattice for the impurity mode when $m_B = 2m_A$, $m_B = 4m$, $\chi = 0$ and $K = 4k$. The impurity mode die out rapidly with distance from the defect. This

Fig. (7). The maximum atomic displacement as a function of position in the lattice for the in-branch (optical) impurity mode when $m_B = 2m_A$,$m_B = 4m$ $\chi = 0$, $K\ 4k$. The solid arrows refer to mass m_A and the dashed arrows to mass m_B .

localized mode is symmetrical about the position of the impurity atom like the localized modes in monatomic crystals with nearest and next-nearest neighbour interactions. One impurity mode with frequency 2.287 which lies into the optical branch of the normal modes frequencies allowed to the perfect crystal has been found in this case. The frequency of this type mode is larger when the mass m of the impurity atom is lighter, but always is localized inside of the optical branch of the normal modes of the perfect crystal. The most important finding is the existence of two different type of localized modes that can be classified according whether or not their frequencies lie inside or outside the allowed branch of frequencies of the perfect crystal.

In the case of the effect of the clean surface on the vibrational properties in a diatomic crystal with nearest and next-nearest interactions we can assume that the system has semi-infinite extension. It provides the two boundary conditions of the problem namely

$$-m_B\,\omega^2 U_{2N} = -K\,(U_{2N}-V_{2N-1})-\chi\,(U_{2N}-U_{2N-2}) \qquad (13)$$

$$-m_A\,\omega^2 V_{2N-1} = K\,(U_{2N}+U_{2N-2}-2V_{2N-1})-k\,(V_{2N-1}-V_{2N-3}) \qquad (14)$$

The localization of the solution and the boundary conditions eqs. (13) and (14) define the characteristic value problem. The solution and analysis will be given elsewhere.

In the Fig. (8) we have ploted the maximum atomic displacements of the surface mode as a function of position of the atoms in the lattice when $2m_B=m_A$, $K=8\chi$ and $K=8k$. The displacements of the atoms decrease towards the interior of the crystal. This

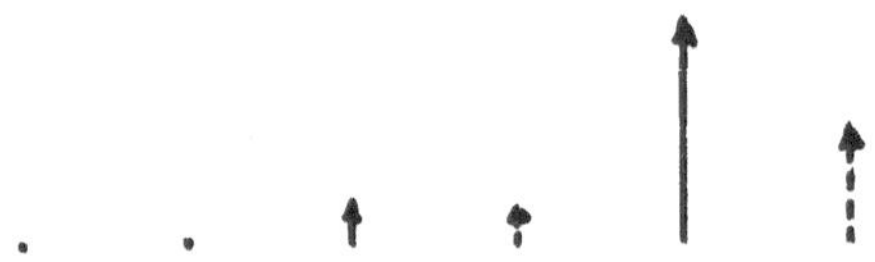

Fig. (8) The maximum atomic displacements of the surface mode as a function of position in the lattice when $2m_B=m_A$, $K=8\chi$ and $K=8k$. The solid arrows refer to mass m_A and the dashed arrows to mass m_B.

mode is a in-branch (accoustical) localized mode and it is a very localized. The frequency of the surface mode is .38 and this frequency lies in the accoustical branch of the normal modes of the perfect crystal. The maximum frequency of the optical branch is 3 and the minimum frequency of the optical branch is 2.375. The maximum accoustical frequency of the normal modes of the infinite crystal is 1.375. These frequencies are normalized with respect to the square of the maximum accoustical frequency of the nomal modes in a crystal with nearest neighbour interactions.

Gazis and Wallis (19) proposed a general numerical method for the solution of this problem and they found surface modes in the forbidden region for the normal modes of the infinite crystal only. We can't compare with their results because they used different boundary conditions but the author think that their method is inable to give the solution of this problem.

4. DISCUSSION AND CONCLUSIONS

The in-branch and out-branch localized modes that we have found are infinitely long-lived and they are a consequence of the more complex solution of the fourth order difference equation that it results when we include nearest and next-nearest neighbour interactions. They are not of resonant type like the Brout,Visscher (20) modes. The analysis outlined in the preceeding sections leads to some basic conclusions about the methods employed for the solution of the consequences of the imperfections on the lattice vibrations when we include nearest, next-nearest and more complex interactions. Because of the general solution of the difference equation in those cases is more complex its transformation to normal coordinates must be done appropriately. Like the methods used to solve this kind of problem i.e. the Green function technique, M- transformation matrix, transfer matrix, scattering matrix and others, utilize expansions in wave planes only,these methods are unable to predict long-lived localized modes with frequencies in-branch of the nomal modes of the perfect crystal and the more complicated localized modes with frequencies in the forbidden branchs of the nomal modes of the perfect crystal in those systems when we include more complex interactions. The advantage of the method of the difference equations is that it provides at least in principle and often in fact, an exact solution to the defect problem under consideration. Finally some remarks may be regarding the possibility of application of the present discussion to the consequences of the imperfections on electrons in solids, localized spin waves in crystal lattices and in general in a number of problems in physics where we find difference equations.

REFERENCES

1. The Fourth International Conference on Solid Surfaces and The Third European Conference on Surface Science. Proceedings, 1980. Edited by D.A. Degras and M. Costa.
2. A.A. Maradudin "Theoretical and Experimental Aspects on the Effects of Point Defects and Disorder on the Vibrations of the Crystals. "1" Solid State Phys. 18, 273, (1966)
3. I.M. Lifshitz. Dokl. Akad. Nauk. S.S.S.R. 48, 83, (1945)
   I.M. Lifshitz. Zu. Eksper. Teor. Fiz. 17, 1017, (1947); 17 , 1076 (1947); 18, 293, (1948)
   I.M. Lifshitz. J. of Phys. U.S.S.R. 7, 211, (1949); 7, 249, (1949)
   I.M. Lifshitz. Ups. Matemat. Nauk. 7, 170, (1952)
   E.W. Montroll and R.B. Potts. Phys. Rev. 100, 525, (1955); P.Mazur. E.W. Montroll, R.B. Potts. J. Wash. Acad. Sci. 46, 2, (1956)
   J. Hori and T. Asahi. Prog. Theor. Phys. 17, 523, (1957)
   R.F. Wallis. Phys. Rev. 105, 540 (1957); Phys. Rev. 116, 302, (1959)
   M. Lax. Phys. Rev. 94, 139, (1954)
4. I.M. Lifshitz. Nuov. Cim. Suppl. 3, 716 (1956)
5. A.A. Maradudin, E.W. Montroll and G.W. Weiss. " Theory of the Lattice Dynamics in the Harmonic Approximation ". Acad. Press. N.Y. (1963)
6. A.A. Maradudin. " Theoretical and Experimental Aspects on the Effects of Point Defects and Disorder on the Vibrations of Crystals. 2 " Solid State. Phys. 19, 2, (1966)
7. R.F. Wallis. Surf. Sci. 2, 146 (1964). Lattice Dynamics of Crystal Surfaces. Progr. in Surf. Sci. Vol. 4
8. L. Andrade. To be published.
9. J. Hori and T. Asahi. Progr. Theoret. Phys. 17, 523, (1957)
10. L. Andrade . Fac. Ciencias. UNAM. Reporte Interno 1979 (unpublished)
11. L. Andrade. Fourth International Conference on Solid Surfaces and Third European Conference on Surface Science. Cannes. France. Sep. (1980).
12. M. Fukushima Sci. Rep. 16, 1 (1969)
13. J. Slater. M. I. T. Report. No. 5, 1953 (unpublished)
14. J.L. Synge. J. Math. Phys. 35, 323 (1957)
15. L. Andrade. To be published
16. R.F. Wallis Phys. Rev. 105, 540 (1957)
17. L. Andrade. Fac. Ciencias. UNAM. Reporte Interno (unpublished)
18. L. Andrade. Fourth International Conference on Solid Surfaces and Third European Conference on Surface Science. Cannes. France. Sep. (1980).
19. D.C. Gazis and R.F. Wallis J. Math. Phys. 3 , 190 (1962)
20. R. Brout and W. Visscher . Phys. Rev. Letters 9, 54 (1962)

SPIN-ALIGNMENT IN CONDENSED ATOMIC HYDROGEN

M.L. Ristig and P.M. Lam[+]
Institut für Theoretische Physik
Universität zu Köln, 5 Köln 41, Germany

Abstract. - The variational approach is adopted to explore equilibrium properties of the condensed phase of atomic hydrogen in high magnetic fields. The calculations are based on optimized trial wave functions which take account of intra-atomic hyperfine mixing and interatomic spatial correlations. We present numerical results on the associated density matrices, the energy expectation value, the pressure, the magnetic equation of state and the static stability.

Introduction. - The recent preparation /1-3/ of stable atomic hydrogen at densities of order 10^{16}-10^{17} atoms/cm^3 has been a major step forward in exploring this exciting new many-body system. Although these densities are presumably still two orders of magnitude too small to detect macroscopic quantum properties of this fluid, one is fascinated by the prospect of testing experimentally predictions in the near future which are based on microscopic theories of degenerate Bose gases /4-6/. At the moment the critical wall-effects seem to be the most serious obstacle for producing the condensed phase /7-9/.

Two ground state hydrogen atoms interact via a potential

$$v(12) = v_T(r)P^3 + v_S(r)P^1 \tag{1}$$

differing in the electronic triplet and singlet channel. This potential is well known from variational calculations /10-12/. The triplet part is predominantly repulsive in contrast to the strongly attractive singlet portion which leads to the H_2 molecule. The hydrogen atoms are generally considered as composite bosons /13-15/. When an ensemble of H atoms is forced into triplet states by applying a large magnetic field we therefore expect Bose condensation to occur at low temperatures. The perfectly spin-aligned condensed ground state of H atoms has been studied within perturbation theory /16,17/, variational theory /18-20,12,7/ and the Monte Carlo approach /21-23/.

In this contribution we address ourselves to the numerical exploration of some ground state properties of the condensed hydrogen phase at large but finite magnetic fields. Our study focusses on the energy per particle, the pressure, as functions of density and the magnetic equation of state which permits information on the static stability of this phase. To learn something about the spatial correlations generated in bulk hydrogen at various densities and fields we also present some numerical results on the one- and two-body density matrices or, equivalently, on the momentum distribution

$$n(k) = \langle \psi_A | a^+_{\underline{k}} a_{\underline{k}} | \psi_A \rangle , \tag{2}$$

the structure function

$$S(\underline{k}) = \frac{1}{A} \langle \psi_A | \rho_{\underline{k}} \rho_{-\underline{k}} | \psi_A \rangle , \quad \rho_{\underline{k}} = \sum_{i=1}^{A} e^{i\underline{k}\cdot\underline{r}_i} \tag{3}$$

and the pairing function

$$\chi(\underline{k}) = \langle \psi_A | a_{\underline{k}} \, a_{-\underline{k}} | \psi_{A+2} \rangle . \tag{4}$$

The analysis (and notation) is based on the variational approach as described in Ref. /24/ employing a (unit-normalized) optimal ground state $|\psi_A\rangle$ for a homogeneous system of A hydrogen atoms embedded in a constant magnetic field. The operator $a^+_{\underline{k}}$ ($a_{\underline{k}}$) creates (destroys) a boson with wave number $\underline{k}$, $\rho_{\underline{k}}$ represents the density fluctuation operator. The present study should be considered as a first step in the right direction of an adequate description of finite field effects in bulk atomic hydrogen. In the next step one should, for example, incorporate the effects of spin-density fluctuations.

2. Microscopic description. - Let us begin our theoretical analysis of an extended system of A hydrogen atoms embedded in a sufficiently large uniform magnetic field $\underline{B}$ (say, of 1-10 Tesla depending on the particle density ρ) with the total hamiltonian

$$H_{tot} = H + H_Z + H_H + H_R + H_W + \ldots . \tag{5}$$

The various terms in eq. (5) represent the kinetic operator and the interatomic potentials (1),

$$H = T + V = \sum_{i=1}^{A} t(i) + \sum_{i<j}^{A} v(ij), \tag{6}$$

the Zeeman term

$$H_Z = -\sum_{i=1}^{A} \underline{\mu}_i \cdot \underline{B} , \tag{7}$$

the intra-atomic hyperfine interaction

$$H_H = \sum_{i=1}^{A} \alpha \underline{\sigma}_i \cdot \underline{\tau}_i , \tag{8}$$

the reactive hamiltonian H_R representing, among more complex reactions, the ionization of the composite boson into a nuclear proton and an electron /15/, the experimentally crucial hydrogen-wall interactions H_W, etc. The vector $\underline{\mu}$ describes the magnetic moment $\underline{\mu} = \underline{\mu}_e + \underline{\mu}_p$ of a H atom with the electron- (proton-) component being antiparallel (parallel) to the Pauli spin-vector $\underline{\sigma}(\underline{\tau})$, $\underline{\mu}_e = -\mu_e \underline{\sigma}$ and $\underline{\mu}_p = +\mu_p \underline{\tau}$. The strength of the hyperfine field is measured by the constant α.

At the present stage we shall only explore the effects arising from the first to third contribution to the total hamiltonian dropping the terms H_R, H_W, ... in eq. (5). The truncated problem could be further drastically simplified if it would be permitted to neglect the interatomic potential part V. In this case the hamiltonian would reduce to a superposition of single-particle operators which can be easily diagonalized /25/. For given momentum there are four single-particle eigenstates characterized by the spin-projections ↑ or ↓ for the electron and ⇑ or ⇓ for the proton. The energetically lowest hyperfine state is of the form

$$|0\rangle = (1+\varepsilon^2)^{-1/2}\{|\uparrow\Downarrow\rangle - \varepsilon|\downarrow\Uparrow\rangle\} \tag{9}$$

assuming that the external field $\underline{B}$ is directed along the negative z-direction. The positive parameter ε describes the hyperfine mixing of ↑ and ↓ spins and is related to the total spin-polarization of the hydrogen gas in its ground state, $P_z \equiv \langle \underline{\psi}_A | \mu_z | \underline{\psi}_A \rangle = \mu(1-\varepsilon^2)(1+\varepsilon^2)^{-1}$ with $\mu = \mu_p + \mu_e$. For the independent boson system represented by $H_{tot} = T + H_Z + H_H$ the admixture parameter ε is determined by the ratio of external field and internal hyperfine field /25/,

$$\varepsilon = \{1 + (\frac{\mu B}{2\alpha})^2\}^{1/2} - (\frac{\mu B}{2\alpha}) \quad . \tag{10}$$

In the limit $B \to \infty$ the parameter ε vanishes and the system is completely polarized. Inclusion of the interaction (1) into the treatment leads to important modifications of the simple form (10) which represent the influence of the medium on the magnetic equation of state. The incorporation of such effects will be done in two steps. We shall first deal with the completely polarized system and then, in a second step, proceed to the problem defined by the hamiltonian $H_{tot} = H + H_Z + H_H$.

3. Spin-aligned state. - When a system of hydrogen atoms is forced into the completely spin-aligned state, the effective portion of the interaction (1) is the triplet potential v_T. Under this condition we may treat the ensemble of H atoms by standard methods designed for one-component Bose systems such as the ^{4}He fluid. Numerical studies within perturbation theory have been reported in Refs. /16,17/. The variational approach has been adopted for the spin-aligned system in Refs. /18-20/.

At low densities, i.e. densities less than 10^{-4} Å^{-3}, the behavior of the spin-aligned system is characterized by the H-H triplet scattering at small momentum transfer. Thus, the equilibrium properties are determined by the triplet scattering length a being about 0.72 Å /16/. In particular, the energy per particle is given, to lowest order, by

$$E \to E_0 = \frac{\hbar^2}{m} \cdot 2\pi a\rho \tag{11}$$

where m is the mass of the H atom. Eq. (11) may be derived by employing many-body perturbation theory or the variational approach. At densities $\rho > 10^{-4}$ Å^{-3} many-body effects must be taken into account. This may be achieved most appropriately within variational theory.

Adopting this approach we may begin with the Bijl-Dingle-Jastrow ansatz for the ground state Bose wave function

$$\psi_A = N^{-1/2} \prod_{i<j}^{A} f(r_{ij}) \quad , \tag{12}$$

(N being the norm) and evaluate the structure function (3) and its Fourierinverse g(r) in hypernetted-chain approximation /24/. Subsequently, we minimize the energy expectation value associated with function (12),

$$E = \frac{1}{2} \rho \int v^{*}(r)\, g(r)\, d\underline{r} \quad , \tag{13}$$

where $v^{*}(r) = v_T(r) - \frac{\hbar^2}{2m} \Delta \ln f(r)$ is an effective potential /26/. The resulting Euler-Lagrange equation /26,27/ for the radial distribution function g(r) may be written as a zero-energy Schrödinger equation /28/

$$-\frac{\hbar^2}{m} \Delta g^{1/2}(r) + \{v_T(r) + w(r)\} g^{1/2}(r) = 0 \tag{14}$$

The induced potential w(r) is generated by the structure of the medium and must be evaluated selfconsistently,

$$w(r) = -\frac{1}{(2\pi)^3} \frac{\hbar^2}{4m\rho} \int k^2 \{ 2\, S(k)+1 \}\{ S(k)-1\}^2\, S^{-2}(k)\, e^{i\underline{k}\cdot\underline{r}} d\underline{k} \ . \tag{15}$$

Thereupon, the pressure p may be expressed by the optimal structure function /27/,

$$p/\rho = E - \frac{1}{(2\pi)^3} \frac{\hbar^2}{8m\rho} \int k^2 \{S(k)-1\}^3\, S^{-2}(k)\, d\underline{k} \quad . \tag{16}$$

Once the radial distribution function is available the one-body density matrix n(r), the condensate fraction n and the pairing function $\chi(r)$ — n(r), $\chi(r)$ are the Fourier-inverse of functions (2) and (4), respectively — may be calculated by employing established hypernetted-chain procedures /24,20/.

Some of our numerical results on various quantities for a system of completely spin-aligned H atoms interacting via the K-W triplet potential /10-12/ are presented in Figures 1-3. Figs. 1 and 2 depict the radial distribution function and the pairing function normalized to unity as $r \to \infty$ at $\varepsilon = 0$ and densities $\rho = 0.5 \cdot 10^{-4}$ Å^{-3} and $\rho = 2 \cdot 10^{-3}$ Å^{-3}, respectively. In the dilute system the radius r_o of the specific volume being large compared with the range of the potential, the effects of the medium are small and, consequently, the square of the pairing function agrees very well with the radial distribution function (Fig. 1). In contrast, significant differences between both functions develop at intermediate (and high) densities (Fig. 2).

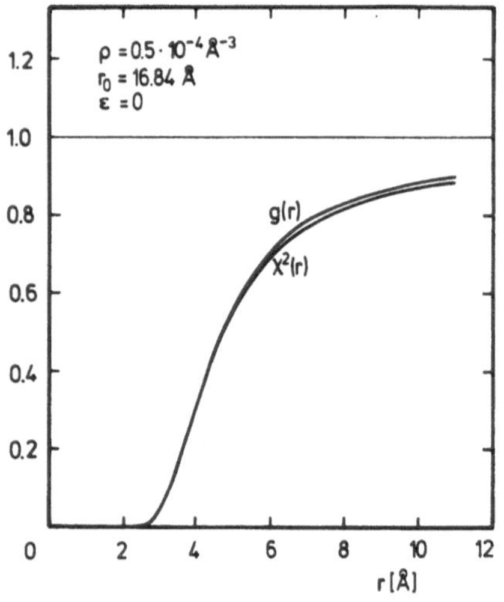

Figure 1: Radial distribution function and square of the pairing function for dilute spin-aligned hydrogen.

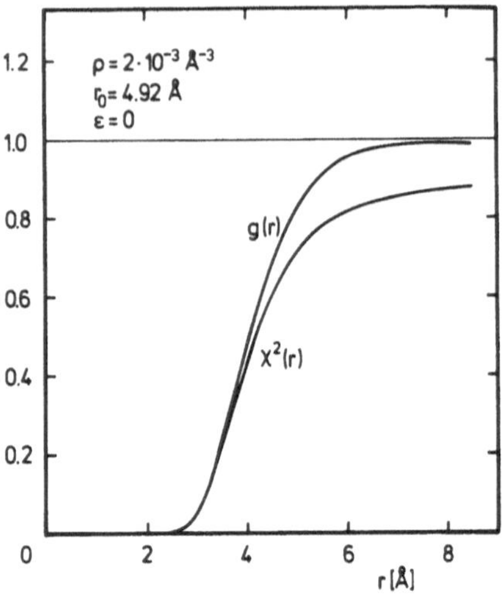

Figure 2: Same as Figure 1 but at intermediate density.

With the function g(r) as input we have calculated the equation of state. Figure 3 displays the energy per particle E, the pressure divided by density p/ρ, and the pairing energy E_p associated with the pairing function /24/ at intermediate densities. For the dilute gas, $\rho < 10^{-4}$ Å^{-3}, the numerical results on these quantities merge correctly into the linear behavior E_o described by eq. (11) with a = 0.72 Å. The dashed line E_1 indicates the energy per particle in a next order of perturbation theory taken into account in Ref. /16,17/. There is rough numerical agreement between the pairing energy and the energies E_o or E_1 over the density range considered. However, the variational result E on the energy lies considerably above the approximations E_1 and E_o indicating the breakdown of the approach adopted in Refs. /16,17/ at such densities. To check the numerical accuracy of our calculations we have exploited the virial theorem which provides an alternative formula for evaluating the pressure

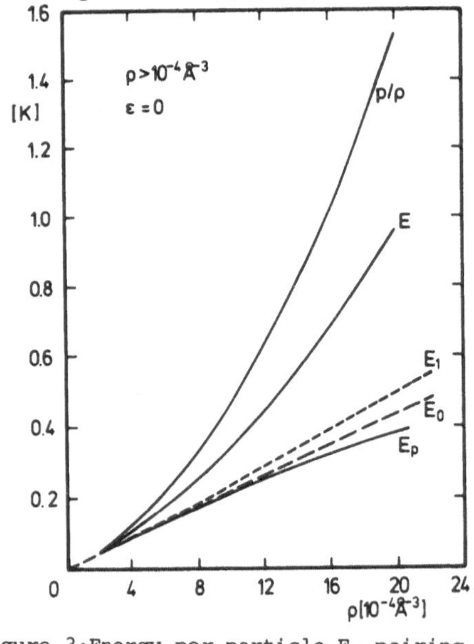

Figure 3: Energy per particle E, pairing energy E_p and pressure per density as functions of density for spin-aligned hydrogen.

/27/. Adopting the analytic Lennard-Jones potential /12/ we found excellent agreement between the results of the two independent methods.

4. Finite magnetic fields.- At finite magnetic fields $\underline{B}$ two hydrogen atoms can also interact in the singlet state, its probability being controlled by hyperfine mixing in the lowest single-particle state (9). To take account of the spatial correlations between and among the hydrogen atoms induced by the potential (1) we generalize the wave function (12),

$$\underline{\psi}_A = N^{-1/2} \prod_{i<j}^{A} f(r_{ij}) \prod_{i=1}^{A} \psi_o(i) \quad , \tag{17}$$

where the single-particle spin function $\psi_o(i)$ represents state (9) occupied by atom i. Ansatz (17) may be considered as the Bose analog of the Jastrow-Slater ansatz /29,20/ for fermions of spin 1/2. In a next step we could improve expression (17) by allowing spin-dependent spatial correlations /30,31/.

Taking the expectation value of the total hamiltonian with respect to the correlated state (17) we get the total energy $E_{tot} = E + E_Z + E_H$. Zeeman- and hyperfine-energy are of the form

$$E_Z = \langle O|-\underline{\mu}\cdot\underline{B}|O\rangle = -\mu B(1-\varepsilon^2)(1+\varepsilon^2)^{-1}$$

$$E_H = \langle O|\alpha\underline{\sigma}\cdot\underline{\tau}|O\rangle = -\alpha(1+4\ \varepsilon+\varepsilon^2)(1+\varepsilon^2)^{-1} \quad . \tag{18}$$

The internal energy E may be evaluated just as described in Section 3. However, the triplet potential v_T in eqs. (13),(14) must be replaced by an averaged potential,

$$v(r,\varepsilon) = \langle OO|v(12)|OO\rangle$$

$$= v_T(r) + \{v_s(r) - v_T(r)\}\varepsilon^2(1+\varepsilon^2)^{-2} \quad , \tag{19}$$

where the admixture ε plays the role of an auxiliary parameter.

Figure 4 displays our results on the optimal energy per particle E, pairing energy E_p and pressure p/ρ as functions of the parameter ε at given density, $\rho = 0.5\cdot 10^{-4}\ \text{Å}^{-3}$. As we expect for a dilute system the curves almost coincide. We are therefore permitted to describe the properties of state (17) at such densities within elementary effective range theory /16/. This statement is supported by the (almost) linear dependence of the energy E (and E_p) on the density at any fixed value ε in the dilute region. Our numerical results are very well represented by eq. (11) with an effective scattering length a = a(ε) being a function of polarization (Figure 5). At intermediate densities quantities E, E_p and p/ρ are of differing magnitude,

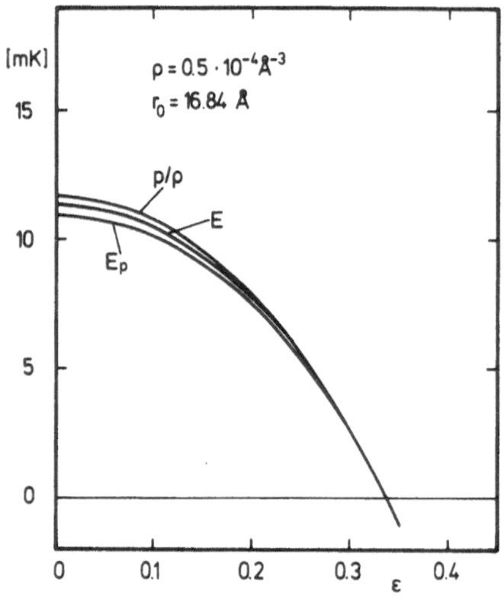

Figure 4: Energy per particle E, pairing energy Ep and pressure per density as functions of admixture parameter ε for a dilute system of atomic hydrogen.

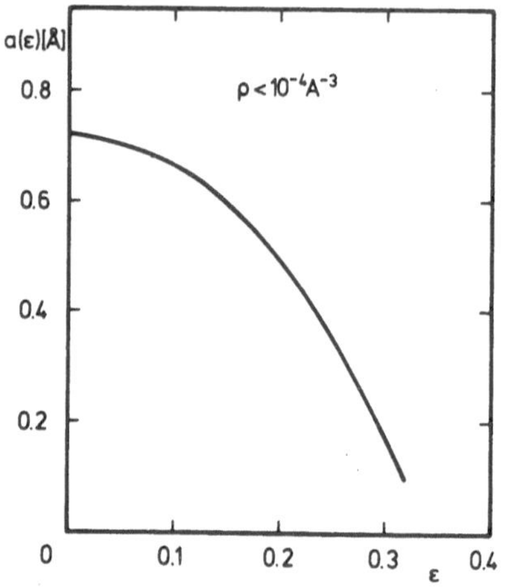

Figure 5: Effective scattering length as function of admixture parameter ε (polarization).

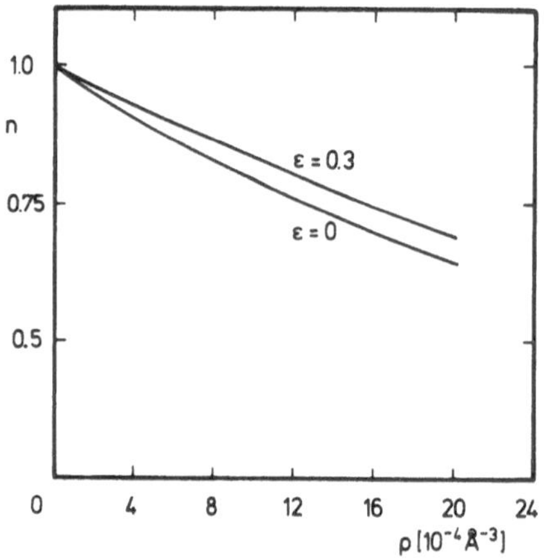

Figure 6: Condensate fraction for atomic hydrogen as function of density at given spin-polarization.

although they change sign at about the same value of ε. The condensate fraction n may be extracted from the distribution (2) at zero momentum or from function n(r) /24/. Numerical results based on the optimal choice (17) are displayed in Figure 6.

In equilibrium the polarization is determined by the magnetic field and the density of the system. This magnetic equation of state may be derived from the condition $dE_{tot}/d\varepsilon = 0$. It is elementary to evaluate the derivatives of the analytic expressions (18). For the derivative of the internal energy we get

$$dE = -\varepsilon(1-\varepsilon^2)(1+\varepsilon^2)^{-3} I(\rho,\varepsilon)d\varepsilon \, . \tag{20}$$

To arrive at eq. (20) we exploit eq. (13) and the variational property $\delta E/\delta g = 0$ which holds for the optimal function $g(r)$. Quantity $I(\rho,\varepsilon)$ defines the exchange integral

$$I(\rho,\varepsilon) = \rho \int \{v_T(r) - v_S(r)\} g(r) d\underline{r} \tag{21}$$

depending linearly on density ρ for sufficiently dilute systems, $I(\rho,\varepsilon) \sim I_o(\varepsilon)\rho$. Numerically we find that the linear approximation is excellent at densities $\rho < 10^{-4}$ Å^{-3} and admixtures $\varepsilon < 0.3$. It breaks down quickly for increasing densities above 10^{-4} Å^{-3}. Quantity $I_o(\varepsilon)$ increases rapidly with increasing parameter $\varepsilon > 0.1$ (Figure 7). Employing the explicit expressions for the derivatives of functions E, E_Z, E_H with respect to the parameter ε we find the following magnetic equation of state:

$$I(\rho,\varepsilon) = 4\alpha(1+\varepsilon^2)(1-\varepsilon^2)^{-1} \left\{\frac{\mu B}{\alpha} - \frac{1}{\varepsilon}(1-\varepsilon^2)\right\} \, . \tag{22}$$

For small values of the parameter ε eq. (22) reduces to the expression

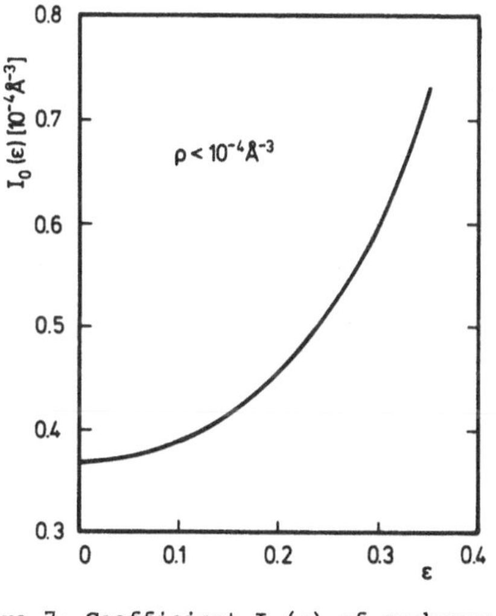

Figure 7: Coefficient $I_o(\varepsilon)$ of exchange integral as function of admixture parameter ε for a dilute gas of hydrogen atoms.

$\alpha/\varepsilon = \mu B - I(\rho,0)/4$. To achieve high spin-alignment we therefore need external fields $\underline{B}$ with $B >> I(\rho,0)/4\mu$. We may further conclude from eq. (22) that $\mu B/\alpha \geq (1-\varepsilon^2)\varepsilon^{-1}$ since $I \geq 0$ ($I = 0$ at $\rho = 0$). Consequently,

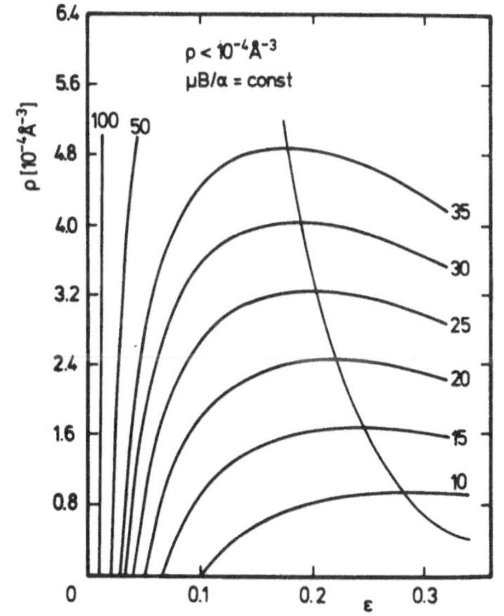

Figure 8: Phase diagram of atomic hydrogen. The density is plotted versus the admixture parameter ε at given external magnetic field B. The system is unstable in the region to the right of the separation curve where $\delta\rho/\delta\varepsilon \leq 0$.

for given magnetic field the maximum spin-alignment in the interacting medium is determined by eq. (10) which is realized at low densities. The general properties of the magnetic equation of state may be conveniently grasped by plotting the density as a function of polarization at given external field (Figure 8). The various curves are parametrized by the ratio $\mu B/\alpha$ employing the hyperfine constant α = 0.017K taken from Ref. /25/. Let us select, for discussion, the results at $\mu B/\alpha = 30$. At zero density the polarization P_Z assumes its highest possible value. With increasing density the admixture of "wrong" spins measured by ε increases very slowly up to densities $\sim 3\cdot 10^{-4}$ Å^{-3}. Then the equation of state becomes "weak" and we reach rapidly a point where $\partial\varepsilon/\partial\rho$ is singular (the maximum of this curve). Beyond this point the equilibrium data represent only the results corresponding to an unstable state (17). Thus, the curve connecting these points separates the region of static instability from a region characterized by $\partial\rho/\partial\varepsilon > 0$ (or by $\partial B/\partial\varepsilon < 0$). This inequality is a necessary condition for the stability (metastability) of the trial state adopted.

Acknowledgements:- We thank T. Chakraborty and S. Fantoni for several discussions and valuable assistance at various stages of the numerical computation. The financial support of the Deutsche Forschungsgemeinschaft under Grant Ri 267/5 is gratefully acknowledged.

Footnote and References:

+present address: Institute of Theoretical Physics, Academia Sinica, P.O. Box 2735, Peking, Peoples Republic of China.

/1/ J.F. Silvera and J.T.M. Walraven, Phys. Rev. Lett. 44 (1980) 164.
/2/ J.T. Walraven, J.F. Silvera and A.P.M. Matthey, Phys. Rev. Lett. 45 (1980) 449.
/3/ R.W. Cline, D.A. Smith, T.J. Greytak and D. Kleppner, Phys. Rev. Lett. 45 (1980) 2117.
/4/ W.C. Stwalley and L.H. Nosanow, Phys. Rev. Lett. 36 (1976) 910.
/5/ L.H. Nosanow, J. Phys. (Paris) 41 (1980) C7-1.
/6/ L.H. Nosanow, this conference.
/7/ J.F. Silvera and V.V. Goldman, Phys. Rev. Lett. 45 (1980) 915.
/8/ J.F. Silvera, Phys. Rev. Lett. 45 (1980) 1268.
/9/ M. Morrow, R. Jochemsen, A.J. Berlinsky and W.N. Hardy, Phys. Rev. Lett. 46 (1981) 195.
/10/ W. Kolos and L. Wolniewicz, J. Chem. Phys. 43 (1965) 2429.
/11/ W. Kolos and L. Wolniewicz, Chem. Phys. Lett. 24 (1974) 457.
/12/ M.D. Miller, L.H. Nosanow, Phys. Rev. B 15 (1977) 4376.
/13/ E.D. Siggia and A. E. Ruckenstein, Phys. Rev. Lett. 44 (1980) 1423.
/14/ J.H. Freed, J. Chem. Phys. 72 (1980) 1414.
/15/ M.D. Girardeau, J. Math. Phys. 16 (1975) 1901.
/16/ D.G. Friends and R.D. Etters, J. Low Temp. Phys. 39 (1980) 409.
/17/ Y.H. Uang and W.C. Stwalley, J. Phys. (Paris) 41 (1980) C7-33.
/18/ L.J. Lantto and R.M. Nieminen, J. Phys. (Paris) 39 (1978) C6-106.
/19/ L.J. Lantto and R.M. Nieminen, J. Low Temp. Phys. 37 (1979) 1.
/20/ M.L. Ristig, in From nuclei to particles, International School of Physics 'Enrico Fermi', Varenna 1980.
/21/ R.D. Etters, J.V. Dugan and R.W. Palmer, J. Chem. Phys. 62 (1975) 313.
/22/ R.L. Danilowicz, J.V. Dugan and R.D. Etters, J. Chem. Phys. 65 (1976) 498.
/23/ P. Entel and J. Anlauf, preprint.
/24/ M.L. Ristig and P.M. Lam, Nucl. Phys. A328 (1979) 267.

/25/ W.J. Mullin, Phys. Rev. Lett. 44 (1980) 1420.
/26/ E. Feenberg, Theory of quantum fluids (Academic, New York, 1969).
/27/ K. Hiroike, Prog. Theor. Phys. 27 (1962) 342.
/28/ L.J. Lantto and P.J. Siemens, Phys. Lett. 68B (1977) 308.
/29/ J.W. Clark, in Progress in Particle and Nuclear Physics, edited by D.H. Wilkinson (Pergamon, Oxford, 1979) Vol. II.
/30/ M.L. Ristig, K.E. Kürten and J.W. Clark, Phys. Rev. B19 (1978) 3539.
/31/ A.J. Berlinsky, Phys. Rev. Lett. 39 (1977) 359.

THE BORN-GREEN-YVON THEORY OF FERMION QUANTUM FLUIDS

C. E. Campbell and K. E. Kürten
School of Physics and Astronomy
University of Minnesota
Minneapolis, Minnesota 55455, U. S. A.
and
E. Krotscheck
Department of Physics, SUNY
Stony Brook, New York 11794, U. S. A.

I. INTRODUCTION

Much of the recent progress in the microscopic theory of fermion quantum fluids has been accomplished using the Feenberg-Jastrow-Correlated Basis Function theory [1-3]. Implementation of this theory was advanced significantly by the development of the fermion generalization of the hypernetted chain method (FHNC) [4] for calculating matrix elements between Slater-Jastrow functions Ψ_n, which are written in the form

$$\Psi_n = \prod_{i<j}^{N} e^{\frac{1}{2}u(r_{ij})} \, \Phi_n(r_1\sigma_1 \cdots r_N\sigma_N) \tag{1}$$

where Φ_n is some Slater determinant. The most common task is to estimate the ground state energy by choosing Φ_o to be in non-interacting ground state and then minimizing the expectation value of the Hamiltonian with respect to u(r).

Recently we pointed out that one can obtain some new insights into the problem of calculating this expectation value by developing the fermion generalization of the Born-Green-Yvon method (FBGY) as an alternative to the FHNC method. [5,6] The most illuminating result obtained in this way is that the FBGY method is the "natural" way to find approximations which preserve the identity of the several alternative expressions frequently employed for the kinetic energy. Moreover we find that the numerical accuracy of the simplest approximation within this scheme is better than the FHNC/0 approximation while being no more difficult to implement. There are some disadvantages to the FBGY method, however, which are already familiar from work using HNC and BGY on comparable boson problems.

II. FERMION BGY EQUATIONS

The potential and kinetic energy of the Slater-Jastrow trial ground state Ψ_o can

---

* Research supported in part by NSF grant DMR-7926447 and by the Deutsche Forschungsgemeinschaft.

be expressed in terms of the two- and three-body distribution functions [1-3]:

$$E = \langle\Psi_o|H|\Psi_o\rangle/\langle\Psi_o|\Psi_o\rangle = \langle T\rangle + \langle V\rangle \tag{2}$$

where

$$\langle V\rangle = N\rho\int g(r)V(r)d^3r \tag{3}$$

$$\begin{aligned}\langle T\rangle = T_f + N\rho\int g(r)\frac{\hbar^2}{8m}[\vec{\nabla}_1 u(r)]^2 d^3r \\ + \rho^3\int g_3(\vec{r}_1\vec{r}_2\vec{r}_3)\frac{\hbar^2}{8m}\vec{\nabla}_1 u(r_{12})\cdot\vec{\nabla}_1 u(r_{13})d^3r_1 d^3r_2 d^3r_3 \equiv T_{cw}\end{aligned} \tag{4}$$

where ρ is the number density, V(r) is the two-body potential, $g(r_{12}) = g(\vec{r}_1,\vec{r}_2)$ is the radial distribution function, T_{cw} is the Clark-Westhaus (CW) expression for the kinetic energy, T_f is the free Fermi gas energy, and $g_n(\vec{r}_1...\vec{r}_n)$ is the n body distribution function defined for Ψ_o:

$$g_n(\vec{r}_1\cdots\vec{r}_n) = \frac{N!}{(N-n)!\,\rho^n\langle\Psi_o|\Psi_o\rangle}\int d^3r_{n+1}\cdots d^3r_N\,|\Psi_o|^2 \tag{5}$$

In Eq. (5) and throughout the remainder of this paper we suppress spin and isospin variables.

These equations also apply to the boson system, except that Ψ_o should then be a boson function so that the Slater determinant in Ψ_o is replaced by a constant. Thus T_f vanishes in that case, and <u>the relationship between g_n and u(r) differs from the fermion case.</u> It is an accurate representation of this relationship (particularly for n = 2) in the fermion case which is the subject of both the FHNC and FBGY models.

The BGY hierarchy of equations is produced by operating with $\vec{\nabla}_1$ on the definitions of g_n in Eq. (5). In the boson-Jastrow case, where $\Psi_o^2 = \exp\sum_{i<j}u(r_{ij})$, the equation for $\vec{\nabla}_1 g_n$ depends in a simple way on $\vec{\nabla}_1 u$, g_n, and g_{n+1}, the simplicity of the result resting on the fact that Ψ_o^2 has the simple exponentiated form. Though this simplicity does not occur in the fermion case, it was pointed out by Gaudin, Gillespie and Ripka [7] that the square of the Slater-Jastrow function has a relatively simple dependence on one additional two-body function $\ell(r_{12})$:

$$|\Psi_o|^2 = \rho^N\prod_{i<j}^{N} e^{u(r_{ij})}\,\mathrm{Det}\,\ell(\vec{r}_m,\vec{r}_n) \tag{6}$$

where

$$\ell(\vec{r}_m,\vec{r}_n) = \rho^{-1}\sum_{\alpha=1}^{N}\varphi_\alpha^*(\vec{r}_m)\varphi_\alpha(\vec{r}_n) \tag{7}$$

the sum over α being over the N occupied orbits ϕ_α in the Slater determinant. Consequently, $\vec{\nabla}_1 g(r_{12})$ can only have terms proportional to $\vec{\nabla}_1 u(r_{12})$, $\vec{\nabla}_1 \ell(r_{12})$, $\vec{\nabla}_1 u(r_{1j})$, and $\vec{\nabla}_1 \ell(r_{1j})$, $j \neq 1, 2$:

$$\vec{\nabla}_1 g(r_{12}) = g(r_{12})\vec{\nabla}_1 u(r_{12}) + \rho\int d^3r_3\, g_3(\vec{r}_1\vec{r}_2\vec{r}_3)\vec{\nabla}_1 u(r_{13})$$
$$-2\Gamma_{cc}(r_{12})\vec{\nabla}_1 \ell(r_{12}) - 2\rho\int d^3r_3\, Z_{cec}(\vec{r}_1\vec{r}_2;\vec{r}_3)\vec{\nabla}_1 \ell(r_{13}) \quad . \tag{8}$$

The first line of Eq. (8) is the boson BGY equation for g(r), while the second line introduces a new two-body function Γ_{cc} and three-body function Z_{cec}. (The choice of subscripts is explained below.) Aside from the usual question of how to obtain a closed expression for g from Eq. (8) by expressing g_3 (and now Z_{cec}) as a functional of g, this fermion equation requires at least one more equation to produce the two-body function Γ_{cc}. Both of these tasks are accomplished by making use of some results obtained from the fermion generalization of the Ursell-Mayer diagrammatic analysis of $g(r_{12})$. The diagrammatic elements are dynamical bonds $h(r_{ij}) \equiv \exp u(r_{ij}) - 1$ and statistical (or exchange) bonds $\ell(r_{ij})$. The diagram rules for the correlation lines are identical to the boson (or classical) case. The rules for the exchange bonds follow from the fact that Det $\ell(r_{mn})$ can be written as the sum of the product of all exchange polygons, with every coordinate $\vec{r}_i$ having exactly one exchange bond entering and one (possibly the same) exchange bond leaving. Each ℓ carries a minus sign, and every exchange cycle carries a minus sign, including $(\vec{r}_i, \vec{r}_i)$. Finally, because of the Pauli exclusion principle, $\ell(r_1,r_2)$ satisfies a convolution property

$$\rho\int \ell(\vec{r}_1,\vec{r}_3)\,\ell(\vec{r}_3,\vec{r}_2)\,d^3r_3 = \ell(\vec{r}_1,\vec{r}_2) \tag{9}$$

which has the diagrammatic consequence that no diagram appears which contains an internal point (integration point) which is <u>only</u> exchange correlated with other particles (known as the reduction to equivalent diagrams). Then $g(r_{12}) - 1$ can be shown to be the sum of all diagrams constructed from dynamical and statistical bonds containing two external points (1 and 2), in which all internal points are at least singly connected to the remainder of the diagram. [2] It is then most convenient to regroup these diagrams in accord with the exchange character of their external points [2]:

$$g(r_{12}) - 1 = \Gamma_{dd}(r_{12}) + \Gamma_{de}(r_{12}) + \Gamma_{ed}(r_{12}) + \Gamma_{ee}(r_{12}) \tag{10}$$

where an external point labelled d is a direct point (no exchange bond with any other point) and one labelled e is connected by an exchange bond to some other point.

The function $\Gamma_{ee}(r_{12})$ includes among others all of the terms in $g(r_{12})$ which are proportional to $\ell(\vec{r}_1,\vec{r}_2)$ or $\ell(\vec{r}_2,\vec{r}_1)$, which are convenient to accumulate as

$$-\ell(\vec{r}_1,\vec{r}_2)\Gamma_{cc}(\vec{r}_{21}) - \ell(\vec{r}_2,\vec{r}_1)\Gamma_{cc}(\vec{r}_{12}) + \ell(\vec{r}_2,\vec{r}_1)\ell(\vec{r}_1,\vec{r}_2)g_{dd}(r_{12}) \tag{11}$$

where $\Gamma_{cc}(\vec{r}_{12})$ can be thought of as a dressed exchange line between 1 and 2, $g_{dd} = \Gamma_{dd} + 1$, and the last term in (11) cancels the overcounting of the terms proportional to $\ell(\vec{r}_1,\vec{r}_2)\ell(\vec{r}_2,\vec{r}_1)$ in the first two terms of (11).

The $\nabla_1\ell(r_{12})$ term in the FBGY equation for $g(r_{12})$ (Eq. (8)) is obtained from Eq. (11) by operating with ∇_1 on only the factors $\ell(\vec{r}_1,\vec{r}_2)$ and $\ell(\vec{r}_2,\vec{r}_1)$. The factor of 2 in Eq. (8) is due to the numerical equivalence of the two directions around an exchange cycle (the first two terms in (11) are also equivalent), and the effect of overcounting cancels the contributions of the last term in (11).

The FBGY equation for Γ_{cc} is obtained by operating with ∇_1 on $\Gamma_{cc}(r_{12})$, the general structure of which is

$$\begin{aligned}\vec{\nabla}_1\Gamma_{cc}(r_{12}) &= \Gamma_{cc}(r_{12})\vec{\nabla}_1 u(r_{12}) + g_{dd}(r_{12})\vec{\nabla}_1\ell(r_{12}) \\ &\quad + \rho\int d^3r_3\left[\Gamma_{cce}(\vec{r}_1\vec{r}_2\vec{r}_3)\vec{\nabla}_1 u(r_{13}) - 2Z_{dcc}(\vec{r}_1\vec{r}_2;\vec{r}_3)\vec{\nabla}_1\ell(r_{13})\right] .\end{aligned} \tag{12}$$

The first term comes from the exponentiated form of the Jastrow factor in Ψ_0^2, while the second term comes from the term proportional to $\ell(r_{12})\ell(r_{21})$ in Γ_{cc}. Note, however, that since the external points in $\Gamma_{dd}(r_{12})$ are not exchange correlated with any point, only $\vec{\nabla}u$ terms appear in the Γ_{dd} equations:

$$\begin{aligned}\vec{\nabla}_1\Gamma_{dd}(r_{12}) &= g_{dd}(r_{12})\vec{\nabla}_1 u(r_{12}) \\ &\quad + \rho\int d^3r_3\left[\Gamma_{ddd}(\vec{r}_1\vec{r}_2\vec{r}_3) + \Gamma_{dde}(\vec{r}_1\vec{r}_2\vec{r}_3)\right]\vec{\nabla}_1 u(r_{13}) .\end{aligned} \tag{13}$$

The three-body functions which appear in these FBGY equations for g,Γ_{cc}, and Γ_{dd} are appropriately defined components of the three-body distribution function g_3. In particular, the functions $\Gamma_{\alpha\beta\gamma}(\vec{r}_1\vec{r}_2\vec{r}_3)$ which multiply $\nabla_1 u(r_{13})$ are those parts of g_3 for which the exchange character of the external points 1, 2, and 3 is denoted by α, β and γ, respectively. Moreover, by the fermion generalization of the Abe-Stell cluster expansion for g_3 [8], it can be demonstrated that these three-body functions can be expressed as functionals of the dressed bonds $\Gamma_{\alpha\beta}$, where $\alpha\beta$ = (dd,de,ed,ee,and cc). As a consequence, the FBGY equations for Γ_{de} and Γ_{ed} are also needed, which by arguments similar to those above have the form:

$$\vec{\nabla}_1\Gamma_{de}(r_{12}) = \Gamma_{de}(r_{12})\vec{\nabla}_1 u(r_{12}) + \rho\int\Gamma_{dee}(\vec{r}_1\vec{r}_2\vec{r}_3)\vec{\nabla}_1 u(r_{13})\,d^3r_3 \tag{14}$$

$$\begin{aligned}\vec{\nabla}_1\Gamma_{ed}(r_{12}) &= \Gamma_{ed}(r_{12})\vec{\nabla}_1 u(r_{12}) + \rho\int\Gamma_{ede}(\vec{r}_1\vec{r}_2\vec{r}_3)\vec{\nabla}_1 u(r_{13})\,d^3r_3 \\ &\quad - 2\rho\int Z_{cdc}(\vec{r}_1\vec{r}_2;\vec{r}_3)\vec{\nabla}_1\ell(r_{13})\,d^3r_3 .\end{aligned} \tag{15}$$

(Since $\Gamma_{de} = \Gamma_{ed}$, only one of these two equations is necessary. For purposes of introducing approximations later, it is best for that single equation to be the sum of these two equations.) Because of Eq. (10), a separate equation for Γ_{ee} is unnecessary.

To define the coefficients $Z_{\alpha\beta c}(\vec{r}_1,\vec{r}_2;\vec{r}_3)$ of $\nabla_1 \ell(r_{13})$, it should be noted that point 3 bears the same (above mentioned) restrictions as an internal point, and thus $Z_{\alpha\beta c}$ must not contain diagrams connected to point 3 only by an exchange line. Moreover, the convolution relationship between $\Gamma_{cc}(r_{12})$ and its non-nodal part, $X_{cc}(r_{12})$:

$$\Gamma_{cc}(r_{12}) = \ell(r_{12}) + X_{cc}(r_{12}) - \rho \int d^3r_3\, X_{cc}(r_{13})\, \Gamma_{cc}(r_{32}) \tag{16}$$

combined with the convolution property of ℓ(Eq. (9)) produces a similar property for this dressed statistical bond

$$\rho \int \Gamma_{cc}(r_{13})\, \ell(r_{32})\, d^3r_3 = \ell(r_{12}) \tag{17}$$

This produces a further reduced property for internal c points such as $\vec{r}_3$ in $Z_{\alpha\beta c}$, namely that they are non-nodal with respect to point 3. Thus, $Z_{\alpha\beta c}$ is that part of g_3 with an $\alpha\beta c$ exhange characteristic in the external points and non-nodal in point 3. Indeed $Z_{\alpha\beta c}$ is related to $\Gamma_{\alpha\beta c}$ by a three-body nodal equation similar to Eq. (16).

This completes the derivation of the FBGY equations, leaving only a specification of a tractable (approximate) expression for the three-body functions in terms of the two-body functions in order to achieve closure. This is discussed below.

III. THE KINETIC ENERGY

The Clark-Westhaus (CW, Eq. (4)), Jackson-Feenberg (JF), and Pandharipande-Bethe (PB) expressions for the kinetic energy are all equivalent through integrations by part [2]. We find another useful expression for the kinetic energy, $\bar{T}$:

$$\begin{aligned}\bar{T} = T_{F} &+ \frac{\hbar^2\rho^2}{8m}\int d^3r_1 d^3r_2 \left\{ g(r_{12})\nabla^2 u(r_{12}) - 2\,\Gamma_{cc}(r_{12})\,\vec{\nabla}_1 u(r_{12})\cdot\vec{\nabla}_1 \ell(r_{12}) \right\} \\ &+ \frac{\hbar^2\rho^3}{8m}\int d^3r_1 d^3r_2 d^3r_3\, Z_{cec}(\vec{r}_1,\vec{r}_2;\vec{r}_3)\,\nabla_1 u(r_{12})\cdot\nabla_1 \ell(r_{13})\end{aligned} \tag{18}$$

which is related to T_{CW} and T_{PB} by

$$\bar{T} = \frac{1}{2}\left(T_{CW} + T_{PB}\right) \tag{19}$$

but has the sometimes useful property that the three-body part is simpler than the three-body part in T_{CW} and T_{PB}, and frequently numerically smaller. It is noteworthy that T_{CW}, T_{PB} and $\bar{T}$ are explicitly equivalent by virtue of the single FBGY equation for $\nabla g(r_{12})$ (Eq. 8)), and this equivalence is preserved for any approximation for Γ_{cc}, g_3 and Z_{cec} used in Eq. (8). However, the Jackson-Feenberg energy depends on quantities not appearing in the FBGY equations:

$$T_{JF} = T_F - \frac{\hbar^2\rho^2}{8m}\int d^3r_1\, d^3r_2 \left[g(r_{12})\nabla^2 u(r_{12}) - 2\Gamma_{dd}(r_{12})\,\vec{\nabla}_1 \ell(r_{12})\cdot\vec{\nabla}_1 \ell(r_{12}) - 2X_{cc}(r_{12})\,\nabla^2 \ell(r_{12}) \right] + \frac{\hbar^2\rho^3}{4m}\int d^3r_1\, d^3r_2\, d^3r_3\, Y_{dcc}(\vec{r}_1;\vec{r}_2\,\vec{r}_3)\,\vec{\nabla}_1 \tag{20}$$

where $Y_{dcc}(\vec{r}_1;\vec{r}_2,\vec{r}_3)$ is the dcc part of $g_3(\vec{r}_1,\vec{r}_2,\vec{r}_3)$ which is non-nodal in points 2 and 3. Thus generally $T_{JF} \neq (T_{PB}, T_{CW}, \bar{T})$ for "natural" approximations within the FBGY method.

The simplest "natural" approximation is the fermion generalization of the Kirkwood superposition approximation, used for all three-body functions in the FBGY equations and the kinetic energies. In addition to producing a T_{JF} which differs from the other kinetic energies, the Γ_{cc} solution to these equations does not satisfy the Γ_{cc} convolution property, Eq. (17), used to derive the equations. More seriously, this latter failure has the numerical consequence that there are solutions to the equations for liquid ^{3}He only at densities $\rho < 0.014\ \text{Å}^{-3}$, somewhat below the physical range of interest.

A nodal analysis of Γ_{cce} and Z_{dcc} shows that the only approximations which satisfy the cc convolution property and also satisfy the "hard" core constraint that $\Gamma_{cc}(r)$ vanish at small r are equivalent to making approximations for the elementary diagrams in the FHNC equation for Γ_{cc}(only). This same analysis shows that the non-nodal cc function X_{cc} satisfies an FBGY equation

$$\vec{\nabla}_1 X_{cc}(r_{12}) = \Gamma_{cc}(r_{12})\vec{\nabla}_1 u(r_{12}) + \Gamma_{dd}(r_{12})\vec{\nabla}_1 \ell(r_{12}) + \rho\int d^3r_3 \left[Z_{cec}(\vec{r}_1\,\vec{r}_3;\vec{r}_2)\vec{\nabla}_1 u(r_{13}) - Y_{dcc}(\vec{r}_1;\vec{r}_2\,\vec{r}_3)\nabla_1 \ell(r_{13}) \right] \tag{21}$$

which is precisely the equation which is needed to demonstrate the equivalence of T_{JF} and the other kinetic energies.

IV. RESULTS AND DISCUSSION

Instead of deducing an elementary diagram approximation for Γ_{cce} and Z_{dcc}, we have used a hybrid approximation whereby the FHNC/0 approximation is used for Γ_{cc} (thus preserving the all-important convolution property), but the superposition approximation is used for all three-body functions which appear in the FBGY equations for g, Γ_{dd}, and $\Gamma_{de} + \Gamma_{ed}$, and also in the evaluation of the kinetic energy. This preserves the equivalence of T_{CW}, T_{PB}, and $\bar{T}$, though they are no longer equal to T_{JF}. The difference between T_{JF} and the others is some indication of the effect of the additional approximation .

This hybrid approximation was tested on liquid ^{3}He by comparing to the recent Monte Carlo variational results of Levesque [9]. The maximum difference between our

T_{JF} and $\bar{T}$ occurred at the highest density calculated, $\rho = 0.0197\ \mathring{A}^{-3}$, and was 0.06K per particle out of 15K total. Our results for the energy per particle are compared to the Monte Carlo results in the Figure, where it is seen that they are rather good at low and intermediate density, but are too negative at the higher densities. However, in comparison to other approximation schemes, the only one which is significantly better than the present results is the FHNC/4 result of Zabolitzky obtained using the PB kinetic energy and an improvement over the superposition approximation for the three-body functions [10]. On the other hand it should also be noted that other BGY based calculations agree rather well with ours when they employ the JF kinetic energy [11,12].

The relationship of the BGY, HNC and Monte Carlo results in these fermion calculations is very similar to the relationship in the boson results, namely, BGY based approximations tend to underestimate the kinetic energy and HNC based approximations using the JF kinetic energy overestimate it. [13] Moreover, the FBGY equations with the superposition approximation almost certainly share with the corresponding boson equations the unfortunate feature of behaving poorly under an Euler-Lagrange analysis, in contrast to HNC-JF analysis. [13]

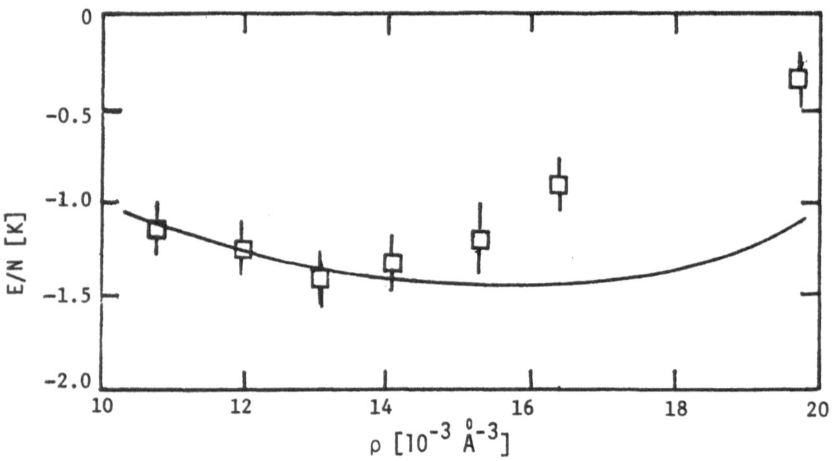

Figure - Energy per particle as a function of density for liquid ^3He obtained using a Jastrow function with the variational parameter determined by Levesque [9]. The Monte Carlo results of Ref. 9 are indicated by the boxes with error bars, while the solid line is the FBGY results obtained using the hybrid approximation discussed in the text.

References

1. E. Feenberg, Theory of Quantum Fluids (Academic Press, 1969); C.-W. Woo, Chapt. 5, The Physics of Liquid and Solid Helium, Part I ed. K.H. Bennemann and J. B. Ketterson (Wiley, 1976); C. E. Campbell, Chapt. 6, Progress in Liquid Physics ed. C. A. Croxton (Wiley, 1978); C. E. Campbell and F. J. Pinski, Nucl. Phys. A328, 210 (1979).
2. J. W. Clark in Progress in Nuclear and Particle Physics ed. D. H. Wilkinson (Pergamon, Oxford, 1979), vol. 2.
3. V. R. Pandharipande and R. B. Wiringa, Rev. Mod. Phys. 51, 821 (1979).
4. S. Fantoni and S. Rosati, Nuovo Cim. 25A, 593 (1975); E. Krotscheck and M. L. Ristig, Nucl. Phys. A242, 389 (1975).
5. E. Krotscheck, Z. Physik B32 395 (1979).
6. C. E. Campbell and K. E. Kürten, Stony Brook Workshop on High Density Matter (1979) (unpublished).
7. M. Gaudin, J. Gillespie and G. Ripka, Nucl. Phys. A176, 237 (1971).
8. R. Abe, Prog. Theor. Phys. 21, 421 (1959); G. Stell, Physics 29, 517 (1963).
9. D. Levesque, Phys. Rev. B12, 5159 (1980).
10. J. G. Zabolitzky, Phys. Rev. A16, 1258 (1977).
11. M. D. Miller, Phys. Rev. B14, 3937 (1976).
12. E. Krotscheck, Z. Physik B33, 403 (1979).
13. R. A. Smith, A. Kallio, M. Puoskari and P. Toropainen, Nucl. Phys. A328, 186 (1979).

STATISTICAL THEORIES OF LARGE ATOMS AND MOLECULES

Elliott H. Lieb
Departments of Mathematics and Physics
Princeton University
POB 708, Princeton, NJ 08544 USA

Since 1972 the author, at first in collaboration with Simon, and then with Benguria and Brezis, [1-10], has undertaken a systematic mathematical investigation of the Thomas-Fermi (TF) theory, and its modifications (von Weizsäcker, Dirac), of atoms, molecules and solids. A rigorous and coherent picture has emerged from the research and, while many problems are still unsolved, a fairly complete account can be given. The following is a thumb-nail sketch which, for brevity, concentrates mostly on atoms. There are three review articles: [3,4] and, more recently [10].

I. A Picture of a Heavy Atom

For a hydrogenic atom, the density is $\rho(x) = \psi(x)^2 = e^{-z|x|}$ in the ground state (unimportant constants will be suppressed). The scale length = $1/z = 1$ if $z = 1$.

What does a large atom ($z = 100$, say) look like? More precisely, what does the single particle density

$$\rho^Q(x) = N \int |\psi(x, x_2, \ldots, x_N)|^2 dx_2 \cdots dx_N \tag{1}$$

look like? (Here, ψ is the normalized ground state eigenfunction and summation on spins is understood).

The question is clear and simple enough, but is is difficult to find a reliable answer to it. Most texts leave their readers with the impression that an atom is something like an onion, with one shell filled after the next. Most physicists would probably guess that $\rho(r)$ is roughly constant, punctuated by ripples indicating the shell structure, out to a distance $R(z)$ which is roughly independent of z. Perhaps $R(z)$ is mildly increasing. Beyond $R(z)$, $\rho(r)$ decreases exponentially fast with a decay constant of order one--the square root of the ionization energy.

Apart from the very last statement, the above picture is wrong! A large atom looks more like a galaxy with a dense core, rather than a uniform ball. The core radius <u>decreases</u> with z as $z^{-1/3}$, yet for large z it contains most of the electrons and the energy. The important fact that will concern us is that TF theory accounts <u>exactly</u> for most of the core, and TFW theory accounts for the remainder of the core. Thus, TF theory is one of the few exact theories we have for many-body systems.

A schematic plot of $\rho(r)$ for a large atom is given in Fig. 1. It is not drawn to scale because to do so would require too large a sheet of paper.

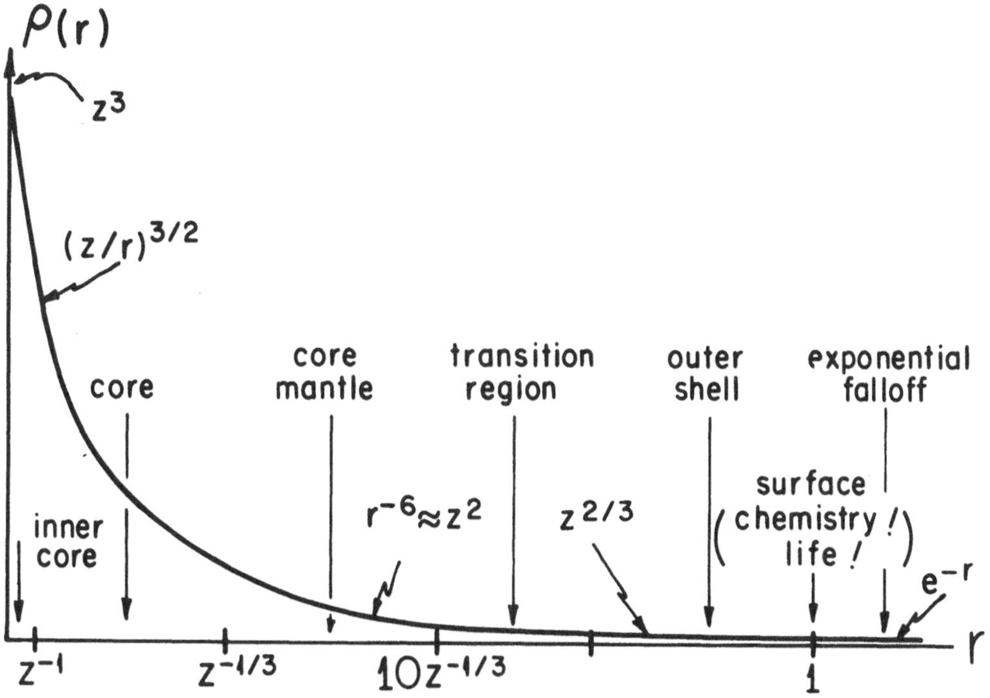

Figure 1

Seven regions can be defined, although the fourth is problematic. TF theory accounts exactly for nos. 2 and 3; TFW theory for 2,3 and probably 1, although the assertions made below about 1 have not yet been proved. 4,5,6 are guesswork. No. 7 has been rigorously established [11]. Units in which the electron charge and $\hbar^2/2m$ is unity is used.

1. <u>The inner core</u>: Distances are $O(z^{-1})$ and ρ is $O(z^3)$. The number of electrons out to $r = \sigma/z$ is $\sim\sigma^{3/2}$, while the energy $\sim z^2\sigma^{\frac{1}{2}}$. If $1 \ll zr \ll z^{2/3}$, $\rho(r)$ is given by the TF expression $(z/\gamma r)^{3/2}$, with $\gamma = (3\pi^2)^{2/3}$. $\rho(r)$ is infinity on a scale of z^2 which is the appropriate scale for the next, or TF, region. ρ satisfies the cusp condition [15]: $(d\rho/dr)(0) = -z\,\rho(0)$.

2. <u>The core</u>: Distances are $O(z^{-1/3})$ and ρ is $O(z^2)$. TF theory is exact to leading order. The energy is $E^{TF} \sim -z^{7/3}$ and almost all the electrons are in this region. Cf. (8), (9).

3. <u>The core mantle</u>: Distances are $\sim\sigma z^{-1/3}$ with $\sigma \gg 1$. $\rho(r) = (3\gamma/\pi)^3 r^{-6}$ which, although it is independent of z, makes $\rho \sim z^2$.

4. A transition region to the outer shell

5. The outer shell: In the Bohr theory $z^{1/3}$ shells are filled. The outer shell, if it can be defined, would then contain $O(z^{2/3})$ electrons, and each would see an unscreened charge $\sim z^{2/3}$. This picture would give a radius unity for the last shell and an average $\rho \sim z^{2/3}$, but a reliable estimate is difficult to obtain. On the same basis the average electron energy would be $O(z^{2/3})$.

6. The surface: The potential is $O(1)$ and so is the energy of each electron. Chemistry takes place here! TF theory, which is unreliable here, nevertheless also predicts a surface radius of $O(1)$.

7. Exponential falloff region: $\rho(r)\exp[2e^{\frac{1}{2}}r] \to 0$ as $r \to \infty$, for any number e which is less than the ionization potential [11].

II. Basic Definitions

The Hamiltonian for N electrons in a molecule of k nuclei is

$$H_N = \sum_{i=1}^{N}\{-\Delta_i - V(x_i)\} + \sum_{1\le i<j\le N} |x_i - x_j|^{-1} + U \tag{2}$$

with

$$V(x) = \sum_{j=1}^{k} z_j\, |x - R_j|^{-1}, \qquad U = \sum_{1\le i<j\le k} z_i z_j\, |R_i - R_j|^{-1}.$$

Here $z_j > 0$ and R_j are the charges and coordinates of the nuclei. $\underline{z}$ and $\underline{R}$ will denote these k-tuples. The R_j are fixed in the approximation of infinite nuclear mass.

The TF energy functional for the electron density $\rho(x) \ge 0$ is

$$\mathcal{E}(\rho) = \frac{3}{5}\gamma \int \rho(x)^{5/3}dx - \int V(x)\rho(x)dx + \tfrac{1}{2}\iint \rho(x)\rho(y)|x-y|^{-1}\,dxdy + U \tag{3}$$

The TFW functional is obtained by adding to $\mathcal{E}$

$$\alpha \int \{\nabla\sqrt{\rho(x)}\,\}^2\, dx\,. \tag{4}$$

For TFD theory we add

$$-\frac{3}{4} C_e \int \rho(x)^{4/3}\, dx\,, \tag{5}$$

and in TFDW theory both corrections are added. (4) is a correction to the kinectic energy while (5) is a correction to the electron-electron repulsion (exchange energy).

α and C_e are adjustable constants; they are not given by any fundamental consideration.

The quantum ground state energy is

$$E^Q(N) = \inf\langle\psi, H_N\psi\rangle / \langle\psi, \psi\rangle \tag{6}$$

Correspondingly

$$E^{TF}(N) = \inf\{\mathcal{E}(\rho) \mid \rho \in L^{5/3} \cap L^1, \int \rho = N\} \tag{7}$$

and similarly for E^{TFW}, E^{TFD}, E^{TFDW}.

The major questions to consider are:

(q1) When does a minimizing ρ (call it ρ^{TF}) exist for E^{TF}? Is it unique and how is it related to ρ^Q?

(q2) How is E^{TF} related to E^Q?

(q3) What are the properties of ρ^{TF} and E^{TF} for fixed R_i?

(q4) How does E^{TF} depend on the R_i? Is there binding?

Similar questions can be asked about the other theories (TFD, TFW, TFDW).

III. Relation of TF Theory to Quantum Theory

The basic scaling of $\underline{z}$,$\underline{R}$ and N (with the number of nuclei, k, fixed) is

$$z_j = a\, z_j^\circ \quad R_j = a^{-1/3} R_j^\circ\ , \quad N = a N^\circ$$

with $a \to \infty$. We can also take $R_j = R_j^\circ$(fixed) but this leads to a less interesting limit of isolated atoms. Neutrality is not assumed.

The main result [1,2] is

$$\lim_{a\to\infty} a^{-7/3} E^Q = \lim_{a\to\infty} a^{-7/3} E^{TF} \tag{8}$$

$$\lim_{a\to\infty} a^{-2} \rho^Q(a^{-1/3} x) = \lim_{a\to\infty} a^{-2} \rho^{TF} (a^{-1/3} x) \tag{9}$$

when ρ^{TF} exists (the existence will be elucidated shortly). The sense of (9) is weak L^1_{loc}. The right sides of (8),(9) are, in fact, independent of a. Thus, TF theory is asymptotically exact.

TFD theory is treated in [8] and reviewed in [10]. The Dirac term (and the true quantum exchange term) modifies E to lower order, namely $a^{5/3}$.

The Weizsäcker term [8,9] is much more interesting for several reasons. As for the energy, in 1952 Scott [12] conjectured that the leading correction beyond TF theory ought to be

$$\Delta E^{\mathrm{Scott}} = \frac{1}{4}\ a^2 \sum_{j=1}^{k} (z_j^{\circ})^2 , \tag{10}$$

independent of N (assuming $N^{\circ} > 0$) and independent of $\underline{R}$. The Scott conjecture has not yet been proved, but it is a fact that TFW theory gives a correction of precisely this kind. To obtain the coefficient 1/4, as in (10), it is necessary to choose [13]

$$\alpha = .186$$

in (4), and not $\alpha = 1/9$ as Kirzhnits and other have suggested.

IV. Properties of TF and Related Theories for Fixed Nuclei

In TF theory there is a minimizing ρ in (7) if and only if $N \le N_c = Z \equiv \Sigma_{j=1}^{k} z_j$. This ρ is unique and satisfies the TF equation:

$$\gamma\, \rho(x)^{2/3} = \max(\phi(x) - \mu,\ 0) \tag{11}$$

and $$\phi(x) = V(x) - \int \rho(y)\ |x-y|^{-7}\ dy . \tag{12}$$

$-\mu$ is the chemical potential $= dE^{TF}/dN$. E^{TF} and μ are convex functions of N, and μ goes from $+\infty$ (as $N^{-1/3}$) at N=0 to zero at N=Z. E^{TF} decreases as N increases. For $N > Z$, $E^{TF} = \text{constant} = E^{TF}(N{=}Z)$. There is no minimizing ρ (and hence no solution to (11)) if $N > Z$, i.e. negative ions cannot be supported in TF theory. $\rho(x)$ is continuous and real analytic away from the R_i. $\rho(x)$ has compact support if $N < Z$, while $\rho(x) \sim (3\gamma/\pi)^3\ |x|^{-6}$ as $|x| \to \infty$ when $N = Z$.

In TFD theory [8], there is a (unique) minimizing ρ if and only if $N \le N_c = Z$ as in TF theory. However, ρ always has compact support, even when $N = N_c$. At the boundary of the support, ρ jumps discontinuously from $\rho_o = (5\ C_e/8\gamma)^3$ to zero. This discontinuity in ρ^{TFD} sometimes caused confusion in the past, which sometimes led to doubts that the TFD equation has a solution; it always has one! When $N = N_c$, $\mu = \mu_c \equiv 15\ C_e{}^2(64\gamma)^{-1}$. $\mu = \infty$ when $N = 0$. For $N > N_c = Z$, $E^{TFD} = E^{TFD}(N{=}Z) - \mu_c(N{-}Z)$.

The most important fact about TFW theory [9] is that $Z < N_c < \infty$. (Strictly speaking $N_c > Z$ has been proved so far only for atoms. For molecules, $Z \le N_c < \infty$ has been proved, but it is believed $Z < N_c$ in this case as well.) No rigorous bound is known for $N_c - Z$. Thus, TFW theory supports negative ions. $\mu = 0$ at N_c and at $N = 0$, μ is finite ($= z^2/4\alpha$ for atoms). $\rho(x)$ (which again exists (uniquely) if and only if $N \le N_c$) never has compact support and, except at N_c, $\rho(x)$ decays exponentially as in real atoms. In TFW theory, unlike TF and TFD theories, $\rho(x)$ is finite at the nuclear R_j; $\rho(R_j) \sim z_j^3$ as in the quantum theory.

TFDW theory resembles TFW, except that $\mu > 0$ at $N_c > Z$, as in TFD theory.

The uniqueness of ρ is important. In the true Schrödinger theory, or even in the Hartree-Fock approximation to Schrödinger theory, a minimizing ψ may exist (it certainly exists if $N < Z + 1$), but it need not be unique always.

Fig. 2 shows $E(N)-U$ schematically in all 4 cases. Its slope is $-\mu$ (cf. above). A minimizing ρ exists if and only if $N \le N_c$, but E is defined for all N. Fig. 2 should be read as follows: (i) TF: $\mu(0) = \infty$, $N_c = Z$, $\mu = 0$ for $N > Z$. (ii) TFD: same as TF except that the slope is negative for $N > N_c = Z$. (iii) TFW: $\mu(0)$ is finite, $N_c > Z$, $\mu = 0$ for $N > N_c$. (iv) TFDW: same as TFW except for a negative slope when $N > N_c$.

The negative slope for $N > N_c$ in the Dirac theories is clearly unphysical. It arises from the fact that it is possible to place electron charge (i.e. ρ) in small clumps arbitrarily far from each other and from the nuclei and lower the energy. The lowering comes from the term $-\int \rho^{4/3}$.

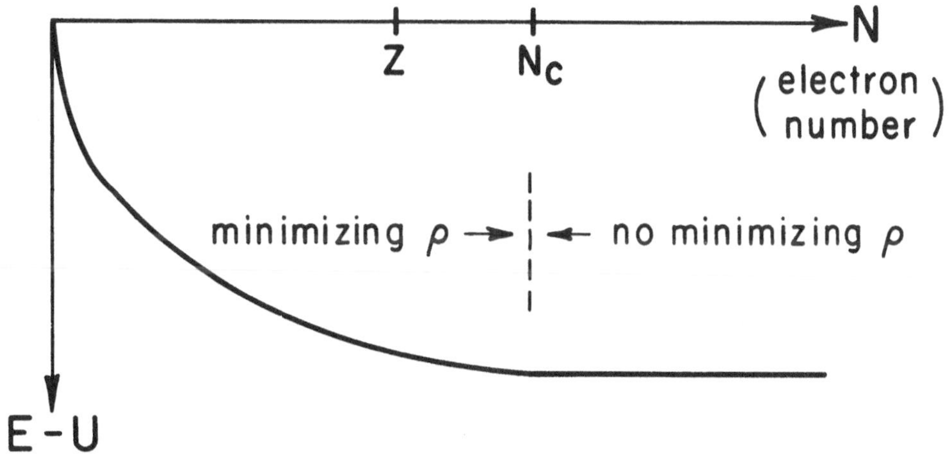

Figure 2

V. Dependence of the Energy on the Nuclear Coordinates

There are four main facts about TF theory. The nuclear repulsion U in (2) is essential for these theorems.

(f1) No binding: In 1962 Teller [14] discovered that a molecule is unstable against every kind of breakup. A rigorous proof is in [2]. To be precise, if $\underline{z}^1$, $\underline{R}^1$ and $\underline{z}^2$, $\underline{R}^2$ are two collections of nuclei then, for any N,

$$E^{TF}(\underline{z}^1 \oplus \underline{z}^2, \underline{R}^1 \oplus \underline{R}^2, N) > \min_{0<N'<N} E^{TF}(\underline{z}^1, \underline{R}^1, N') + E^{TF}(\underline{z}^2, \underline{R}^2, N-N') \tag{13}$$

If U is omitted in (2) then inequality (13) is reversed.

(f2) Positivity of the pressure [6]: If z is fixed and a molecule is dilated by $R_i \to \ell R_i$ then E^{TF} is monotone decreasing and convex in ℓ. Thus there are not even local minima in the energy. This theorem has been proved only for neutral molecules, but is undoubtedly true always. In particular it is trivially true when $N = 0$, for then only the U term remains.

(f3) Sign of the many-body potentials [5]: The n-body energy, $\varepsilon^{(n)}$, is defined by successive differences of the E's.

$$\varepsilon^{(2)} = E(z_1, z_2) - \{E(z_1) + E(z_2)\}, \quad \varepsilon^{(3)} = E(1,2,3)$$

$$- \{E(1,2) + E(1,3) + E(2,3)\} + \{E(1) + E(2) + E(3)\},$$ and so forth.

Only neutral systems are considered. The sign of $\varepsilon^{(n)}$ is $(-1)^n$ for all $R_1, \ldots, R_n$ in TF theory.

(f4) Long range potential [7]: If $R_i \to \ell R_i$ then, for large ℓ, (in the neutral case),

$$E^{TF}(\underline{z}, \underline{R}) - \sum_{j=1}^{k} E^{TF}_{\text{atom}}(z_j) \approx C\,\ell^{-7}$$

where $C \neq 0$ depends on the R_i but not on the z_i. Even more interesting is the fact that all the many-body potentials $\varepsilon^{(n)}$ decay in exactly the same way, namely $C^{(n)}\,\ell^{-7}$. Thus, in TF theory it is not true that three and higher body interactions can be neglected at large distances--as is ordinarily assumed.

The Weizsäcker term (4) seriously changes the above. In TFW theory binding can occur. This has so far been proved for 2 neutral atoms in two cases: (i) when $z_1 \ll z_2$ (cf. [9]), (ii) when $z_1 = z_2$. The existence of binding seems to be intimately connected with the fact that negative ions exist in TFW theory. We conjecture that in TFW theory any two neutral molecules will bind.

The Weizsäcker term thus has a dramatic effect on TF theory. Its original motivation was to improve the approximation to the kinetic energy. This correction is most pronounced near the nuclei and TFW theory correctly gives the Scott correction which comes from a modification of ρ near the nuclei. But (accidentally?) it improves TF theory in two other ways connected with large distances: (i) It gives exponential falloff of ρ; (ii) Negative ions can be supported and binding can occur.

References:

1. E. H. Lieb and B. Simon, Thomas-Fermi Theory revisited, P.R.L. 31, 681-683 (1973).

2. E. H. Lieb and B. Simon, The Thomas-Fermi Theory of Atoms, Molecules and Solids, Adv. in Math. 23, 22-116 (1977).

3. E. H. Lieb, Thomas-Fermi and Hartree-Fock theory, Proc. Int. Cong. Math., Vancouver (1974), vol. 2, pp. 383-386.

4. E. H. Lieb, The stability of matter, Rev. Mod. Phys. 48, 553-569 (1976).

5. R. Benguria and E. H. Lieb, Many-body potentials in Thomas-Fermi Theory, Ann. of Phys. (NY) 110, 34-45 (1978).

6. R. Benguria and E. H. Lieb, The positivity of the pressure in Thomas-Fermi theory, Commun. Math. Phys. 63, 193-218 (1978). Errata 71, 94 (1980).

7. H. Brezis and E. H. Lieb, Long range atomic potentials in Thomas-Fermi Theory, Commun. Math. Phys. 65, 231-246 (1979).

8. R. Benguria, The von-Weizsäcker and exchange corrections in Thomas-Fermi theory, Ph.D. thesis, Princeton (1979).

9. R. Benguria, H. Brezis and E. H. Lieb, The Thomas-Fermi-Von Weizsäcker theory of atoms and molecules, Commun. Math. Phys., in press.

10. E. H. Lieb, Review of Thomas Fermi Theory, Erice 1980 Summer School in Atomic Physics (in preparation).

11. A. J. O'Connor, Commun. Math. Phys. 32, 319 (1973). P. Deift, W. Hunziker, B. Simon, E. Vock, Commun. Math. Phys. 64, 1 (1978). M. Hoffmann-Ostenhof, T. Hoffmann-Ostenhof, R. Alrichs, J. Morgan, Springer Lecture Notes in Physics 116, 62 (1980).

12. J.M.C. Scott, The binding energy of the Thomas-Fermi atom, Phil. Mag. 43, 859-867 (1952).

13. E. H. Lieb and D. Liberman, in preparation.

14. E. Teller, On the stability of molecules in Thomas-Fermi theory, Rev. Mod. Phys. 34, 627-631 (1962).

15. T. Kato, Commun. Pure and Appl. Math. X, 151 (1957). E. Steiner, J. Chem. Physics. 39, 2365 (1963). W. Bingel, Z. Naturf. 18a, 1249 (1963). M. Hoffmann-Ostenhof, T. Hoffmann-Ostenhof and W. Thirring, J. Phys. B11, L571 (1978).

ELECTRONS, PHONONS AND THE OPTICAL PROPERTIES OF SMALL METALLIC PARTICLES

G. Monsivais
Centro de Estudios Nucleares, U.N.A.M.
A. P. 70-543, México 20, D. F., México

and

J. Flores[+]
Instituto de Física, U.N.A.M.
A. P. 20-364, México 20, D. F., México

Abstract

We show that infrared absorption by small metallic particles is due mainly to the excitation of electrons and that electron correlations are essential to obtain an order of magnitude agreement between observed and calculated values of the absorption coefficient.

[+]Work supported in part by the Instituto Nacional de Investigaciones Nucleares, México.

Small metallic particles are atomic aggregates formed by a finite number of atoms for which the thermodynamic limit is not valid and which show a large surface to volume ratio. Due to the presence of the surface, electrons and phonons in these small particles have properties which differ from those in the bulk metal. As a consequence, small metallic particles show anomalous thermodynamic and magnetic behaviour[1]. Other properties, such as the absorption of infrared radiation will also be altered and will depend explicitly on the diameter d of the small particle, as we shall now discuss.

One can think of at least two mechanisms through which infrared radiation of frecuency ω can be absorbed by an isolated small metallic particle. The first is of course through the valence electrons in the particle, which displace to form a dipole[2]. The other mechanism implies the excitation of phonons in the small particle[3]. In what follows we shall restrict ourselves to analyzing the infrared absorption by the small particle itself, excluding the possibility that the absorption takes place at the oxid layer which could surround the particle. This has been taken into account by Simanek[4] and the experimental results can be reproduced qualitatively; however, the experimental situation he assumes might be in disagreement with the actual conditions[3].

Let us discuss the second mechanism first. Near the surface the electric field is not screened completely by the conduction electrons, so that it can couple with the surface ions, creating surface phonons. If the frequencies of these surface modes lie within the spectra of lattice eigenmodes, then bulk phonons modes can be excited. This conversion of electromagnetic into acoustic energy will obviously be more important the smaller the particle is. With suitable approximations, Glick and Yorke[3] obtain for the absorption coefficient due to phonon excitation $A_{ph}(\omega)$ the following expression

$$A_{ph}(\omega) = \frac{K}{d} \int_0^\infty \frac{\mathcal{D}(\omega_\ell)\, \gamma_\ell\, \omega^2 d\omega_\ell}{(\omega_\ell - \omega)^2 + \gamma_\ell^2 \omega^2} \qquad (1)$$

where $\mathcal{D}(\omega_\ell)$ is the density of phonon modes with frequency ω_ℓ, γ_ℓ is the damping factor for mode ℓ and K is a constant which depends on the specific material forming the small particle. The damping factor γ_ℓ is estimated using the ultrasonic absorption coefficient, and it is found that γ_ℓ/ω is small, a fact which allows the integral in (1) to be performed, once $\mathcal{D}(\omega_\ell)$ is known.

Using this expression, one obtains an absorption coefficient which is of the same order of magnitude as the one reported experimentally[5] for Al particles with an average radius of 24 Å. However, it is found experimentally that the absorption increases with the particle radius, whereas expression (1) predicts $A_{ph}(\omega)$ to be proportional to d^{-1}. Furthermore, $A_{ph}(\omega)$ increases with ω, but not quadratically as observed in the experiments. A final criticism to this theory is that Glick and Yorke use a $\mathcal{D}(\omega_\ell)$ appropriate for the bulk metal which is not completely adequate. In fact, experimental data on electrical resistivity[6)] and on X-ray diffraction[7)] by small metallic particles can be explained by assuming that the Debye temperature θ_D decreases with the particle diameter. For silver small particles of radius 75Å θ_D reduces to about three quarters the bulk value. This is mainly due to the existence of soft phonon surface modes[8)].

The absorption coefficient $A_e(\omega)$ due to electrons was calculated by Granqvist et al[5)] using the results of the free-electron model used in reference 2, but modified as suggested by Strässler et al[9)] in order to take into account in some implicit way the electron-electron interaction. They find that $A_e(\omega)$ is indeed quadratic in ω but with a constant of proportionality three orders of magnitude smaller than the one observed experimentally. In what follows we shall take into account explicitly electron-electron correlations as proposed by Lushnikov and Simonov[10)], and show that $A_e(\omega)$ is then of the same order of magnitude as the measured value and with the correct d dependence.

The absorption cross section $\sigma(d,\omega)$ is related to the electrical polarizability $\alpha(d,\omega)$ through

$$\sigma(d,\omega) = \frac{4\pi\omega}{c\,n}\,\mathrm{Im}\;\alpha(d,\omega) \tag{2}$$

where n is the refraction index, c the velocity of light and $\alpha(d,\omega)$ is such that the *induced* dipole moment $\underline{p}$ and the *external* electric field $\underline{E}_0$ are given by

$$\underline{p} = \alpha\,\underline{E}_0 \tag{3}$$

We assume that α is a function only of d and ω.

Now, Migdal et al[11)] show by applying the theory of finite Fermi systems, that α is given by

$$\alpha(d,\omega) = -\frac{e^2}{3}\sum_{\lambda\lambda'} \frac{n_\lambda - n_{\lambda'}}{\varepsilon_\lambda - \varepsilon_{\lambda'} - \omega}\,\underline{r}_{\lambda\lambda'}\cdot\underline{R}_{\lambda'\lambda} \tag{4}$$

where ε_λ and n_λ are the single-electron energies and occupation numbers, respectively, $\underline{r}_{\lambda\lambda'}$ is the matrix element of the position operator $\underline{r}$ taken with respect to the single-electron states λ and λ', and $\underline{R}_{\lambda'\lambda}$ is the corresponding matrix element of the operator $\underline{R}$, which obeys the operator equation

$$\underline{R} = \underline{r} + \Gamma^{\omega} GG\underline{R} \tag{5}$$

Here G is the single-electron Green's function for a system of N electrons and Γ^{ω} is the operator associated with the scattering amplitude.

Expression (4) was used by Lushnikov and Simonov[10] to compute $\alpha(d,0)$, the static polarizability. In particular, they show that the increase in $\alpha(d,0)$ predicted by Gor'kov and Eliashberg[2] is an spurious effect, due to the neglect of electron correlations.

We now separate from $\alpha(d,\omega)$ its static components and arrive to

$$\alpha(d,\omega) = \alpha(d,0) + i\tfrac{2}{3}\pi\, e^2\, \nu^2\, \hbar\omega(\underline{r}_{12} \cdot \underline{R}_{21})_o \tag{6}$$

in much the same way as done in reference (2). Here $\nu = md^3p_F/12\,\pi\hbar^3$, with p_F the Fermi momentum, is the single-electron level density and $(\underline{r}_{12} \cdot \underline{R}_{21})_o$ is an average matrix element, the average being taken over single-particle states around the Fermi surface differing in energy by $\hbar\omega$. Since we know from experiment that $A_e(\omega)$ is proportional to ω^2, it is natural from eqs. (2) and (6) to assume that the average $(\underline{r}_{12} \cdot \underline{R}_{21})_o$ is independent of ω. But then the operator $\underline{R}$ can be taken directly from the calculation for the static case[10]. We then obtain the average $(\underline{r}_{12} \cdot \underline{R}_{21})_o$ using the general method of Shapoval[12]. Combining all this, we have

$$\alpha(d,\omega) = \alpha(d,0) + i\, m\, e^2\nu d^3\omega I(\gamma)\,/24p_F \tag{7}$$

where $I(\gamma)$ is a complicated double integral, which can be evaluated numerically[13]. Here $\gamma = me^2d^2p_F\,/\,3\pi\hbar^3$.

For an ensemble of small particles, whose diameter is distributed around an average diameter $\bar{d}$ according to a log-normal distribution[5], one finds that the absorption coefficient $A_e(\omega)$ grows linearly with $\bar{d}$, which agrees with experiment. However, the magnitude of $A_e(\omega)$

is still smaller than the experimental values, as shown in the following table:

	Theory	Experiment
Al $\bar{d}$=48 Å	$0.083 \times 10^{-3}\ \omega^2$	$0.9 \times 10^{-3}\ \omega^2$
Cu $\bar{d}$=270 Å	$1.08 \times 10^{-3}\ \omega^2$	$3.5 \times 10^{-3}\ \omega^2$

Although our theoretical values for $A_e(\omega)$ are still below the experimental ones, they are three orders of magnitude larger than those calculated by Granqvist et al[5], who only took into account the electron-electron interaction in an implicit way. Therefore, we think that it is more appropriate to include electron correlations using many body theory as we have done, although our approximations might still be too crude to obtain the correct values. In any case, our results show that $A_e(\omega)$ and $A_{ph}(\omega)$ are of the same order of magnitude. Since $A_e(\omega)$ shows the experimentally observed ω and size dependence, it seems that the absorption takes place preferably via the excitation of conduction electrons in the small particle.

REFERENCES

1.- R. Kubo, in "Polarisation, Matière et Rayonnement", Volume Jubilaire en L'honneur du Prof. A. Kastler (Presses Universtaires, Paris, 1969) p. 325

2.- L. P. Gor'kov and G. M. Eliashberg Sov. Phys. (JETP) 21, 940 (1965)

3.- A. J. Glick and E. D. Yorke Phys. Rev. B 18, 2490 (1978)

4.- E. Simanek Phys. Rev. Lett. 38, 1161 (1977)

5.- C. G. Granqvist, R. A. Buhrman, J. Wyns and A. J. Sievers Phys. Rev. Lett 37, 625 (1976)

6.- K. Ohshima, T. Fujita and T. Kuroishi
Jour. de Phys. (Paris) 38 C2, 163 (1977)

7.- Y. Kashiwase, I. Nishida, Y. Kainuma and V. Kimoto
Jour. de Phys. (Paris) 38 C2 , 157 (1977)

8.- J. M. Dickey and A. Paskin
Phys. Rev. Lett. 21, 1441 (1968)

9.- S. Strässler, M. J. Rice and P. Wyder
Phys. Rev. B 6, 2575 (1972)

10. A. A. Lushnikov and A. J. Simonov
Phys. Lett. 44 A, 45 (1973)

11. A. B. Migdal, A. A. Lushnikov and D. F. Zaretsky
Nucl. Phys. 66, 193 (1965)

12. E. A. Shapoval
Sov. Phys. (JETP) 20, 675 (1965)

13. G. Monsivais, Ph. D. Thesis, Universidad de México, 1980
(unpublished)

ATOMIC EXCHANGE ENERGY AS A DENSITY FUNCTIONAL

M. Berrondo†
Instituto de Física, UNAM
Apdo. Postal 20-364,
México 20, D.F.

We give a local expression for the exchange potential and for the exchange energy in terms of the electron density and its gradients. The assumption that the logarithmic derivative of the density varies very slowly with the distance from the nucleus turns out to be surprisingly good in the evaluation of the exchange energy.

1.- Introduction

The outcome of the Thomas-Fermi approximation[1] is an expression for the total energy as a functional of the electron density. In fact, is the prototype of density functionals, and it yields a qualitative global picture of atomic structure. A further correction was introduced by Dirac[1] to include the so-called exchange term, which partially corrects for the electron self-interaction[2] which appears in the Thomas -Fermi expression for the Coulomb repulsion energy. This model can be derived starting from a homogeneous electron gas to compute both the kinetic and enchange-correlation energy as functionals of the electron density. The nuclear attraction and the interelectronic repulsion are given by the usual electrostatic expressions in terms of the density. For inhomogeneous systems like an atom, the assumption is usually made that the density varies slowly from point to point, and then correction terms are added, which depend on the density gradient, both for the kinetic energy[1] and the Dirac term[3].

The formal justificacion for the use of density functionals (for the ground state) is found in the Hohenberg-Kohn theorem[4]. In essence, it states that the ground state energy is a unique functional of the density, and the exact energy is a lower bound. It can be easily proved[5] by expressing the energy in terms of a one-body local potential, and then performing a functional Legendre transform to write the energy as a functional of the density. The problem is, of course, that the exact explicit form of this functional is not known for an interacting electron system. In fact, for an atom, the presence of a central attraction due to the nucleus, makes it a very inhomogeneous system, particularly close to the nucleus, where the density is higher and hence gives the

---

* Consultant at the Instituto Mexicano del Petróleo.

largest contribution to the energy. The Thomas-Fermi expression for the kinetic energy (even with gradient corrections) turns out to be inaccurate for calculating atomic energies. It has hence proven to be much better to solve Schrödinger-like equations for the orbitals and compute the density from these orbitals. The resulting equations[2] for the orbitals are indeed similar to Hartree-Fock's, but the exchange potential is local and, in fact, multiplicative. In this way, the kinetic energy term is computed in the correct fashion both in the orbital equations and in the total energy. A surprisingly good approximation for the exchange potential[2,6] is derived from the Dirac exchange term, particularly if we multiply it by a parameter, which can be determined[7] by imposing the virial theorem.

A better starting point is the assumption that the electron fluid in an atom has an inhomogeneous density whose form is largely dominated by the presence of the nucleus. This inmediately implies[8] a slowly varying logarithmic derivative of the density rather than a locally constant density. In this way, the density gradient appears in a natural fashion, and not as a correction to the homogeneous electron gas. Accordingly there are two different (but related) tasks: to obtain an appropriate local exchange potential, and to compute the approximate exchange energy, given the density.

2.- The local exchange potential

The electron distribution in an atom (or ion) is far from being homogeneous. It presents a cusp at the nucleus and is exponential asymptotically. Actually, it can be better described as piecewise exponential[9]. Although numerically this does not give a perfect fitting, for the (only) purpose of finding the functional form of the energy, let us assume that we can write

$$\rho(\bar{r}) = A\, e^{-2\eta r} \tag{2.1}$$

where A and η are "almost constant". This Ansatz however is not enough to compute the local potential. The exchange energy* is given in terms of the one-density matrix $\gamma(\bar{r}_1,\bar{r}_2)$ as:

$$E_x = -\int \frac{\gamma(\bar{r}_1,\bar{r}_2)\,\gamma(\bar{r}_2,\bar{r}_1)}{r_{12}}\, d\bar{r}_1\, d\bar{r}_2, \tag{2.2}$$

*The correlation energy is disregarded at this stage. For atoms, it represents roughly 1% of the total energy.

where ρ is normalized to half the number of electrons:

$$\mathrm{Tr}\ \gamma = N/2 \tag{2.3}$$

If we assume a functional form for γ :

$$\gamma = B\ e^{-(r_1+r_2)\eta}\ , \tag{2.4}$$

η will still be almost constant, and can be computed directly from the density:

$$\eta = -\frac{1}{2\rho}\ \frac{d\rho(r)}{dr}\ . \tag{2.5}$$

The B appearing in Eq. (2.4) however has to allow for two different conditions, namely normalization, Eq. (2.3) and idempotency:

$$\int \gamma(\bar{r}_1,\bar{r}_3)\ \gamma(\bar{r}_3,\bar{r}_2)\ d\bar{r}_3 = \gamma(\bar{r}_1,\bar{r}_2)\ . \tag{2.6}$$

To calculate the enchange potential, defined as

$$v_x\,[\rho(r)] = \frac{\delta E_x\,[\rho(r)]}{\delta\rho(r)}\ , \tag{2.7}$$

the B-factor "cancels out", so its exact form is not very important. The exchange energy however turns out to be proportional to B, so it cannot be assumed as locally constant.

The explicit form of the local exchange potential is obtained as follows: substitute Eq. (2.4) in (2.2), perform the r_2 integration, reexpress E_x in terms of the density ρ using Eq. (.2.1) and compute the functional derivative in Eq. (2.7). The result is a simple expression[7]:

$$v_x = -\frac{1}{r}\left(1+\frac{1}{2\eta r}\right)\left[1-(1+2\eta r)\,e^{-2\eta r}\right]. \tag{2.8}$$

This potential has the correct behaviour asymptotically

$$v_x \longrightarrow -1/r \tag{2.9}$$

in contrast to the homogeneous gas expression, which decays exponentially. This allows for bound states of negative ions[10], like in the case of F^-, whereas the X_α method does not[11].

Using the local potential (2.8) to determine the orbitals and computing the total energy as the expectation value of the Slater determinant, we obtain the following results[10,12]:

i) the total energy gives a relative error of 10^{-5}, above the Hartree-Fock value for the computed cases of atoms from He to Kr

ii) the one particle energies for the same atoms are accurate to within 0.1% for the deepest levels and a bit worse for the valence shells

iii) taking η as a variational constant[12] in Eq. (2.8), it yields a value close to one for the same series of atoms.

3.- The Exchange Energy

In order to express the exchange energy (2.2) in terms of the density and its gradient, we cannot use the form (2.4) directly, as we mentioned earlier. Instead, we shall extract the main contribution using the idempotency of γ, Eq. (2.6), and then use the Ansatz (2.4) for the remainder.

The ground state density is higher near the nucleus, so we look for a good approximation in this region. If we would expand r_{12}^{-1} in multipoles, we would find that the main contribution[13] comes from the term $r_{>}^{-1}$. Hence we can define

$$I_\circ = \int \frac{|\gamma(\bar{r}_1,\bar{r}_2)|^2}{r_1} \, d\bar{r}_1 \, d\bar{r}_2, \tag{3.1}$$

and the exchange energy becomes:

$$E_x = - I_\circ - \int |\gamma(r_1,r_2)|^2 \left(\frac{1}{r_{12}} - \frac{1}{r_1}\right) d\bar{r}_1 \, d\bar{r}_2 \; . \tag{3.2}$$

But using Eq. (2.6), we can rewrite $I_\circ$ directly in terms of the density:

$$I_\circ \left[\rho(r)\right] = \int \frac{\rho(\bar{r})}{r} \, d\bar{r}. \tag{3.3}$$

In Table I we compare this quantity with Eq. (2.2) using the Hartree-Fock density[14] (double-zeta quality) to show that it indeed gives a sizeable part of E_x. It is also obvious that it improves as Z increases.

Table I

Atom	$I_\circ$ [a]	Exact [b]	Approximate [c]
He	1.687	1.026	1.069
Be	4.204	2.666	2.679
Ne	15.56	12.12	12.679
Mg	19.96	16.00	15.792
Ar	34.86	30.19	29.224

a Equation (3.1) with H.F. density (a.u.)
b Exact (absolute) value, Eq. (2.2) with H.F. orbitals
c Density functional including gradient, Eq. (3.3) with same density

The integral in Eq. (3.2) can now be approximated using Eq. (2.4) and modified by the presence of p electrons, which gives a geometrical factor [13]

$$E_x = - \left(1 + \frac{2}{5} \frac{N_p}{N}\right) \int \frac{\rho(r)}{r} \left\{ 1 - \left[1+r\eta(r)\right] e^{-2r\eta(r)} \right\} d\vec{r} \qquad (3.3)$$

where

$$\eta(r) = - \frac{1}{2\rho(r)} \frac{d\rho(r)}{dr} \qquad (3.4)$$

and N_p is the number of p-electrons. The results for closed shell atoms are shown in Table I, and a more extensive Table can be found in Ref. 13.

Acknowledgements

Most of this work is the result of collaboration with O. Goscinski (Univ. Upsala), J.P. Daudey (Univ. Toulouse), and A. Flores (Inst.Mex. Petróleo).

REFERENCES

1. See e.g. N.H. March, Self-Consistent Fields in Atoms (Pergamon Press, Oxford, 1975).

2. J.C. Slater, The Self-Consistent Field for Molecules and Solids (Mc, Graw Hill, New York, 1974).

3. S.K. Ma & K.A. Brueckner, Phys. Rev. 165, 18 (1968).

4. P. Hohenberg & W. Kohn, Phys. Rev. 136, B864 (1964).

5. M. Berrondo & O. Goscinski, Int. J. Quantum Chem. S9, 67 (1975).

6. J.C. Slater, Phys. Rev. 81, 385 (1951).

7. M. Berrondo & O. Goscinski, Phys. Rev. 184, 10 (1969).

8. M. Berrondo & O. Goscinski, Chem. Phys. Lett. 62, 31 (1979).

9. G. Sperber, Int. J. Quantum Chem. 5, 189 (1971).

10. J.P. Daudey & M. Berrondo, to be published in Int.J.Quantum Chem.

11. K. Schwartz, Chem. Phys. Lett. 57, 605 (1978).

12. M. Berrondo, J.P. Daudey & O. Goscinski, Chem.Phys.Lett.62,34(1979).

13. M. Berrondo & A. Flores, J. Chem. Phys. 72, 6299 (1980).

14. E. Clementi & C. Roetti, At. Data Nucl. Data Tables 14, 177 (1974).

LIOUVILLIAN PROPAGATOR TECHNIQUE FOR PERTURBED WAVE FUNCTIONS, LEVEL SHIFTS AND BROADENINGS OF COMPOSITE PARTICLES IN A MANY-BODY MEDIUM

M.D. Girardeau
Department of Physics and Institute of Theoretical Science, University of Oregon
Eugene, Oregon 97403, U.S.A.

1. Introduction

At this conference there has been a great deal of discussion of methods of calculating various properties of many-body systems in cases where they may be well approximated (or at least, simulated) by systems of structureless particles interacting by potentials. However, many problems in several areas of physics and chemistry involve many-body systems of interacting composite particles, in regimes where their internal transitions and/or reactive collisions (breakup, recombination, rearrangement) are important. Standard many-body Green's function and quantum field theoretic techniques are not well adapted to such situations. I want to discuss generalized representations which allow application of standard techniques to more complicated systems of interacting composite particles and their constituents.

Composite particles in a medium have wave functions and energies which are modified by interactions of the composites with their environment. Consider, for example, spectral line shifts and broadenings, which are related to such distortion of single-composite wave functions, or clusters in nuclear matter, which are again distorted by their environment. There is no single "correct" definition of such distorted wave functions, which are neither truly bound nor even energy eigenstates no matter how their wave functions are chosen. It is really a question of the most convenient choice of basis to optimize calculation of observable properties, the distorted single-composite wave functions being ingredients in the many-particle basis. However, it is possible to give a reasonable criterion for "optimal choice" of these wave functions.

2. Variational criterion for perturbed wave functions, level shifts, and lifetimes

A convenient and rather natural criterion for determination of the perturbed composite particle wave functions ϕ_α is in terms of the appropriate pole z_α of a single-composite Green's function. The energy shift ξ_α and width γ_α are defined by $z_\alpha = \varepsilon_\alpha+\xi_\alpha-i\gamma_\alpha$ where ε_α is the energy of the isolated composite. One method for determination of ϕ_α is by minimization of $\varepsilon_\alpha+\xi_\alpha$, leading to a generalization of Hartree-Fock theory including correlation effects. Another natural criterion is maximization of the lifetime (minimization of γ_α), a composite in a medium being "most likely to be found" in a state of maximal lifetime. These criteria are combined in a natural way by requiring that z_α be stationary under variation of ϕ_α. Our variational principle then consists of the requirement[1] that z_α be stationary under variation of ϕ_α subject to orthonormality, $(\phi_\alpha, \phi_\beta)=\delta_{\alpha\beta}$:

$$[\delta/\delta\phi_\alpha^*(X)][z_\alpha-\Sigma_\beta\lambda_{\alpha\beta}(\phi_\alpha, \phi_\beta)]=0. \tag{1}$$

Here X stands for the position and spin arguments of all of the constituents of the composite state ϕ_α, and the $\lambda_{\alpha\beta}$ are Lagrange multipliers for the orthonormality constraint.

3. Composite particle Green's functions

Several inequivalent definitions of such Green's functions have been proposed.[2-4] The analysis here will be based on the Soulet-Gomes Green's function[4] in Fock-Tani representation,[5-10] within which composite particles are consistently described by field operators satisfying elementary-particle commutation rules but nevertheless incorporating inelastic and reactive effects such as breakup and recombination of composites. The causal single-composite Green's function is[4]

$$g(\alpha,t;\alpha)=-(i/\hbar)\mathrm{Tr}[\hat{a}_\alpha(t)\hat{a}_\alpha^\dagger(0)\hat{\rho}]\theta(t) \tag{2}$$

with $\hat{\rho}$ the statistical ensemble density operator and $\hat{a}_\alpha(t)$ the composite particle annihilator in Heisenberg picture and Fock-Tani representation. The energy Green's function is the Laplace transform

$$\tilde{g}(\alpha,z;\alpha)=\int_0^\infty g(\alpha,t;\alpha)e^{izt/\hbar}dt. \tag{3}$$

We are interested in that pole z_α which approaches the isolated-composite energy ε_α as the interaction $\hat{V}$ is switched off; in this representation[5-10] the internal binding is included in $\hat{H}_0$.

4. Fock-Tani representation

This representation has been described in considerable detail;[5-10] here only the salient features will be indicated.

The problem to be overcome is that physical composite annihilation and creation operators $\hat{A}_\alpha$ and $\hat{A}_\alpha^\dagger$ are of higher than first degree in the constituent fields and hence satisfy complicated commutation relations. Following the lead of Bohm-Pines theory of plasmons[11] and Dyson's theory of ideal spin waves,[12] one performs a unitary transformation $\hat{U}$ which changes the physical single composite states $\hat{A}_\alpha^\dagger|0\rangle$ into ideal composite states $\hat{a}_\alpha^\dagger|0\rangle$ where the $\hat{a}_\alpha$ and $\hat{a}_\alpha^\dagger$ satisfy elementary Bose or Fermi commutation or anticommutation relations. The required transformation is[5-7]

$$\hat{U}=\exp(\tfrac{\pi}{2}\hat{F}),\ \hat{F}=\Sigma_\alpha(\hat{A}_\alpha^\dagger\hat{a}_\alpha-\hat{a}_\alpha^\dagger\hat{A}_\alpha). \tag{4}$$

The same transformation applied to the Fock space Hamiltonian transfers all internal binding of the composites into an unperturbed Hamiltonian $\hat{H}_0$ of free particle form, whereas the interaction Hamiltonian $\hat{V}$ contains the effects of decay[9] and collisional interaction[5-8] of the composites and their free constituents, explicitly exhibiting elastic, inelastic, and reactive collision channels.

5. Fock-Tani Hamiltonian

Although the method is general, I shall now, for pedagogical reasons, specialize to the case of a many-body system of hydrogen atoms and their constituent free protons and electrons. Since the method has so far been developed only nonrelativistically, it will also be assumed that relativistic effects are negligible. The Fock-Tani Hamiltonian is[5-7] $\hat{H}=\hat{H}_0+\hat{V}$ with unperturbed part (including internal binding)

$$\hat{H}_0=\Sigma_\alpha \varepsilon_\alpha \hat{a}_\alpha{}^\dagger \hat{a}_\alpha + \Sigma_K \varepsilon_K \hat{\psi}_K^\dagger \hat{\psi}_K + \Sigma_k \varepsilon_k \hat{\psi}_k^\dagger \hat{\psi}_k \tag{5}$$

in which $\varepsilon_\alpha=(\alpha|H|\alpha)$ is the single atom (translational plus internal) energy and ε_K and ε_k the free proton and electron energies, K standing for the proton wave vector and spin quantum numbers and k for those of the electron. The interaction Hamiltonian $\hat{V}$ is an infinite series exhibiting all decay and collision channels, of the general form

$$\begin{aligned}\hat{V}= &\sum_{\alpha\neq\beta} \hat{a}_\alpha{}^\dagger (\alpha|H|\beta)\hat{a}_\beta + \sum_{Kk\alpha} [\hat{\psi}_K{}^\dagger \hat{\psi}_k^\dagger (Kk|H|\alpha)\hat{a}_\alpha + h.c.] \\ &+ \tfrac{1}{2} \sum_{\alpha\beta\alpha\delta} \hat{a}_\alpha{}^\dagger \hat{a}_\beta{}^\dagger (\alpha\beta|H|\alpha\delta)\hat{a}_\delta \hat{a}_\gamma \\ &+ H_{pp}+H_{ee}+H_{pe} \\ &+ \sum_{\alpha\beta KK'} \hat{a}_\alpha{}^\dagger \hat{\psi}_K{}^\dagger (\alpha K|H|\beta K')\hat{\psi}_{K'}\hat{a}_\beta + \sum_{\alpha\beta kk'} \hat{a}_\alpha{}^\dagger \hat{\psi}_k{}^\dagger (\alpha k|H|\beta k')\hat{\psi}_{K'}\hat{a}_\beta \\ &+ \sum_{\alpha Kkk_1k_2} [\hat{a}_\alpha{}^\dagger \hat{\psi}_{k_1}{}^\dagger (\alpha k_1|H|Kkk_2)\hat{\psi}_{k_2}\hat{\psi}_k \hat{\psi}_K + h.c.] \\ &+ \ldots\end{aligned} \tag{6}$$

The off-diagonal $(\alpha|H|\beta)$ and the $(Kk|H|\alpha)$ vanish identically[6] if the ϕ_α are free-atom wave functions. However, we are now considering ϕ_α perturbed by the many-body medium. H_{pp} and H_{ee} are formally identical with the standard proton-proton and electron-electron Fock interaction Hamiltonians, although the $\hat{\psi}_K$ and $\hat{\psi}_k$ operators now refer only to unbound protons and electrons. However, H_{pe} is modified[6] due to the fact that p-e binding effects are already included in $\hat{H}_0$. Only the few-body collision terms relevant to the application we wish to discuss here are exhibited in (6); for more details and explicit expressions for the matrix elements, see the references.[5-9]

6. Liouvillian propagator diagram expansion for the Green's function

In order to implement our variational method for perturbed wave functions, level shifts, and broadenings, we need a systematic procedure for calculating the position of the relevant pole of the single-composite Green's function. The Heisenberg equation for $\hat{a}_\alpha(t)$ in (2) is

$$i\hbar d\hat{a}_\alpha(t)/dt=[\hat{a}_\alpha(t),H] \tag{7}$$

which has the formal solution

$$\hat{a}_\alpha(t)=\exp(-i\mathcal{L}t/\hbar)\hat{a}_\alpha \tag{8}$$

in terms of the Liouvillian superoperator[13-16] which operates on operators $\hat{A}$ according to $\mathcal{L}\hat{A}=[\hat{A},\hat{H}]$. The Laplace transform (3) is then

$$\tilde{g}(\alpha,z;\alpha)=\mathrm{Tr}\left\{[\mathcal{G}(z)\hat{a}_\alpha]\hat{a}_\alpha^{\dagger}\hat{\rho}\right\}, \tag{9}$$

$\mathcal{G}$ being the Liouvillian propagator $(z-\mathcal{L})^{-1}$, which has a perturbation expansion

$$\mathcal{G}(z)=\mathcal{G}_0(z)+\mathcal{G}_0(z)\mathcal{L}'\mathcal{G}_0(z)+\ldots \tag{10}$$

with $\mathcal{G}_0(z)=(z-\mathcal{L}_0)^{-1}$. The Liouvillian is decomposed into $\mathcal{L}_0+\mathcal{L}'$ in accordance with the decomposition $\hat{H}_0+\hat{V}$ of the Hamiltonian.

Any product of annihilation and creation operators is an eigenoperator of $\mathcal{L}_0$. For example,

$$\mathcal{L}_0\hat{\psi}_K^{\dagger}\hat{\psi}_k^{\dagger}\hat{a}_\beta=(\varepsilon_\beta-\varepsilon_K-\varepsilon_k)\hat{\psi}_K^{\dagger}\hat{\psi}_k^{\dagger}\hat{a}_\beta. \tag{11}$$

$\mathcal{L}'$ converts such a product into a sum of products with factors of matrix elements from $\hat{V}$. The diagram rules are illustrated by Fig.1, giving a contribution

$$-(z-\varepsilon_\alpha)^{-1}\sum_{\substack{\beta\gamma\delta\lambda\\(\beta\neq\lambda)}}(\alpha\beta|H|\gamma\delta)(z+\varepsilon_\beta-\varepsilon_\gamma-\varepsilon_\delta)^{-1}(\lambda|H|\beta)$$
$$\times(z+\varepsilon_\lambda-\varepsilon_\gamma-\varepsilon_\delta)^{-1}\langle\hat{a}_\lambda^{\dagger}\hat{a}_\delta\hat{a}_\gamma\hat{a}_\alpha^{\dagger}\rangle \tag{12}$$

to $\tilde{g}$, with $\langle\cdots\rangle$ the ensemble average $\mathrm{Tr}(\cdots\hat{\rho})$. In general (a) the vertices are ordered from left to right in accordance with the order of operation of factors $\mathcal{L}'$ in a product $\cdots\mathcal{G}_0\mathcal{L}'\mathcal{G}_0\mathcal{L}'\mathcal{G}_0\hat{a}_\alpha$ (rightmost factor $\mathcal{L}'$ corresponding to leftmost vertex). (b) The "α-stump" to the left of the first (leftmost) vertex stands for the operator $\hat{a}_\alpha$ on which the superoperator product operates. (c) Of the lines present "just after" (to the right of) a given vertex, forward lines (those directed toward the right) stand for annihilation operators and backward lines (those directed toward the left) for creation operators. Not all of these lines need be connected to the given vertex; some may originate at previous vertices. However, all diagrams are necessarily connected as a result of their origin through commutation. The rules for the contribution of such a diagram to $\tilde{g}$ are: (d) a factor $(z-\varepsilon_\alpha)^{-1}$ for the α-stump on the left. (e) A matrix element factor $(\alpha|H|\beta)$, $(\alpha\beta|H|\gamma\delta)$,

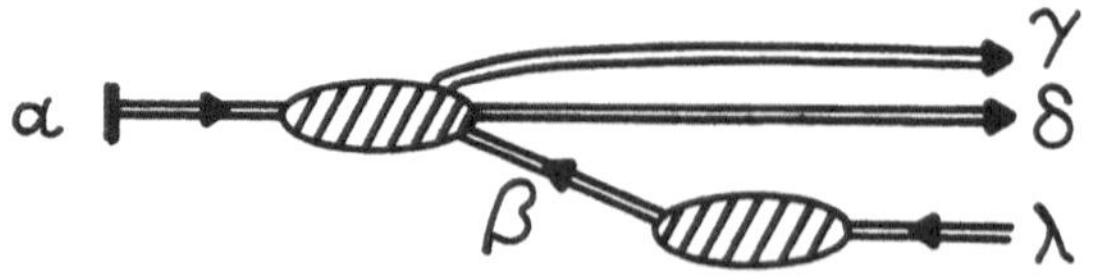

Fig.1. Example of a diagram contributing to $\tilde{g}(\alpha,z;\alpha)$.

etc. for each vertex from the expression for $\hat{V}$, Eq. (6), labels on the left standing for lines entering the vertex and those on the right for lines leaving it; the left-right order of these labels corresponds to the counterclockwise ordering of lines entering the vertex (left-side matrix element labels) and to the clockwise ordering of lines leaving the vertex (right-side matrix element labels). (f) A factor $(z+E_b-E_f)^{-1}$ for each vertex, where E_b (resp. E_f) is the sum of the energies [according to $\hat{H}_o$, Eq. (5)] of the backward (resp. forward) lines present just after the vertex. (g) A factor $\langle\hat{\Pi}_f^{\dagger}\hat{\Pi}_b\hat{a}_\alpha^{\dagger}\rangle$ where $\hat{\Pi}_f^{\dagger}$ (resp. $\hat{\Pi}_b$) is the product of creation (resp. annihilation) operators for all forward (resp. backward) lines present after the last vertex. (h) An overall sign factor as follows: Plus (resp. minus) sign for each pair of vertices connected by a forward (resp. backward) line or lines; plus or minus signs for even or odd permutations of fermion (here proton and electron) operators. In this connection, note that a vertex connected with a vertex to its left by forward lines arises from contraction of the term $\hat{A}\hat{V}$ in the commutator $\mathcal{L}'\hat{A}=[\hat{A},\hat{V}]$, whereas a vertex connected with a vertex to its left by backward lines arises from contraction of the term $-\hat{V}\hat{A}$. Although the particular example shown in Fig.1 contains only atomic lines, diagrams containing proton or electron lines also occur, corresponding to processes involving free (unbound) protons or electrons [see $\hat{V}$, Eq. (6)].

7. α-self-energy diagrams

The pole z_α of $\tilde{g}$ needed in (1) arises from "α-self-energy" diagrams satisfying (i) one outgoing external line is an α-stump matching the incoming α-stump; (ii) any other external lines present after the last vertex are "forward-backward paired," i.e. any such forward line is paired with a backward line with the same index, and vice versa. The contribution of such a diagram is $(z-\varepsilon_\alpha)^{-1}\chi(z-\varepsilon_\alpha)^{-1}f$ where χ is the contribution of the vertices and of all lines except the initial and final ones, and

$$f=\langle\hat{a}_\beta{}^{\dagger}\hat{a}_\gamma{}^{\dagger}\cdots\hat{a}_\gamma\hat{a}_\beta\hat{a}_\alpha\hat{a}_\alpha{}^{\dagger}\rangle \tag{13}$$

with $\beta,\gamma,\cdots$ the labels of the paired final lines; paired proton or electron lines may also occur. f can be decomposed as

$$f=n_\beta n_\gamma\cdots(1+n_\alpha)+f' \tag{14}$$

with $n_\alpha=\hat{a}_\alpha{}^{\dagger}\hat{a}_\alpha$ the statistical average in the ensemble $\hat{\rho}$ (which can be either an equilibrium or nonequilibrium ensemble). $n_\beta\cdots(1+n_\alpha)$ is the statistically uncorrelated contribution whereas f' vanishes in the ideal gas approximation, coincidence of two or more $\beta,\gamma,\cdots$ being a set of measure zero in the thermodynamic limit. The exact f' does not vanish, but only the uncorrelated part of f contributes to the relevant pole.

A self-energy diagram is "α-irreducible" if it cannot be separated into two by cutting an internal forward α-line plus any forward-backward paired lines present with it. Any α-self-energy diagram can be built by joining α-irreducible ones at their α-stumps and extending forward-backward paired lines to the right without connection. A minor generalization of the usual argument[17,18] implies

$$\tilde{g}(\alpha,z;\alpha)=[z-\varepsilon_\alpha-\varepsilon_\alpha(z)]^{-1}(1+n_\alpha)+\tilde{g}'(\alpha,z;\alpha) \qquad (15)$$

where $\Sigma_\alpha(z)$ is the proper self energy and $\tilde{g}'$, the contribution of diagrams other than the α-self-energy ones and of the f', is analytic in the neighborhood of the zero of $z-\varepsilon_\alpha-\Sigma_\alpha(z)$. The diagram rules for Σ_α are the same as those for $\tilde{g}$ except for (i) omission of the initial and final $(z-\varepsilon_\alpha)^{-1}$; (ii) Replacement of the factor $\langle\cdot\cdot\cdot\rangle$ of rule (g) by a product of factors n_α, n_K, or n_k for each forward-backward pair present after the final vertex. Some examples of α-irreducible diagrams and their contributions to Σ_α have already been given[1] and others will be discussed in Sec. 9.

8. Nonlinear Schrödinger equation for ϕ_α

In evaluating the explicit form of the variational equation (1) it is necessary to use the dependences of the matrix elements in $\hat{V}$ [Eq. (6)] on ϕ_α as well as that of[7]

$$\varepsilon_\alpha=(\alpha|H|\alpha)=\int\phi_\alpha{}^*(x_p x_e)H(x_p x_e)\phi_\alpha(x_p x_e)dx_p dx_e \qquad (16)$$

where $H(x_p x_e)$ is the single atom Schrödinger Hamiltonian, including translational and internal kinetic energy and internal Coulomb interaction. Note that ε_α contributes directly to $z_\alpha=\varepsilon_\alpha+\xi_\alpha-i\gamma_\alpha$, whereas ξ_α and γ_α are determined in principle by substitution of this expression for z_α into $z_\alpha-\varepsilon_\alpha-\Sigma_\alpha(z_\alpha)=0$. Carrying out the functional differentiation in (1) then leads to an equation of the general form

$$H(X)\phi_\alpha(X)+\int\Sigma_\alpha(X,X')\phi_\alpha(X')dX'=\sum_\beta\lambda_{\alpha\beta}\phi_\beta(X) \qquad (17)$$

where

$$\Sigma_\alpha(X,X')=\Sigma_\alpha{}^{(r)}(X,X')-i\Sigma_\alpha{}^{(i)}(X,X') \qquad (18)$$

and $\Sigma_\alpha{}^{(r)}$ and $\Sigma_\alpha{}^{(i)}$ are nonlocal hermitian kernels depending implicitly on the ϕ_β. Here we have reverted to the general notation of (1); in the hydrogen case $X=(x_p x_e)$. Some contributions to $\Sigma_\alpha(x_p x_e, x_p' x_e')$ (statistical Hartree-Fock and a few correlation corrections) have already been exhibited.[1]

The inner product of (17) with ϕ_α gives $\lambda_{\alpha\alpha}=\varepsilon_\alpha+\xi_\alpha-i\gamma_\alpha=z_\alpha$. Leaving $\lambda_{\alpha\alpha}=z_\alpha$ on the right and transferring the terms off-diagonal in $\lambda_{\alpha\beta}$ to the left, one obtains a nonlinear eigenvalue equation for ϕ_α and z_α, with a generalized complex optical potential operator. Expressions for the off-diagonal $\lambda_{\alpha\beta}$ follow from the inner product of (17) with ϕ_β. Translational invariance implies $\lambda_{\alpha\beta}=0$ unless ϕ_α and ϕ_β have the same wave vector $\underset{\sim}{k}$. One is interested in ϕ_α which are <u>discrete</u> in that z_α is isolated (even in the thermodynamic limit) from other z_β <u>with the same</u> $\underset{\sim}{k}$.

Variational approximations can be found by the Rayleigh-Ritz method, inserting ansätze for the ϕ_α into $\Sigma_\alpha(z)$ and requiring that z_α be stationary under variation of parameters in the trial ϕ_α. Such a procedure is the same as that of the complex stabilization method,[19] a generalization of the complex coordinate method.

9. Example: Atomic partition function divergence problem

It is well known that the partition function of an isolated atom diverges due to the fact that the series limit is an accumulation point of an infinite sequence of bound-state energies, because of the long range of the nucleus-electron Coulomb interaction. This divergence is usually treated by various ad hoc cutoff procedures, but it has long been recognized that a satisfactory resolution of the divergence problem requires a consistent treatment of quantum many-body effects of the interaction of the atom with its environment; see, e.g., Jackson and Klein.[20] The representation we have been discussing is particularly well suited to a systematic treatment of such effects.

Any mechanism limiting the number of atomic bound states will remove this partition function divergence. In fact, it is easy to think of several candidates for such a mechanism, all of which may be important in some circumstances: (i) Screening of the internal electron-nucleus interaction by interpenetration of "free" charges. This mechanism is the motivation for some phenomenological treatments in which the internal atomic Coulomb interaction is replaced by a screened Coulomb interaction. It is only important when the temperature or density are high enough that an appreciable fraction of the atoms are ionized. (ii) For highly excited states, $\varepsilon_\alpha+\xi_\alpha$ can be positive when $\varepsilon_\alpha<0$. There may then no longer be a discrete state; however, this requires detailed investigation since the z_α are complex. (iii) For a highly excited state, even a weak collision can promote an outer electron into the continuum. Note that these three mechanisms are not disjoint.

As an example of α-irreducible diagrams which may be important for mechanism (i), consider the series shown in Fig.2. The screening arises through summation of the usual ring diagrams, although these have a different appearance in our diagram notation. For ease of physical interpretation, these have been drawn as left-right time-

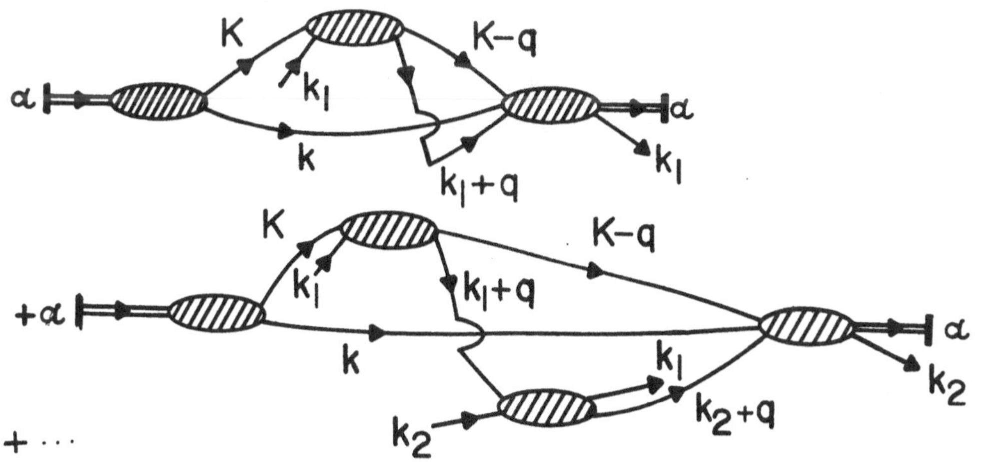

Fig.2. Diagrams representing screening of the internal proton of a hydrogen atom by interaction with free electrons in the medium. Capital K's refer to protons and small k's to electrons.

ordered diagrams. They are converted into Liouvillian diagrams (which involve integration over all time orders) by rotation of each free electron line entering from the left counterclockwise about its respective vertex, so that it becomes a backward line entering from the right, and extension of each such line and its forward partner (with the same k_j) to the right extremity of the diagram.

An example of diagrams contributing to mechanism (iii), through virtual dissociation of the atom followed by collision of its electron and proton with other atoms in the medium, is shown in Fig.3. Note that although the virtual dissociation matrix element $(\alpha|H|Kk)$ vanishes identically for an isolated atom,[6] this is not true for an atomic state ϕ_α perturbed by the medium [solution of Eq. (17)]. As before, the diagrams of Fig.3 are to be converted into Liouvillian diagrams by rotation of the incoming β line so that it becomes a backward β line, extended along with its forward partner to the right extremity of the diagram.

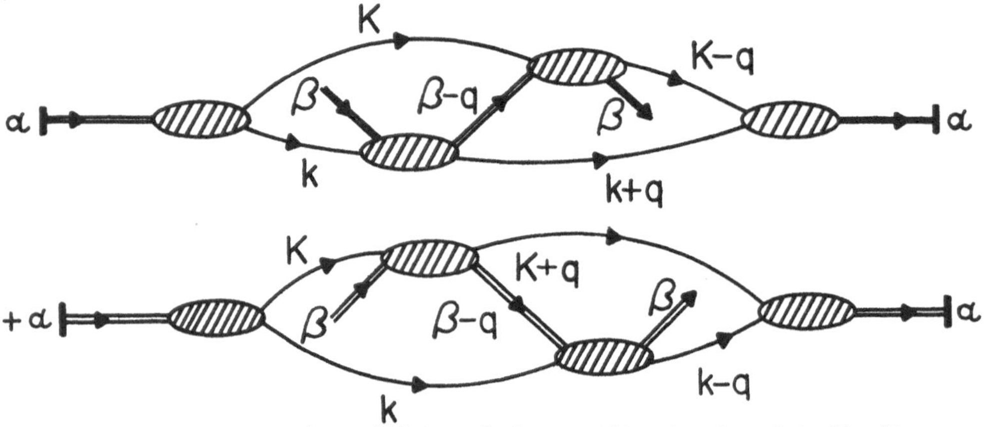

Fig.3. Diagrams representing collision of the constituents of a virtually dissociated atom with atoms of the medium.

Acknowledgement

This work was partially supported by a grant from the M.J. Murdock Charitable Trust through the Chemical Physics Program, University of Oregon.

References

1. M.D. Girardeau: Phys. Rev. Lett. (submitted).
2. H.L. Sahlin, J.L. Schwartz: Phys. Rev. 138, B267 (1965)
3. A. Goldberg, R.D. Puff: Phys. Rev. A10, 323 (1974)
4. Y. Soulet, A. Gomes: J. Stat. Phys. (to be published)
5. M.D. Girardeau: Phys. Rev. Lett. 27, 1416 (1971)
6. M.D. Girardeau: J. Math. Phys. 16, 1901 (1975)
7. J.D. Gilbert: J. Math. Phys. 18, 791 (1977)
8. M.D. Girardeau, J.D. Gilbert: Physica 97A, 42 (1979)
9. M.D. Girardeau, Physica 98A, 139 (1979)
10. M.D. Girardeau: Int. J. Quant. Chem. XVII, 25 (1980)
11. D. Bohm, D. Pines: Phys. Rev. 92, 609 (1953)
12. F.J. Dyson: Phys. Rev. 102, 1217 (1956)

13. I. Prigogine: Non-Equilibrium Statistical Mechanics (Interscience, New York, 1962)
14. R. Balescu: Statistical Mechanics of Charged Particles (Interscience, New York, 1963)
15. R. Balescu: Equilibrium and Nonequilibrium Statistical Mechanics (Wiley, New York, 1975)
16. R. Zwanzig: Physica 30, 1109 (1964)
17. R.D. Mattuck: A Guide to Feynman Diagrams in the Many-Body Problem (McGraw-Hill, London, 1967), Chaps. 10, 14
18. A.L. Fetter, J.D. Walecka: Quantum Theory of Many-Particle Systems (McGraw-Hill, New York, 1971), Chaps. 3, 7, 9.
19. B.R. Junker: Phys. Rev. Lett. 44, 1487 (1980)
20. J.L. Jackson, L.S. Klein: Phys. Rev. 177, 352 (1969)

DENSITY FUNCTIONALS FROM MODELS OF THE ELECTRONIC CHARGE DENSITY.

Jaime Keller, Cristina Keller and Carlos Amador
Facultad de Química, Universidad Nacional Autónoma
México 20, D.F.

ABSTRACT.

The well known Thomas-Fermi theory with its subsequent development, provides a local density functional in terms of powers of the free electron gas parameter $\rho^{1/3}$. For spherically symmetric charge densities, additional terms including $1/r$, $1/r^2$, $\partial/\partial r$ and $\partial^2/\partial r^2$ should be studied. The way they arise from the study of charge densities and the results that could be expected with their use, are analyzed in the present paper.

INTRODUCTION.

Physical considerations and models of the electron charge density allow the derivation of electronic density functionals for the total energy. For the free electron gas case there is only one parameter in the theory: the density ρ, then the dimensional quantities that should be formed (x^{-1} and x^{-2}) can be obtained from $\rho^{1/3}$ and $\rho^{2/3}$ for the potential and kinetic energy parts respectively. Corrections to these terms are obtained either from dimensionless functionals of ρ or from an appropriate use of the operator $\partial/\partial x$ or $\partial^2/\partial x^2$. For spherically symmetric charge densities (a sum over angular momenta components) the parameters in the theory are $\rho^{1/3}$, $\rho^{2/3}$, $\partial/\partial r$ and $\partial^2/\partial r^2$, besides the obvious $1/r$ and $1/r^2$.

The present paper shows how this can be done with illustrations for spherically symmetric atomic charge densities.

The kinetic energy functional for the central symmetric problem is

$$E = \int \left\{ -\frac{1}{4}\,\partial^2/\partial r^2 - \frac{1}{r}\,\partial/\partial r + A/r^2 + 0.5\,\partial^2 \ell n\rho/\partial r^2 \right\} \rho(r)\, 4\pi r^2 dr \qquad (1)$$

Here A should not be strictly constant but depends on Z. The use of the value A = - .0519 + .0041 Z gives results which agree with the total kinetic energy within 3% for the first 45 atoms of the periodic table. But as the difference with the Hartree-Fock total kinetic energy depends smoothly with Z, an optimized value of A could be used for a given range of atomic numbers to obtain the best results in practice.

A potential energy functional including accurate terms for exchange and correlation which has been obtained in previous work, is also shown.

DENSITY FUNCTIONAL.

The search of a local density functional and in general of a density functional within the Hohenberg-Kohn-Sham theory[1,2] to study the electronic structure of atoms, molecules and condensed matter should, desirably, be undertaken from a first principles approach. In practice the procedure requires the introduction of auxiliary conditions from symmetry and statistics. The Thomas-Fermi theory performs a free electron gas analysis of the various terms. It could therefore be possible to introduce the same information for other systems in the construction of the density functionals. This is made step by step in the present paper for spherically symmetric atoms. The procedure can be improved systematically and extended to molecules and condensed matter.

To develop a density functional we shall start from the Schrödinger equation for an atom (in Rydberg units)

$$\sum_i \left\{ -\nabla_i^2 - \frac{2Z}{r_i} + \sum_{j\neq i} \frac{2}{r_{ij}} \right\} \psi(\{\underline{r}_i\}) = E\psi(\{\underline{r}_i\}) \tag{2}$$

The usual Hartree electronic wave function for an atom ψ_H

$$\psi_H(\{\underline{r}_i\}) = \prod_i \phi_i(\underline{r}_i) \tag{3}$$

could be substituted in equation (2) if, as usual in local density theory, an exchange-correlation effective potential is introduced.

The choice of the set of monoelectronic wave functions ϕ_i contains the long known rules to find either the lowest energy state for an atom or a fixed <u>ad hoc</u> electronic configuration. We call this set of rules the "rule of the diagonals"[3] which states

a) The electrons enter into the unoccupied subshells with lowest value for $n + \ell$. The subshells which share a common value of $n + \ell$ are called diagonals.
b) Within each diagonal the subshell with lowest n is occupied first.
c) Within each subshell the electron configuration is chosen to be that of the highest possible multiplicity.

These rules, in practice, fix the set of ℓ_i needed to compute the expectation value of the angular part of the kinetic energy

$$K_\Omega = \sum_i \ell_i(\ell_i + 1) < \frac{1}{r_i^2} > \qquad (4)$$

A useful way of splitting the total energy is into radial kinetic energy, angular kinetic, nuclear-electron potential and electron-electron potential contributions. This has been done to several degrees of approximation by previous authors. When electron gas theory is used the total kinetic energy is directly found.

But, and this is the main idea of this paper, if an electronic configuration and a form for the monoelectronic functions ϕ_i are given or, equivalently, of the density matrices, an equation for the total energy can be written explicitly. The resulting functional can be parametrized to optimize it a posteriori.

Let us assume in the present example that (neglecting orthogonality for different n of a given ℓ) each ϕ_i is of the type

$$\phi_i(\underline{r}_i) = C r^{\ell_i} e^{-ar_i} Y_{\ell_i m_i}(\hat{r}_i) \qquad (5)$$

C and a depend on n_i and ℓ_i.

Then the n-electron wave function (3) can be substituted in (2). The angular integrations are performed, the Hartree-Coulomb potential is introduced together with the exchange-correlation energy density $\varepsilon_{xc}(r)$, and the result is multiplied by the left by the wave function; the result of these manipulations is

$$\sum_i \left\{ -\frac{1}{4}\left\{ \frac{\partial^2}{\partial r^2} + \frac{4}{r}\frac{\partial}{\partial r} \right\} + \frac{\ell_i}{2r^2} + \frac{\ell_i(\ell_i+1)}{r^2} - \frac{2Z}{r} + \int_0^r 2\frac{\rho(r')}{r} 4\pi r'^2 dr' + \right.$$
$$\left. + \int_r^\infty 2\rho(r')4\pi r' dr' + \varepsilon_{xc}(r) \right\} |\psi|^2 = E|\psi|^2 \qquad (6)$$

Finally, from this expression, a density functional can be obtained averaging over the ℓ_i.

For each summation in (6) the integral over the $j \neq i$ has to be made; this is straightforward for the first and the last four terms. If a suitable A could be found, the final functional would be

$$\int \left\{ -\frac{1}{4}\left[\frac{\partial^2}{\partial r^2} + \frac{4}{r}\frac{\partial}{\partial r} \right] + A\frac{1}{r^2} - \frac{2Z}{r} + \varepsilon_{coul}(r) + \varepsilon_{xc}(r) \right\} \rho(r) d^3r = E \qquad (7)$$

Use of the one-particle density matrix $\rho(r)$ is allowed by the fact that the local electron-electron potential $\varepsilon_{coul}(r) + \varepsilon_{xc}(r)$ has been introduced.

In equation (7) the (Z dependent)

$$A = \overline{\ell(\ell + 3/2)} = \left(\sum_i \langle \ell_i(\ell_i + 3/2) r_i^{-2} \rangle \right) \left(\sum_i \langle r_i^{-2} \rangle \right)^{-1} \tag{8}$$

stands only for the form (5) of the wave function, a more general (orthogonal) set would not allow this definition. The term $\ell_i/2r^2$ can also be obtained from (5) using $\partial^2 \ell n\rho/\partial r^2$ as an extra term in (7).

The functional (7) should be used with a density which avoids the inaccuracies introduced. This can be accomplished if the density is constructed as

$$\rho(r) = \sum_i \phi_i^2(r) \tag{9}$$

with improved forms for ϕ_i (spherical averages performed to avoid cross terms, as usual in practice for density functional theory). The parameters used to construct the final ϕ_i can, otherwise, be optimized from the functional itself. The free electron gas density can be expressed as a sum over spherical harmonics and used with (7).

The Coulomb, exchange and correlation energy densities may be given in a local density functional approximation also. In previous papers[4,5,6] we already modeled the exchange and correlation part of the potential. This can be extended to the Coulombic term

$$V_{coul}(r) = \frac{4}{3} D\rho^{1/3}(r) = \frac{\partial \varepsilon_{coul}(r)}{\partial \rho(r)} \tag{10}$$

if a form is given for the electron gas pair-correlation function. A suitable first approximation for the second-order density matrix could be

$$\Pi(1,2) = \rho^2(1)\, e^{-r_{12}/r_o} \left\{ 1 - f_{xc}(1,2) \right\} \quad , r_{12} \leq r_o ; \tag{11}$$

with r_o adjusted in such a way that the total charge n (in units of the electron charge) is

$$\rho(1) \int_o^{r_{max}} e^{-r_{12}/r_o}\, 4\pi r_{12}^2 dr_{12} = n \tag{12}$$

$r_{max} = mr_o$, m is an integer we fix so that the model obeys the virial theorem.

and $f_{xc}(1,2)$ is described in references [4], [5] and [6], including the spin polarized case. The energy density $\varepsilon_{coul}(r)$ resulting from the Coulombic potential of the electron gas is

$$\varepsilon_{coul}(r) = 1.7921\ Z^{2/3}\ \rho^{4/3} = D\rho^{4/3}, \tag{13}$$

A similar approach for the Coulomb energy was introduced by Parr, Gadre, Bartolotti and Handy[7] in the Hohenberg-Kohn formalism and by Gázquez[8] in the Kohn-Sham formalism.

The exchange-correlation part has also the form

$$V_{xc}(r) = B\left(1 + G(\rho(r))\right)\rho^{1/3}(r) \tag{14}$$

with B and the functional G given in [4], [5] and [6].

The density (9) can be fixed or given a form with parameters which will be optimized by conventional methods.

In conclusion, we have shown a systematic procedure to model the density functional and simultaneously the electron density for atoms. In the case of molecular and solid state physics, the functionals here discussed can be used directly in the renormalized atom approach, the one center and some cellular methods, allowing for straightforward evaluation of total energies.

We should mention that there are alternative forms of writing the functional (7). The numerical analysis which follows should be useful in this respect.

NUMERICAL RESULTS.

Table 1 shows the contributions to the kinetic energy obtained from (near) Hartree-Fock charge densities using (1) and, as a comparison, the Hartree-Fock values (HF) and the Thomas-Fermi (TF) values ($E_k = \frac{3}{5}(3\pi^2)^{2/3}\int \rho^{5/3}d\tau$). The first (gradients) term in (1) is insufficient but good for light atoms. The $\partial^2 \ell n\rho/\partial r^2$ (logarithmic) term brings the results closer to HF for medium atoms but overestimates the kinetic energy for the light atoms. The third term contains then possitive as well as negative contributions. Equation (1) gives kinetic energies very close to HF with a r.m.s. percentage error of 1.8. This is better than Thomas-Fermi plus the original Weizacker correction for non-homogeneity but not as good as improved Thomas-Fermi functionals[9].

TABLE I. KINETIC ENERGY COMPONENTS FOR HARTREE-FOCK DENSITIES.

Z	(1)	(2)	(3)	% error	HF	TF
2	5.5075	6.2986	5.8169	- .0260	5.9720	5.121
3	15.1450	16.8082	15.6083	.0337	15.1000	13.359
4	29.0181	31.9527	29.8923	.0216	29.2593	26.256
5	47.1706	53.1643	50.2025	.0218	49.1310	43.930
6	69.9268	80.9363	77.1199	.0219	75.4653	67.296
11	257.7361	328.1413	324.6415	.0026	323.8063	297.56
12	311.3269	399.2921	397.6140	- .0041	399.2643	368.00
13	370.4601	478.4927	479.5279	- .0087	483.7402	446.74
15	505.5525	661.1085	670.7906	- .0156	681.4457	631.08
16	581.6776	765.0021	780.8960	- .0178	795.0112	737.22
17	663.6967	877.5827	901.1391	- .0194	918.9805	853.36
20	946.6783	1268.4628	1325.2371	- .0209	1353.593	1260.14
25	1544.3070	2115.9185	2271.3919	- .0126	2300.4271	2143.26
28	1985.1040	2754.1175	3002.3203	- .0042	3014.8838	2816.00
30	2316.4070	3238.7435	3565.9148	.0025	3557.1981	3331.44
33	2875.5720	4059.5875	4535.5042	.0146	4470.0241	4197.58
35	3291.5430	4675.2415	5273.1244	.0210	5164.5627	4840.38

(1) $= -\frac{1}{4}\,\partial^2/\partial r^2 - \frac{1}{r}\,\partial/\partial r$ (2) $=$ (1) $+ 0.5\;\partial^2 \ln\rho/\partial r^2$ (3) $=$ (2) $+ A\langle 1/r^2\rangle$

$A = -0.0519 + 0.0041Z$ r.m.s. = 0.0181

Figure 1 shows the above results more clearly and suggests that the scaling of any one of the terms (alone or with a second one) could be useful. Figure 2 shows the value of A in equation (1) that should be used, if the "logarithmic" term is suppressed, to obtain the HF kinetic energies.

On the other hand the total HF kinetic energies E_k could have been obtained from the parametrized formulae

$$E_k = 0.7722\, Z^{1/5} \left\langle -\frac{1}{4}\frac{\partial^2}{\partial r^2} - \frac{1}{r}\frac{\partial}{\partial r}\right\rangle \quad (\mathrm{Ry}) \tag{15}$$

$$E_k = 0.3861\, Z^{1/5} \langle 1/r^2 \rangle \quad (\mathrm{Ry}) \tag{16}$$

(the first coefficient is the double of the second!)

These formulae should be useful to obtain other type of relations, for example from the Parr and Gadre relation[10]

$$E_T = -E_k = -1.0398\, Z^{2.3947} \quad (\mathrm{Ry}) \tag{17}$$

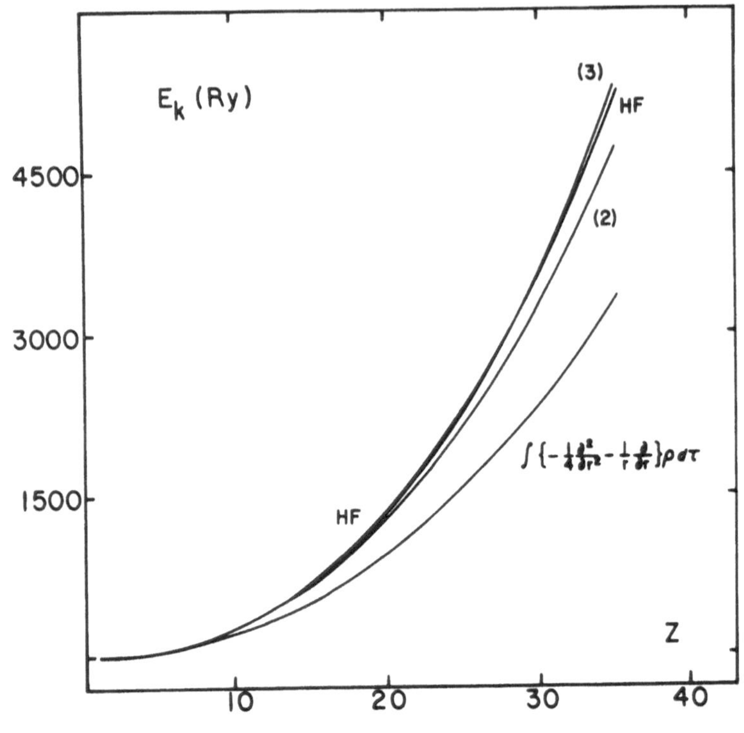

Figure 1

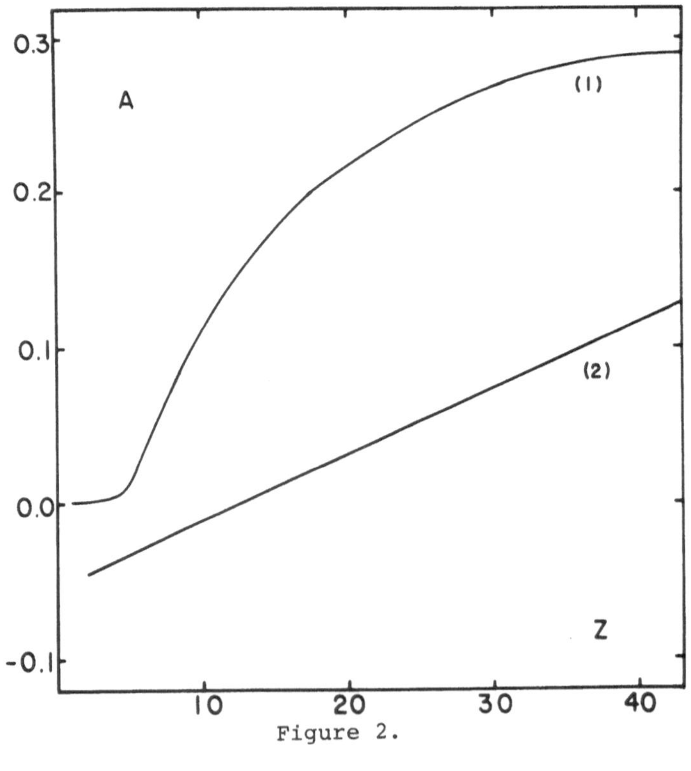

Figure 2.

we obtain

$$\langle \frac{1}{r^2} \rangle = \frac{1.0398}{0.3753} Z^{2.1947} \quad \text{(a.u.)} \tag{18}$$

Finally, Table II shows the values obtained with (13) for the electron-electron Coulomb interaction.

TABLE II.

Electron-Electron Coulomb Energy for Hartree-Fock Atomic Charge Densities Using the $(2/r_{ij})$ Functional and the $dZ^{2/3}\rho^{4/3}$ Approximate Functional. d = 1.7921 (Rydberg units).

Z	$1 = \langle 2/r_{ij} \rangle$	$2 = \langle dZ^{2/3}\rho^{4/3} \rangle$	$\frac{2-1}{1}$
2	2.5741	3.1541	0.2253
3	8.2017	7.6045	-0.0728
4	14.4067	13.9227	-0.0336
5	23.2581	22.5667	-0.0297
6	35.7043	34.4717	-0.0345
7	52.3522	49.8783	-0.0473
8	73.2312	69.3661	-0.0528
9	99.6268	93.4348	-0.0622
11	160.1051	152.9021	-0.0450
12	191.7887	185.4778	-0.0329
13	225.8359	221.2618	-0.0203
15	307.1687	304.9702	-0.0072
16	354.1731	353.3905	-0.0022
17	406.2906	406.5086	-0.0005
19	515.4579	523.6388	0.0159
20	570.8364	585.2978	0.0253
23	791.5412	809.5511	0.0228
25	972.4858	988.4735	0.0164
26	1072.396	1089.8655	0.0163
28	1295.166	1304.4992	0.0072
30	1549.740	1549.5837	-0.0001
33	1924.141	1937.3470	0.0069
35	2199.101	2223.4977	0.0111
38	2619.405	2685.0758	0.0251
42	3282.344	3386.9152	0.0319
47	4266.963	4423.8140	0.0368

References.

1. P. Hohenberg and W. Kohn, Phys. Rev. 136,B864 (1964).
2. W. Kohn y L.J. Sham, Phys. Rev. 140, A1133 (1965).
3. J. Keller, Ciencia (Mex.) XVI, 86 (1956).
4. J.L. Gázquez and J. Keller, Phys. Rev. A16, 1358 (1977).
5. J. Keller and J.L. Gázquez, Phys. Rev. A20, 1289 (1979).
6. J.L. Gázquez, E. Ortiz and J. Keller, Int. J. Quantum Chem. Symp. 13, 377 (1979).
7. R.G. Parr, S.R. Gadre, and L.J. Bartolotti, Proc. Natl. Acad. Sci. U.S.A., 76, 2522 (1979). S.R. Gadre, L.J. Bartolotti and N. C. Handy, J. Chem. Phys. 72, 1034 (1980).
8. J.L. Gázquez, paper presented at the Seventh Canadian Symposium on Theoretical Chemistry, 1980, to be published.
9. D.R. Murphy and W.P. Wang, J. Chem. Phys. 72, 429 (1980).
10. R.G. Parr and S.R. Gadre, J. Chem. Phys. 72, 3669 (1980).

EFFECTIVE HAMILTONIAN THEORY: RECENT FORMAL RESULTS AND NON-NUCLEAR APPLICATIONS*

B. H. Brandow
Theoretical Division
Los Alamos National Laboratory
University of California
Los Alamos, NM 87545

I. Introduction

Effective Hamiltonian theory is actually quite an old subject, dating back to a 1929 paper of Van Vleck,[1] and a subsequent refinement by Kemble.[2] Their approach, the so-called canonical transformation or unitary transformation method, is widely known throughout all branches of quantum physics, from high-energy theory to quantum chemistry. The alternative is to use degenerate perturbation theory. This is available in several different forms, most notably those of Kato,[3] Bloch,[4] and des Cloizeaux.[5] The past 15 years have seen an intensive development of the subject by nuclear physicists,[6-8] based on one of these forms of degenerate perturbation theory.

At first sight, the various degenerate perturbation formalisms all seem more complicated than the unitary approach, their interrelations seem rather obscure, and moreover they seem totally unrelated to the unitary method. It is now recognized, however, that most of the perturbative formalisms which lead to Hermitean effective Hamiltonians are, in fact, completely equivalent, and are connected by simple identities.[9-11] We will focus here on a point which is far less obvious: the fact that, subject to an important caveat, these Hermitean perturbation formalisms are actually identical, term-by-term, to the unitary transformation approach, when the latter's effective Hamiltonian is expanded as a perturbation series. The study of this connection also has the important benefit of revealing the relative merits of these approaches for practical applications. The conclusion is that the approach familiar to nuclear physicists is by far the most powerful and efficient one, especially for many-body applications.

Another important recent development concerns the coupled-cluster formalism for many-body systems. This form of many-body theory was originally developed by Coester and Kümmel[12] for closed-shell systems. Over the years, both Coester[13] and Kümmel and coworkers[14] have worked on extensions to open-shell systems, and Zabolitzky and Ey[15] have done some highly sophisticated nuclear calculations using this approach. Quite recently, however, Lindgren[16] has developed a beautifully clean formulation of the coupled-cluster method for open-shell systems. Although not essentially different from the previous work, it has the important pedagogical advantages of being clear, concise, and quite general. We shall outline the main features of Lindgren's formula-

*Work supported by US Department of Energy

tion. This alternative to perturbation theory may well be advantageous for certain systems, as it suggests different classes of approximations;[17] it certainly deserves much further development.

Although the many-body linked-cluster form of degenerate perturbation theory was first developed for nuclear physics applications,[6-8] it is clear that this is, in fact, a very general technique for deriving effective Hamiltonians for the low-energy excitations of any many-fermion system. (This formalism has also been extended to handle the elementary excitations of the boson system of liquid ^{4}He.[18]) I shall briefly describe three recent applications where this formalism has contributed significantly to the understanding of other many-body systems. These applications are (a) the derivation of effective spin Hamiltonians in magnetic insulator systems; (b) derivation and ab initio calculation of effective π-electron Hamiltonians for planar conjugated hydrocarbon molecules, and (c) understanding the so-called valence fluctuation phenomenon exhibited by certain rare-earth compounds.

The present formal matters are reviewed in considerably more detail in Ref. 11, together with application (b). Application (a) is covered in depth in Ref. 19, whereas application (c) is quite new, and is yet only partially published.[20]

II. Degenerate Perturbation Theory

We shall first outline what we regard as the most simple and efficient formulation of degenerate perturbation theory. We present only the key equations, and refer the reader to Ref. 11 for further details.

The simplest version of degenerate perturbation theory is the Brillouin-Wigner form, where the effective interaction matrix is $P\mathcal{V}P$, and $\mathcal{V}$ is determined from

$$\mathcal{V} = VP + \frac{Q}{e}\mathcal{V} = V\sum_{n=0}^{\infty}\left(\frac{Q}{e}V\right)^n P \equiv V\Omega \quad . \tag{2.1}$$

Here P and Q are the usual projection operators onto the "model" and "virtual" subspaces $(P + Q = I)$, $H = H_0 + V$, and $e \equiv E-H_0$. The effective Hamiltonian of this formulation is $P(H_0 + \mathcal{V})P$. Its eigenvalues E are identical to certain eigenvalues of the complete Hamiltonian H, and its eigenvectors represent the "model projections" $P\Psi$ of the corresponding complete eigenvectors Ψ. For later use we have also introduced the wave operator Ω, which has the property that $\Psi = \Omega(P\Psi)$.

From a practical standpoint, this formulation has three serious drawbacks. Most obvious is that $\mathcal{V}$ depends on the (initially unknown) eigenvalue E, and furthermore this operator must be re-determined for each of the desired eigenstates with different eigenvalues E. Second, for many-body systems the Brillouin-Wigner expansion (2.1) lacks the important linked-cluster property. (Even for relatively few-body systems this property remains vital; see Ref. 11.) Finally, the above effective Hamiltonian has a non-Hermitean aspect, since its eigenvectors are not mutually orthogonal. (They are merely the projections $P\Psi$ of the complete eigenvectors Ψ.) This is undesireable because the various phenomenological effective Hamiltonians which one would like to explain are invariably Hermitian.

The first two of these deficiencies can be removed by expanding the energy dependence of $\mathcal{V}$ in a Taylor series. This eventually leads to the implicit equation

$$\mathcal{W} = \sum_{r=0}^{\infty} \mathcal{V}_r [-\mathcal{W}]^r , \qquad \mathcal{V}_r = \frac{(-1)^r}{r!} P \left. \frac{d^r \mathcal{V}}{dE^r} \right|_{E_o} P \quad . \tag{2.2}$$

This $\mathcal{W}$ replaces $P\mathcal{V}P$ as the effective interaction matrix. The Rayleigh-Schroedinger (RS) expansion for $\mathcal{W}$, i.e. the ordinary power series in V, can now be obtained by first solving (2.2) recursively, in terms of matrix products of the various $\mathcal{V}_r$'s, then replacing the latter by their perturbation series, thus:

$$\begin{aligned}\mathcal{W} &= \mathcal{V}_o + \mathcal{V}_1 [-\mathcal{V}_o] + \mathcal{V}_2 [-\mathcal{V}_o]^2 + \mathcal{V}_1 [-\mathcal{V}_1][-\mathcal{V}_o] + \mathcal{O}(\mathcal{V}_r)^4 \\ &= VPV + PV\frac{Q}{e_o}VP + PV\frac{Q}{e_o}V\frac{Q}{e_o}VP + PV\frac{Q}{e_o}\frac{Q}{e_o}V(-P)\,VP + \mathcal{O}(V^4) \quad .\end{aligned} \tag{2.3}$$

Here $e_o \equiv E_o - H$, and E_o comes from PH_oP, assuming H_o exactly degenerate within P. (This restriction is only for simplicity; it can easily be removed.) Finally, $\mathcal{W}$ may be replaced by a "Hermitized" effective interaction matrix,

$$\mathcal{K} = \tfrac{1}{2}[(I+\theta)^{\frac{1}{2}}\mathcal{W}(I+\theta)^{-\frac{1}{2}} + \text{h.c.}] = \tfrac{1}{2}[\mathcal{W}+\mathcal{W}^{\dagger}] + \mathcal{O}(\theta) \quad , \tag{2.4}$$

where $\theta = \Omega^{\dagger} Q \Omega$ has a well-defined RS expansion which follows from the preceeding equations. This $\mathcal{K}$ operator has the same eigenvalues as $\mathcal{W}$, but its eigenvectors are now precisely orthogonal.

The RS expansion for this $\mathcal{K}$ operator can be generated by several alternative methods,[4,5,11] but the present procedure has significant practical advantages. The expansion (2.1) is obviously a geometric series, and (2.2) also has a geometric-like character [since one is expanding the denominators $(E_o + \Delta E - H_o)^{-1}$]. One finds, therefore, that (2.1)-(2.3) present many opportunities for infinite partial summation of the series, a technique of great importance for practical applications. (See for example Ref. 8.) The last step (2.4), on the other hand, is by far the most complicated one, from the standpoint of its effect on the structure of the perturbation series. But in all applications to date that we are aware of, the lack of Hermiticity in $\mathcal{W}$ has turned out to be quantitatively quite minor. It should usually, therefore, be quite adequate to use just the "zeroth order" approximation, $\mathcal{K} \approx \tfrac{1}{2}(\mathcal{W}+\mathcal{W}^{\dagger})$. This is a <u>major</u> simplification. The alternative methods for generating the RS expansion for $\mathcal{K}$ have this complicated "Hermitization aspect" inextricably mixed with the other aspects, which therefore severely restricts the possibilities for efficient partial summation.

III. The Unitary Transformation Method

Van Vleck[1] introduced the idea of a unitary transformation of H,

$$\mathcal{H} = U^{-1} HU \quad , \tag{3.1}$$

where U is to be chosen such that

$$Q\mathcal{H} P = 0 \quad . \tag{3.2}$$

The desired effective Hamiltonian is then $P\mathcal{H}P$. Unfortunately, (3.2) does not suffice to determine U or $P\mathcal{H}P$ uniquely, since arbitrary unitary transformations within the P subspace are still allowed. It seems most reasonable to add a requirement that U should have as little effect as possible within the P subspace (and likewise for the Q subspace). Kemble[2] suggested that U should be expressed in a matrix-exponential form,

$$U = e^G, \quad G^\dagger = -G \quad , \tag{3.3}$$

whereby this somewhat vague "minimal effect" requirement can be incorporated via the simple subsidiary conditions

$$PGP = 0, \quad QGQ = 0. \tag{3.4}$$

One may then express G as a formal expansion in powers of V, and collect the resulting terms in $\mathcal{H}$ for each order in V. The condition (3.2) can then be imposed separately for the terms of each order in V [subject also to $G^\dagger = -G$ and (3.4)] to determine the successive terms in the G expansion.

The net result of this procedure is to generate a Rayleigh-Schroedinger expansion for the effective Hamiltonian $P\mathcal{H}P$. In common with the preceeding $P(H_o + \mathcal{K})P$ this should generate some subset of the exact eigenvalues E, but it is not at all obvious whether these two effective Hamiltonians should have the same sets of model eigenvectors. These effective Hamiltonians might well differ by a unitary transformation within P, in which case their perturbative expansions would also be different.

IV. Formal Equivalence of the Perturbative and Unitary Approaches

It turns out that the effective Hamiltonians of Sections II and III are not merely unitarily equivalent; they are actually identical. That is, their respective Rayleigh-Schroedinger expansions are identical. This welcome result was first recognized by Klein,[9] and was later proven in a quite different manner by Jørgensen.[31] (Klein's proof unfortunately contains some errors; a corrected proof consists of two parts, given in appendices in Refs. 11 and 19.) In retrospect, one can see that both proofs are based on the idea (Section III) that the transformation from H to the effective Hamiltonian should have "minimal effect within P." The proofs also share a common strategy: a precise definition is given for this "minimal effect," this requirement is shown to have a unique solution, and then each of the effective Hamiltonians is shown to satisfy this requirement.

Jørgensen's proof[21,11] is based on the requirement that PUP should be Hermitean. (To motivate this choice, consider the one-dimensional case where $U = e^{i\theta}$. Here Hermiticity requires that $U = \pm 1$. The -1 possibility is then eliminated by requiring continuity as $V \to 0$.) It turns out that the subsidiary conditions (3.4) are sufficient (although not necessary) to make PUP Hermitean.[11] It is less obvious how to apply this requirement to $P(H_0 + \mathcal{K})P$, since no U is visible here. There are, however, some simple identities[11] which show that

$$P(H_0 + \mathcal{K})P = (\Omega^\dagger\Omega)^{-\frac{1}{2}} (\Omega^\dagger H\Omega)(\Omega^\dagger\Omega)^{-\frac{1}{2}}, \tag{4.1}$$

whereby $UP = \Omega(\Omega^\dagger\Omega)^{-\frac{1}{2}}$, and thus $PUP = (\Omega^\dagger\Omega)^{-\frac{1}{2}}$, which is now obviously Hermitean.

Klein's proof is based on the following variational problem: Let $\{\alpha\}$ be the set of d eigenstates which are described by the d-dimensional model Hamiltonian (d = dimension of P), and let $\{\Psi_\alpha\}$ be the corresponding set of complete eigenvectors (eigenvectors of H). Let $\{B_\alpha\}$ be a set of d vectors which lie entirely within P. These B_α's are required to be orthonormal, but are otherwise arbitrary; the infinity of possible choices for $\{B_\alpha\}$ are therefore related by unitary transformations within P. The Ψ_α's are also required, here, to have unit norms, but of course they do <u>not</u> lie entirely within P. The problem is to find the basis set $\{B_\alpha\}$ such that the quantity $\Sigma_\alpha \langle B_\alpha - \Psi_\alpha \; B_\alpha - \Psi_\alpha \rangle$ attains its absolute (i.e. global) minimum. The solution of this vector variational problem is known to be unique,[22,19] and the sets of model eigenvectors of the perturbative and unitary effective Hamiltonians both satisfy this condition (as proven, respectively, in the appendices of Refs. 19 and 11). Finally, since the eigenvalues and eigenvectors of these model Hamiltonians are identical, the operators themselves much be identical.

Jørgensen's proof leads to important insights about the relative merits of the perturbative and unitary approaches. As described in Section III, the unitary approach has a simplicity and elegance which has appealed to generations of physicists. In reality, however, this simplicity is only an illusion. Beyond the lowest orders the recursive procedure for determining G and $P\mathcal{H}P$ becomes exceedingly tedious, and offers no general insights of the type needed for infinite partial summations. Two reasons for this complexity can now be seen. One is that the unitary approach must necessarily include the complicated "Hermitisation aspect" of (2.4). The other is that the perturbation series for $UP = \Omega(\Omega^\dagger\Omega)^{-\frac{1}{2}}$, as determined by the methods of Section II, does not have an exponential-like character, thus it is "unnatural" (i.e. inefficient) to focus on the (matrix) logarithm of U, as is done in the Van Vleck-Kemble approach.

We must not leave this subject without mentioning an important caveat. The original works did not fully specify how the unitary approach is to be implemented in higher orders. We have presumed a <u>single</u> unitary transformation, whose G contains all orders in V. There have been a number of applications, however, which employ a succession of unitary transformations,

$$U' = U_1 U_2 U_3 \cdots = e^{G_1} e^{G_2} e^{G_3} \cdots , \tag{4.2}$$

where each transformation enforces (3.2) for one higher order in V. The subsidiary conditions (3.4) are commonly imposed for each of the G_n's. (A well-known example of this procedure is the work of Foldy and Wouthuysen,[23] whose object was to eliminate the small components of the Dirac equation.) Explicit calculation shows that this gives different results from the methods of Sections II and III; specifically, PU'P - PUP ~ $\mathcal{O}(V^3)$ and $P\mathcal{W}P - P\mathcal{W}P \sim \mathcal{O}(V^4)$. Such differences have sometimes led to confusion, as pointed out by Friar.[24]

V. Coupled-Cluster Formalism for Open-Shell Systems

Elementary manipulation of the Schroedinger equation leads to the operator identity

$$[\Omega, H_o] = V\Omega - \Omega V\Omega \quad . \tag{5.1}$$

(It is to be understood that $\Omega = \Omega P$, i.e., that Ω acts only on the P subspace.) This is one of the two basic ingredients of Lindgren's formulation.[17] His other ingredient refers explicitly to the many-body nature of an open-shell many-fermion system, as follows.

For closed-shell systems, it is well known[6] that the wave operator Ω can be expressed as

$$\Omega = e^{\hat{W}} \quad , \tag{5.2}$$

where $\hat{W} = \Sigma_{n=1}^{N} \hat{W}_n$, N being the total number of particles. Each $\hat{W}_n$ corresponds to the sum of all linked but open perturbation diagrams which lead to the creation of n particle-hole pairs (starting from the closed-shell configuration Φ_o). The caret symbol is a reminder that $\hat{W}$ is a second-quantized operator, with a particle creation or annihilation operator attached to the end of each outgoing particle or hole line associated with an individual $\hat{W}_n$ component amplitude. Lindgren noted that this representation is inadequate for open-shell systems, and that it should be replaced by

$$\Omega = \left\{e^{\hat{W}}\right\} = \sum_{r=0}^{\infty} \frac{1}{r!}\left\{\hat{W}^r\right\} , \tag{5.3}$$

where { } indicates normal-ordering of the various creation and annihilation operators. Failure to do this would lead to many spurious terms.[11] (This point was also recognized by Kümmel and coworkers,[14] but was not clearly stated.)

It is easy to see that systematic use of (5.1) as a recursion formula will generate the Rayleigh-Schroedinger perturbation expansion for Ω. Following this procedure, Lindgren was able to prove by induction that the perturbation-theoretic Ω does indeed have the form (5.3), where each term in $\hat{W}$ is fully connected, and is also "open" in the sense of always leading to states in the Q subspace. (It then follows that $\mathcal{W} = PV\Omega$ is fully linked.) In a similar manner, Lindgren then obtained a formal equation for $\hat{W}$ itself. This translates into an inhomogeneous set of equations for the various cluster amplitudes within $\hat{W}$. If one adopts some suitable (physically motivated) truncation of these equations, it becomes possible to obtain the "most relevant" amplitudes directly, without using perturbation theory. This is the open-shell analog of the coupled-cluster technique.

VI. Recent Non-Nuclear Applications

A. Effective Spin Hamiltonians for Magnetic Insulator Materials

Magnetic insulator materials include nearly all halides, most oxides, and a number of sulphides of the 3d (transition) and 4f (rare earth) metals, as well as some of the 5f (actinide) metals, plus many other ionic compounds of these metals; thousands of examples are known. Their magnetic behaviors can generally be described by effective Hamiltonians of the form

$$\mathcal{H} = -\sum_{ij} J_{ij}\vec{S}_i\cdot\vec{S}_j + \text{small corrections}, \qquad (6.1)$$

where the couplings J_{ij} are typically found to be antiferromagnetic and of fairly short range. This is the so-called Heisenberg spin Hamiltonian, and efforts to understand its microscopic origin date back to the late 1920's. The so-called superexchange theory of Anderson[25] is the standard in this field, and gives a good qualitative and semi-quantitative account of the physics. But this theory is restricted to an isolated pair of magnetic ions in a non-magnetic host crystal [two Ni's in MgO, two Cr's in Al_2O_3 (= "ruby"), etc.]. All previous attempts to extend this (or any other) theory to a crystal with a macroscopic number N of magnetic ions had met with difficulties of the unlinked-cluster type: terms involving high powers of N. This is known historically as the nonorthogonality catastrophe, first observed by Slater[26] in 1930, and it is quite possibly the first unlinked-cluster problem to be recognized since the development of wave mechanics. It was, therefore, quite gratifying to find that the folded-diagram expansion resolves this problem in a clean, general, and complete manner.[19] A curious feature of this application is that the appropriate H_0 now contains two-body as well as one-body terms. Apart from some minor refinements, however, this was simply a matter of embedding Anderson's two-site perturbation theory into the full many-body formalism.

B. π-Electron Hamiltonians

In planar hydrocarbon molecules with double bonds, molecules such as ethylene, benzine, anthracene, etc., the two bonds of a double bond are not equal. One is a strong bond composed of so-called σ orbitals (hybrids of carbon 2s, $2p_x$ and $2p_y$ orbitals) which lie in the molecular plane, while the other is a weak bond involving carbon $2p_z$ orbitals oriented perpendicular to the plane, the so-called π orbitals. Pariser, Parr, and Pople[27] showed in 1953 that the lowest few electronic excited states of these molecules can be described fairly accurately by attributing all of the action to just the π electrons alone, with their interactions described by a small number of phenomenological parameters. Moreover, these parameters are quite transferable -- those determined from the optical absorptions of benzine provide good predictions for the corresponding spectra of napthaline, anthracine, and the other "chickenwire" compounds. This scheme has since been extended to much wider classes of molecules, and it is now a standard textbook subject for organic chemists. Nevertheless, many theoretical chemists have rejected this as "dirty phenomenology, with no theoretical justification," and they will have nothing to do with this scheme.

There is, of course, another school of theoretical chemists who have been attacking this problem with various formal techniques, and in recent years their efforts have been evolving towards the folded-diagram expansion of Ref. 6. I am convinced that the latter (or its coupled-cluster counterpart) really is the optimum formalism for the π-electron problem, and I have therefore written some pedagogical reviews[28,11] directed towards these chemists. At the least this provides a sound formal justification for the phenomenology, and efforts are also underway by several investigators to calculate the parameters from "first principles."

C. Valence Fluctuations in Rare Earth Compounds

"Valence fluctuations" is the name of a many-body phenomenon first recognized about ten years ago.[29] It is seen in a number of rare earth compounds, some "classic" examples being SmS, SmB_6, and TmSe. At low temperatures essentially all electronic properties become quite anomalous, indicating a novel type of many-body ground state. The subject is complex and still poorly understood. Suffice it to say that this is closely related to the Kondo effect, and the most popular model for theoretical study is a dense lattice of Kondo-like ions (actually Anderson-Hamiltonian ions) embedded in a simple metal. It is also somewhat analogous to the BCS problem, to the extent that a "zeroth order" description involves an enormous degeneracy, whereby strong cooperative effects can result from a weak residual interaction.

Pursuing this BCS analogy, we constructed simple variational wavefunctions for the ground states of various model systems.[20] The central problem was to evaluate the necessary many-body expectation values, so that parameters could be optimized and the physics extracted. The graphology for the above spin-Hamiltonian problem turned out to be well-suited for this task, and the various expectation values were found to have simple analytic forms. The resulting physical output is consistent with much of the observed phenomenology.

These examples suggest that the effective Hamiltonian formalism has much potential for other fruitful applications.

References

1) J. H. Van Vleck, Phys. Rev. 33, 467 (1929).
2) E. C. Kemble, The Fundamental Principles of Quantum Mechanics (McGraw-Hill, New York, 1937), p. 394.
3) T. Kato, Prog. Theor. Phys. 4, 514 (1949).
4) C. Bloch, Nucl. Phys. 6, 329 (1958).

5) J. des Cloizeaux, Nucl. Phys. 20, 321 (1960).

6) B. H. Brandow, Rev. Mod. Phys. 39, 771 (1967).

7) Effective Interactions and Operators in Nuclei, B. R. Barrett, ed. (Springer-Verlag, Berlin, 1975).

8) P. J. Ellis and E. Osnes, Rev. Mod. Phys. 49, 777 (1977).

9) D. J. Klein, J. Chem. Phys. 61, 786 (1974).

10) B. H. Brandow, in Ref. 7.

11) B. H. Brandow, Int. J. Quantum Chem. 15, 207 (1979).

12) F. Coester and H. Kümmel, Nucl. Phys. 17, 477 (1960).

13) F. Coester, in Lectures in Theoretical Physics, Vol. 11B, K. T. Mahanthappa and W. E. Brittin, eds. (Gordon & Breach, New York, 1969), p. 157.

14) R. Offermann, W. Ey, and H. Kümmel, Nucl. Phys. A 273, 349 (1976); R. Offermann, Nucl. Phys. A 273, 368 (1976); W. Ey, Nucl. Phys. A 296, 189 (1978).

15) J. Zabolitsky and W. Ey, Nucl. Phys. A328, 507 (1979).

16) I. Lindgren, Int. J. Quantum Chem., Symp. Vol. 12, 33 (1978).

17) J. Paldus, J. Cizek, and I. Shavitt, Phys. Rev. A5, 50 (1975).

18) B. H. Brandow, Ann. Phys. (NY) 64, 21 (1971).

19) B. H. Brandow, Adv. Phys. 26, 651 (1977); see §6.5.

20) B. H. Brandow: Int. J. Quantum Chem. 13, 423 (1979); in Crystalline Electric Field and Structural Effects in f-Electron Systems, J. E. Crow, R. P. Guertin, and T. W. Mihalisin, eds. (Plenum, New York, 1980), p. 353; and preprint.

21) F. Jørgensen, Mol. Phys. 29, 1137 (1975), see pp. 1144-46; also private communication.

22) B. C. Carlson and J. M. Keller, Phys. Rev. 105, 102 (1967).

23) L. L. Foldy and S. A. Wouthuysen, Phys. Rev. 78, 29 (1950).

24) J. L. Friar: in Mesons in Nuclei, M. Rho and D. H. Wilkinson, eds. (North-Holland Publ. Co., Amsterdam, 1979), see §3; Phys. Rev. C 22, 796 (1980).

25) P. W. Anderson, Phys. Rev. 79, 350 (1959); Solid State Phys. 14, 99 (1963).

26) J. C. Slater, Phys. Rev. 35, 509 (1930).

27) R. Pariser and R. G. Parr, J. Chem. Phys. 21, 466 and 767 (1953); J. A. Pople, Trans. Faraday Soc. 49, 1375 (1953).

28) B. H. Brandow, in Advances in Quantum Chemistry, P.-O. Löwdin, ed. (Academic Press, New York, 1977), p. 188; B. H. Brandow, in Quantum Theory of Polymers, J.-M. André, J. Delhalle, and J. Ladik, eds. (D. Riedel Publ. Co., Dordrecht, 1978).

29) C. M. Varma, Rev. Mod. Phys. 48, 219 (1976).

APPLICATIONS OF MOMENT METHODS TO FINITE NUCLEI

James P. Vary
Physics Department, Ames Laboratory
Iowa State University
Ames, IA 50011/USA

ABSTRACT

We summarize approaches to a number of many-body problems utilizing moment methods. We develop some tests with valence particle systems and some tests with the soluble Lipkin-Meshkov-Glick Hamiltonian in order to assess accuracy of the moment method approach for finite nuclei. We present results for the binding energy and for the elastic electron scattering cross section of ^{16}O using a realistic microscopic no-core effective Hamiltonian.

I. INTRODUCTION AND MOTIVATION

All many-body theories invoke some truncation procedure in order to obtain solutions with a general Hamiltonian, H. The basic question is: which many-body theory is the most practical, flexible and accurate for a given physical system and for a given set of physical observables? I characterize the theories discussed at this conference by the five major groupings:

1.) Variational
2.) Perturbational
3.) Coupled Cluster
4.) Green's Function Monte-Carlo
5.) Moment

The first three assume a particular reference state $|\phi_0>$ as a starting point for calculations of ground-state properties while the last two do not.

Since our primary goal is to obtain the ground state and excited state properties of nuclei with a realistic Hamiltonian and without assuming a particular $|\phi_0>$ we have chosen moment methods[1] for our many-body technique. A primary motivation for this choice stems from the flexibility of working in a no-core space which solves the problems discovered in the "Perturbational" approach to finite nuclei with realistic effective Hamiltonians.[2]

II. DEFINITIONS AND TESTS

a.) Method of Moments-Brief Sketch

We define a set of N single-particle orbits in which all of the m Fermions of the system are free to move. For the nucleus ^{16}O we have taken up to $N_p = 110$ and $N_n = 110$ so that the dimensionality D of the many-particle states is

$$D = \binom{N_p}{8}\binom{N_n}{8} \simeq 10^{22} \qquad (1)$$

In this no-core model space, the problem to solve is

$$H\,|\Psi_i\rangle = E_i|\Psi_i\rangle \qquad i = 1,\ldots,D \qquad (2)$$

where H is a realistic microscopic effective (N_p, N_n dependent) Hamiltonian which is described below. In principle, the solutions to (2) are obtained exactly by diagonization for a general H. Moment methods afford a systematic approximation scheme for the physical observables and these methods are applicable to situations such as ours where diagonalization is unfeasible.

The pioneering developments of French and co-workers[3] and the trace reduction formulae of Ginocchio and co-workers[4] have brought moment methods to the stage of a practical method for the nuclear many-body problem. Another primary goal of this research is to assess the accuracy of the method.

Briefly then, the experience of diagonalizing valence nucleon Hamiltonians has demonstrated that the results, even for a fixed total J and total T, as well as the summed results yield on eigenvalue distribution, $\rho(E)$, which is remarkably close to Gaussian. This suggests that we may accurately represent the results of diagonalizing an effective Hamiltonian by a few terms of a Gram-Charlier series for $\rho(E)$

$$\rho(E) = \frac{D}{\sqrt{2\pi}\,\sigma}\exp\left[-\frac{(E-E_o)^2}{2\sigma^2}\right]\left\{1+\sum_{\mu=3}\frac{a_\mu}{\mu!}H_\mu\left(\frac{E-E_o}{\sigma}\right)\right\} \qquad (3)$$

For eq'n (3) the ingredients are calculated directly, without diagonalization, as traces in the multi-particle space and employed with the Hermite polynomials, H_μ, to provide a continuous distribution approximating the exact eigenvalue distribution. The moment of an operator K is signified by $\langle K\rangle$ which equals the normalized trace of K. That is, we employ the notation

$$\langle K\rangle = D^{-1}\,\mathrm{tr}(K) \equiv D^{-1}\,\langle\langle K\rangle\rangle\ . \qquad (4)$$

Then, the lowest four moments and their nomenclature are summarized in Table 1.

Table 1.

Moment	Name	Value
Zeroeth	dimensionality	$D = \langle\langle 1 \rangle\rangle$
First	centroid	$E_o = \langle H \rangle = D^{-1}\langle\langle H \rangle\rangle$
Second	width	$\sigma = [\langle H^2 \rangle - E_o^2]^{1/2}$
Third	skewness	$a_3 = \langle (H-E_o)^3 \rangle/\sigma^3$

The rate of convergence with the number of moments is expected to depend upon the physical property sought (ground state energy, excited state level density, spin properties, etc.,) and upon the size of the model space and number of particles. We concentrate on tests which help assess this convergence.

The general approach to the distribution of eigenvalues of a physical observable is cast as an orthogonal polynomial expansion[3,4]. For example

$$\rho(E) = \langle \delta(H-E) \rangle = w(E)\sum_{\mu} a_{\mu} P_{\mu}(E) \tag{5}$$

where $w(E)$ is some chosen weight function and $a_{\mu} = \langle P_{\mu}(H) \rangle$. The polynomials, P_{μ}, satisfy the equations

$$\langle P_{\mu}(H) P_{\nu}(H) \rangle = \delta_{\mu\nu} \tag{6}$$

and are utilized to evaluate the expectation value of an operator K in the state E through

$$\langle E|K|E \rangle = \frac{1}{N(E)} \langle\langle K\, \delta(H-E) \rangle\rangle = \sum_{\mu} \langle K\, P_{\mu}(H) \rangle P_{\mu}(E) \tag{7}$$

or

$$\langle E|K|E \rangle = \langle K \rangle + [\langle KH \rangle - \langle K \rangle E_{o}] \frac{[E-E_{o}]}{\sigma^{2}} + \ldots \tag{8}$$

where w is chosen unity here and N(E) is the degeneracy of $|E\rangle$.

In this paper we concentrate on $\rho(E)$ and the radial moments of the one-body density distribution. Specifically, we employ $\rho(E)$ to obtain a prediction of the binding energy of ^{16}O and we employ the radial moments to evaluate the elastic electron scattering cross section. The operator for the kth moment of the one-body density distribution is simply

$$K = R^{(k)} = \sum_{ij} \langle i|r^{k}|j \rangle a_{i}^{\dagger} a_{j} \tag{9}$$

where $a_{i}^{\dagger}$ and a_{j} are Fermion creation and destruction operators in the chosen single particle space whose states are labelled by the subscripts.

There are a number of special moment method techniques necessary to carry out our investigations. In particular we employ the trace reduction technique of Ginocchio[4] to reduce the sum over D many-body states to statistical factors times "basic diagrams." We also employ the "fixed J, T expansions"[3,4] in our determination of $\rho(E)$ and this is summarized in Ref. (5). In order to obtain a prediction of the ground state energy of ^{16}O we employ the Ratcliff procedure[6] or "predictor method." The most likely position of the ground state E_{g} is given by

$$\int_{-\infty}^{E_g} \rho(E)dE = 1/2 \tag{10}$$

Of course, such a method must be extensively tested to verify its accuracy. For shell model problems we have performed a number of tests where we limit the model spaces so that we can compare with the results of exact diagonalization. These comparisons proved very favorable.[5]

b.) Tests with a Soluble Model

Motivated by the need to test the moment methods and the predictor method for the ground state energy in very large model spaces, we have employed the Lipkin-Meshkov-Glick[7] soluble model in model spaces up to $D \sim 10^{20}$ in size. The results of these tests will be published in detail elsewhere.[8] One conclusion is that for moderate values of the coupling constants the predictor method, with $\rho(E)$ determined by the lowest three moments, is remarkably accurate in the very large spaces. On the other hand, with additional moments used to determine $\rho(E)$, the progress towards a more accurate estimate of the ground state energy is very slow. Two conclusions are obtained: first, with the predictor method we gain little advantage by going to higher moments while the three moment results are remarkably accurate; second, we must obtain more powerful methods to utilize the higher moment information. We are making progress in developing better methods[9] but further work is necessary to employ them with realistic Hamiltonians.[10]

c.) Realistic Effective Hamiltonians

For our no-core studies we have developed a set of effective Hamiltonians for a sequence of harmonic oscillator ($\hbar\Omega$ = 14 MeV) model spaces including up to six major shells. The details are available in Ref. (11). Briefly, we solve for

$$H = T_{rel} + G(\omega) = H_o + (G-U) - T_{cm} \tag{11}$$

where T_{rel} is the relative kinetic energy operator, H_o is the harmonic oscillator one-body operator, U is the harmonic oscillator potential and T_{cm} is the center of mass kinetic energy operator for the m-Fermion system. In eqn. (11), $G(\omega)$ is the Brueckner G-matrix which we solve for the Reid soft-core potential V in the oscillator basis. Thus, we solve

$$G(\omega) = V + V \frac{Q}{\omega - H_o} G(\omega) \tag{12}$$

where Q is the two particle Pauli operator permitting scattering to intermediate two particle states with at least one particle outside the model space. We treat it exactly in a single particle representation. We neglect the Coulomb potential for the present. It is clear that H is appropriate for a treatment by full diagonalization and, therefore, with moment methods. Since we employ a no-core basis space there are no higher-order particle-hole processes to contribute to H. Thus, the first correction to eqn. (11) is an effective three-body force. The importance of this correction and the importance of other effective many-body forces depend upon the size of the model space. Therefore, we study our results as a function of model space size and compare with results of alternative approaches.

The quantity ω, the starting energy, is defined as the average energy of two particles interacting in the model space. We may employ an added shift, C, in the energy gap between the last shell of the model space and first shell of the "particle" space. In this case, we have a spectrum for H_0 with $\hbar\Omega$ spacing everywhere except on $\hbar\Omega$ + C spacing between the two sections of the single-particle space. We study the dependence on C as a gauge of our overall convergence in a given physical quantity. Independence of C would be a significant indicator of convergence.

III. APPLICATIONS

a.) Binding Energy of ^{16}O

We have calculated $\rho(E)$ in a fixed J, T moment method expansion for ^{16}O utilizing our realistic Hamiltonians. Initial results for two, three and four major oscillator shells are reported in Ref. (5). More recently,[11] we have added two major shells so that we now treat the 16 nucleons in a model space consisting of 220 single particle states. Substantial independence of C is achieved in that the total binding energy of ^{16}O changes by about 40 MeV over a comparable range of change in C. More importantly, the results lie between the experimental results and those of the coupled cluster[12] method when corrections for the Coulomb effects are applied. Currently, we conclude there is coarse agreement with the coupled cluster results pending results with further independence of C achieved by adding two more major oscillator shells. If residual discrepancies are found at convergence we will investigate the possibility that many-particle correlations permitted by the moment methods approach are providing significant contributions to the binding energy. Concurrently, we are investigating improved methods to obtain the binding energy from the moments[9] which will utilize more completely our higher moment calculations.[10]

b.) Moments of the Ground State Charge Density and Electron Scattering

We have recently evaluated[13] radial moments $< E|R^{(k)}|E > \equiv < R^{(k)}(E) >$ the ground state density distribution of ^{16}O, ^{40}Ca and ^{58}Ni. Specifically we have employed the terms displayed in eqn. (8) with the definition of eqn. (9) for k = 0 through k = 5. We note that the first term of eqn. (8) is independent of m and E and therefore depends solely on the model space and choice of $\hbar\Omega$. Other terms build in the specific nucleus dependence. Consequently, we expect certain deficiencies when we include only the terms displayed in eqn. (8). In Table 2 we show the values of the first term and its square root in eqn. (8) for k = 2 as a function of the number of major oscillator shells included in the model space. After including the additional terms of eqn. (8) for ^{16}O we obtain an rms radius of 2.66 fm for the three major shell space which is sufficiently close to the empirical value that we adopt this model space for extended studies of the higher radial moments. In the future we will employ more terms in eqn. (8) in order to release the constraint of a fixed model space.

We next obtain the k = 3,4,5 moments with eqs. (8) and (9). The results are found remarkably similar to the corresponding moments of the density obtained from the phenomenologically successful Density Dependent Hartree-Fock approach[14] with the Skyrme III Hamiltonian for ^{16}O.

Within the Born approximation we may then proceed to evaluate the elastic electron scattering cross section from ^{16}O. To do this,[13] we employ the radial charge moments $< R^{(k)}(E) >$ to fix the parameters $\{\rho_i, \xi_i\}$ of a weighted delta distribution

$$\bar{\rho}(r) = \sum_i \rho_i \; \delta(r-\xi_i) \tag{13}$$

$$F(q) = \frac{4\pi}{M} \sum_i \rho_i \frac{\sin(q\xi_i)}{q\xi_i} \tag{14}$$

Utilizing the k = 0 thru k = 5 moments obtained in this fashion we successfully describe the ^{16}O elastic scattering data out to q = 1 fm^{-1}. In order to extend the predictions to higher q values with accuracy we must evaluate higher radial moments. However, our primary goal should be to obtain model space independence. It is indeed encouraging that the higher radial moments are in concert with existing data.

Table 2. Values of the second moment of the radial operator as function of model space Ω = 14 MeV

Major oscillator shells	$< R^{(2)} >$(fm^2)	$< R^{(2)} >^{1/2}$(fm)
3	8.89	2.98
4	11.11	3.33
5	13.33	3.65
6	15.55	3.94

IV. SUMMARY AND FUTURE PROSPECTS

Moment methods provide a practical, flexible and potentially accurate approach to the nuclear many-body problem with realistic Hamiltonians. These methods allow us to perform calculations in extremely large model spaces and, therefore, to circumvent difficulties obtained in the perturbative treatment. They also admit calculations independent of an assumed unperturbed ground state $|\phi_o>$.

Since our primary goal is to obtain the properties of finite nuclei with realistic Hamiltonians we have learned from the present investigations that three improvements are necessary

(1) We must obtain H in even larger model spaces to insure that convergence has indeed been achieved.

(2) We must develop efficient methods to evaluate higher moments of observables and products of observables.

(3) We should obtain methods that more effectively employ the higher moments.

We are encouraged by the results obtained to date to pursue these goals. Progress on all three improvements has been made and will be reported in the future.

It is a pleasure to acknowledge my collaborators R. H. Belehrad, B. J. Dalton, A. Klar and F. Margetan for their individual and collective insights on these challenging problems. This research was supported by the Department of Energy, contract number W-7405-Eng-82, Division of High Energy and Nuclear Physics, budget code No. KB-03-0000.

REFERENCES

[1] For a recent review of moment methods and a variety of applications see: Moment Methods in Many-Fermion Systems, B. J. Dalton, S. M. Grimes, J. P. Vary and S. A. Williams, eds., Plenum Press (N.Y.) 1980.

[2] For a brief review of the difficulties with perturbation theory see: J. P. Vary, R. H. Belehrad and R. J. McCarthy, Phys. Rev. C21, 1626 (1980).

[3] J. B. French, in Nuclear Structure (North Holland, Amsterdam, 1967), p. 85; F. S. Chang, J. B. French and K. F. Ratcliff, Phys. Lett. 23, 251 (1965); J. B. French and K. F. Ratcliff, Phys. Rev. C3, 94 (1971).

[4] J. N. Ginocchio, Phys. Rev. C8, 135 (1973); S. Ayik and J. N. Ginocchio, Nucl. Phys. A221, 285 (1974).

[5] J. P. Vary, R. Belehrad and B. J. Dalton, Nucl. Phys. A328, 526 (1979).

[6] K. F. Ratcliff, Phys. Rev. C3, 117 (1971).

[7] H. J. Lipkin, N. Meshkov and A. J. Glick, Nucl. Phys. 62, 188 (1965).

[8] A. Klar and J. P. Vary, to be published.

[9] M. C. Cambiaggio, A. Klar, F. Margetan and J. P. Vary, to be published.

[10]F. Margetan, J. P. Vary and B. J. Dalton, to be published.

[11]J. P. Vary, ibid. p. 423.

[12]H. Kümmel, K. H. Lürhmann and J. G. Zabolitzky, Physics Reports 36C, 1 (1978).

[13]R. H. Belehrad, B. J. Dalton and J. P. Vary, to be published.

[14]D. Vautherin and D. M. Brink, Phys. Rev. C5, 626 (1972).

VARIATIONAL CALCULATIONS ON LIGHT NUCLEI

R. A. Smith
Department of Physics
SUNY at Stony Brook
Stony Brook, NY 11794

The nuclear interaction is a bridge between physical systems and our underlying beliefs about the nature of matter. Based on phenomenology, meson field theory, QCD, or whatever, nuclear force models provide a means for calculating two-body scattering, studying finite nuclei, and investigating the nearly infinite matter of which neutron stars are made. Although mean-field meson and quark-matter calculations neatly sidestep the nuclear interaction, they do not address directly the two-body data or light nuclei. The two-body problem is very important, because the non-relativistic Schrödinger equation can be solved for a given potential; the great body of two-body scattering data and the deuteron then place great constraints on the two-body potential. There is considerable theoretical justification for three-body forces as well; for these to be seen and studied requires the study of finite nuclei or infinite matter. While infinite matter is computationally much simpler than most finite systems, interpretation of results requires extrapolation from real nuclei. Light nuclei, where one may keep a firm grip on the (relatively few) degrees of freedom, provide a useful testing ground for potentials and few-body methods. Besides the ground state energy, excited state energies and form-factors may also be computed. Several approaches to the few-body problem are possible.

The Schrödinger equation is easily solved for the deuteron, and suitable juggling of integral equations leads to the Faddeev equations [1] for three particles and the Yakubovsky equations [2] for four. An independent approach, solution by the Greens-function Monte-Carlo method, gives impressive results for central potentials [3]. The coupled-cluster method, truncated at a suitable level [4,5], allows approximate solution of the Schrödinger equation.

Variational calculations for light systems may be based on a single particle (e.g. oscillator) basis [6], or with correlated wave-functions. The former suffers from the number of basis states needed to screen effectively the short-range part of the interaction and the accompanying difficulty of using physical insight in constructing the wavefunction. Lomnitz and Pandharipande (LP) [7] have suggested a simple form for a Jastrow-correlated wavefunction and shown that it gives good results for the triton.

LP adopted a trial wavefunction of the form

$$\psi_v = S \{ \prod_{i<j}^{A} f_{ij} \} \Phi \qquad (1)$$

where Φ is a spin-isospin determinant and the symbol S denotes symmetrization of the (non-commuting) operator product. The operator product may be written symbolically as

(2)
$$F_c (1 + \bullet\!\!-\!\!\bullet + \bullet\!\!-\!\!\bullet\!\!\diagup^{\bullet} + \triangle\hspace{-0.3em}\bullet\ )$$

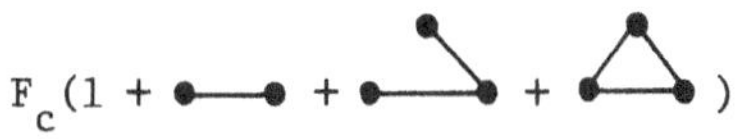

where the F_c includes the central correlations and the lines indicate non-central correlations. The LP calculation, using calculational methods originally devised for the infinite problem, obtained exact variational bounds when the first two (independent pair) terms in the parentheses in eq. (2) were used; approximate calculations with more of the wavefunction suggested that the full symmetrized product could make a substantial improvement. More recently, I have joined LP in extending the calculations to the full symmetrized wavefunction for both the triton and the alpha particle [8]. The methods used to generate the wavefunction and calculate the energy, along with possible extensions, will be described in the remainder of this lecture.

Let's begin by considering the deuteron with a non relativistic Hamiltonian

(3)
$$H = - (\hbar^2/2m) \ (\nabla_1^{\ 2} + \nabla_2^{\ 2}) + V_{12}$$

Since the V has operators $(1, \sigma_1 \cdot \sigma_2, S_{12}, \vec{L} \cdot \vec{S}) \times (1, \tau_1 \cdot \tau_2)$, the correlation function f may have these operators as well. For the deuteron, the Slater determinant is particularly simple: $\Phi = n \uparrow p \uparrow - p \uparrow n \uparrow$; all of the operators acting on this state have the effect of either central or tensor correlations. Hence,

(4)
$$f = f^c + f^t \ S_{12}$$

is the most general form of the correlation operator. Variation of the energy expectation $\langle \psi_v | H | \psi_v \rangle / \langle \psi_v | \psi_v \rangle$ with respect to the functions f^c and f^t leads to the usual deuteron equations. The f^c and f^t are associated with the $\ell = 0$ and $\ell = 2$ components of the wavefunction. Note that while the bare Φ has the particles in relative s-states, the correlation operator does excite the d-state admixture.

For the triton and alpha particle, the Slater determinant may still avoid spacial dependence. Since it must then be antisymmetric under the combined spin-isospin exchange operators, any of the operators acting on Φ has the same effect as some combination of $1, \sigma_i \cdot \sigma_j, S_{ij}$. A second f, however, no longer acts on Φ, and the form used for the single-particle f affects the computation of the whole wavefunction. This ambiguity in choice is broken in favor of the operators $1, \sigma_i \cdot \sigma_j$, and $S_{ij} \tau_i \cdot \tau_j$. The most important effects result from the tensor correlations induced by pion exchange; this motivates the last form. The quadratic effects from the $\sigma_i \cdot \sigma_j$ term should be less important. With this choice, the correlation function f has the form

(5)
$$f_{ij} = f_{ij}^c \ (1 + X_{ij} (u^\sigma \ \sigma_i \cdot \sigma_j + u_{ij}^{t\tau} S_{ij} \tau_i \cdot \tau_j))$$

(6) $$X_{ij} = \prod_{k\neq i,j} \{1 - t_1(r_{ij}/(r_{ij}+r_{ik}+r_{jk}))^{t_2}\}$$

The function X acts as a many-body correlation factor which suppresses state-dependent correlations for one pair whenever a third particle is close to them. It gives an improvement in energy by discouraging repulsive configurations. The functions f^c, u^σ, and $u^{t\tau}$ are obtained as projections of the correlated two-body wavefunctions in the (T,S)= (1,0) and (0,1) channels in external potentials $\varepsilon_{T,S}$ which simulate effects of the other particle(s).

(7) $$[-(\hbar^2/m)\nabla^2 + v({}^1S_o) - \varepsilon_{1,0}]\, f_{1,0} = 0$$

(8) $$[-(\hbar^2/m)\nabla^2 + v({}^3S_1) - \varepsilon_{0,1}]\, f_{0,1} + 8v^t({}^3S-{}^3D)\, f^t_{0,1} = 0$$

(9) $$[-(\hbar^2/m)\nabla^2 + 6\hbar^2/(mr^2) + v({}^3D_1) - \varepsilon_{0,1}]\, f^t_{0,1} + v^t({}^3S-{}^3D)\, f_{0,1} = 0$$

Eq. (7) is the equation for a bound dineutron (given some extra attraction), while eqs. (8) and (9) correspond to a deuteron. The equations (7)-(9) are similar in spirit, if not in detail, to the Euler-Langrange equations which can be derived for infinite systems, or, with less ease, for the general few-body problem. The projections give

(10) $$f^c = (3f_{0,1} + f_{1,0})/4$$

(11) $$f^c u^\sigma = (f_{0,1} - f_{1,0})/4$$

(12) $$f^c\, u^{t\tau} = -\frac{\beta}{3}\, f^t_{0,1}\ ,$$

and β is a parameter which may be used to adjust the strength of the tensor correlation.

The asymptotic form of the f's is rather different from that adopted for Jastrow theories of infinite matter, since the f's asympotically vanish rather than going to unity. The f is adjusted so that if one particle is pulled far from the others, the wave function factors roughly into a product of an (A-1) body wave-function with an exp(-Kr)/r factor for the single body at a distance r from the (A-1) body cluster. The parameter K is related to the separation energy of the particle, and is a variational parameter in the wavefunctions used. It determines the long-range part of the induced potentials $\varepsilon_{T,S}$; the short-range part is parametrized with a Woods-Saxon form [9].

As outlined above, the wavefunction has been parametrized by choosing the induced potential. This results in a few parameters which are, for the most part, physically meaningful; hence reasonable guesses for most parameters are rather easily made. The parameters t_1 and t_2 which occur in the many-body correlation factor X are less obvious, but also not critical. The wavefunction has four parameters for the induced potential, the two t's, and the β of eq. (12).

Given the correlation function, one must then construct the expectation values. In the present case, this was implemented by first constructing the wavefunction. This is suitable at least for light nuclei, as Table 1 indicates.

Table 1.
The number of spin-isospin components of the wavefunction

Nucleus	${}^{A}_{N}Z$	d	t	a	${}^{2Z}_{Z}Z$
Spin	2^A	4	8	16	2^A
Isospin	$\binom{A}{Z}$	2	3	6	$2^A/\sqrt{\pi Z}$
Total	$2^A\binom{A}{Z}$	8	24	96	$2^{2A}/\sqrt{\pi Z}$

More spin than isospin components are necessary, because the interaction conserves T_z but not S_z; the tensor force couples the angular momentum associated with the spin and spacial parts of the wavefunction.

The operator products needed to compute the wavefunction from the correlation function are most conveniently carried out when the spin-isospin components of the wavefunction are arranged as a matrix, in which the rows correspond to the different spin components and the columns to the isospin components. With this convention, the action of any spin operator on a wavefunction is equivalent to a premultiplication of the wavefunction matrix by a spin-operator matrix; isospin operators correspond to postmultiplication. Using conventional matrix methods, each element in the matrix product (for spins) would require 2^A products and additions. For wavefunctions of the type chosen, where the correlations are successive factors involving only pairs of particle spins, the situation is greatly improved. Since a pair of particles can have only four possible spin configurations, only four elements in each row of the operator matrix will be non-zero. By representing the spin components as A-digit binary numbers, and using these numbers as indexes, it is easy to determine which four products are necessary. Similar improvements may be made for the isospin calculations.

There remains the problem of symmetrizing the product of correlation factors. Since there are altogether 720 ways of arranging the 6 correlation factors of the alpha particle, it is obvious that complete symmetrization would be extremely time-consuming. It is preferable to take an average over independent choices for the left-hand and right-hand wavefunctions. The independent-pair component of the wavefunction is, in any event, independent of the ordering, so that the left and right wavefunctions will not be grossly different.

The spin and isospin operators in the Hamiltonian are handled in the same fashion. The differential operators in the kinetic energy and spin-orbit terms are handled by recomputing the wavefunction for slightly different coordinate values and explicitly computing the partial derivatives. This feature of the calculation makes the inclusion of three-body correlations extremely simple.

The energy expectation value is the ratio of the matrix elements of the Hamiltonian and the unit operator between the left and right wavefunctions. These are sampled in a form of the classic random-walk procedure of MRRTT [10]. Two forms of the importance-sampling function have been tried. In one, the sampling function is the IP approximation to the square of the wavefunction, and the orders of the operators in the left and right wavefunctions are chosen independently and uniformly. In the second case, the sampling function is the real part of the overlap of the left and right wavefunctions. In the second case, this weighting function is non-negative for the parameters used. Each step of the random walk involves trial displacements of each of the particles.

Pandharipande [11] has brought to the conference some results obtained by adding a three-body force to the interaction model [12]. In order to take some advantage of this force, he has included an additional factor in the wavefunction of $(1 + p\, V_{ijk})$, where

$$V_{ijk} = \sum \{.003T_\pi^2(ij)T_\pi^2(jk) + (-3U_2)T_\pi(ij)T_\pi(jk)P_2(ij\cdot jk)\} \tag{13}$$

The $T_\pi(x)$ represents the form $(1 + 3/x + 3/x)$, and $U_2 = -1.5$ MeV. The first term represents repulsion due to the diminished binding of an intermediate Δ resonance in matter, while the second term gives the attraction obtained when a Δ resonance is excited by one particle and de-excited by another. His talk described a more elaborate form for the variational wavefunction, for which calculations have not yet been made.

Table 2 shows the results for the triton, which is bound by 8.54 MeV. The potentials used are the Malfliet-Tjon model V (MTV) [13], Reid [14], Urbana V_{14}[15], and the Reid plus three-body potential. In the last case, the strength of the three-body term has been adjusted to obtain the experimental binding energy.

For the last entry in the table, Pandharipande has also shown indications of a hole forming in the center of the triton as the result of the three-body correlations induced by the three-body potential. The change in energy associated with this three-body correlation is 1 MeV, which is larger than (and in addition to) the .6 MeV which comes from the three-body force without three-body correlations.

Table 2.

Triton results

Potential	Method	Energy	Reference
MTV	Faddeev (approx)	-7.3	13
	Correlated var.	-8.03	8
Reid	Faddeev	-7.0	16,17
	Sing. part. var.	-6.7	18,19
		-7.0±.1	20
	Correlated var.	-6.86±.08	8
Urbana V_{14}	Correlated var.	-8.46±.06	11
Reid + V_{ijk}	Correlated var.	-8.53±.15	11

Table 3 shows corresponding results for the alpha particle. The experimental binding energy, for comparison, has been adjusted to 29 MeV to correct for the Coulomb energy neglected in the calculations listed.

Table 3.

The alpha particle

Potential	Method	Result	Reference
MTV	GFMC	-31.3±.3	3
	Coupled cluster	-31.24	5
	Correlated var.	-30.7±.1	8
Reid	Coupled cluster	-24.4	3
	Approx. Yakubovsky	-20	21
	Correlated var.	-22.9±.5	8
Urbana V_{14}	Correlated var.	-25.1±.4	11
Reid V_8 + V_{ijk}	Correlated var.	-30.7±1.	11

The results are again in good agreement with the trial problems, and the method well suited to the more complicated potentials.

The use of the correlated variational method is still somewhat in its infancy. There remain a number of interesting extensions. For the case of simple central potentials, it is possible to derive an Euler-Langrange equation for the f. This equation for the triton is presently being studied. With a practical technique for solving this equation, it should be possible to optimize calculations of this type for reasonable numbers of particles. In addition to improving on energies, this optimization would reduce uncertainties in calculated form factors by generating an f with no parameters. In addition, the energy is stationary at excited states, as well as the ground state, so that it might be possible to investigate excited states with the same angular momentum and parity as the ground state. The form of the induced potential may also be useful in considering more complicated problems.

Another attractive use of the method would be for the larger state space which includes delta as well as nucleon components. The nucleons may be excited to deltas by correlation functions with the operator properties of the transition potentials.

In a similar vein, one could study light hypernuclei, in which the lambda may be excited to a sigma by means of transition potentials. Interesting problems include investigation of the charge asymmetry [22] and the magnitude of the binding of the lambda to an alpha particle. Again, correlation functions can be used to connect the lambda and sigma states.

For looking at larger nuclear systems, it would be very useful to not have to compute all of the components of the wavefunction matrix. Since many of them will be small, it would be nice to be able to use importance sampling and evaluate only some of the terms in the product. In addition, there remains open the question of how to best handle particles in the p shells and beyond.

While the future development of the method is not yet clear, we may expect substantial progress along at least some of these lines in the next few years.

Preparation of this contribution was supported, in part, by USDOE contract DE-AC02-76ER13001.

References

1. L.D. Faddeev, ZhETF (USSR) 39 (1960) 1459.
2. O.A. Yakubovsky, Yad. Fiz. 5 (1967) 1312.
3. J.G. Zabolitzky, M.H, Kalos, preprint.
4. J.G. Zabolitsky, Nucl. Phys. A228 (1974) 285; H. Kummel, K.H. Lührmann, J.G. Zabolitzky, Phys. Reports 36C (1978)1.
5. J.G. Zabolitzky, preprint.
6. A.D. Jackson, A. Lande, P. Sauer, Nucl. Phys. A156 (1970)1.
7. J. Lomnitz-Adler, V.R. Pandharipande, Nucl. Phys. A342 (1980)404.
8. J. Lomnitz-Adler, V.R. Pandharipande, R.A. Smith, Nucl. Phys., in press.
9. R.D. Woods, D.S. Saxon, Phys. Rev. 95 (1954) 577.
10. N. Metropolis, A.W. Rosenbluth, M.N. Rosenbluth, A.H. Teller, E. Teller, J. Chem. Phys. 21 (1953) 1087.
11. V.R. Pandharipande, this volume.

12. I.E. Lagaris, V.R. Pandharipande, preprint.

13. R.A. Malfliet, J.A. Tjon, Nucl. Phys. A127 (1969)161.

14. R.V. Reid, Ann. Phys. 50 (1968) 411.

15. I.E. Lagaris, V.R. Pandharipande, preprint.

16. A. Laverne, C. Gignoux, Nucl. Phys. A203 (1973) 597.

17. R.A. Brandenburg, Y.E. Kim, A. Tubis, Phys. Rev. C12 (1975) 1368.

18. P. Nunberg, D. Prosperi, E. Pace, Nucl. Phys. A285 (1977) 58.

19. M.R. Strayer, P.V. Sauer, Nucl. Phys. A231 (1974)1.

20. M.A. Hennell, L.M. Delves, Nucl. Phys. A246 (1975) 490.

21. J.A. Tjon, Phys. Rev. Lett. 40 (1978) 1239.

22. S.A. Coon, P.C. McNamee, Nucl. Phys. A322 (1979) 267.

STUDY OF LIGHT NUCLEI FROM ^{4}He TO ^{40}Ca WITH THE FAHT CLUSTER EXPANSION

R.Guardiola
Departamento de Fisica Nuclear
Universidad de Granada
(Spain)

1. INTRODUCTION

The purpose of this work is the study of the behaviour with the mass number A of the binding energy and radii of 4n nuclei in the presence of short range correlations. These correlations are described <u>via</u> a Jastrow ansatz in the state independent form, and the expectation value of the energy and radii is computed according to the FAHT cluster expansion [1]. Practical studies on the convergence conditions of this and other cluster expansions have shown [2] that the FAHT cluster expansion is very appropriate for the calculation of expectation values, and that one can safely stop at third order.

The complexity of the computation of three-body matrix elements prevents for the application of this method to heavy nuclei, the computing time being roughly proportional to A^3, so that our systematics will stop at ^{40}Ca.

The potentials used to describe the nucleon-nucleon interaction are the B1 of Brink and Boeker [3] and the S3 potential of Afnan and Tang [4] (appropriately modified, see section 2). These potentials have been recently used in a comparison between the Brueckner and Jastrow approaches in the doubly-magic nuclei ^{4}He, ^{16}O and ^{40}Ca [5] and some of that results will be commented here.

To my knowledge, there exists only another systematic study of finite nuclei in the Jastrow framework [6]: this work used the Reid soft core potential and state dependent correlations. The behaviour of the binding energy pernucleon in the range ^{16}O-^{40}Ca in Ref. [6] is very different of the behaviour we have found. We should however point out that there are several corrections and approximations in that work not fully justified, and probably that work shouldbe improved.

2. THE NUCLEON-NUCLEON POTENTIALS CONSIDERED

The calculations presented below have been carried out for the B1 potential of Brink and Boeker [3] and for the modified S3 potential of Afnan and Tang [4].

The Brink-Boeker B1 force was devised as an "effective interaction". This means that actual calculations with this force should be limited to the case of uncorrelated nuclear states. However, the large amount of work concerning this force (Hartree-Fock and Hartree-Fock-Bogoliubov [7] , alpha-particle model [8,9] , deformations [10,11], hyperspherical harmonics method [12] ,...) justifies their use in the present context. The B1 force is given by

$$
\begin{aligned}
V_W &= 595.55\exp(-2.041r^2) - 72.212\exp(-0.5102\, r^2) \\
V_M &= -206.04\exp(-2.041r^2) - 68.388\exp(-0.5102\, r^2)
\end{aligned}
\qquad (1)
$$

where r is measured in fm, and the Wigner (V_W) and Majorana (V_M) components are given in MeV. The B1 force presents a small core at the origin, but of a rather long range ($R_c \simeq 1/\sqrt{2.04} = 0.7$fm), so that short range correlations may have a significative role in the energy.

The S3 potential was adjusted to the even nucleon-nucleon phase shifts and to static properties (radius and energy) of ^{4}He. The work of Afnan and Tang [4] deals with a very general function for ^{4}He, much more general than ours, and one may consider their calculation as <u>exact</u>. Further work on this nucleus with the same potential [13] confirms the calculations of Afnan and Tang. The only problem for the use of this potential is that only even waves are defined (odd waves do not enter in the determination of the ^{4}He structure) and to extend the range of applicability to heavier nuclei we have added a repulsive part in the singlet- and triplet-odd components. The modified S3 potential reads:

$$
\begin{aligned}
V_W &= 1000\, e^{-3r^2} - 41.5\, e^{-0.8r^2} - 5.75 e^{-0.4r^2} - 81.675 e^{-1.05r^2} - 10.75\, e^{-0.6r^2} \\
V_M &= 41.5\, e^{-0.8r^2} + 5.75 e^{-0.4r^2} + 81.675 e^{-1.05r^2} + 10.75\, e^{-0.6r^2} \\
V_B = -V_H &= 41.5\, e^{-0.8r^2} + 5.75 e^{-0.4r^2} - 81.675 e^{-1.05r^2} - 10.75\, e^{-0.6r^2}
\end{aligned}
\qquad (2)
$$

As above, r is measured in fm and V is given in MeV. The central part of the potential shows a strong core, being both high and wide (R_C=.58f) so that one expects in this case short range correlations to be very important.

There is a further remark on this potential: as far as we will consider only spin/isospin saturated nuclei, the Bartlett and Heisenberg components can be taken as null, and still obtain the same values for the energy of those nuclei.

3. NUCLEAR WAVE FUNCTIONS

Nuclei are described according to the Jastrow form

$$\Psi = \prod_{i<j} f(r_{ij}) \quad \Phi \tag{3}$$

where $f(r_{ij})$ is the state independent Jastrow correlation factor, and Φ is the model state. Practical calculations require a specific functional structure for both correlations and model wave functions [2], namely gaussian forms or polynomials times gaussians. This means that the model state must be constructed with single particle states from the harmonic oscillator potential. On the other hand, the correlation factor considered has the simple form

$$f(r) = 1 - a \exp(-\beta r^2) \tag{4}$$

so that $\underline{a}$ controls the depth of the correlation and $1/\sqrt{\beta}$ is the range of the correlation.

The fact of constructing the model function with the help of Slater determinants of the harmonic oscillator potential determines strictly the model function in some cases, particularly in magic nuclei. In the cartesian notation (n_x n_y n_z) for the single-particle states, the configurations for magic nuclei are given by:

^{4}He: $(0\ 0\ 0)^4$

^{16}O : $(0\ 0\ 0)^4\ (1\ 0\ 0)^4\ (0\ 1\ 0)^4\ (0\ 0\ 1)^4$

^{40}Ca: $(^{16}\mathrm{O})\ (1\ 1\ 0)^4\ (1\ 0\ 1)^4\ (0\ 1\ 1)^4\ (2\ 0\ 0)^4\ (0\ 2\ 0)^4 (0\ 0\ 2)^4$

these configurations being unique and without any angular momentum couplig problem.

This is not the situation in the case of open shell nuclei, i.e., all remaining nuclei. Extensive work in these nuclei has shown the need of including deformations in the model function: the alpha-particle model, for example, assigns to ^{8}Be a prolate shape and to ^{12}C an oblate form [8], but does not offer so clear indications for the s-d shell nuclei [9]. The deformation of these nuclei has a double origin: first the true deformation related to the fact of having different harmonic oscillator constants in two or three directions of the space. Second, there is a deformation related to the form the valence particles are distributed in the (major) shell. This statement is clarified with an example: if ^{12}C is described according to the SU_3 model by filling the $1p_x$ and $1p_y$ orbitals with the four spin/isospin orientations there results also an oblate shape for this nucleus, and the angular momentum projection (without the true deformation) gives three levels 0^+, 2^+ and 4^+ satisfying exactly the $\ell(\ell+1)$ law [14].

To simplify the counting of the possible states for open shell nuclei we have made the following assumptions:

i. The spatial orbitals are saturated, i.e., each orbital is filled up with the four spin/isospin orientations.

ii. The spatial orbitals are taken in the cartesian representation

iii. No true deformation is considered.

Accordingly, we have enumerated all possible configurations with the above restrictions and we have chosen the best one after a variational search in the absence of correlations. In this search no angular momentum projection has been carried out, but the center-of-mass motion has been appropriately considered. The result of this search is shown in Table 1 for the nuclei of interest and for the B1 force. Note that ^{8}Be and ^{12}C are not included because, according to our hypothesis, there is a unique configuration, namely $(0\ 0\ 0)^4(0\ 0\ 1)^4$ for ^{8}Be and $(0\ 0\ 0)^4$ $(1\ 0\ 0)^4(0\ 1\ 0)^4$ for ^{12}C. All remaining possibilities result from a rotation of the above configurations. This statement also applies to the results of Table 1. With regard to the numbers of Table 1 is has to be noticed the importance of the configuration, sometimes giving differences of the order of 1 MeV/A. Note also that the best configuration for ^{24}Mg has not axial symmetry, and is the only exception: probably the ^{24}Mg

	B1 POTENTIAL					MODIFIED S3					EXP.DATA	
	α	β	a	E/A	$<r^2>^{\frac{1}{2}}$	α	β	a	E/A	$<r^2>^{\frac{1}{2}}$	E/A	$<r^2>^{\frac{1}{2}}$
	fm^{-2}	fm^{-2}		MeV	fm	fm^{-2}	fm^{-2}		MeV	fm	MeV(a)	fm(b)
^4He	0.82	1.6	.475	9.15	1.58	0.82	2.1	.725	6.04	1.58	7.07	1.63-1.71
^8Be	0.68	1.4	.500	5.85	2.21	0.68	2.1	.725	2.83	2.19	7.06	
^{12}C	0.69	1.2	.525	7.48	2.33	0.68	2.0	.725	4.28	2.32	7.68	2.44-2.46
^{16}O	0.72	1.1	.550	10.1	2.34	0.71	1.8	.725	6.73	2.32	7.98	2.67-2.73
^{20}Ne	0.62	1.6	.525	8.67	2.72	0.65	2.0	.725	5.39	2.61	8.03	3.00-3.04
^{24}Mg	0.60	1.6	.525	8.87	2.89	0.64	2.0	.725	5.55	2.73	8.26	2.99-3.08
^{28}Si	0.61	1.6	.550	9.63	2.92	0.64	1.9	.725	6.22	2.80	8.45	3.08-3.14
^{32}S	0.61	1.6	.550	10.3	2.97	0.64	1,9	.725	6.84	2.86	8.49	3.24-3.26
^{36}Ar	0.60	1.7	.550	11.1	3.04	0.64	1.9	.725	7.63	2.89	8.52	
^{40}Ca	0.60	1.6	.525	12.0	3.08	0.62	2.0	.725	8.39	2.99	8.55	3.42-3.48
N.M.	1.45^c	0.8	.550	31.8		1.8^c	1.6	.700	23.6		≃16	$\simeq 1.36^c$

(a) Ref.[15] (b) Ref.[16] (c) k_F in fm^{-1}

Table 2. Energies, radii and parameters of the wave function (defined in eq.(4); $\alpha=\sqrt{m\omega/\hbar}$) for the B1 and S3 interactions corresponding to 4n nuclei and nuclear matter.

the results poorer, but the variational principle still applies(i.e. we still have an upper bound for the energy)

v. There are three variational parameters: the harmonic oscillator constant $\alpha=\sqrt{m\omega/\hbar}$ and the parameters of the correlation, eq. (4). All three parameters have been independently varied.

vi. The calculation of the rms radius includes the center-of-mass correction and the finite size of the proton: is the charge radius.

Table 2 collects all interesting quantities corresponding to 4n nuclei :energies, rms radius and the wave function parameters. We have included also the nuclear matter results with the same correlation function (4) computed in the FHNC/0 theory.

There are some interesting consequences from Table 2:

i. The two-body correlation function is almost independent on the nucleus. This statement applies exactly to the depth a of the correlation, but that regularity is not so strong with regard to the range β but in S3 potential is still manifestly constant(probably due to the rather strong core of this force).

Nucleus		Configuration	Energy(MeV)
^{20}Ne		(110)	116.0
	→	(002)	116.8
^{24}Mg	→	(101)(002)	141.6
		(101)(110)	138.4
		(110)(002)	134.1
		(200)(020)	136.7
^{28}Si		(110)(200)(002)	166.7
		(110)(200)(020)	174.1
		(110)(101)(020)	170.0
	→	(101)(011)(002)	177.4
		(200)(020)(002)	161.9
		(110)(101)(011)	171.7
^{32}S	→	(110)(101)(011)(002)	216.3
		(101)(011)(200)(020)	207.8
		(101)(011)(200)(002)	215.2
		(110)(200)(020)(002)	205.5
^{36}Ar	→	(110)(101)(011)(200)(020)	266.7
		(101)(011)(200)(020)(002)	259.1

TABLE 1. Determination of the configuration in absence of correlations. Levels are filled up in the cartesian representation of the harmonic oscillator. All nuclei listed have the oxygen core. The energy corresponds to the B1 force. The arrow indicates the best configuration.

wave function should be improved. To end up this section we would like to stress that the model wave functions considered are not the most general possible within the p-shell or the sd-shell. However, in most of the cases the resulting model functions correspond to the accepted alpha-particle model in the limit in which the distance of the clusters shrinks to zero.

4. CALCULATION OF THE CORRELATED ENERGY

The energy in the presence of correlations has been calculated for the correlated functions given by eq. (3) by considering always the simple form eq.(4) for the two-body correlation and with the model wave functions determined in the previous section.The energy and radius has been computed according to the conditions specified below (further details may be found in Ref. [2]):

i. The Coulomb force is not included

ii. Center of mass effects are apropriately taken into account by substracting $3/2\ \hbar\omega$.

iii. The FAHT cluster expansion up to third order has been used, i.e. four particle clusters have been disregarded.

iv. No angular momentum projection has been carried out. This makes

ii. The results for the energy per particle are, however, dissappointing: open shell nuclei have an energy per particle manifestly smaller than closed shell nuclei. This difference is of the order of 4MeV in ^{8}Be and 3 MeV in ^{12}C, but is smaller in heavier nuclei, as expected.

iii. The charge r.m.s. radii vary smoothly with A, but the resulting values are much smaller than the experimental values.

iv. The gain in energy due to correlations is impressive.

To stress this last point we have included in Table 3 the values of E/A for the B1 potential obtained from several methods. These values

	Jastrow	Alpha-model	HFB (c)	BHF (d)	Various
^{12}C	7.48	5.17[a]			10.6[e] 7.8[f]
^{16}O	10.10	5.90[a]	5.80	10.23	5.82[g]
^{20}Ne	8.67	5.62[b]	5.50		
^{24}Mg	8.87	5.49[b]	5.02		
^{28}Si	9.63	5.61[b]	5.67		
^{32}S	10.28	5.58[b]	5.41		
^{36}Ar	11.08	5.83[b]	5.99		
^{40}Ca	11.95	6.26[b]	6.29	12.68	6.28[g]
N.M:	31.8			26.6	15.70[h]

(a): Ref. [8]
(b): Ref. [9]
(c): Ref. [7] some cases HF
(d): Ref. [5]
(e): Ref.[17];deformations
(f): Ref.[17];no deformation
(g): Ref.[12];hypersph.Harm.
(h): Ref.[3];uncorrelated

TABLE 3: Energies for the B1 potential from several methods and corresponding to various nuclei and nuclear matter.See the text for further details.

are written as they appear in the original works, and before the comparison with our results it should be considered that the alpha-model, HFB and hyperspherical harmonics calculations include the Coulomb potential. Nevertheless, the gain in energy is always much larger than the Coulomb part.

Table 3 includes the results of a BHF calculation of the energy [5]. The BHF results are very close to the Jastrow results, but they are slightly lower in finite nuclei. This is in contrast with the numbers of the last row corresponding to nuclear matter, the Jastrow method giving 5 MeV more binding energy per nucleon than the BHF method (note that the Jastrow calculation has been carried out with the same simple correlation function eq. (4) than finite nuclei). The conclusion is that BHF is a good approximation in finite nuclei (low density systems)[5].There remains still the question that the rms charge radii (not reported here) are systematically lower in Jastrow method (and, correspondingly, the Fermi momentum in nuclear matter is higher). The same qualitative behaviour is found in the S3 potential.

We now comment the column labelled "various" in Table 3. The values for ^{16}O and ^{40}Ca correspond to the hyperspherical harmonics method [12] and they compare very well with other uncorrelated calculations. The two numbers corresponding to ^{12}C are prelimary results corresponding to a Jastrow wave function [17]. The value of E/A=7.8 MeV corresponds to the full angular momentum projection of the wave function obtained before: the gain due to the angular momentum is then of some 0.3 MeV per particle. The value E/A=10.6 corresponds to a variational calculation of the ^{12}C energy in an axially deformed harmonic oscillator potential, including short range correlations and full angular and center-of-mass projection. We consider that result very important, because it puts the ^{12}C value in the same range than the results for doubly magic nuclei, restoring the systematics of E/A vs. A. This suggests the convenience of doing analogous calculations for the other open shell nuclei.

We would finally answer the question: how good the trial function is? A partial answer for this question results from the comparison of our results for ^{4}He with the S3 potential with those of Refs. [4] and [13]: the energy per particle in [4] is 6.84 Mev, and the result of [13] is 6.63 MeV against our result of 6.04.MeV. This means that we lose about 1 MeV per particle (note however that our wave function has three parameters against the seven parameters of Ref. [13]). The improvement of the wave function in those works is double: in the correlation function and in the single particle wave functions. We believe that second aspect to be the most important in the gain in energy, and this should be our first improvement in future work.

Acknowledgements

This work is part of a contract with the Comisión Asesora Cientifica y Técnica (Spain). Finantial support from the INAPE, Dirección General de Universidades and Vicerrectorado de Extensión Universitaria is also acknowledged.

In this work I have included some results obtained in collaboration with A. Faessler, H. Müther, A. Polls and E.Buendía, and fruitful suggestions of S.Rosati. All them are also acknowledged.

References

[1]J.W.Clark and P.Westhaus, Jour.Math.Phys. 9 (1986) 131

[2]R.Guardiola, Nucl.Phys A328 (1979) 490
R.Guardiola and A.Polls, Nucl.Phys. A342 (1980) 385
R.Guardiola, A.Polls and J.Ros, Nuovo Cim., in press

[3]D.M.Brink and E.Boeker, Nucl.Phys. A91 (1967) 1

[4]I.R. Afnan and Y.C. Tang , Phys.Rev. 175 (1968) 1337

[5]A.Faessler, R.Guardiola, H.Müther and A.Polls. To be published

[6]C.S. Warke and M.R. Gunye, Jour.Phys. A7 (1974) 718

[7]P.U. Sauer, A.Faessler,H.H.Wolter and M.M. Stingl, Nucl.Phys. A125 (1969) 257

[8]D.M. Brink, H.Friedrich, A.Weiguny and C.W.Wong, Phys.Lett.33B (1970) 143

[9]W.Bauhoff "The alpha-particle model of sd shell nuclei" Proc.18th Int. Winter Meeting on Nucl.Phys. BORMIO (Italy) 1980

[10]Y.Abgrall,G.Baron,E.Caurier and G.Monsonego Nucl.Phys. A131 (1969) 609

[11]J.Garcia-Roger and R.Guardiola, Nucl.Phys. A267 (1976) 137

[12]J.Navarro. Proc. Second Topical School GRANADA (Spain) 1979. Anales de Física (Madrid) 1980

[13]S.Fantoni, L.Panattoni and S.Rosati, Nuovo Cim. 69A (1970) 80

[14]J.P. Elliot, Proc.Roy.Soc. (London) A245(1958) 128

[15]A.H.Wapstra and K.Boss, At. and Nucl. Data Tables 19(1977)177

[16]C.W.de Jager,H de Vries and C. de Vries, At. and Nucl. Data Tables 14 (1974) 479

[17]E.Buendia and R.Guardiola, in preparation.

THREE-BODY FORCES IN NUCLEI

SIDNEY A. COON
Department of Physics, University of Arizona
Tucson, Arizona 85721/USA

Three-body forces (TBFs) may play a non-negligible role in such old problems of nuclear physics as i) the saturation properties of nuclear matter, ii) the binding energies and charge form factors of light nuclei,[1-5] and iii) the effective interaction of the nuclear shell model. The possible importance of TBFs for problems i) and ii) has been discussed at this meeting. It is perhaps not so well known, however, that the contributions of TBFs to problem iii) have been estimated for one-body,[6] two-body,[7] and three-body[8] terms in the effection interaction. After a general discussion of the structure of TBFs of the two-meson exchange type, and a brief description of the two-pion-exchange three-body potential (2πTBF) based on Refs. 4 and 9, we will present some new results with that potential on problem iii), the single particle spin-orbit splitting[6] in ^{16}O and ^{40}Ca, and on problem i), nuclear matter saturation. We will close by discussing recent constructions[10,11] of πρTBFs and ρρTBFs.

Structure of two-meson exchange TBFs

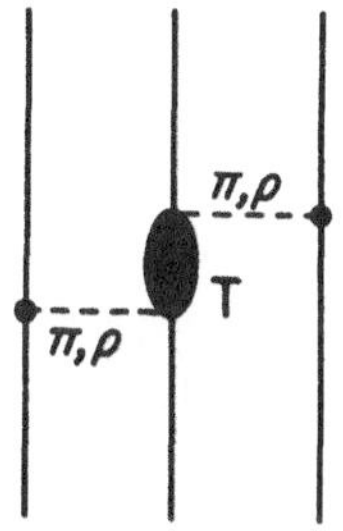

Fig. 1

The shaded oval of Fig. 1 represents a meson-nucleon scattering amplitude with the term corresponding to the iterated one-meson exchange (or forward propagating Born term, FPBT) subtracted. The two mesons are off-mass-shell and space-like. The dots at the meson-nucleon-nucleon vertices represent the damping of the coupling constant in the space-like region, usually parameterized by a form factor normalized to 1 on the meson mass shell. The three-body potential W is defined in terms of the non-relativistic reduction of the three-body S-matrix of Fig. 1 as $(S-1)_{NR} = -i(2\pi)\delta(E_f-E_i)\,W$. Therefore W has the form of a product of simple Yukawas multiplied by a single off-mass-shell scattering amplitude.

In contrast, the two-meson-exchange (i.e., 2π) two-nucleon potential is derived from an integral over the πN amplitude for time-like values of the momentum transfer variable t so that the potential is a superposition of Yukawas. In addition, the dispersion theoretic two-nucleon forces utilize πN amplitudes in which all four particles are on-mass-shell. Given these structural differences, it is a subtle business to relate meson-nucleon amplitudes used in two-body and three-body forces. Certainly one cannot simply substitute the amplitudes discussed here for the amplitudes appearing in the two-body force. We will confine our presentation to TBFs.

Determination of the meson-nucleon amplitude

We first outline a general approach to this problem and then relate to it two alternate methods often seen in the literature. One can begin with Compton-like amplitudes of axial-vector current-nucleon scattering or vector current nucleon scattering and use PCAC or the current field identities to derive πN amplitudes or ρN amplitudes, respectively. The appropriate off-mass-shell amplitude is written as

$$T = T_B - T_{FPBT} + \Delta T + q\cdot C\cdot q' \qquad (1)$$

where T_B is the nucleon pole (Born) term and ΔT is added to T_B so that $T_B + \Delta T$ satisfy low energy theorems (LETs). The LETs impose a certain amount of model independence to the amplitude as they require T to be expressed in terms of coupling constants and the target mass as q and $q'\to 0$. The LETs constrain TBFs because a practical potential contains the low order terms of an expansion of the amplitude in powers of q and q' around the nucleon pole. The axial-vector LETs lead to the soft ($q^2\to 0$) pion theorems associated with the current algebra (CA) Ward identity program[12] for πN scattering. The importance of nucleon pole dominance and the soft pion theorems for the 2πTBF was early recognized by Brown and Green.[13] To include higher order terms in T, one must consider the background $q\cdot C\cdot q'$ which is certainly model-dependent. In the 2πTBF case, however, the sum $\Delta T + q\cdot C\cdot q'$ is given by the empirical on-mass-shell expansion coefficients (Höhler coefficients) of the non-pole background πN amplitude,[14] so that the entire off-shell amplitude needed is model-independent up to $O(q^2)$.[15] Corrections of $O(q^4)$ in T are known to be small from the on-shell data, and such corrections in $q\cdot C\cdot q'$ are scaled by the isobar mass so the truncation at $O(q^2)$ terms is justified in the 2πTBF case.[9] (Some recent off-shell πN amplitudes[17,18] were also based on the Höhler coefficients, but the truncation was not justified in those papers.)

An often used alternative to the Compton-like amplitude just described is the representation of T of (1) by a sum over isobars (Δ, N*, etc.). Often the emphasis then shifts from T to an expansion of nonrelativistic nuclear states to include isobars, with a concomitant shift from TBFs to three-body clusters of nucleons and isobars.[18-20] Viewed as a πN amplitude generator, this approach has some difficulties. Firstly, it generates only the $q\cdot C\cdot q'$ part of (1) and misses the nucleon pair term $T_B - T_{FPBT}$ and the non-pole ΔT. It is known[14] that a relativistic $q\cdot C\cdot q'$ from $\pi N\to\Delta\to\pi N$ alone cannot reproduce the Hohler coefficients. Secondly, the nonrelativistic method actually used reproduces the Lorentz invariant term $q\cdot C\cdot q'$ only when normal dependent terms are included in the Hamiltonian,[21] and may not even contain all of this term. Thirdly, the PCAC-CA amplitude already includes the Δ, N*, and higher mass isobars; the recent suggestion of a separate TBF due to an N* state[20] would amount to double counting of isobars if added to the PCAC-CA amplitude and missing other more significant components if added to a Δ-based amplitude. This approach allows one to include ρ-exchange easily,[19,10] but the neglect of the nucleon pair term in the resulting TBF then becomes a serious deficiency,[11] which will be discussed later.

Another alternative amplitude is the one pioneered for the 2πTBF by Fujita and Miyazawa[22] and still in use.[23] In effect, the low energy amplitude is converted into a sum rule via a dispersion relation of a static theory of πN interactions, and the resulting strength parameter is given by an integral over total cross sections. This approach doesn't account for ΔT, and $T_B - T_{FPBT}$ was incorrectly set to zero in Ref. 22. The πN cross section near threshold is dominated by the (3,3) resonance, so the resulting 2πTBF is nearly identical with one obtained by summing over isobars.

Many-body forces versus many-body clusters

Friman and Nyman[24] have shown how to sum the ring diagrams of a model many-body system of nucleons and Δ isobars interacting via meson exchange. They point out that their results include, in a systematic way, many diagrams which are usually classified as due to many-body forces. So we now have two theoretical schema, many-body forces and many-body clusters, each containing parts of the other as a substructure. It will take more work to sort this out, but some implications can be drawn for the nuclear structure problems mentioned in the introduction.

A three-body potential has the following single exchange (Fig. 2) and double exchange (Fig. 3) diagrams of first-order perturbation theory in nuclear matter. These

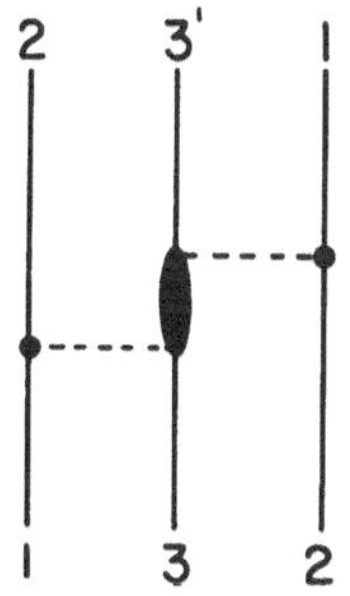

Fig. 2

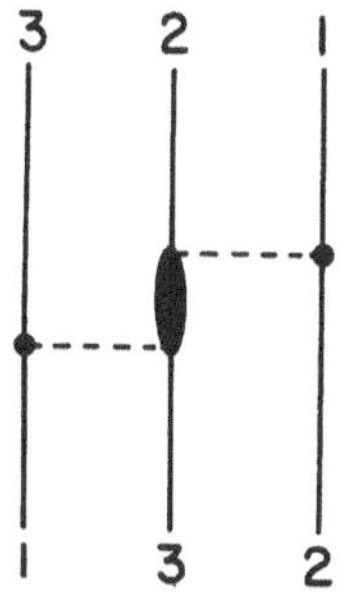

Fig. 3

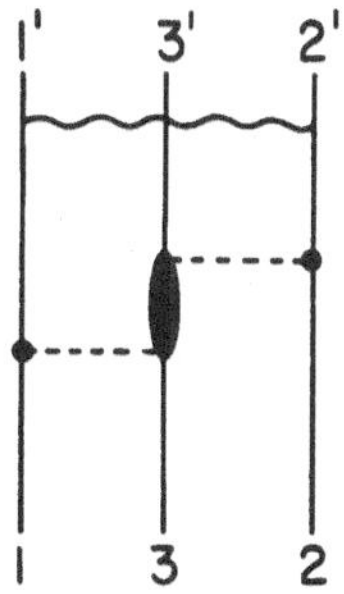

Fig. 4

diagrams are sometimes neglected three-body clusters,[19] but are not negligible (see row $B^{(1)}$ of the Table) compared to the second order diagram of Fig. 4. In addition, the smallest nucleus with a three-body cluster is ^{3}He, which seems to have a pronounced depression in the one-body density of point nucleons as deduced from charge form factors. The origin of such a hole can be seen rather transparently from a 2πTBF,[1,2] but it seems harder to get such a deep hole from a Hamiltonian that includes isobar degrees of freedom.[25] The many-body cluster approach might, however, lend itself better to studies of the Landau parameters of nuclear excitations or extreme situations like pion condensation (see, however, Ref. 26).

The 2π three-body potential

The potential[4,9] is (with the labels of Fig. 2 and $f \equiv g\mu/2m \approx 1$)

$$\begin{aligned} W(\vec{r}_1,\vec{r}_2,\vec{r}_3) = {} & (f/4\pi)^2\, \vec{\tau}_1\cdot\vec{\tau}_2\; \vec{\sigma}_1\cdot\vec{\nabla}_1\; \vec{\sigma}_2\cdot\vec{\nabla}_2 \Big\{ b\, \vec{\nabla}_1\cdot\vec{\nabla}_2\, Z_1(r_{31})Z_1(r_{32}) \\ & + (a-2\mu^2 c)Z_1(r_{31})Z_1(r_{32}) + c\big[Z_0(r_{31})Z_1(r_{32}) + Z_1(r_{31})Z_0(r_{32})\big]\Big\} \\ & + (f/4\pi)^2\, \vec{\tau}_2 x \vec{\tau}_1\cdot\vec{\tau}_3\; \vec{\sigma}_1\cdot\vec{\nabla}_1\; \vec{\sigma}_2\cdot\vec{\nabla}_2\; \vec{\sigma}_3\cdot\vec{\nabla}_1 x \vec{\nabla}_2\; e\, Z_1(r_{31})Z_1(r_{32}) \\ & + \text{cyclic permutations} \end{aligned} \qquad (2)$$

where

$$Z_n(r_{ij}) = \frac{4\pi}{\mu}\int \frac{d^3\vec{q}}{(2\pi)^3}\ \frac{e^{-i(\vec{r}_i-\vec{r}_j)\cdot\vec{q}}\ H(\vec{q}^{\,2})}{(q^2+\mu^2)^n} \quad . \tag{3}$$

The quantities a, b, and c are the coefficients of a low energy expansion of the isospin even, forward πN amplitude for virtual pions which satisfies the program of Eq. (1):

$$T = F_{\pi NN}(-\vec{q}^{\,2})\ F_{\pi NN}(-\vec{q}^{\,\prime 2})\left[a+b\vec{q}\cdot\vec{q}^{\,\prime}+c(\vec{q}^{\,2}+\vec{q}^{\,\prime 2})+d\vec{p}_3\cdot\vec{p}_3^{\,\prime} + \ldots\right] \tag{4}$$

where

$$\begin{aligned}
a &= \sigma/f_\pi^2 &&= +1.13\ \mu^{-1}\\
b &= -2/\mu^2\left[\sigma/f_\pi^2 - \bar{F}^{(+)}(\nu=0,t=\mu^2)\right] &&= -2.58\ \mu^{-3}\\
c &= \sigma/(\mu^2 f_\pi^2) - g^2/4m^3 + F'_{\pi NN}(0)\sigma/f_\pi^2 &&= +1.00\ \mu^{-3}\\
d &= -g^2/(2m^3) &&= 0.29\ \mu^{-3}\ .
\end{aligned} \tag{5}$$

In these equations, σ is the πN sigma term, f_π the pion decay "constant," $\bar{F}^{(+)}(0,\mu^2)$ the non-pole background amplitude, and $g^2/4\pi \approx 14.3$. We have dropped the small non-local term $p_3 \cdot p_3'$. The final coefficient $e = -0.75\ \mu^{-3}$ is the leading term in the expansion of the isospin-odd, spin flip πN amplitude. The forward even and spin-flip odd amplitudes don't contribute to the TBF to $O(q^2)$.

This potential is distinguished from that in Refs. 22-23 by being non-static and hermitian and including (small) contributions from nucleon pair terms. Those TBFs have the parameters $a=c=d=0$, $b=-1.4\ \mu^{-3}$, $e=-0.35\ \mu^{-3}$; that is, no contribution from s-wave πN scattering and a contribution from p-wave scattering about half as large as that determined by the PCAC-current algebra program of Eq. (1). Small $a=-0.003\ \mu^{-1}$ and $c=-0.002\ \mu^{-3}$ terms were once added to the isobar-based potential but were dropped already in the same paper.[27]

Recently there has been progress made on understanding of the form factor $g(q^2) \equiv gF_{\pi NN}(q^2)$ which appears in Eq. (3) as $H(\vec{q}^{\,2}) = F^2_{\pi NN}(-\vec{q}^{\,2})$. Theoretical calculations[28] and phenomenology[29] agree that there should be a variation of about 3% from $q^2 = m_\pi^2$ (the on-shell normalization point) to $q^2 = 0$, so that $F_{\pi NN}(0) \approx 0.97$. The Goldberger-Treiman discrepancy $\Delta = 1 - (mg_A(0)/f_\pi g) \approx 0.06\pm0.01$, however, implies that if the entire discrepancy is blamed on $F_{\pi NN}$, then $F_{\pi NN}(0) \approx 0.94$. The resolution of this apparent paradox would seem to be the realization that f_π is momentum-dependent also. A simple constituent quark loop calculation[30] yields an additional 3% variation in $f_\pi(q^2)$ from $q^2=m_\pi^2$ to $q^2=0$. Therefore, $H(\vec{q}^{\,2})$ should reflect the value $F_{\pi NN}(0) = 0.97$; for example, $H(\vec{q}^{\,2}) = (\eta^2-\mu^2)/(\vec{q}^{\,2}+\eta^2)$ where $\eta = 4\ \mu$. This form factor is shorter ranged in coordinate space than the one used in Refs. 3, 7, and 9. We shall see this apparently benign change has rather dramatic consequences.

New results I: Spin-orbit splitting

The origin of the large spin orbit (s.o.) term in the nuclear shell model has been a classical unsolved problem in nuclear physics for 30 years. The theoretical spin-orbit splitting ($\Delta_{s.o}$) in spin- and isospin-saturated nuclei evaluated microscopically from the two-body force is generally too small compared with experiment. Andō and Bandō have calculated the contribution of the 2πTBF in Eqs. (2)-(4) to the splitting of the 0p state of ^{16}O and the 0d state of ^{40}Ca. They find $\Delta_{s.o.}(0p, {}^{16}O)$= 1.55 MeV, and $\Delta_{s.o.}(0d, {}^{40}Ca)$ = 2.12 MeV, which are comparable to the values of 3.2 MeV (3.4 MeV) from the two-body force in $^{16}O(^{40}Ca)$. Although the sum of two-body force and TBF doesn't reach the experimental $\Delta_{s.o.}$ = 6.2 (7.0) MeV for $^{16}O(^{40}Ca)$, Andō and Bandō suggest that the investigation of $\Delta_{s.o.}$ provides a good testing ground for TBFs.

New results II: Saturation of nuclear matter

The 2πTBF contribution to nuclear matter at different densities has been estimated once before,[15] but in an approximate way which is not reliable.[3] This time we use the effective potential approximation of Ref. 27 in which one averages over spin and isospin of nucleon 3 in Figs. 2 and 4 and integrates over position to get an effective potential

$$V_{eff}(r_{12}) = \rho \int W(\vec{r}_1\vec{r}_2\vec{r}_3)\phi^2(r_{31})\phi^2(r_{23})d^3\vec{r}_3 \tag{6}$$

where ρ is the density and ϕ is a correlation function taken from nuclear matter calculations.[31] This approximation allows contributions of the single exchange Fig. 2 and of Fig. 4 to the binding energy per particle to be expressed as

$$E^{(1)} = \langle W\rangle = \langle V_{eff}\rangle, \quad E^{(2)} = 2\langle V_2 Q/e\ V_{eff}\rangle, \quad E = E^{(1)} + E^{(2)} , \tag{7}$$

where a sum over occupied states is implied on the right hand side. One cannot calculate the double exchange diagram (Fig. 3) nor other exchange diagrams of higher order in the two-body potential V_2 and W with this approximation. At low densities, double exchange (Fig. 3) is about 10-20% of the single exchange (Fig. 2) in finite nuclei[6] and about 10% in nuclear matter.[27] Higher order exchange diagrams are perhaps more important.

The technical details of the calculation are given in Ref. 9; it is a minor extension to find the result at different densities. It turns out that V_{eff} scales with ρ to three significant figures; a scaling suggested but not demonstrated in Ref. 7. The contribution E, as we shall see, does not scale as ρ ($\rho=2\pi^2k_F^3/3$) but as $\rho^{1.4}$. This is in part due to the density dependence of the operator Q/e which is evaluated by a generalized Eular function,[32] thus avoiding an angle average approximation. The single particle energies in e are taken from Ref. 23.

We first tabulate the coefficients which give $E^{(1)}$ and $E^{(2)}$ in terms of a, b, and c:

$$\begin{aligned} E^{(i)} &= A^{(i)}\ a + B^{(i)}\ b\mu^2 + C^{(i)}\ c\mu^2 \text{ for } i=1,2 \\ E &= E^{(1)} + E^{(2)} = A\ a + B\ b\mu^2 + C\ c\mu^3 . \end{aligned} \tag{8}$$

This table permits the construction of the contributions to the energy of any potential of the form we have used, given the values of a, b, and c. A negative contribution represents attraction. The forms in parentheses represent a least squares fit to the k_F dependence of the coefficients.

TABLE - Expansion coefficients in MeV μ

		$k_F = 1.0\ fm^{-1}$	$k_F = 1.36\ fm^{-1}$	$k_F = 1.7\ fm^{-1}$
$A^{(1)}$	$(-.32\ k_F^{4.4})$	-0.324	-1.289	-3.345
$A^{(2)}$	$(+1.0\ k_F^{4.1})$	+0.101	+0.381	+0.869
A	$(-.22\ k_F^{4.5})$	-0.223	-0.908	-2.476
$B^{(1)}$	$(-.28\ k_F^{5.9})$	-0.284	-1.802	-6.631
$B^{(2)}$	$(+.92\ k_F^{4.5})$	+0.926	+3.601	+10.172
B	$(+.68\ k_F^{3.9})$	+0.642	+1.799	+3.541
$C^{(1)}$	$(-.52\ k_F^{5.8})$	-0.516	-3.157	-10.968
$C^{(2)}$	$(+1.4\ k_F^{4.3})$	+1.412	+5.263	+14.047
C		+0.896	+2.106	+3.079

The results of the 2πTBF are also plotted in Fig. 5 along with the results of the Fujita-Miyazawa force (FMTBF) which has coefficients a=c=0, b=-1.39. The box encloses the empirical saturation point. The topmost curve is the energy per particle from the RSC potential in the two-hole line approximation and is taken from Ref. 10. The curves labeled with an asterisk correspond to $F_{\pi NN}(0) = 0.94$, called form factor III in the literature. The perturbation calculation agrees well with a procedure[23] in which V_{eff} is simply added to V_2 except at the higher densities where perturbation theory (dashed line) gives a larger contribution. At $k_F = 1.7\ fm^{-1}$, Grangé et al.[23] found that the single particle energies had to be readjusted when V_{eff} was added to the Hamiltonian, so perturbation theory might not be expected to agree with Ref. 23 at such a high density. For this form factor the complete 2πTBF gives a

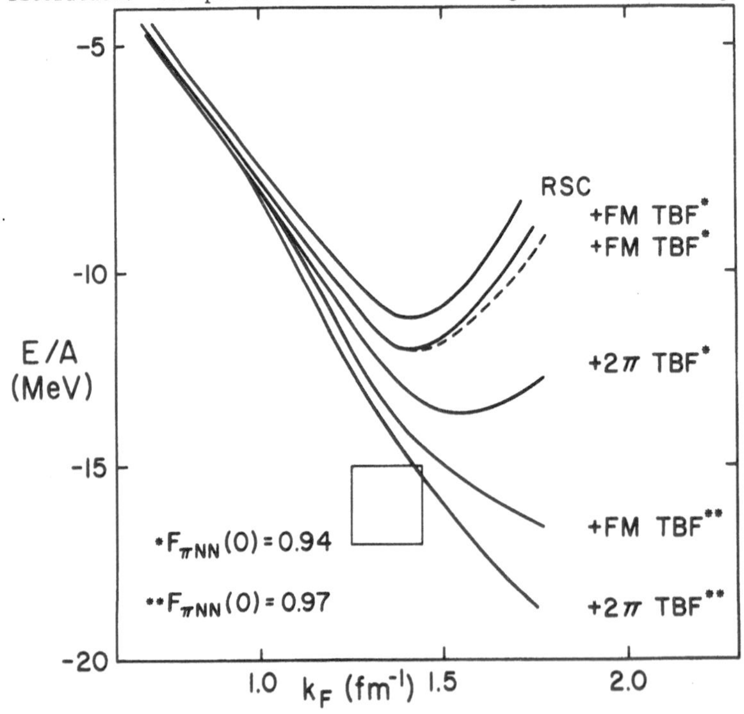

Fig. 5

significantly higher contribution than the earlier FMTBF.

Changing only the form factor in Eq. (3) to the realistic case $F_{\pi NN}(0) = 0.97$ yields the two lower curves of Fig. 5. The 2πTBF plus RSC doesn't saturate. It is often remarked[6,9] that the 2πTBT (especially the c part) is very dependent upon the form factor. This dependence would lead to an uncertainty in the predictions of the 2πTBF if $F_{\pi NN}$ were free, but its low energy behavior is fixed, thus substantially reducing the uncertainty. The 2πTBF, when treated as a perturbation to a two-hole line calculation (and including only Figs. 2 and 4), contributes to the energy of nuclear matter (E in MeV, k_F in fm^{-1})

$$E \approx -.97\ k_F^{4.3}\ . \tag{9}$$

This is a relatively unambiguous prediction because all aspects of the 2πTBF have been determined by experiment. The reader is cautioned against immediately adding this attractive contribution to alleged repulsive contributions from presumed "Δ-resonances in two-body clusters."

Nuclear matter does saturate, however. If we wish to continue to think of nuclear potentials in the traditional way, we must examine the caveats in the sentence culminating in Eq. (9). First the effective potential approximation can be checked by taking as a benchmark Faddeev[34] or variational[34] calculations of light nuclei with a Hamiltonian that contains a TBF, and redoing the same calculation with a V_{eff}. We know already that V_{eff} does not produce a hole in the one-body density,[3] but the FM TBF does.[1,2] Evidently the V_{eff} approximation loses some physics; whether it has a strong effect on energies remains to be seen.

The second caveat is the restriction of the TBF to 2π-exchange. One suggestion, revived at this conference,[35] is 3π-exchange potentials. Early work treated separately nucleon pair terms $(T_B\text{-}T_{FPBT})$[36] and isobar terms.[37] The full amplitude should be put together, subjected to the current-algebra soft pion constraints (ΔT), and compared with πN data before undertaking a necessarily elaborate calculational program. Preliminary work[35] does suggest that these 3πTBFs may play a more important role in nuclear saturation and nuclear structure than anyone has suspected for a long time. More work in this direction is desirable.

Potentials due to πρ and ρρ exchange

From the implications of the many-body cluster results,[18,19,24,25] one would propose that a useful supplement to the 2πTBF would be a shorter ranged TBF with the structure of Fig. 1, but with ρs (or possibly ωs) replacing one or both πs. Such a TBF based on i) the Δ-isobar, ii) the static approximation, and iii) only the largest ρNN coupling does give, in the V_{eff} approximation, a repulsive contribution which overwhelms the line labeled FM TBF** in Fig. 5 and yields an overall saturating effect (see Fig. 4 of Ref. 10). This is a very encouraging result, but is it the entire story? We will conclude by explaining why it is not.

The pair term $T_B - T_{FPBT}(\propto g^2/m^3)$ plays a small role in the 2πTBF (15% of c, all of d, and 20% of e in Eq. (5)) because of "pair suppression" enforced by the soft pion theorem ("Adler zero"). In the πρ and ρρ case, no such suppression occurs and the pair terms give a major contribution which, to the order given, dominates the contribution to T from the Δ-isobar. This can be seen intuitively by appealing to the Kroll-Ruderman theorem[38] for pion photo production and the Thirring[39] LET for Compton scattering. The Kroll-Ruderman theorem states that the isospin odd amplitude obeys

$$\lim_{k_\gamma, q_\mu \to 0} T_{\gamma\pi} = \frac{egi\vec{\varepsilon}\cdot\vec{\sigma}}{2m} . \tag{10}$$

But this result is readily obtained from the pair term by noting that γ^μ and γ_5 vertices both couple nucleons strongly to antinucleons. Thirring's theorem states that as k and k'→0, $T_{\gamma\gamma} = +e^2/m\vec{\varepsilon}\cdot\vec{\varepsilon}'$, which is again entirely due to the pair term. Because vector dominance is expected to be valid as k^2, the mass of the virtual ρ, approaches zero, we expect the pair terms to dominate the πρTBF and ρρTBF.

They do dominate, even after the covariant amplitudes for ρN→Δ→πN and πN→Δ→ρN have been worked out.[11] Indeed, one of the Beg isospin-odd LETs for "photons with isospin" can be converted into the Cabbibo-Radicati sum rule. It has been shown[40] that the Δ resonance does not saturate this sum rule, but gives a contribution of the wrong sign. The non-resonant πN intermediate states provide a contribution similar in magnitude but opposite in sign to that of the Δ. It appears that any attempt to construct the isospin-odd ρρTBF is doomed to failure if only the Δ, or indeed any set of resonances, is considered.

The leading order πρ and ρρTBFs have been constructed obeying the PCAC-CA plus LET constraints of Eq. (1). We hope that nuclear calculations with these potentials can be made soon. It is interesting that one cannot use a V_{eff} for the dominant part of the πρTBF because, according to Eq. (10), the spin average over the middle nucleon vanishes. Nuclear physicists will be forced to confront this TBF on its own terms.

I gratefully acknowledge the collaboration of B. H. J. McKellar and M. D. Scadron on the latter part of this work and thank B. H. J. McKellar and D. W. E. Blatt for the use and modification of the codes they wrote. This work was, for the most part, supported by the U.S.-Australia Cooperative Science Program administered by NSF Grant PHY-80-09527.

References

1) Fabre de la Ripelle, C. R. Acad. Sc. Paris, Series B, 288 (1979) 325.
2) Y. Nogami et al., McMaster University preprint (1980).
3) S. A. Coon, J. G. Zabolitzky, and D. W. E. Blatt, Z. Phys. A281 (1977) 137.
4) S. A. Coon and W. Glöckle, Phys. Rev. C, Feb. 1981.
5) Y. E. Kim, Muslim, and T. Ueda, Proc. 9th Few Body Conference, Eugene, OR (1980) 48.
6) K. Andō and H. Bandō, Proc. of the International Conference on Nuclear Physics, Berkeley, 1980, p. 6.

7) S. A. Coon, R. J. McCarthy, and C. P. Malta, J. Phys. G4 (1978) 183.
8) T. Lönnroth, J. Blomqvist, I. Bergström, and B. Fant, Physica Scripta 19 (1979) 233.
9) S. A. Coon, M. D. Scadron, P. C. McNamee, B. R. Barrett, D. W. E. Blatt, and B. H. J. McKellar, Nucl. Phys. A317 (1979) 242.
10) M. Martzolff, B. Loiseau, and P. Grangé, Phys. Lett. 92B (1980) 46.
11) S. A. Coon, M. D. Scadron, and B. H. J. McKellar, U. of Az. preprint (1981).
12) See, e.g., M. D. Scadron and L. R. Thebaud, Phys. Rev. D9 (1974) 310.
13) G. E. Brown, A. M. Green, and W. J. Gerace, Nucl. Phys. A115 (1968) 435; G. E. Brown and A. M. Green, Nucl. Phys. A137 (1969) 1.
14) G. Höhler, H. P. Jakob, and R. Strauss, Nucl. Phys. B39 (1972) 237; M. M. Nagels et al., Nucl. Phys. B147 (1979) 189.
15) S. A. Coon, M. D. Scadron, and B. R. Barrett, Nucl. Phys. A242 (1975) 467.
16) T. Ueda, T. Sawada, and S. Takagi, Nucl. Phys. A285 (1977) 429.
17) A. B. Midgal, Rev. Mod. Phys. 50 (1978) 108.
18) See, e.g., A. M. Green, Rep. Prog. Phys. 39 (1976) 1109.
19) T. Kouki, L. E. Smulter, and A. M. Green, Nucl. Phys. A290 (1977) 381.
20) K. Shimizu, A. Polls, H. Müther, and A. Faessler, U. of Tubingen preprint (1980); H. Müther, U. of Tubingen preprint (1981).
21) G. Ball and B. H. J. McKellar, U. of Melbourne preprint.
22) J. Fujita and H. Miyazawa, Prog. Theor. Phys. 17 (1957) 360.
23) P. Grangé et al., Phys. Lett. 60B (1976) 237; T. Kasahara, Y. Akaishi, and H. Tanaka, Prog. Theor. Phys. Suppl. 56 (1974) 76,
24) B. L. Friman and E. M. Nyman, Nucl. Phys. A302 (1978) 365.
25) Ch. Hajduk and P. U. Sauer, Nucl. Phys. A322 (1979) 242; P. U. Sauer, private communication.
26) B. H. Wilde, S. A. Coon, and M. D. Scadron, Phys. Rev. D18 (1978) 4489.
27) B. A. Loiseau, Y. Nogami, and C. K. Ross, Nucl. Phys. A165 (1971) 601.
28) H. F. Jones and M. D. Scadron, Phys. Rev. D11 (1975) 174; A. Cass and B. H. J. McKellar, Nucl. Phys. B166 (1980) 399.
29) C. A. Dominguez and B. J. Verwest, Phys. Lett. 89B (1980) 333.
30) S. A. Coon and M. D. Scadron, Phys. Rev. C, March 1981.
31) P. J. Siemens, Nucl. Phys. A141 (1970) 225.
32) D. W. E. Blatt and B. H. J. McKellar, J. Comp. Phys. 32 (1979) 89.
33) W. Glöckle and S. A. Coon, work in progress.
34) R. A. Smith, this conference; V. R. Pandharipande, this conference.
35) S. A. Moszkowski, this conference.
36) S. Drell and K. Huang, Phys. Rev. 91 (1953) 1527; A. Klein, Phys. Rev. 90 (1953) 1101; K. Brueckner, C. Levinson, and H. M. Mahmoud, Phys. Rev. 95 (1954) 217; E. M. Gelbard, Phys. Rev. 100 (1955) 1530.
37) I. Fujita, M. Kawai, and M. Tanifuji, Nucl. Phys. 29 (1962) 252.
38) N. Kroll and M. A. Ruderman, Phys. Rev. 93 (1954) 233.
39) W. Thirring, Phil. Mag. 41 (1950) 1193.
40) F. J. Gilman and H. J. Schnitzer, Phys. Rev. 150 (1966) 1362.

ROLE OF ISOBARS IN NUCLEAR INDEPENDENT PARTICLE MOTION

Invited Paper presented by

S. A. Moszkowski
University of California, Los Angeles, CA 90024 U.S.A.

at the
2nd International Conference on Recent Progress in Many Body Theories
Oaxtepec, Mexico, Jan. 12-17, 1981

Work supported by the National Science Foundation

Abstract: Multiple meson exchanges proceeding primarily via excitation of isobars can lead to strong many body interactions between nucleons. Three body interactions may be more important in most nuclei than the two body one pion exchange interaction.

I will discuss four simple models of nuclear forces, each of which gives nuclear saturation, but which differ importantly in other respects. In actual nuclei, it appears that all four mechanisms play a role, but I would like to explore at least the possibility that the many body forces due to the isobar mechanism are the most important of the four.

We will also discuss the implication of the isobar model for some low energy properties of finite nuclei. This model, in conjunction with the condition $W = -\frac{1}{2} T_F$ for nuclear matter (which is approximately satisfied) leads to the independent particle model and (approximately) a surface delta residual interaction in finite nuclei.

The isobar model differs from most conventional models of nuclear interactions used at present in that it requires a somewhat weaker tensor force.

1. INTRODUCTORY REMARKS

In view of the title of this conference, "Many Body Theories," I think it might be appropriate to begin this talk with a quotation from Wilkinson[1]:

> "Of course, we have tried very hard to pretend that many-body forces do not exist and have attempted to sweep them under the carpet by, for example, committing ourselves, in our shell modeling of nuclear states, to the idea of two body NN forces only and then adjusting the effective residual NN force to give best overall fit to data, thereby perhaps absorbing into that pragmatic NN force elements that rightly belong to NNN or higher forces. Who knows?"

I think most of us would agree with Wilkinson that many body forces in nuclei have been swept under the rug and should be considered perhaps more seriously than they have been.

I would, however, like to suggest an even more extreme position, namely that many body forces may be much more important than is usually thought, and indeed that they may be a major factor in nuclear saturation. There is a model for this, the Isobar model, which I will discuss in this talk.

Ideas such as this have been considered previously in the literature. Thus Drell and Huang[2] suggested that nuclear saturation is a consequence of repulsive many body forces. However, at the time of their work, the meson nucleon interaction was much more poorly known than now. The particular form of meson theory which Drell and Huang used was shown to be in conflict with the later information, and thus their idea seems to have been abandoned for a time.

2. MESON EXCHANGE MODEL OF TWO AND THREE NUCLEON INTERACTIONS

It has been known for some time that the two pion exchange (2PEP) potential with isobars in intermediate states is of great importance for the nucleon-nucleon interaction. In fact, the 2PEP is much more important than the OPEP contribution to the nuclear matter energy, which is essentially an average over spins and isospins. Up to now, most investigations of three body interactions have stressed two pion exchange contributions. This was discussed, for example, by Coon in the preceding talk. This, however, turns out to be quite small, only about 1 MeV per nucleon, for basically the same reason as for the smallness of the OPEP contribution to the two body interaction energy. In this talk we investigate the 3PEP three body interactions. This diagram was in fact considered previously by Fujita et al.[3] as a source of a strong three body nuclear force, but it seems not to have been followed up. On the other hand, in atoms, the analogous three body force, the so-called triple dipole interaction, seems to be of some importance in the interaction between helium atoms (as discussed in this conference). For the 3PEP three body interaction (again involving isobars in intermediate states), just as for the 2PEP two body interaction, the S,T averages tend to add, so that one gets a large repulsive contribution to the energy.

We briefly review some results for the 2PEP interaction. The OPEP interaction (including also ρ contributions) is:

$$V_{OPEP} = \frac{1}{3} (\tau_1 \cdot \tau_2) [(\sigma_1 \cdot \sigma_2)\ V_c(r) + S_{12}\ V_T(r)]$$

where

$$V_c = V_c^{\pi} + 2V_c^{\rho}\ (> 0)\ , \qquad V_T = V_T^{\pi} - V_T^{\rho}$$

The ρ-coupling tends to reduce the strength of the tensor interaction.

It is interesting to note that on the basis of an extreme version of the quark models, in which the spin dependence of the quark quark interaction is neglected entirely, we find that

$$V^{\pi}_{c\ \mathrm{or}\ T} = V^{\rho}_{c\ \mathrm{or}\ T}\ , \qquad V_c = 3V_c^{\pi}\ , \qquad V_T = 0\ ,$$

i.e., no tensor interaction. Of course, such a model would also give no splitting between π and ρ nor between N and Δ. Some spin dependence (color magnetic interaction) is required to obtain the required hadron splittings. This leads to a non-vanishing tensor interaction which is required anyway by the finite quadrupole moment of the deuteron.

However, the quark model at least suggests that the tensor force (especially at short and intermediate distances) might well be smaller than thought previously. In

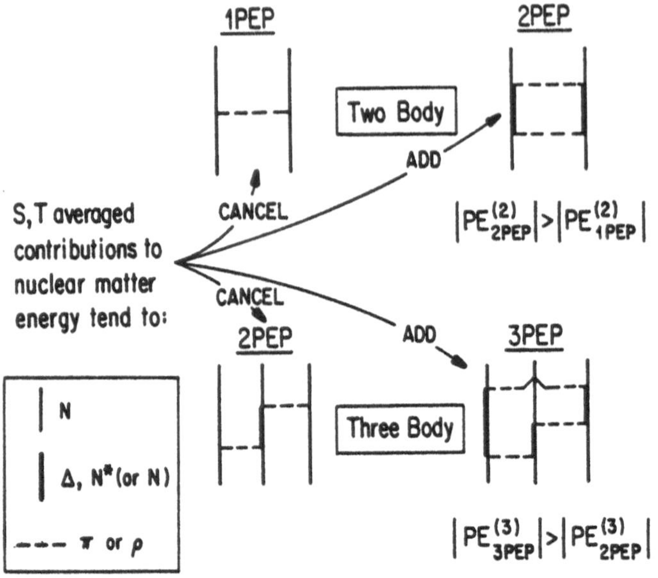

Fig. 1. Meson Exchange Model of Two and Three Nucleon Interactions.

most of the rest of this talk, we will disregard the tensor force.

The contribution of the central interaction to the 2PEP potential can be estimated using the closure approximation. This gives:

$$V_{2PEP} = -\frac{V_c^{\,2}(r)}{\Delta E}\Big[(1+\tfrac{4}{9}R)^2 - \tfrac{2}{3}(1-\tfrac{2}{9}R)^2\big((\sigma_1\cdot\sigma_2)+(\tau_1\cdot\tau_2)\big)$$

$$+\tfrac{4}{9}(1+\tfrac{1}{9}R)^2(\sigma_1\cdot\sigma_2)(\tau_1\cdot\tau_2)\Big]$$

where ΔE is an average energy denominator, and

$$R = (f^2_{\pi N\Delta}/f^2_{\pi NN}) = (f^2_{\rho N\Delta}/f^2_{\rho NN}) \quad .$$

As is well known, for $R = 0$, i.e., no isobar contribution, the 2PEP interaction is strongly (and wrongly) spin-dependent. This is why the 2PEP was not taken seriously as the main ingredient of the intermediate range attraction until the introduction of the Δ-isobar (which had been discovered by Fermi in 1951) into nuclear physics about 1970 by A. M. Green and his collaborators.[4] The best empirical value of R

$$R = (0.35/0.081) = 4.3$$

the constituent quark model gives a lower value:

$$R = (72/25) = 2.88 \ .$$

Interestingly enough, for the special value:

$R = 9/2$

(close to the experimental one), most of the spin dependence disappears, but the scalar-isoscalar contribution is greatly enhanced.

$$V_{2PEP} = -\ [V_c^{\ 2}(r)/\Delta E]\ [9 + (\sigma_1\cdot\sigma_2)(\tau_1\cdot\tau_2)] \quad .$$

This effect was first noted by Smith and Pandharipande,[5] in more accurate calculations of the nucleon-nucleon interaction. It was also studied by Durso et al.[6] and by Holinde et al. (talk presented at this conference). It is not surprising that a similar effect holds for the 3PEP three body interaction, and indeed, neglecting V_T, we obtain

$$V_{3PEP} \sim O(V_c^{\ 3}) > 0 \quad ,$$

a repulsive three body contribution to the energy. Earlier calculations using a smaller $f_\rho^{\ 2}$ and R (e.g., with Reid soft core interaction) give larger tensor interactions. For this case our previous argument must be modified, and, in fact, the three body contribution to the energy was generally found to be attractive. Note that $V_c(r) > 0$. (Of course, the OPEP is attractive in even states due to presence of the factor $(\sigma_1\cdot\sigma_2)(\tau_1\cdot\tau_2)$, which is negative in this case.)

We can investigate the characteristics of the three body interaction. For mathematical simplicity, we use a Gaussian rather than a Yukawa interaction form for V_c. This is

$$V_c(r) \sim \exp(-r^2/a^2) \quad ;$$

We find:

$$V_{2PEP}(r) = V_\sigma(r) = -V_c^{\ 2}(r)/\Delta E \sim \exp(-2r^2/a^2)/\Delta E$$

and a 3PEP interaction equivalent to a density dependent two body interaction of the form:

$$V_{3PEP} \sim 2\rho V_c(r_{12})\left(\int V_c(r_{13})\ V_c(r_{23})\ d^3r_3\right)/(\Delta E)^2 \sim \rho\ \exp(-3r_{12}^2/2a^2)/(\Delta E)^2$$

i.e., nearly the same range as V_{2PEP}. The comparatively long range of the three body interaction is not surprising when one considers that it arises from ring diagrams. Thus we are dealing with a "giant isobar resonance," which is propagated through the nuclear medium. We can, of course, also have higher order ring diagrams as well, but these are not likely to be important.

Most calculations using density dependent interactions (usually of the Skyrme form) assume a zero range interaction for the sake of mathematical simplicity. However, some authors, for example Onishi and Negele,[7] used essentially finite range Skyrme interaction, though in a different connection.

3. TWO AND THREE BODY ENERGIES IN NUCLEAR MATTER

Taking into account terms of higher order in ρ, we find, very roughly,

$$V(\rho,r) \sim V_o\, e^{-\alpha\rho} \exp(-2r^2/a^2) \quad .$$

We can make a crude fit to empirical saturation properties of nuclear matter

$$T_F = 37 \text{ MeV} , \quad T_{Av} = 22 \text{ MeV} , \quad PE_{Av} = -38 \text{ MeV} , \quad W = -16 \text{ MeV at } \rho_o \sim 0.16 \text{ fm}^{-3} .$$

Replacing the $\rho^{2/3}$ dependence of the kinetic energy by a linear one, yields simple analytic results. We find:

$$PE_{Av}(\rho) = -63\, \hat{\rho} \exp(-\frac{1}{2}\hat{\rho}) = -63\, \hat{\rho}(1 - \frac{1}{2}\hat{\rho} + \frac{1}{8}\hat{\rho}^2 \cdots)$$

where $\hat{\rho} = \rho/\rho_o$. Thus the two body and three body and higher contributions are

$$W^{(2)} = -63 \text{ MeV}, \quad W^{(3)} = -63x - \frac{1}{2} = 31.5 \text{ MeV}, \quad W^{(\geq 4)} = -63\left(\frac{1}{8} - \frac{1}{48} \cdots\right) = -6.5 \text{ MeV},$$

i.e., the three body contribution must be half as large as the two body energy (in this simple model). This model gives a reasonable value of the compression modulus

$$K = 9|W|(2+\phi)/(1+\phi)$$

where

$$\phi = 2|W|/T_F \quad ,$$

an important dimensionless parameter, which will come up frequently in this work. The empirical value of ϕ in nuclear matter is ~ 0.86 which is close to 1. For $\phi = 1$ we obtain

$$K \simeq 220 \text{ MeV}$$

in the range of the empirical values of K deduced from electric monopole giant resonance energies. In order to reproduce these values of $W^{(2)}$ and $W^{(3)}$ in the framework of a meson exchange model, using a Gaussian interaction with $\Delta E = 700$ MeV (a reasonable value for the average intermediate state energy involving isobar excitation), we need a transition interaction with a volume integral of about 6000 MeV fm^3. On the other hand, the volume integral of the empirical one pion exchange N-Δ transition interaction turns out to be only about 4300 MeV fm^3, which is close but not quite large enough. Of course, we can also have excitation to other isobaric states besides the Δ; for example, N*(1470), $N^*_{3/2}(1520)$. In fact, the results of calculations of the 2PEP three body interaction by Shimizu et al.[8] suggests that the N*(1470) gives as large a contribution to the nuclear matter energy as the Δ.

Presumably, this will hold for the 3PEP as well, so there should be no difficulty in accounting for the nuclear matter saturation properties using the isobar excitation model. We would roughly guess that the Δ alone accounts for about half of the density dependence required, maybe of order of 10 MeV/A for the three body term, and that all other isobars together contribute about the same amount. This rough guess is supported by the apparent circumstance that the low energy Δ-nucleon optical potential seems to be about half as large as the low energy nucleon-nucleus potential (Pandharipande, talk at this conference). These results are in fact consistent if indeed the low energy interactions and optical potentials are mainly due to an isobar excitation mechanism as discussed here.

4. FOUR MODELS FOR NUCLEAR SATURATION

We consider four extreme models of nuclear interactions, each of which can account for nuclear saturation, but not necessarily other features of nuclear structure. These are:

A. EXCHANGE INTERACTION, which is, in fact, a consequence of the OPEP without isobar excitation, but relatively unimportant in nuclear matter.

B. MOMENTUM DEPENDENT INTERACTION, which arises from relativistic corrections to OPEP.

C. SHORT RANGE REPULSION, which arises from ω exchange (also from π-ρ two meson exchange). Using many body theory, this alone leads to a short range density dependent two body repulsion plus a longer range attraction, and finally the model considered here:

D. ISOBAR MODEL, which is approximately equivalent to a density dependent interaction of the same range as the bare two body interaction. As we have seen, this results from multiple meson exchange via isobars. We would like to suggest the possibility that, in fact, model D may be more important than models A, B, or C for nuclear matter and for the low energy properties of finite nuclei.

5. BINDING AND RESIDUAL INTERACTIONS IN NUCLEAR MATTER

The existence of important many body effects leads to an ambiguity in the definition of the effective two body interaction. We may consider a "binding" interaction (often referred to as the reaction matrix). In the low density approximation where

$$V^{Bind} = - V_2 + \frac{1}{3} \hat{\rho} V_3$$

V_2 may be momentum or state dependent. Note that both V_2 and V_3 are two body interactions. The potential energy is just a sum of the binding interaction energies over all pairs of nucleons. On the other hand, also of importance is the "residual" interaction, which for a simple local interaction used here (and neglecting exchange) is given by the Landau prescription:

$$V^{Resid} = \frac{d^2}{d\rho^2} \left(\frac{1}{2} \rho^2 \, V^{Bind} \right) .$$

Thus, at low density,

$$V^{Resid} = - V_2 + \hat{\rho} V_3 .$$

The ratio 3 in the strength of the V_3 term was first noted by Sharp and Zamick.[9] Note that model C (short range repulsion) is equivalent to a short range density dependent interaction, i.e., the range of V_3 is much less than that of V_2. On the other hand, in model D (Isobar model), which is equivalent to a finite range density dependence, the range of V_3 is about the same as that of V_2. Supposing that $V_3 \sim V_2$, which turns out to be not a bad approximation, we find that at $\hat{\rho} = 1$,

$V^{Resid} = 0$, but $V^{Bind} = -\frac{2}{3} V_2 < 0$. Thus, we obtain overall binding, but also independent particle motion in nuclear matter.

Consider a Skyrme interaction, i.e., short but not necessarily zero range, of arbitrary exchange mixture, and of linear but finite range density dependence. Again we replace $\rho^{2/3}$ by ρ, which is equivalent to using a two dimensional model. The results in a realistic three dimensional model are not very different numerically; however, the two dimensional model has the great advantage that the calculations can be done analytically.

We can show that, for a Skyrme interaction in two dimensions, the average binding interaction of a pair in nuclear matter is

$$g_{Av} = \langle V^{Bind} \rangle_{Av} = -(T_F/A)(1+\phi)$$

while the average residual interaction energy is

$$F_o = \langle V^{Resid} \rangle_{Av} = -(T_F/A)[1 - (\rho \cdot \alpha_3/3T_F)]$$

where α_3 is the volume integral of the V_3 part of the interaction. Thus the density dependence reduces F_o relative to g_{Av}. In this approximation, the effective mass is

$$m^*/m = [1 + \phi - (\rho \cdot \alpha_3/3T_F)^{-1} \quad .$$

We give here some simple results for the four models. For a state dependent but density independent interaction (models A or B) we find that

$$m^*/m = F_o/g_{Av} = 1/(1+\phi)$$

for nuclear matter with $W = -\frac{1}{2} T_F$, i.e., $\phi = 1$. Both m^* and F_o are reduced by a factor 2. On the other hand, for a purely density dependent interaction (models C or D) we have

$$\rho \cdot \alpha_3 = 3T_F \phi \quad .$$

Thus, $m^* = m$, and $F_o/g_{Av} = (1-\phi)/(1+\phi)$. For $\phi = 1$ we find that F_o vanishes. Indeed, the spin average of the Landau parameter F_o in nuclear matter seems to be close to zero.[10]

It is also of interest to consider the monopole pairing matrix element, G_o, which in three dimensional nuclear matter can be defined by

$$G_o = \int V^{Resid}_{even}(r)\ j_o^{\,2}(k_F r)\ d^3r \quad .$$

For a Skyrme interaction, G_o is related to F_o by a simple equation,

$$G_o - F_o = (\rho_o/3A) \int [V_3(r) - V_2(r)]_{even} \times r^2\ d^3r \quad .$$

For a short range three body repulsion (model C), we find that $G_o > F_o$, but in the isobar model with $\phi = 1$, we have $V_3 = V_2$ and $G_o = F_o$. We summarize key nuclear properties predicted by the four extreme models in the table below.

Table 1. Key Nuclear Properties (for $\phi = 1$).

	m^*/m	F_o	G_o
(a) EXCHANGE	1/2	$1/2\ g_{Av}$	$1/2\ g_{Av}$
(b) MOM. DEP.	1/2	$1/2\ g_{Av}$	$> 1/2\ g_{Av}$
(c) SHORT RANGE REP.	1	0	> 0
(d) ISOBAR MODEL	1	0	0
(e) EXPERIMENT	0.7 to 1	~ 0	~ 0

Thus experiment favors the isobar model. Note that the effective mass right at the Fermi surface (which determines nuclear level densities) is close to 1. I think this is the relevant value for low energy properties. The lower value of about 0.7 applies to the energy dependence of the optical potential. The connection between these two values has been considered by several groups (as discussed by Friman at this conference). We have not considered this point here. In order to explain the energy dependence of the optical potential it is necessary to take into account the exchange terms arising from the OPEP.

6. FINITE NUCLEI

Using a two dimensional harmonic oscillator shell model, and taking the limit of large quantum numbers, we can then derive simple explicit expression for the higher multipole terms F_L as well as for the average pairing matrix elements. We find, for models A and B (state dependent, density independent interaction)

$$F_o = -\frac{2}{3}(T_F/A)(1+\frac{2}{3}\phi) \ , \qquad F_2 = -\frac{1}{3}(T_F/A)(1+\frac{4}{3}\phi)$$

while for models C and D (density dependent but state independent interaction)

$$F_o = -\frac{2}{3}(T_F/A) \ , \qquad F_2 = -\frac{1}{3}(T_F/A)(1+\phi) \quad .$$

In order to obtain $F_2 = F_o$, we need $\phi = \infty$ for models A and B, but $\phi = 1$ for models C and D. The latter is close to the empirical nuclear matter value. In either case, we find $F_4/F_o = 7/8$, also close to 1.

7. SURFACE DELTA INTERACTION

The interaction predicted by the isobar model is close to the surface delta interaction. This can be seen as follows: The well known SΔI conditions are[11]: (1) Zero range interaction. (2) All radial integrals are equal. We are, of course, primarily concerned with the matrix elements of this interaction. In two dimensions this gives the very simple result:

$$F_o = F_2 = F_4 \cdots = G_o = G_2 = G_4 \cdots$$

i.e., all particle-particle and particle-hole multipole strengths are equal.

Indeed model D (isobar), with $\phi = 1$, gives $F_2 = F_o$, $F_4 = \frac{7}{8} F_o$, $G_L = F_L$. These are very close to the SΔI matrix elements.

8. CONCLUDING REMARKS

The interactions used in the idealized models discussed in this talk do not fit experimental nuclear force parameters such as nucleon-nucleon scattering data, and thus they are not realistic, strictly speaking. On the other hand, they are chosen so as to fit nuclear saturation properties, and so in a sense they are self consistent, besides being very easy to handle. As I see it, the main advantage to using simple interactions such as Skyrme and of using a two dimensional model is that it makes it possible to explore the connection between different models in a much simpler way than is possible for realistic forces. As we have seen, for the four simple models used here we can obtain analytic relations between such quantities as the average binding and residual matrix elements.

The convergence of the perturbation series used in this paper is supported by estimates from the quark model (lecture by S. Moszkowski, Workshop Mexico City, Jan. 5-9 1981). In fact, it probably holds quite generally, as a consequence of the finite size of the nucleon. When two nucleons approach closer than about 0.8 fm, a repulsive transition interaction operates which couples the nucleons to other channels (such as the Δ). (Note that this is not quite the same as conventional short range repulsion without other channels.) Such a repulsive transition potential automatically leads to an attractive two body and repulsive three body interaction and to nuclear saturation.

We conclude by discussing some interesting problems for the future. First, it has been pointed out by many people that in order to pin down the nuclear interaction, we must not only fit (besides the two and three body data) nuclear ground state properties, but also low energy properties, such as level densities, pairing matrix elements, etc., indeed low energy spectra. (An interesting problem here is the possible need for three body effective interactions to fit nuclear spectra.) In this way, we can distinguish between various nuclear forces much better than by looking only at nuclear matter or only ground state properties. This is clearly seen in the two dimensional model discussed here. However, it is not easy to actually implement such calculations. In practice, it is difficult to make reliable and convergent shell model calculations, using the usual realistic interaction (such as, e.g., Reid Soft Core). This is, of course, to some extent unavoidable, given the complexity of nuclear systems.

However, I have a specific suggestion that might make things at least a little easier. That is the idea of a "Minimal tensor interaction" (named after the idea of minimal relativity suggested by G. Brown about 20 years ago). This means that we use only the minimum tensor interaction (especially at short and intermediate distances) required not to do violence to fits of experimental data such as the deuteron quadrupole moment. This probably means less saturation due to the higher order effects of the tensor force, which might have seemed disturbing if one wants to use only two body interactions. On the other hand, as we have pointed out, there is every reason to believe that the mainly spin independent 3PEP three body interaction will be very important, and this interaction can easily take up the slack due to its strongly repulsive density dependence. I think it is actually easier to work with the latter than with the strongly spin dependent tensor force. Thus we expect to get better convergence of the perturbation expansion. The relative role of the short range part of the tensor force and of the three body interaction is a problem that still needs to be considered further.

We have suggested that the isobar effect, as manifested by repulsive three body forces, actually might be the main ingredient in nuclear saturation. This is admittedly an extreme position (clearly the OPEP will have some effect even at low energies). Nevertheless, it is interesting that with the isobar model we can obtain (at least for a Skyrme interaction in two dimensions and using harmonic oscillator single particle wavefunctions) simple relations between quantities of interest for finite nuclei, such as average binding, average residual and pairing interactions. These relations seem to be approximately satisfied in actual finite nuclei.

REFERENCES

1. Excerpt from D. Wilkinson, Lectures at 1977 Summer School on Heavy Ions and Mesons in Nuclear Physics, Les Houches, France.
2. S. Drell and K. Huang, Phys. Rev. 91, 1527 (1953).
3. J. Fujita, M. Kawai, and M. Tanifuji, Nucl. Phys. 29, 252 (1962).
4. A. M. Green and P. Haapakoski, Nucl. Phys. A221, 429 (1974).
5. R. A. Smith and V. R. Pandharipande, Nucl. Phys. A256, 327, (1976).
6. J. Durso, M. Saarela, G. E. Brown, and A. D. Jackson, Nucl. Phys. A278, 445 (1977).
7. N. Onishi and J. W. Negele, Nucl. Phys. A301, 336 (1978).
8. K. Shimizu, A. Polls, H. Muther, and A. Faessler, U. of Tubingen preprint (1980).
9. R. W. Sharp and L. Zamick, Nucl. Phys. A208, 130 (1973).
10. P. Ring and J. Speth, Nucl. Phys. A235, 315 (1974).
11. A. Plastino, R. Arvieu, and S. A. Moszkowski, Phys. Rev. 145, 837 (1966).

HYPERONS IN NUCLEAR MATTER - AN IMPURITY PROBLEM *

J. Dąbrowski
Institute of Nuclear Research
Hoża 69, PL-00-681 Warsaw, Poland

Hypernucleus is a system composed of a hyperon bound to a nucleus. Usually the hyperon is the Λ particle. These Λ hypernuclei have been known since 1953 when the first hypernucleus was observed in nuclear emulsion exposed to cosmic rays.[1] Today, hypernuclei are produced in a controlled manner in the (K,π) strangeness exchange reaction on nuclei. It is this (K,π) reaction which led to the first observation of Σ hypernuclei about two years ago.[2]

Instead of discussing finite hypernuclei, we consider the theoretically simpler problem of a hyperon in nuclear matter (NM). Most of the talk is devoted to the much better known case of the Λ hyperon. The problem of a Σ hyperon in NM is considered at the end.

B_Λ - Reaction-Matrix Method

The quantity we want to discuss is the binding energy of a Λ particle in NM, B_Λ, whose empirical value, B_Λ^{EMP}, is about 30 MeV. The investigation of B_Λ is of considerable interest as it enables us to gain valuable information on the ΛN interaction $v_{\Lambda N}$. Furthermore, the Λ + NM system, i.e., NM with a Λ "impurity" is an interesting testing ground for nuclear many-body theories.

We defined B_Λ by

$$-B_\Lambda = E_\Lambda = E_{NM+\Lambda} - E_{NM} \quad , \tag{1}$$

where E_{NM} and $E_{NM+\Lambda}$ are the ground state energies of NM and of the Λ + NM system. In calculating these energies, we use a periodicity box of volume Ω, and apply the limit $\Omega, A \to \infty$, with the density $\varrho = A/\Omega$ kept constant at its equilibrium value.

Most of the existing calculations of B_Λ with realistic $v_{\Lambda N}$ were done with the Brueckner reaction-matrix method (for a review, see ref. 3). By applying the low-order-Brueckner (LOB) theory expressions to E_{NM} and $E_{NM+\Lambda}$, we obtain (to simplify the notation, spins and isospins are suppressed):

---

* Research supported by the Polish-U.S. Maria Skłodowska-Curie Fund under Grant No P-F7F037P

$$-B_\Lambda^{LOB} = V_\Lambda = \sum_{\vec{k}_N}^{<k_F} (\vec{k}_N\, \vec{k}_\Lambda = 0 | \mathcal{K}_{N\Lambda} | \vec{k}_N\, \vec{k}_\Lambda = 0), \qquad (2)$$

where the $N\Lambda$ reaction matrix is determined by the equation

$$\mathcal{K}_{N\Lambda} = v_{N\Lambda} + v_{N\Lambda} \{ Q / [e_N + V_\Lambda - e_N^* - e_\Lambda^*] \} \mathcal{K}_{N\Lambda}, \qquad (3)$$

where Q is the exclusion principle operator. Notice that in ground state of the non-interacting Λ + NM system, the Λ particle has zero momentum, $\vec{k}_\Lambda = 0$, and in this state the single-particle (s.p.) energy of the Λ particle is equal to the s.p. potential V_Λ. By ϵ_N and ϵ_Λ, we denote the nucleon and Λ particle kinetic energies. For the states in the Fermi sea, the nucleon s.p. energy $e_N = \epsilon_N + V_N$, where V_N is the nucleon s.p. potential. By $e_N^* = \epsilon_N + V_N^*$, and $e_\Lambda^* = \epsilon_\Lambda + V_\Lambda^*$, we denote the nucleon and Λ s.p. energies of the excited states. All the calculations of B_Λ reported here were performed with the "standard choice" of s.p. energies: $e_N^* = \epsilon_N$, $e_\Lambda^* = \epsilon_\Lambda$. Since V_Λ, eq. (2), appears in eq. (3), it has to be determined self consistently.

Eqs (2) and (3) were used to calculate $V_\Lambda = -B_\Lambda$ in ref.4 and by Rote and Bodmer[5]. The two calculations differ only in technical details and their results agree very well. Here, we shall follow ref.4.

It turns out that V_Λ is not sensitive to changes in the shape of $e_N(k_N)$, $k_N < k_F$, as long as the average value of $e_N(k_N)$ in the Fermi sea, $\bar{e}_N$, remains unchanged. Consequently, we may replace $e_N(k_N)$ in eq. (3) by $\bar{e}_N$. In the LOB theory of NM, the expression for the energy per nucleon in NM, $-\epsilon_{vol}$, is

$$-\epsilon_{vol} = \bar{\epsilon}_N + \bar{V}_N/2 = (\bar{e}_N + \bar{\epsilon}_N)/2, \qquad (4)$$

i.e., we have for $\bar{e}_N$:

$$\bar{e}_N = -[3\epsilon_N(k_F)/5 + 2\,\epsilon_{vol}]. \qquad (5)$$

In the LOB calculation of V_Λ at a given value of k_F, it is sufficient to assume a given value of ϵ_{vol}. There is no need to specify v_{NN}. The tacit assumption concerning v_{NN} is here that in the LOB theory of NM it leads to the assumed value of ϵ_{vol}. As in ref.4, we assume for ϵ_{vol} the empirical value of 15.8 MeV, and for k_F the value of 1.35$\,fm^{-1}$.

A number of spin-dependent ΛN potentials were used in calculating B_Λ. As a typical example, we consider the best potential (H) of Herndon and Tang[6], adjusted to the Λ binding in A = 3,4 hypernuclei and to Λp scattering data. This potential H has an exponential attraction and a hard core of radius c_Λ = 0.6 fm, and its strength in odd-angular-momentum states is equal 0.6 times the strength in even states. The result obtained for B_Λ with potential H is: $B_\Lambda^{LOB}(H)$ = 43.8 MeV, much bigger than $B_\Lambda^{EMP} \cong 30$ MeV.

To solve this overbinding problem, i.e., to reduce the calculated value of B_Λ to its empirical value, one has to consider the strong coupling of the ΛN channel with the ΣN channel. This coupling is expected to be suppressed in NM as was first suggested by Bodmer [7]. The origin of this suppression is the following. The contribution of $\Lambda\Sigma$ coupling to B_Λ is at least of second order in the coupling, and is reduced by the exclusion principle and the higher excitation energy in the intermediate states (as visualized in the reaction-matrix approach by the gap in the s.p. spectrum, $-(\bar{V}_N + V_\Lambda) \gtrsim 100$ MeV) in NM than in an isolated ΛN system. Since the one-pion exchange contributes to $\Lambda\Sigma$ coupling, the expected long range of the coupling potential may lead to a substantial reduction in B_Λ.

To investigate the $\Sigma\Lambda$ conversion, we introduce a 2x2 potential matrix

$$\hat{v} = \begin{pmatrix} v(\Lambda N \to \Lambda N) & v(\Sigma N \to \Lambda N) \\ v(\Lambda N \to \Sigma N) & v(\Sigma N \to \Sigma N) \end{pmatrix} = \begin{pmatrix} v_{\Lambda N} & v_{\Lambda\Sigma} \\ v_{\Sigma\Lambda} & v_{\Sigma N} \end{pmatrix} . \qquad (6)$$

The reaction-matrix $\hat{\mathcal{K}}$ for YN interaction $(Y = \Lambda, \Sigma)$ in NM becomes also a 2x2 matrix, with components which correspond to those of $\hat{v}$. Expression (2) for V_Λ remains unchanged, and reaction-matrix equation (3) is replaced by a system of two coupled equations for $\mathcal{K}_{\Lambda N}$ and $\mathcal{K}_{\Sigma\Lambda}$ (in the energy denominator in the ΣN channel, the mass difference $\Delta = (M_\Sigma - M_\Lambda)c^2 \cong 78$ MeV is added to e^*_Σ).

Because of the scare experimental information on the YN system, determining a reliable YN interaction matrix $\hat{v}$ is a difficult problem. Here, an essential progress has been done recently by the Nijmegen group[8,9]. The authors apply the OBE model and assume SU(3) relations for the coupling constants. Free parameters are determined from a combined analysis of the available NN and YN scattering data.

Two recent forms of the Nijmegen interaction, model D[8] and model F[9], have been used in calculating B_Λ^{LOB} with the following results: $B_\Lambda^{LOB}(D) = 37.6$ MeV and $B_\Lambda^{LOB}(F) = 31.4$ MeV.[10] These results are in a satisfactory agreement with B_Λ^{EMP}, especially as there are negative corrections (to be discussed later) to the LOB results.

B_Λ - Comparison of Reaction-Matrix and Correlated-Wave-Function Methods

In the Jastrow correlated-wave-function method, we approximate expression (1) by

$$B_\Lambda \cong \langle \Psi | H | \Psi \rangle / \langle \Psi | \Psi \rangle - \langle \mathbf{\Psi} | \mathbf{H} | \mathbf{\Psi} \rangle / \langle \mathbf{\Psi} | \mathbf{\Psi} \rangle , \qquad (7)$$

where H , Ψ , and $\mathbb{H}$, $\mathbf{\Psi}$ are the hamiltonians and trial wave functions of the ground state of NM , and of the Λ + NM system. For Ψ and $\mathbf{\Psi}$, we make the following Ansatz:

$$\Psi(1,\dots,A) = \prod_{j<k} f_{NN}(r_{jk})\,\Phi(1,\dots,A), \tag{8}$$

$$\mathbf{\Psi}(\Lambda,1,\dots,A) = \prod_{i} f_{N\Lambda}(r_{i\Lambda})\,\varphi_\Lambda(\Lambda)\,\Psi(1,\dots,A), \tag{9}$$

where Φ is the NM Slater determinant, φ_Λ is the Λ s.p. wave function, and f_{NN}, $f_{N\Lambda}$ are the NN, NΛ correlation functions.

With the help of the NN and NΛ radial distribution functions in the Λ + NM system, $\mathfrak{g}_{NN}$ and $\mathfrak{g}_{N\Lambda}$, and the NN radial distribution function in NM , g_{NN} , we may write expression (7) as:[11]

$$-B_\Lambda = \tilde{V}_{N\Lambda} + \Delta\tilde{V}_{NN} + \frac{1}{4}(\hbar^2/2M_N)\Delta\tau , \tag{10}$$

where

$$\tilde{V}_{N\Lambda} = \varrho\int d\vec{r}\,\mathfrak{g}_{N\Lambda}(r)\,\tilde{v}_{N\Lambda}(r) , \tag{11}$$

$$\Delta\tilde{V}_{NN} = \frac{1}{2}\varrho^2\Omega\int d\vec{r}\,\hat{g}_{NN}(r)\,\tilde{v}_{NN}(r) , \tag{12}$$

where $\hat{g}_{NN} = \mathfrak{g}_{NN} - g_{NN}$, and $(X = N,\Lambda)$

$$\tilde{v}_{NX}(r) = v_{NX}(r) + \frac{\hbar^2(M_N+M_X)}{4M_NM_X}\left[(\nabla f_{NX}(r)/f_{NX})^2 - \Delta f_{NX}(r)/f_{NX}\right]. \tag{13}$$

The Jackson-Feenberg form of the kinetic energy applied here, leads to the additional term $\Delta\tau$ which turns out to be very small and is neglected.

Most of the existing Jastrow type calculations of B_Λ apply a low order cluster (LOC) expansion of $\mathfrak{g}_{N\Lambda}$ and $\hat{g}_{NN}$ (see,e.g., refs 12, 13). The LOC results for B_Λ turn out to be much bigger than the LOB results. Looking for a possible source of the discrepancy between B_Λ^{LOC} and B_Λ^{LOB} , we have improved the Jastrow method by applying the FHNC approximation,i.e., by calculating $\mathfrak{g}_{N\Lambda}^{FHNC}$, and $\hat{g}_{NN}^{FHNC}$. The FHNC approximation in a general impurity problem is described in ref.11. Here, we outline the calculation of B_Λ^{FHNC} of ref.14.

As a typical case, we consider the ΛN potential HNX which is identical with potential H of ref.6, except that it is not suppressed in odd states. For v_{NN} , we take the OMY6 potential[15]. For the NN and NΛ correlation functions, we assume the following form:

$$f_{NX}(r) = \begin{cases} 0 & , r \leq c_X , \\ [1 - e^{-\alpha_X(r-c_X)}][1 - \beta_X e^{-\gamma_X(r-c_X)}] & , r > c_X , \end{cases} \tag{14}$$

where c_X is the hard core radius of v_{NX}. The parameters of f_{NN} ($\alpha_N = 2.5\ fm^{-1}$, $\beta_N = 0.884$, $\gamma_N = 2.0\ fm^{-1}$) were determined by the Pisa group[16] in the NM calculation at $k_F = 1.366\ fm^{-1}$ with v_{NN}(OMY6), in which E_{NM}^{FHNC}/A was minimized with respect to the parameters α_N, β_N, γ_N, uder the subsidiary "average Pauli condition". The parameters of $f_{N\Lambda}$ were determined by maximizing B_Λ^{FHNC} at $k_F = 1.366\ fm^{-1}$. To avoid unreasonably large computations, we introduced the restriction $\alpha_\Lambda = \gamma_\Lambda$, and obtained: $\alpha_\Lambda = 6.0\ fm^{-1}$, $\beta_\Lambda = -0.05$. Our final result is: $B_\Lambda^{FHNC}(HNX) = 77$ MeV, very close to earlier LOC results[12,13].

To make a meaningful comparison with the LOB results, we have used in eq.(5) for ϵ_{vol} the value 8.2 MeV which follows from the LOB calculation of NM at $k_F = 1.366\ fm^{-1}$ with v_{NN}(OMY6).[17] In this way, for $k_F = 1.366\ fm^{-1}$, we get $B_\Lambda^{LOB}(HNX) = 60$ MeV.

Although the Jastrow expression for B_Λ, eq.(7), does not have any strict lower bound character, the discrepancy between $B_\Lambda^{FHNC} = 77$ MeV and $B_\Lambda^{LOB} = 60$ MeV is disturbing. It appears that the discrepancy increases if we go in the reaction-matrix method beyond the LOB approximation. The first correction B_Λ' to B_Λ^{LOB} consists of two parts: the rearrangement energy[18] $B_{\Lambda R} = -\varkappa B_\Lambda^{LOB}$ ($\varkappa$ is the NN wound integral), and the three-body cluster energy $B_{\Lambda 3}$. Now, $B_{\Lambda R}$ is obviously negative, and the existing estimates[19] of $B_{\Lambda 3}$ indicate that $B_{\Lambda 3}$ is also negative. The best way of resolving the problem of the discrepancy would be to calculate B_Λ with both methods in a model case of simple central ΛN and NN hard core potentials of pure Wigner type. In the Brueckner method, one should calculate accurately the three-hole-line contributions, and analyze the form of the s.p. potentials in the excited states. In the Jastrow method, one should determine carefully the optimal correlations. A comparison of the radial distribution functions in both methods would also be valuable.

Σ Hyperon in NM

The recent observation of Σ hypernuclei[2] indicates that the nuclear potential well depth of Σ, $-V_\Sigma \sim 20 - 30$ MeV, i.e., $V_\Sigma \cong V_\Lambda$, and the width Γ of the Σ states is surprisingly small, $\Gamma \lesssim 10$ MeV, although these states are expected to udergo a fast decay via the strong conversion process $\Sigma N \to \Lambda N$. Here, we review a reaction-matrix calculation[20] of V_Σ and Γ for the ground state of Σ in NM.

The expression for the energy $E_\Sigma = V_\Sigma - i\Gamma/2$ is similar to that for V_Λ in the presence of $\Lambda\Sigma$ conversion:

$$E_{\Sigma} = \sum_{\vec{k}_N}^{<k_F} (\vec{k}_N\ \vec{k}_{\Sigma} = 0 \,|\, \mathcal{K}_{N\Sigma} \,|\, \vec{k}_N\ \vec{k}_{\Sigma} = 0) \quad , \tag{15}$$

where $\mathcal{K}_{N\Sigma}$ is determined, together with $\mathcal{K}_{\Lambda\Sigma}$, by the system of coupled equations:

$$\mathcal{K}_{N\Sigma} = v_{N\Sigma} + v_{N\Sigma} \frac{Q}{e_N + V_{\Sigma} - e_N^* - e_{\Sigma}^*} \mathcal{K}_{N\Sigma} + v_{\Sigma\Lambda} \frac{Q}{e_N + V_{\Sigma} + \Delta - e_N^* - e_{\Lambda}^* + i\epsilon} \mathcal{K}_{\Lambda\Sigma} ,$$

$$\mathcal{K}_{\Lambda\Sigma} = v_{\Lambda\Sigma} + v_{\Lambda\Sigma} \frac{Q}{e_N + V_{\Sigma} - e_N^* - e_{\Sigma}^*} \mathcal{K}_{N\Sigma} + v_{N\Lambda} \frac{Q}{e_N + V_{\Sigma} + \Delta - e_N^* - e_{\Lambda}^* + i\epsilon} \mathcal{K}_{\Lambda\Sigma} . \tag{16}$$

Due to the energy release Δ in the $\Sigma N \to \Lambda N$ process, real energy conserving transitions $\Sigma N \to \Lambda N$ may occur in NM. The infinitesimal parameter $+ i\epsilon$ guarantees that only outgoing waves appear in the ΛN channel.

By applying the identity $1/(x + i\epsilon) = \mathcal{P}(1/x) - i\pi\delta(x)$, we get

$$\Gamma = \text{const} \sum_{\vec{k}_N}^{<k_F} k'_{N\Lambda}\, Q \int d\hat{k}'_{N\Lambda} \,|\, (\vec{k}'_N\ \vec{k}'_{\Lambda} \,|\, \mathcal{K}_{\Lambda\Sigma} \,|\, \vec{k}_N\ \vec{k}_{\Sigma} = 0) \,|^2 , \tag{17}$$

where $\vec{k}'_{N\Lambda} = (M_{\Lambda}\vec{k}'_N - M_N\vec{k}'_{\Lambda})/(M_N + M_{\Lambda})$, $\vec{k}'_N$ and $\vec{k}'_{\Lambda}$ are final nucleon and Λ momenta in the $\Sigma N \to \Lambda N$ process, the exclusion principle operator $Q = 0$ for $k'_N \leq k_F$, and const involves reduced masses μ_{NY} and numerical constants. Because of the energy conservation,

$$\hbar^2 k'^2_{N\Lambda}/2\mu_{N\Lambda} = \hbar^2 k^2_{N\Sigma}/2\mu_{N\Sigma} + \Delta + V_N(k_N) + V_{\Sigma} - V_N^*(k'_N) - V_{\Lambda}^*(k'_{\Lambda}) , \tag{18}$$

the magnitude of $k'_{N\Lambda}$, and consequently of k'_N, is sensitve to dispersive effects, i.e., to momentum dependence of the s.p. potentials. These dispersive effects reduce the k'_N momenta to such a degree that an essential part of them are smaller than k_F, and are excluded by the exclusion principle. This suppression of the $\Sigma N \to \Lambda N$ process in NM leads to a narrow width Γ.

The results of our calculation[20] at $k_F = 1.35\ \text{fm}^{-1}$ with model D of the Nijmegen interaction $\hat{v}$ [8] (with acceptable choices of the s.p. potentials),

$$-35\ \text{MeV} \lesssim V_{\Sigma} \lesssim -10\ \text{MeV} , \qquad 0 \lesssim \Gamma \lesssim 12\ \text{MeV} , \tag{19}$$

are consistent with the CERN experiment[2]. The big range of the results, in particular of our Γ values, reflects uncertainities in the choice of the s.p. potentials V_N^*, V_Y^*. This makes the problem of Γ particularly interesting from the point of view of the theory of NM.

References

1 M.Danysz, J.Pniewski, Phil.Mag.44,348/1953/.

2 R.Bertini et al., Phys.Lett.90B,375/1980/.

3 J.Dąbrowski, Nukleonika 23,875/1978/.

4 J.Dąbrowski, M.Y.M.Hassan, Phys.Rev.C 1,1883/1970/.

5 D.M.Rote, A.R.Bodmer, Nucl.Phys.A148,97/1970/.

6 R.C.Herndon, Y.C.Tang, Phys.Rev.159,853/1967/.

7 A.R.Bodmer, Phys.Rev.141,1387/1966/.

8 M.M.Nagels,T.A.Rijken, J.J.de Swart, Phys.Rev.D 12,744/1975/; ibid.15,2547/1977/.

9 M.M.Nagels, T.A.Rijken, J.J.de Swart, Phys.Rev.D 20,1633/1979/.

10 J.Rożynek, J.Dąbrowski, Phys.Rev.C 20,1612/1979/.

11 J.Dąbrowski, W.Piechocki, Ann.Phys./N.Y./126,317/1980/.

12 G.Mueller,J.W.Clark, Nucl.Phys.B7,217/1969/.

13 S.Ali,M.E.Grypeos,B.Kargas, Phys.Rev.C 14,285/1976/.

14 W.Piechocki, J.Dąbrowski, Acta Physica Polonica - in press.

15 T.Ohmura,M.Morita,M.Yamada,Progr.Theor.Phys.15,222/1956/.

16 S.Fantoni, private communication.

17 S.O.Backman,J.W.Clark,W.J.Ter Louw,D.A.Chakkalakal,M.L.Ristig, Phys.Lett.41B,247/1972/.

18 J.Dąbrowski, H.S.Köhler, Phys.Rev.136,B162/1964/.

19 A.Daniluk, J.Dąbrowski, Acta Physica Polonica B6,317/1957/; ibid. B11,675/1980/.

20 J.Dąbrowski, J.Rożynek, Raport "P" IBJ Nr 5/VII/80/P, Phys.Rev.C - submitted for publication.

PION CONDENSATION, EQUATION OF STATE OF DENSE MATTER AND NEUTRON STARS *

P. Haensel and M. Prószyński
Polish Academy of Sciences, N. Copernicus Astronomical Center,
Bartycka 18, PL-00-716 Warszawa, Poland

1. Introduction

The possibility of pion condensation in dense nucleon matter was pointed out by Migdal[1] in 1971 and independently by Sawyer and Scalapino[2] in 1972. Up to now the problem of pion condensation has been studied in numerous papers (see Refs. 3-5 and references therein). Recent calculations of the critical baryon density, n_c , for the appearence of pion condensate in nuclear matter indicate that n_c is higher than the normal nuclear matter density, $n_{NM} = 0.16\ \mathrm{fm}^{-3}$ (corresponding to mass density $\rho_{NM} = 2.7 \cdot 10^{14}\ \mathrm{gcm}^{-3}$). This theoretical result seems to be consistent with negative results of experimental search for direct evidence for the presence of pion condensate in atomic nuclei. On the other hand, theoretical calculations of n_c in neutron matter leads to values corresponding to baryon densities that are expected to exist in the interiors of sufficiently massive neutron stars. One of the main effects implied by the pion condensation would be a significant softening of the equation of state of dense cold matter at the densities following n_c. Such an effect could be important for neutron star structure[6] and this aspect of pion condensation will constitute the topic of the present talk. The possible presence of pion condensate may strongly influence the process of cooling of young neutron stars[7] but this important problem will not be discussed here. It is clear, that in view of a rapid growth of the quantity and quality of the observational data on neutron stars, the studies of possible astrophysical consequences of pion condensation in dense matter are of particular interest.

2. Equations of state with pion condensate

An essential input for the calculation of a neutron star model is the equation of state of dense cold matter. At $T = 0K$ the equation of state is determined by the functional dependence of the energy per baryon, $\mathcal{E}$, on the baryon density, n : $\mathcal{E} = \mathcal{E}(n)$. The pressure, P and the mass density, ρ , can be calculated from $\mathcal{E} = \mathcal{E}(n)$ using

$$P = n^2 \frac{d}{dn} \mathcal{E}(n) \quad , \quad \rho = n m_n + n \frac{\mathcal{E}}{c^2} \quad , \qquad (1)$$

where m_n is the neutron mass. For $n > n_{NM}$ the simplest model of normal (non pion-condensed) neutron star matter is that of pure neutron matter. If one allows for pion condensation, then for $n < n_c$ neutron matter is in its normal state described by $\mathcal{E} = \mathcal{E}_o(n)$ and $P = P_o(n)$, $\rho = \rho_o(n)$. For $n > n_c$ neutron matter is in the pion-condensed state with

$$\mathcal{E}(n) = \mathcal{E}_o(n) + \mathcal{E}_{cond}(n), \qquad (2a)$$

$$P(n) = P_o(n) + P_{cond}(n), \qquad (2b)$$

$$\rho(n) = \rho_o(n) + \rho_{cond}(n), \qquad (2c)$$

where $\mathcal{E}_{cond}$, P_{cond} and ρ_{cond} are the pion condensate contributions to corresponding quantities. For the description of the condensate contributions we shall use an approximate analytical formula of Migdal and collaborators[4]

$$\rho_{cond}(n) = -3.594\, n_c^2\, (x-1)^2 \left(A + \frac{B}{x} + \frac{C}{x^2}\right) \cdot 10^{14}\, g\, cm^{-3}, \qquad (3)$$

which implies corresponding analytical formulae for $\mathcal{E}_{cond} = \rho_{cond} c^2/n$ and $P_{cond} = n^2 \frac{d}{dn}\mathcal{E}_{cond}$. Here $x = n/n_c$, n_c is expressed in fm^{-3} and A, B and C are dimensionless parameters. The formula (3) implies that pion condensation has character of a second order phase transition. The uncertainties in the numerical values of n_c, A, B and C stem from the lack of a precise knowledge of the strong interactions as well as from the deficiences and approximations of the many body theory of dense matter. In the present talk we shall use three different models of pion condensate. The models a and d were derived by Migdal and his collaborators[4] while our BW model is an approximate representation of a pion condensate model derived by Brown and Weise[3], extrapolated using Eq.(3) for $n > 1.1\ fm^{-3}$. The parameters of these pion condensate models are given in Table 1.

Table 1.

Condensate model	n_c (fm^{-3})	A	B	C
a	0.2362	0.91	-0.20	0.09
BW	0.300	0.97	-1.41	2.15
d	0.1850	1.26	-0.08	-0.07

For $\rho < 1.59 \cdot 10^{14} gcm^{-3}$ we shall use a standard equation of state of cold catalyzed matter[8]. The uncertainty in theoretical knowledge of equation of state of normal neutron star matter for $\rho \gtrsim \rho_{NM}$ grows rapidly with increasing density. In view of this, we consider three different equations of state for normal phase. The softest of them is that of Pandharipande for pure neutron matter[9]. The model I of Bethe and Johnson[10] may be considered as a medium stiff one. As an example of a very stiff equation of state we shall use a model calculated by Kutschera[11] using the relativistic mean field theory developed by Serot[12]. Combined with the standard equation of state for $\rho < 1.59 \cdot 10^{14}$ gcm^{-3} they will be hereafter referred to as the PN, BJI and MFT equations of state. The equations of state with pion condensation will be obtained in a very simplified procedure by adding the pion condensate contributions to the quantities corresponding to the normal phase.

Let us consider a slow compression of a small element of cold matter. For our models given by Eq.(3) the appearence of pion condensate leads to a discontinuous decrease of the compression modulus, $(\partial P_o/\partial n)_{n=n_c} > (\partial P/\partial n)_{n=n_c+0}$. For $n > n_c$ two situations are then possible (Fig.1):

(i) If the pion condensate is too weak to produce negative compression modulus of matter for some $n > n_c$, then the $P(n)$ dependence corresponds to the schematic plot in Fig.1a.

(ii) For a sufficiently strong pion condensate, for which $(\partial P/\partial n)_{n=n_c+0} \lesssim 0$, we have a characteristic van der Waals type P vs. n curve (Fig. 1b,c). For $P < P_{NS}$ (i.e. $n < n_N$) matter is in its normal phase. Under a further increase of n matter can either undergo a first order phase transition at constant pressure $P = P_{NS}$ or remain in metastable supercompressed states (dashed part NC of the $P(n)$ curve). For $n > n_c$ matter becomes thermodynamically unstable $(\partial P/\partial n < 0)$. The parameters of the first order phase transition (Fig.1b) can be calculated from the Maxwell construction.

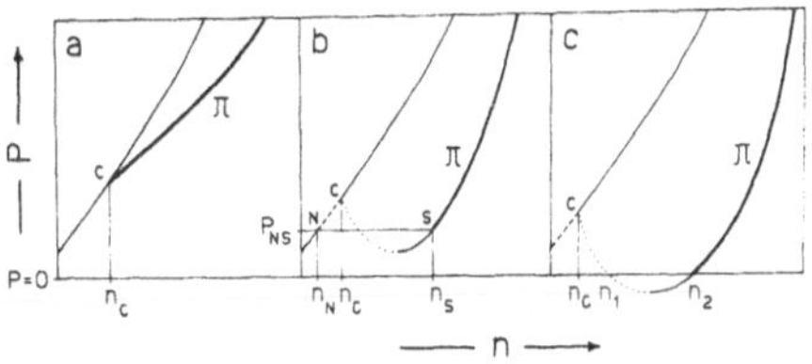

Fig.1. —— normal phase; ▬▬ stable pion-condensed phase with $P \geq 0$; – – – metastable normal phase; ······ unstable pion-condensed phase.

A suitable thermodynamic variable for the neutron star interior is the local pressure, $P(r)$, which remains strictly decreasing continuous function of the radius, r. Treated as functions of P the baryon density n and the mass density ρ are both discontinuous at $P = P_{NS}$. At $P = P_{NS}$ the normal phase of matter of baryon density n_N and mass density $\rho_N = \rho(n_N)$ can coexist with pion-condensed phase of baryon density n_S and mass density $\rho_S = \rho(n_S)$. The parameters of a possible first order phase transition to a pion-condensed state are presented in

Table 2. In the last column of this table we give the values of parameter $\lambda \equiv \rho_S/(\rho_N + P_{NS}/c^2)$ which will be relevant for the discussion of the stability of configurations with a developing small pion-condensed core. The equations of state with pion condensation which are not given in Table 2 lead to <u>second order</u> phase transition to a pion condensed state (Fig.1a).

Table 2. The parameters of the first order phase transition implied by the pion condensation

Equation of state	ρ_N (10^{15}gcm^{-3})	ρ_S (10^{15}gcm^{-3})	P_{NS} (10^{34}dyncm^{-2})	λ
PN + a	0.3538	0.8688	0.6036	2.41
PN + BW	0.4701	0.7028	1.1829	1.45
BJI + BW	0.5179	0.5857	2.537	1.07
BJI + d	0.2835	0.7178	0.5853	2.47

Our results[17] show, how strongly the parameters of the phase transition to a pion-condensed state depend on the stiffness of the equation of state for the normal phase and on the condensate model itself. The first order phase transition with a large density jump occuring for the BW model combined with the PN equation of state becomes much less dramatic when the BW model is combined with the BJI equation of state and reduces to the second order phase transition with no density jump in the MFT+BW case.

The case of the PN+d model is a very particular one. The combination PN+d leads to negative values of P for $n_1 < n < n_2$ (Fig.1c) with n_1 = 0.2372 fm^{-3} and n_2 = 0.9592 fm^{-3}. Dense catalyzed matter described by this equation of state has, apart from a standard self-bound ($P = 0$) state of the ^{56}Fe crystal at $n_0 = 0.473 \cdot 10^{25}$cm^{-3}, an additional stable ($\partial P/\partial n > 0$) self-bound state at n_2 . Moreover, this stable self-bound state at $n = n_2$ is energetically preferred over that of a ^{56}Fe crystal because $\mathcal{E}(n_2) = -0.1776 \cdot 10^{-4}$erg $< \mathcal{E}(n_0) = -0.1185 \cdot 10^{-4}$erg. The PN+d equation of state could thus allow (in principle) for the existence of very unusual systems like neutron nuclei[4,13] and "golf ball" neutron stars with arbitrarily small mass and radius[6]. Pion-condensed neutron star for the PN+d equation of state would have a liquid surface of density $\rho_2 = \rho(n_2) = 1.587 \cdot 10^{15}$gcm^{-3}. We shall show, however, that the confrontation with existing neutron star data seems to rule out such a peculiar equation of state as the PN+d one.

3. Neutron star models

The most important macroscopic parameters characterizing a model of a neutron star are: the total gravitational mass M, the stellar radius R, the total number of baryons A, and the moment of inertia (for a slow rigid rotation) I. The models of spherically symmetric configurations of cold matter representing non-rotating neutron stars can be obtained through the numerical integration of the relativistic equations of hydrostatic equilibrium (i.e. the Tolman-Oppenheimer-Volkoff equations),

$$\frac{dP}{dr} = -G\,\frac{[\rho(r)+P(r)/c^2][m(r)+4\pi r^3 P(r)/c^2]}{r^2 - 2Gm(r)r/c^2},$$
$$\frac{dm}{dr} = 4\pi r^2 \rho(r), \qquad \frac{da}{dr} = 4\pi r^2\left[1-2G\frac{m(r)}{c^2 r}\right]^{-1/2} n(r), \tag{4}$$

Here $m(r)$ and $a(r)$ are, respectively, the mass and the number of baryons within a sphere of a radius r and $\rho(r)$, $n(r)$ and $P(r)$ are mass density, baryon density and pressure profiles. The integration starts from the center of configuration with $P(0) = P_{centr}$, $m(0) = 0$ and $a(0) = 0$. The boundary of the equilibrium configuration, $r = R$, is given by a condition $P(R) = 0$; the total gravitational mass M and the total number of baryons A are correspondingly given by $M = m(R)$ and $A = a(R)$. The corrections to these parameters, induced by a slow rigid rotation of the star, may then be calculated by using a perturbation technique (Ref.8 and references therein).

The equilibrium configuration is completely determined by the value of the central pressure P_{centr} (or correspondingly by ρ_{centr} or n_{centr}) and the equation of state of cold matter. A family of the equilibrium configurations, obtained for the same equation of state but different values of the central pressure, is usually visualized as a curve in the (M, R)-plane because the stability of configurations against small radial perturbations (oscillations) can then be easily checked out (method 2-A from the catalogue of Bardeen et al.[21]). The form of the $M-R$ curve for the configurations (stars) containing pion-condensed matter in their interiors depends on the character of the phase transition between normal dense matter and the pion-condensed one. The schematic plots of the $M-R$ curves in the three possible cases are shown in Fig.2. Let us discuss these possibilities. Fig.2a is obtained in the case of the second order phase transition (no density jump, equation of state represented schematically in Fig.1a). The derivative dM/dR is continuous at S_1 where $n_{centr} = n_c$. Configurations lying on the π branch contain a pion-condensed core (pion-condensed neutron stars). Figs.2b

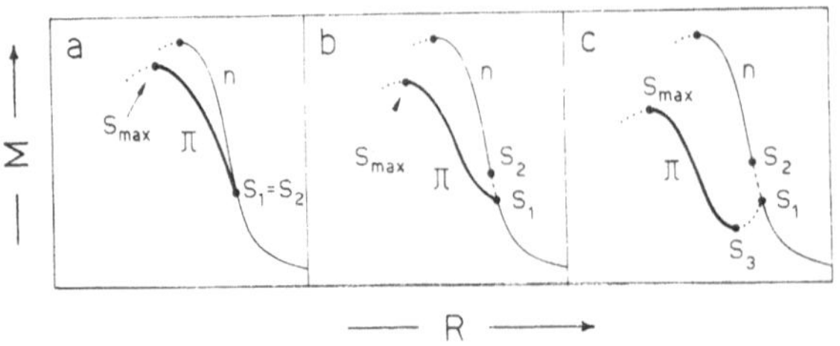

Fig.2. The $M-R$ relation for the normal (n) and pion-condensed (π) neutron stars. Each curve represents a family of equilibrium configurations parametrized by P_{centr}, i.e. each point on the curve corresponds to one neutron star model. —— stable configurations; - - - configurations metastable with respect to nucleation of the pion-condensed phase; ······· configurations unstable with respect to small radial perturbations.

and 2c are obtained for the equations of state represented schematically in Fig.1b. Configurations lying to the right of S_1 are normal, stable neutron stars. Normal configurations between S_1 and S_2 are <u>metastable</u> with respect to the nucleation of the pion-condensed phase in their central regions with $n>n_N$. The lifetime of these configurations is closely related to the characteristic time of nucleation of pion-condensate in the supercompressed states of cold matter with $n_N<n<n_C$ (this lifetime was unknown up to now, its estimates being extremely model dependent). In Fig.2b the configuration S_1 is not an <u>extremal</u> one; all pion-condensed configurations lying between S_1 and the configuration corresponding to the maximum allowable mass, S_{max}, are <u>stable</u>. This occurs for the first order phase transition when $\lambda \equiv \rho_S/(\rho_N+P_{NS}/c^2) < \frac{3}{2}$ (Ref.15). This is the case of the PN+BW and BJI+BW equations of state. The most interesting, from the point of view of possible astrophysical consequences, is the case of the first order phase transition with $\lambda>\frac{3}{2}$, represented by the π branch in Fig.2c. The configuration S_1 is there an <u>extremal</u> one. The pion-condensed configurations lying between S_1 and S_3 are <u>unstable</u> with respect to small radial perturbations. A sufficiently strong pion condensate with $\lambda>\frac{3}{2}$ leads then to existence of two <u>distinct</u> families of <u>stable</u> neutron stars. The normal neutron stars lie to the right to S_1 while stable pion-condensed neutron stars lie on the π branch between S_3 and S_{max}.

Pion-condensed neutron stars have in general smaller radii and moments of inertia than normal neutron stars consisting of the same number of baryons. This is illustrated in Table 3 where we present some macroscopic parameters of the normal and pion-condensed neutron stars consisting of $A = 1.675\cdot 10^{57}$ baryons. Such configurations may be considered as models of a "canonical" neutron star (CNS) which has the atomic mass $M_A = A\cdot m_A = 1.4\,M_\odot$, where m_A is the mass unit based on the ^{12}C atom ($m_A = m\,(^{12}C)/12 = 1.6604\cdot 10^{-24}$g , Ref.14).

Table 3. Models of "canonical" neutron stars (configurations containing $A = 1.675\cdot 10^{57}$ baryons). The radius of pion-condensed core and its mass are denoted by R_{core} and M_{core}, respectively.

	units	Equation of state PN	PN+BW	BJI	BJI+BW	MFT	MFT+BW
R_{core}	km	-	6.872	-	7.66	-	same as for MFT
M_{core}	$M_\odot$	-	1.162	-	0.929	-	
R	km	10.16	7.86	12.08	10.71	14.50	
M	$M_\odot$	1.267	1.222	1.288	1.280	1.301	
I	10^{45}gcm^2	0.91	0.57	1.24	0.96	1.84	

In the case of the PN+BW and BJI+BW equations of state the pion-condensed CNSs consist of a large pion-condensed core of radius R_{core} and mass M_{core} and an outer envelope of normal matter. The density jump at the core boundary is $\Delta\rho = \rho_S - \rho_N$. Pion condensed neutron stars are more tightly bound than the normal ones (larger mass defect $M_A - M$). The most dramatic difference between normal and pion-condensed neutron stars concerns their moments of inertia.

The case of a very stiff MFT equation of state deserves a comment. Because of the stiffness of the equation of state for the normal phase the central baryon density of CNS, n_{centr}, is lower than the threshold density for the pion condensation. Hence, for the MFT+BW equation of state CNS is normal.

4. Collapsing neutron stars [16 - 18]

Let us consider a neutron star of the central density slightly below n_N. Let us assume that due to a slow, spherically symmetric accretion of matter onto its surface, the number of baryons, A , the star mass, M and consequently, the central baryon density, n_{centr}, increase. This would correspond to moving upwards along the $M - R$ curve in Fig. 3.

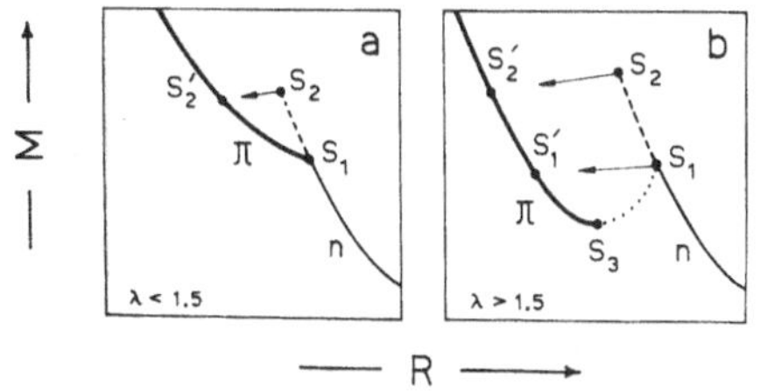

Fig.3. —— stable, - - - - - metastable and ······· unstable configurations

Eventually, the star reaches the configuration S_1 with $n_{centr} = n_N$. Let us assume that the nucleation of condensate is rapid as compared to the mass accretion rate. Then the configurations between S_1 and S_2 cannot be reached. If at the same time $\lambda < \frac{3}{2}$ (the case presented in Fig.3a) then the star starts to move on the π branch. If, however, $\lambda > \frac{3}{2}$ (the case of Fig.3b) then neutron star in S_1 with $n_{centr} = n_N$ and $A = A_1$ is unstable under further increase of A. A small pion-condensed core appearing at the star center destroys the equilibrium of the star. The star collapses. Assuming no baryon ejection during the collapse and complete cooling of the final configuration, the final state of the collapsed star is the configuration S_1'.

Let us assume now that the nucleation of pion condensed phase is slow as compared to the rate of change of the neutron star parameters caused by accretion. Then, neutron star can reach the supercompressed normal configuration S_2 with $n_{centr} = n_c$ and $A = A_2$. Such a configuration is unstable under further increase in A. A small increase in A leads to the appearence of an unstable ($\partial P/\partial n < 0$) pion-condensed phase; this implies pion-condensation of the metastable supercompressed core and collapse of the whole star. Assuming no baryon loss and complete cooling the final stable configuration is S_2' with $A = A_2$.

Even without studying the dynamics of collapsing neutron star we may estimate the collapse time as $\sim 10^{-4}$s. Also, the upper bound, E_{max}, for the energy released during the collapse of an unstable (or nonequilibrium) configuration S can be calculated without entering the complexities of the collapse dynamics as

$$E_{max}(S) = [M(S) - M(S')]c^2 . \qquad (5)$$

Here, $M(S')$ is the mass of the stable, pion-condensed configuration of the same number of baryons as the normal (collapsing) configuration S. Two further effects would characterize the neutron star transition to a pion-condensed configuration: decrease of the stellar radius, R, and of the moment of inertia, I. These effects are described by the quantities $\Delta R = R(S') - R(S)$ and $\Delta I = I(S') - I(S)$.

Table 4. Examples of collapsing configurations and of the values of E_{max} ΔR and ΔI [17,18]. For the BJI+BW equation of state the collapse is possible only in the case of a slow nucleation of pion-condensed phase.

	Eq. of state / units	BJI+BW	BJI+d	
		S_2(slow nucl.)	S_2(slow nucl.)	S_1(rapid nucl.)
M	$M_\odot$	0.697	0.355	0.301
R	km	12.86	14.11	14.70
I	10^{45}gcm^2	0.59	0.25	0.20
ΔR	km	-0.009	-4.58	-4.37
ΔI	%	-0.2	-56	-53
E_{max}	erg	$0.5\cdot10^{48}$	$2.5\cdot10^{51}$	$0.9\cdot10^{51}$

The numerical values of E_{max}, ΔR and ΔI as well as the parameters of collapsing configurations depend very strongly on the model used. This is visualized by the results presented in Table 4.

5. Confronting theoretical results with observations

Because of many uncertainties, concerning both the physics of strong interactions and the many-body theories of dense matter, the results and models presented in preceding sections should be regarded as only illustrating the range of (theoretical) possibilities.

At the time when the Refs.3 and 6 were published (1975-76) the theoretical results could be confronted with only a few observational estimates of neutron star parameters. The most important constraint was an estimation of the moment of inertia of the Crab pulsar ($I_{Crab} > 0.15 \cdot 10^{45}$ gcm^2), from the luminosity of the Crab Nebula. This gave a lower limit on the maximum moment of inertia for a neutron star, $I_{max} > I_{Crab}$. More careful considerations of the energetics of the Crab Nebula yield for I_{Crab} a value of at least $0.8 \cdot 10^{45}$gcm^2 (pp.60, 67, 68 and 172 of Ref.22) and probably $I_{Crab} \approx 1 \cdot 10^{45}$gcm^2. At present, optical and X-ray observations of binary X-ray pulsars provide us with a possibility to estimate the masses of neutron stars in these binaries[20,26]. It turns out, however, that masses of these objects are not yet sufficiently precisely determined to be used to constraint the present models of dense matter.

Analysis of the timing data collected during last six years on the first radio pulsar discovered to be in a binary system, PSR 1913+16, provide us with much better determination of the nature and masses of both orbiting objects. In this close binary system the general relativistic effects play an important role and, in consequence, the orbital parameters characterizing dynamics of this system are now over-determined. With an indirect evidence for the emission of gravitational waves at the level predicted by general relativity, the analysis leads to a self-consistent estimation of masses of the pulsar and its companion[24]. The most recent values[23] are $M_{PSR} = 1.43 \pm 0.05 M_{\odot}$ and $M_{comp} = 1.40 \pm 0.05 M_{\odot}$. On the other hand the only model accepted now for a radio pulsar is that of a highly magnetized neutron star. The companion star is most likely also a neutron star[25] (though there is no observational evidence, except for its low mass, to exclude a black hole). If the companion star <u>is</u> a neutron star, then, due to the fact that the sum of masses of both objects is a quantity known with a high accuracy, $M_{PSR} + M_{comp} = 2.83 M_{\odot}$, the maximum allowable mass of neutron stars, M_{max} should satisfy $M_{max} > 1.415 M_{\odot}$.

The pion condensation leads to a significant lowering of M_{max} and of a maximal moment of inertia, I_{max}. Our results for M_{max} obtained using

Table 5

Equation of state	$M_{max}/M_\odot$	can be ruled out?
PN	1.66	
PN+a	1.38	yes
PN+BW	1.38	yes
PN+d +	1.32	yes
BJI	1.86	
BJI+a	1.54	
BJI+BW	1.58	
BJI+d	1.44	yes (?)
MFT	2.84	
MFT+a	2.70	
MFT+BW	2.75	
MFT+d	2.59	

Based on Ref.17,18.
+ Pion-condensed stars with a liquid surface (see Section 2)

various equations of state with pion condensation are shown in Table 5. Using the constraint stated above we can rule out equations of state which yield M_{max} lower then $1.415 M_\odot$, that is PN+a, PN+BW and PN+d. All these equations yield $I_{max} \leq 0.61 \cdot 10^{45} gcm^2$, to be compared with $I_{Crab} \gtrsim 0.8 \cdot 10^{45} gcm^2$. It should be stressed that the estimate for I_{Crab} is, however, not so certain as the constraint on M_{max}.

Summarizing, using observational data we are able - at least within our rather broad set of pion-condensate models - to rule out a possibility of existence of neutron nuclei advanced in Ref.13 and of neutron stars with liquid surface discussed in Ref.6. The BJI+d equation of state yields M_{max} which is dangerously close to the observational limit of $1.415 M_\odot$ and $I_{max} = 0.67 \cdot 10^{45} gcm^2$ which is clearly inconsistent with the estimation of I_{Crab}. Therefore we think that this equation of state can also be disqualified. Finally, let us mention that the observational constraint $I_{max} > 0.8 \cdot 10^{45} gcm^2$ is satisfied for all other models of dense matter considered here.

* Presented by P. Haensel. Supported in part by the Polish-U.S. Maria Sklodowska-Curie Fund, Grant No. P-F7F037P.

References

1) A.B.Migdal, JETP 61, 2210 (1971).
2) R.F.Sawyer, Phys.Rev.Lett. 29, 382 (1972); R.F.Sawyer and D.J. Scalapino, Phys.Rev. D7, 953 (1972).
3) G.E.Brown and W.Weise, Phys.Rep. 27, 1 (1976).
4) A.B.Migdal, Rev.Mod.Phys. 50, 107 (1978).
5) S.-O.Bäckman and W.Weise, Mesons in Nuclei, (M.Rho and D.Wilkinson, ed.) vol.III, p.1097 (North-Holland, 1979).
6) J.B.Hartle, R.F.Sawyer and D.J.Scalapino, Ap.J. 199, 471 (1975).
7) O.Maxwell, G.E.Brown, D.K.Campbell, R.F.Dashen and J.T.Manassah, Ap.J. 216, 77 (1977).
8) G.Baym, C.Pethick and P.Sutherland, Ap.J. 170, 299 (1971).
9) V.R.Pandharipande, Nucl.Phys. A178, 123 (1971).

10) R.C.Malone, M.B.Johnson and H.A.Bethe, Ap.J. 199, 741 (1075).
11) M.Kutschera, unpublished, private communication.
12) B.D.Serot, Phys.Lett. 86B, 146 (1979).
13) A.B.Migdal, D.A.Markin, I.A.Mishustin, and G.A.Sorokin, Phys.Lett. 65B, 423 (1976); Zh.Eksp.Teor.Fiz. 72, 1247 (1977).
14) W.D.Arnett and R.L.Bowers, Ap.J.Suppl. 33, 415 (1977).
15) B.Kaempfer, Proceedings of International Conference on Extreme States in Nuclear Systems, Dresden, GDR, 1980 February, Ch.1, p.12.
16) A.B.Migdal, A.J.Chernoutsan and I.N.Mishustin, Phys.Lett. 83B, 158 (1979); Phase transition in neutron matter and neutron star dynamics, Preprint Saclay DPhT/79/741.
17) P.Haensel and M.Prószyński, to be submitted for publication.
18) P.Haensel and M.Prószyński, Phys.Lett. 96B, 233 (1980).
19) M.A.Ruderman, Ann.Rev.Astron.Astrophys. 10, 427 (1972).
20) J.N.Bahcall, Ann.Rev.Astron.Astrophys. 16, 24 (1978).
21) J.M.Bardeen, K.S.Thorne, D.W.Meltzner, Ap.J. 145, 505 (1966).
22) R.N.Manchester, J.H.Taylor, Pulsars , Freeman (1977).
23) J.H.Taylor, talk presented at the IAU Symposium No95 "Pulsars" (1980).
24) J.H.Taylor, L.A.Fowler, P.M.McCulloch, Nature 277, 437 (1979).
25) K.H.Elliot et al., M.N.R.A.S. 192, 51P (1980).
26) S.Rappaport, in Compact Galactic X-Ray Sources, F.K.Lamb and D.Pines, Eds. (Physics Dept. University of Illinois, Urbana, 1979) p.181.

STRUCTURE OF BARYONIC SYSTEM WITH PION CONDENSATION AND ITS IMPLICATION IN NEUTRON STAR PROBLEMS

R. Tamagaki
Department of Physics, Kyoto University, Kyoto 606, JAPAN

1. Introduction

Studies of pion condensation are closely related to the problems on structures of nucleon (baryon) system under pion condensation. In the 1978 Trieste Conference, I reported our studies based on a model of the Alternating-Layer-Spin (ALS) structure in nuclear medium.[1)] This paper deals with recent works made from the viewpoint of the ALS model by developing these studies:

(1) In order to take into account the OPE effect enhanced by Δ-mixing leading to pion condensation and short-range effect on equal footing, a reaction matrix theory in the ALS structure by Tamiya and myself[2)] is presented. The calculated results in neutron matter enable us to understand the mechanism of the structure change by using the same notions in the usual nuclear matter theory.

(2) Study on the ALS structure in neutron matter based on the σ-model made by Tatsumi[3)] is reported, by putting emphasis on his recent result that π^c condensation is realized in a limited range of density upon π^0 condensation, namely under the ALS structure.

(3) The effects of the ALS structure and its associated aspects of pion condensation on neutron star problems are briefly discussed.

2. Alternating-Layer-Spin (ALS) structure of baryonic system under π^0 condensation

The nucleon structure under a typical π^0 condensation with the standing-wave type, among various possible pion condensates,[4),5)] has been shown to be well described by the ALS model of nucleon system.[6)] For the characteristic aspects of the ALS structure, refs.6) and 1) should be referred to. Based on the standpoint of the equivalence between the field description (with explicit use of π^0 field) and the potential description (with use of nucleonic states and potentials only), we notice that there is a way to understand the π^0 condensed phase in the potential description, where the same notions used in the ordinary nuclear matter theory are applicable.

Total potential energy of neutron matter in the normal phase calculated by the reaction matrix theory[7)] is almost given by the $^1S_0$ contribution at the density $\rho \lesssim 5\rho_0$ (ρ_0 being the nuclear density), although the $^3P_1$ and $^3P_2$ waves give large contributions beyond $\rho \sim 2\rho_0$. This is because the statistically weighted sum of the $^3P_{J=0,1,2}$ contributions are almost cancelled even in the reaction matrix calculations; as we have $\Sigma_J(2J+1)\langle V(^3P_J)\rangle = 9\langle V_C\rangle$ in the Fermi gas model because of the tensor operator S_{12} and the spin-orbit (LS) operator $\vec{L}\cdot\vec{S}$, where $V(^3P_0)=V_C-4V_T-2V_{LS}$, $V(^3P_1)=V_C+2V_T-V_{LS}$, $V(^3P_2)=V_C-0.4V_T+V_{LS}$ with C, T and LS denoting central, tensor and LS forces. Occurrence of the ALS structure means that such "cancellation balance" of non-central forces is broken to obtain energy gain overwhelming the kinetic energy increase.[6),8)] Takatsuka[9)] studied this problem in the developed ALS structure by using the OPEP only in the Born approximation and found such tendency brought about by its tensor part. Here this problem is discussed by extending the reaction matrix theory so as to be applicable to the ALS structure in a way that the change of every partial-wave contribution can be traced with variation of localization including the limit of no localization.

2A. Reaction matrix formulation in the ALS structure

For our aim the Bloch-type basis of one-particle wave functions is constructed from the Wannier-type basis $\{\phi^W\}$ used previously in the developed (ALS) structure; $\phi^W_\alpha = \phi_{\ell\vec{q}_{\perp\alpha}}(\vec{\xi}_i) \propto \exp(i\vec{q}_{\perp\alpha}\cdot\vec{r}_{\perp i})\phi(z_i)\chi_{\sigma_\ell}(i)$, which represents the localized wave around $z=\ell d$ approximated by $\phi_\ell(z_i) \cong (a/\pi)^{1/4}\exp[-a(z-\ell d)^2/2]$ with layer distance d and layer number ℓ, the two-dimensional Fermi gas state occupied to $|\vec{q}_{\perp\alpha}| \leq q_{\perp F} = (4\pi\rho d)^{1/2}$ and the spin state with a spin component $\sigma_\ell/2 = (-)^\ell/2$, where $\vec{q}_{\perp\alpha} = \{q_{x\alpha}, q_{y\alpha}\}$, $\vec{r}_{\perp i} = \{x_i, y_i\}$, $\vec{\xi}_i = (\vec{r}_i, \text{spin})$. The Bloch-type basis is obtained by converting $\phi_\ell(z_i)$ into $\phi_{\kappa_\alpha\sigma_\alpha}(z_i) \propto \Sigma_{\ell(\sigma_\alpha)}\exp(i\kappa_\alpha \ell d)\phi_\ell(z_i)$ with $\ell(\sigma_\alpha)$ =even(odd) for $\sigma_\alpha=1(-1)$.[10),1),11)] After the Fourier transformation of $\phi_\ell(z_i)$, we have the Bloch-type basis $\{\phi_\alpha\}$:

$$\phi_\alpha(i) \equiv \phi_{\vec{q}_\alpha\sigma_\alpha}(\vec{\xi}_i) = 1/\sqrt{\Omega}\cdot\sum_n(\sigma_\alpha)^n U_\alpha(n)\exp[i\vec{q}_\alpha(n)\cdot\vec{r}_i]\chi_{\sigma_\alpha}(i)$$

where $\vec{q}_\alpha \equiv \vec{q}_{\perp\alpha} + \kappa_\alpha\hat{z}$, $U_\alpha(n) \equiv \exp[-\kappa_\alpha^2(n)/2a]/\sqrt{N_{\kappa_\alpha}}$ with $N_{\kappa_\alpha} = \Sigma_n \exp[-\kappa_\alpha^2(n)/a]$ and $\kappa_\alpha(n) = \kappa_\alpha + nk_0$, $\vec{q}_\alpha(n) = \vec{q}_\alpha + nk_0\hat{z}$ and the three-dimensional volume Ω. $\vec{k}_0 = k_0\hat{z}$ is the momentum of condensed π^0 and related to the layer distance as $k_0 = \pi/d$. The occupied states are restricted within the cylindrical Fermi surface given by $|\vec{q}_{\perp\alpha}| \leq q_{\perp F}$, $|\kappa_\alpha| \leq q_{zF} = k_0/2$. When the localization disappears ($a\to 0$), $\phi_\alpha(\vec{\xi}_i)$ tends to the plane wave $\Omega^{-1/2}\exp(i\vec{q}_\alpha\cdot\vec{r}_i)X_{\sigma_\alpha}(i)$ and the ALS state $|\Phi_{ALS}\rangle$ becomes the Fermi gas state. Two-particle wave functions such as $\phi_\alpha(1)\,\phi_\beta(2)$ are composed of the center of mass part and the relative plus spin part to which the partial wave (p.w.) expansion is applied. Basic p.w. contributions essentially depend on the azimuthal quantum numbers of angular momenta for such spin-spatial correlated configuration as the ALS one.

After the summation over them we have the p.w. contributions specified by ${}^{2S+1}L_J$ (spin S, orbital L and total J) as in the spin-spatial non-correlated spherical Fermi surface.

Under the ALS structure, a single-particle (s.p.) potential U in the state ϕ_α is the diagonal matrix whose elements are the same for the up and down spins; $\langle\alpha|U|\alpha\rangle=U(\vec{q}_\alpha)=U(q^2_{\perp\alpha},\ \kappa^2_\alpha)$. These matrix elements are obtained by the reaction matrix G. The kinetic energy matrix elements is diagonal; $\langle\alpha|t|\alpha\rangle=t(\vec{q}_\alpha)=1/2M\cdot\{q^2_{\perp\alpha}+\Sigma_n\kappa^2_\alpha(n)U^2_\alpha(n)\}$. (The $\hbar=c=1$ unit is used throughout.) The total energy is given as

$$\overset{\circ}{E}_{ALS}\equiv\langle T\rangle+\langle G_{ALS}\rangle=\sum_\alpha^{(OCC)}\{\langle\alpha|t|\alpha\rangle+\langle\alpha|U|\alpha\rangle/2\}=\sum_\alpha^{(OCC)}\{t(\vec{q}_\alpha)+U(\vec{q}_\alpha)/2\}.$$

The treatment of the Pauli operator is more complex than in the usual case, and we can dealt with it only approximately.[2)]

2B. Partial wave contributions leading to the realization of the ALS structure

By applying the formulation mentioned above to the OPEP only in the Born approximation, we have confirmed the results previously obtained without p.w. expansion[6)] and those in ref.9, where the realization of the ALS structure takes place due to the lowering of all the ${}^3P_J$ contributions as the growth of localization specified by increase of a. The half-value width of the Gaussian localization is given by $1/\sqrt{a}$ and $\Gamma\equiv ad^2$, the squared ratio of the layer distance to this width, is the measure indicating the growth of the ALS structure. Thus we can say that the ALS structure is realized as a result of breaking of the "cancellation balance" among the ${}^3P_J$ waves, for the OPEP-Born case. Can this aspect persist even for realistic case?

The RSC potential[12)] V^{RSC} as well as other realistic nuclear forces does not favor the ALS structure because the tensor component is damped in the intermediate region from the OPEP value, mainly due to the ρ-meson exchange. Actually we found the clear energy minimum at a=0 near the values obtained by the LOBT (Lowest-Order Bruckner Theory) in the normal phase for $\rho=(1\sim3)\rho_0$ by using the formalism in 2A.

For the realization of the ALS structure, the enhancement of the OPEP tensor component due to Δ-mixing is indispensable. The reaction matrix theory under the ALS structure is still applicable by making the following minimal modification based on the results obtained previously for a simple model with the (πNN, $\pi N\Delta$, $\pi\Delta\Delta$) P-wave interaction.[14)] $|\Phi_{ALS}\rangle$ is extended to the Slater-determinantal state built up on the basis $\{\phi_\alpha\}$ of quasineutrons ($\tilde{n}$) represented as $|\tilde{n}\rangle=u|n\rangle-v|\Delta\rangle$ with real u and v satisfying $u^2+v^2=1$. This prescription is similar to that in the three-dimensional solid model by Pandharipande and

Smith.[13] When the developed ALS structure is dealt with, we have the effective Δ-$\tilde{n}$ coupling constant f_{eff}; $f_{eff}=fu^2+4/3\cdot guv+1/4\cdot hv^2$, where f, g and h are the πNN, $\pi N\Delta$ and $\pi\Delta\Delta$ coupling constants, respectively. The expression of the total energy obtained in the field description for the developed phase is rewritten by the potential description for the OPEP with the enhanced factor ENF $\equiv f_{eff}^2/f^2$ and the increase in self-energy. To put forward this description in more realistic way, we replace the enhanced OPEP by the reaction matrix G_{ALS} under the ALS structure calculated from a modified realistic potential such as

$$V_{ij}=(f_{eff}^2/f^2)\tilde{V}_{ij}^{(OPE)}+(V_{ij}^{(RSC)}-\tilde{V}_{ij}^{(OPE)})$$

which tends to $V_{ij}^{(RSC)}$ when $v^2\to 0$, i.e. ENF$\to$1. Here $\tilde{V}_{ij}^{(OPE)}$ means the OPEP regularized at small distances in its tensor component from $Z(x)/3=(1/3+1/x+1/x^2)e^{-x}/x$ with $x=m_\pi r_{ij}$ to $Z(x)/3-(4/x+1/x^2)e^{-4x}/x$. The term with e^{-4x} damping reduces the tensor potential at $r_{ij}=1.4\sim 1.0$ fm and simulates partially the ρ-meson effect. Thus our program is to obtain $E_{ALS}/N=\overset{\circ}{E}_{ALS}/N+\delta Mv^2$ with $\delta M=M_\Delta-M\cong 300$ MeV, where $\overset{\circ}{E}_{ALS}/N=\langle t\rangle+\langle G_{ALS}\rangle/N$ is calculated with V_{ij}. E_{ALS}/N is to be compared with $E_{FG}/N=\langle t\rangle+\langle G_{FG}\rangle/N$ in the LOBT.

For ENF$\lesssim$2, we have energy minimum at a=0 for any d in the density range of $(1\sim 3)\rho_0$, as in the original RSC case. For ENF=3 and $\rho\gtrsim 2\rho_0$, we find the minimum at finite a. We find in Fig.1 that (i) $\rho=2\rho_0$ is just close to the critical point because we have a local minimum at a=0 and d=1.0 fm, close in magnitude to the minimum at a=3.0 fm^{-2} and d=1.4 fm and (ii) the ALS structure is surely realized at $\rho=3\rho_0$ as we have the clear minimum at a=3.8 fm^{-2} and d=1.2 fm.

The a-dependence of each p.w. contribution in $\langle G_{ALS}\rangle/N$ is shown in Fig.2 for a typical case of ENF=3 and $\rho=3\rho_0$ with d=1.2 fm. When a increases, the $^3P_2+^3F_2$ attractive contribution grows signigicantly, the $^3P_1$ repulsive one decreases and the $^1S_0$ plus $^1D_2$ one slightly favors the localization. Contrary to the OPEP-Born case, the attractive $^3P_0$ contribution becomes relatively small and even acts against the localization, because the short-range LS force produces strong repulsion. The ALS phase is realized by breaking the "cancellation balance" in such a manner that two dominating p.w. ($^3P_1$ and $^3P_2+^3F_2$) in the normal phase favors the localization.

The comparison in total energy with the normal phase should be done with the addition of δMv^2 which depends sensitively on the $\pi N\Delta$ coupling constant g. For ENF=3, case I; δMv^2=56.0 MeV (v^2=0.188) for $g^2=72/25\ f^2$ (the quark model prediction) and case II; δMv^2=31.6 MeV

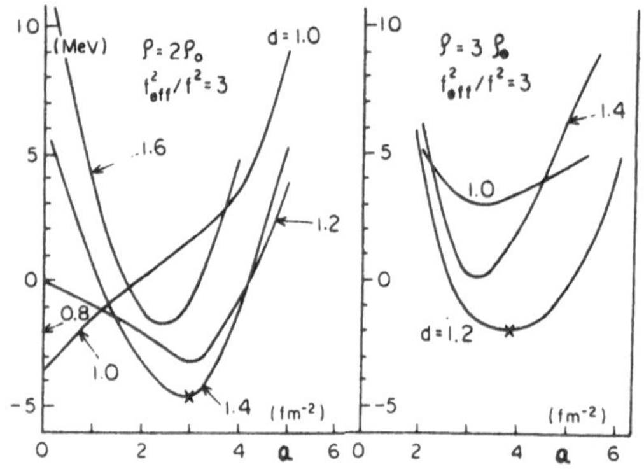

Fig.1. Dependence of $\mathring{E}_{ALS}/N$ on the localization parameter a for ENF=3 and $\rho/\rho_0=2$ and 3. The numbers attatched are d in fm.

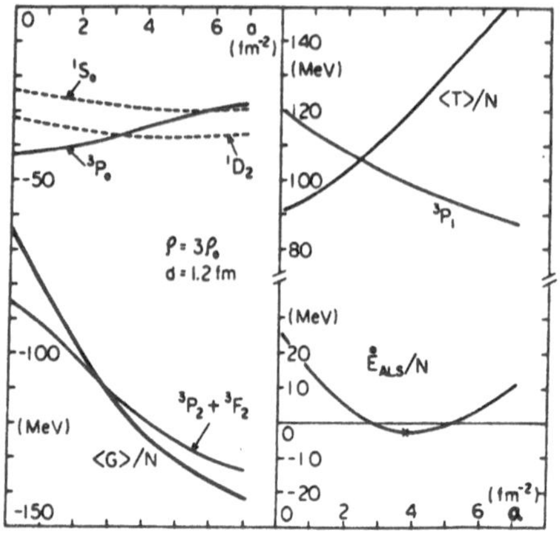

Fig.2. a-dependence of $\langle T\rangle/N$ $\langle G_{ALS}\rangle/N$, $\mathring{E}_{ALS}/N$ and the partial wave contributions for ENF=3, $\rho=3\rho_0$ and d=1.2 fm.

(v^2=0.106) for g^2=4 compatible with the decay width of Δ. The results are shown in Table 1. E_{ALS}/N including the p.w. with $J\leq 2$ becomes lower than E_{FG}/N at $\rho \gtrsim 3\rho_0$ ($2\rho_0$) for case I (II),as a result of the enhanced attractive interaction term still larger than the enlarged kinetic and self-energy terms. The higher p.w. effect due to the OPEP-tail in the ALS phase, which becomes attractive at higher density as seen in (b)-(a), even strengthens this tendency. Thus we can say

Table 1. Energies of quasi-neutron matter under the ALS structure at $\rho=2\rho_0$ and $3\rho_0$. Column (a) shows $E_{ALS}/N=\mathring{E}_{ALS}/N(J\leq 2)+\delta M v^2$, (b) E_{ALS}/N with the higher p.w. contributions and (c) E_{ALS}/N supplemented by the $N\Delta$-channel correction[15] in addition. The values in the parentheses are the corresponding ones of E_{FG}/N.[7]

ρ (fm^{-3})	d (fm)	a (fm^{-3})	Γ $=ad^2$	$\langle T\rangle/N$ (MeV)	$\langle G\rangle/N$ (MeV)	$\mathring{E}_{ALS}/N$ (MeV)	E_{ALS}/N(MeV) case	(a)	(b)	(c)	$m_\perp^*$
0.34	1.4	3.0	5.88	93.1 (58.0)	-103.4 (-24.5)	-4.6 (33.5)	I II	51.4 27.0	54.4 30.0	72.4 48.0	0.92
0.51	1.2	3.8	5.47	119.1 (76.0)	-121.9 (-23,0)	-2.8 (53.0)	I II	53.2 28.8	47.4 23.0	82.4 59.0	0.86

that the ALS structure equivalent to the typical π^0 condensation is realized at $\rho \gtrsim 2\rho_0$ even under the influence of short-range (s.r.) correlation, if the OPEP enhancement due to Δ-mixing comes into play.

Inclusion of s.r. correlation stabilizes the matter; firstly the resulting moderate values of $\Gamma = ad^2 = 5 \sim 6$ mean the stability against the collapse to $\Gamma \to \infty$ and secondly the ρ-dependence of E_{ALS}/N is so moderate that the pressure remains possitive if some repulsive correction increasing with ρ is taken into account such as quoted in (c). Such s.r. effect for stabilizing the matter has been found in other energy calculations in the ALS phase, in the OBEP approach with Δ-mixing effect[16] and in the LOBT approach with the 2π-exchange three-body force mediated by Δ in addition to the G-matrix in the normal phase.[17] The extracted single-particle asepcts in the ALS phase are characterized (i) by the oblate cylindrical Fermi surface and (ii) by the deep single-particle potential with very weak κ_α (the z-component of $\vec{q}_\alpha$) dependence and moderate $q_{\perp\alpha}$ dependence represented by large effective mass $m_\perp^* \equiv (1 + M/q_{\perp\alpha} \cdot \partial U/\partial q_{\perp\alpha})^{-1}_{q_{\perp F}} \sim 0.9$ as shown in Table 1.

3. Charged pion condensation under the ALS structure — Combined condensation of neutral and charged pions realized in the σ-model—

As one of realistic approaches to pion condensation, studies taking into account the dynamics of pion+nucleon system implied in the chiral symmetry such as the π-N S-wave and the π-π interaction have been made, in particular for the charged pion (π^c) condensation following the line initiated by Campbell, Dashen and Mannasah.[18] It is an important study to examine the ALS structure and its associated aspects from such viewpoint. Previously Tatsumi[3] showed in neutron matter that the ALS structure is the phase firstly coming into existence around ρ_0 which corresponds to the π^0 condensation natually realized in the σ-model, while the phase suggested by Dautry and Nyman[19] may be possible at higher density. His recent study in the σ-model extended to ($\pi^0+\pi^c$) combined pion condensation (CPC) is briefly reported here. Starting point is to introduce the unitary operator U transforming a normal state of nuclear matter $|N.M.\rangle$ to a pion condensed state;

$|PC; \varphi, \chi, \theta\rangle = U|N.M.\rangle$ with

$$U = \exp(i\int \varphi(x) A_0^3 d^3x)\exp(i\int \chi(x) V_0^3 d^3x)\exp(iQ_1^5\theta),$$

where V_μ^i and A_μ^i are the vector and axial-vector currents, respectively and $Q^5{}_1 = \int A^1{}_0(x)d^3x$.[18] A limiting case for $\chi=\theta=0$ describes the pure π^0 condensation, another case for $\varphi=0$ the pure π^c one and the case with nonvanishing (φ, χ, θ) the combined one. A suitable choice to describe the CPC realized upon the ALS structure is to take ($\varphi = A\sin k_0 z$, $\chi = \vec{k}_c \cdot \vec{r}_\perp - \mu_\pi t$, θ=const.) with the two condensed momenta being perpendicular $\vec{k}_0 = k_0 \hat{z} \perp \vec{k}_c$.

The total energy is the expectation value of the Hamiltonian with $|PC; \varphi, \chi, \theta>$, whose ρ-dependence is shown in Fig.3. The results are characterized by three density regions; (1) the pure π^0 condensation corresponding to the ALS structure firstly appears at $\rho_1 \cong 0.8 \rho_0$, (2) next the π^c condensation appears under the ALS structure at $\rho_2 \cong 1.6\rho_0$ but disappears at $\rho_3 \cong 5.5\rho_0$ and (3) again the pure π^0 condensation persists at $\rho \gtrsim \rho_3$. The notable points are; (i) the π^c condensation is realized only in the limited region ($\rho = \rho_2 \sim \rho_3$), (ii) the feature at $\rho = \rho_2 \sim 3.5\rho_0$ is similar to the one previously obtained in the conventional π-N P-wave interaction only[20] except higher ρ_2 due to the π-N S-wave effect and (iii) decrease in energy gain by the combined condensation at $\rho = 3.5\rho_0 \sim \rho_3$ is due to the repulsive effect arising from the interaction between condensed π^0 and π^c. It is a future problem to examine how this feature persists when the effects of short-range correlation and Δ-mixing.

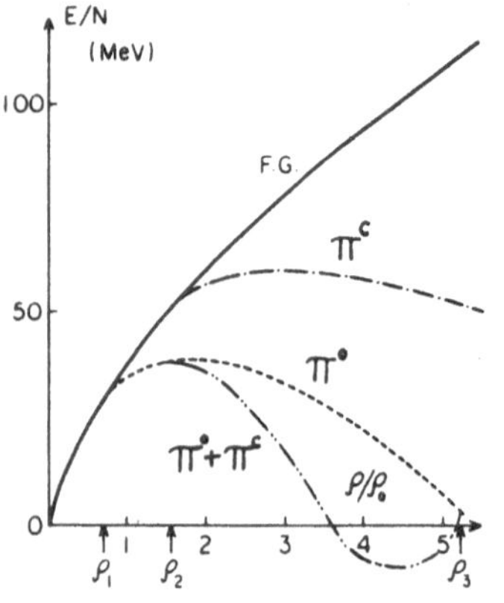

Fig.3. Energies per particle of pion-condensed phases in the σ-model. π^c, π^0 and $\pi^0+\pi^c$ indicate the pure π^c, pure π^0 and the combined consensations, respectively.

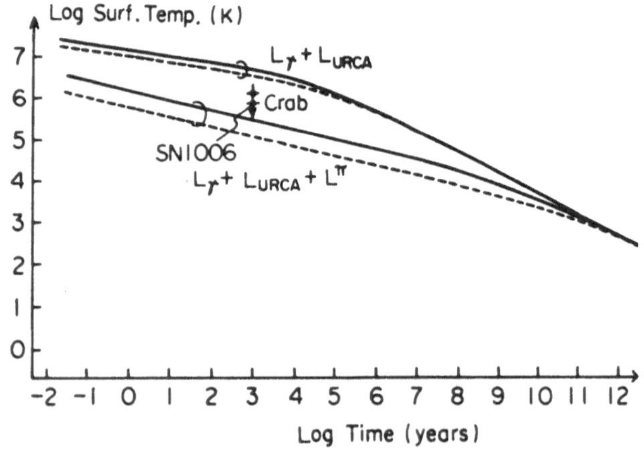

Fig.4. Cooling curves for a model neutron star with surface radius=10 km and the pion condensed core of $2\rho_0$ and 8 km radius; solid (dashed) lines are with (without) the ALS structure. $L\gamma$ means the photon luminocity.

4. Implication in neutron star problems

Aspects of baryonic structure significantly related to neutron star problems are the equation of state (EOS), the single-particle (s.p.) aspects near the Fermi surface and the effects associated with pion condensation. Since our reaction matrix calculation under the ALS structure has been done only in the limited density range ($\rho = (1\sim3)\rho_0$), we can not say anything about EOS more than that the transition to the ALS phase from the Fermi gas one necessarily makes EOS softer but recovering is possible due to the suppresion of $^1S_0$ attraction brought about by the channel coupling effect, which increases with ρ.[15] As for the latter two points a few remarks are given.

In the developed ALS structure indicated by $\Gamma=5\sim6$, a large band gap appears in the z direction of the oblate cylindrical Fermi surface.[10] Therefore the transition between s.p. states near the Fermi surface is restricted to that between the two-dimensional Fermi gas states. As is reported by Takatsuka,[21] the effect of such restriction on nucleon superfluidity is suppressive but moderate enough to be restored by larger effective mass ($m_\perp^* > m^*$ in the normal phase), and the theoretical results of superfluid effects on neutron star phenomena remain without serious modification. Another phenomenon affected by such restriction is the ν-cooling of neutron stars. For the pure π^0 condensate, luminosity of the modified URCA process ($n+n \to n+p+e^- +\bar{\nu}_e$ and its inverse one) L_{ALS}^{URCA} becomes lower than that in the normal phase L^{URCA} by one order of magnitude.[23]

If the π^c condensation takes place, the pion cooling as the β-decay of quasinucleons (η) without by-stander nucleons ($\eta(\vec{q}) \to \eta(\vec{q}\,') + e^-(\vec{p}_e) + \bar{\nu}_e(\vec{p}_\nu)$) is the most efficient process at early stage. For the pure π^c condensation, the luminosity given by Maxwell et al.[22] is $L_\nu^{\pi n}/\Omega \sim 3.4\times10^{27}$ erg sec^{-1} cm$^{-3}\theta^2/4\ m^*\ T_9^6$, where m^* is the effective mass parameter, θ the chiral angle and T_9 temperature/10^9 K. The luminosity for the combined ($\pi^0+\pi^c$) condensation under the ALS structure shown in 3 is estimated as $L_{ALS}^{\pi c} \simeq (1.2\times10^{26}/3.4\times10^{27})\ L_\nu^{\pi n}$, where the core matter with constant density $2\rho_0$ and radius 8 km is assumed.[23] The ALS structure brings about the reduction of pion cooling luminosity by one order of magnitude but this process is still dominant ($L_{ALS}^{\pi c} \gg L_{ALS}^{URCA}$). At later stage photon emission from the surface is dominating. To illustrate relative importance of these effects, cooling curves are shown in Fig.4.

Based on the recent calculations[24] on neutron star cooling, the "standard" cooling senario (without presuming any new phases in neutron star interior such as pion condensates) is consistent with the recent observations which possibly indicate the actual surface temperatures T_{surf} for the Crab and Vela. For the stiff EOS case, where no π^c condensate exists because of relatively low central density, this statement is valid as it stands. For the soft EOS case, where central density reaches about 10 ρ_0, the pion cooling gives T_{surf} with one order of magnitude lower than the "standard" cooling curves, which contradicts with the data mentioned above. However, if π^c condensate is realized as the combined condensate only in the region $\rho_2 \sim \rho_3$ as shown in 3, the π^c condensate exists only in a small portion just below the core-crust boundary. Therefore the pion cooling is much reduced, and the soft EOS model is also consistent with the recent cooling data.

Acknowledgements

The author would like to thank Dr. T. Takatsuka, Dr. T. Tatsumi, Mr. T. Kunihiro and Mr. H. Frukawa for their valuable discussions. He is especially grateful to Dr. T. Tatsumi for his cooperation in preparing this report.

References

1) R. Tamagaki, Nucl. Phys. A328 (1979) 352.
2) K. Tamiya and R. Tamagaki (in preparation)
3) T. Tatsumi, Prog. Theor. Phys. 63 (1980) 1252 and private communication.
4) A.B. Migdal, Rev. Mod. Phys. 50 (1978), 107.
5) T. Matsui, T. Otofuji, K. Sakai and M. Yasuno, Prog. Theor. Phys. 63 (1980) 1665.
6) T. Takatsuka, K. Tamiya, T. Tatsumi and R. Tamagaki, Prog. Theor. Phys. 59 (1978) 1933.
7) As a paper by the use of the RSC, P.J. Siemens and V.R. Pandharipande, Nucl. Phys. A173 (1971) 561.
8) F. Calogero and F. Palumbo, Lett. Nouvo Cimento, 6 (1973), 663
9) T. Takatsuka, Prog. Theor. Phys. 61 (1979) 1564.
10) T. Matsui, K. Sakai and M. Yasuno, Prog. Theor. Phys. 60 (1978) 442, 61 (1979) 1093.
11) F. Palumbo, preprint LNF-80/41(P).
12) R.V. Reid, Ann. of Physics 50 (1968) 411.
13) V.R. Pandharipande and R.A. Smith, Nucl. Phys. A237 (1975) 507.
14) T. Kunihiro and R. Tamagaki, Prog. Theor. Phys. 61 (1979) 1107.
15) A.M. Green and P. Haapakoski, Nucl. Phys. A221 (1974) 429.
16) T. Kunihiro and T. Tatsumi, Prog. Theor. Phys. 65 (1981) No.2.
17) T. Takatsuka, Y. Saito and J. Hiura, private communication.
18) D.K. Campbell, R.F. Dashen and J.H. Manassah, Phys. Rev. D12 (1975) 979, 1010. G. Baym and D.K. Campbell, "Mesons in Nuclei" Vol.III, ed. by M. Rho and D. Wilkinson (North Holland Pub. Comp., 1979), chapter 27.
19) F. Dautry and E. Nyman, Nucl. Phys. A319 (1979) 323.
20) K. Tamiya and R. Tamagaki, Prog. Theor. Phys. 60 (1978) 1753.
21) T. Takatsuka, Invited Talk in this Conference.
22) O.V. Maxwell, Astrophys. J. 231 (1979) 201.
23) T. Tatsumi, H. Frukawa and R. Tamagaki (in preparation).
24) S. Tsuruta, Invited Talk at the IAU Symposium No.95, on Pulsars, in Bonn, 1980 August and the references cited therein.

Nucleon Superfluidity under Pion Condensation

T. Takatsuka

College of Humanities and Social Sciences,
Iwate University, Morioka 020

1. Introduction

In this talk, on the basis of investigations by Tamagaki and myself, we present our studies on the nucleon superfluidity under pion condensation.[1)] It was previously shown that both of neutrons and protons in neutron star interior are in the superfluid states at the densities $\rho \simeq (1-3)\rho_0$ (ρ_0: nuclear density), where the neutron $^3P_2$-superfluid and the proton $^1S_0$-one are coexistent.[2-4)] On the other hand, pion condensation, another interesting phase of nuclear medium, has been recognized to set in or develope in the same region of densities, causing a remarkable structure change of nucleon system.[5)] Then, there arises the important question whether the superfluidities of nucleons, shown to be realizable from the ordinary Fermi gas, persist or not when pion condensation comes into play. This problem is of particular interest from the observational viewpoint since the existence of superfluids largely affects the bulk properties of neutron stars, such as cooling processes[6)] and glitch phenomena,[7)] and also from the viewpoint of many-body quantum theory.

In order to study the problem, it is essential to find out the new single-particle basis describing the nucleon system under pion condensation, together with the ground state configuration and the single-particle spectra. In this report, we consider two typical cases for pion condensation which enable us to give insight into the basic aspects of the superfluidities. The one is the neutral pion (π^0) condensation of standing wave mode $\varphi_{\pi^0} \propto \sin k_0 z$, with the condensed momentum $\vec{k}_0$ in z-direction. For this case, the nucleon system undergoes a drastic structure-change well described by the Alternating Layer Spin (ALS) model.[8-10)] In this model nucleons localize one-dimensionaly forming a layer structure with a particular spin- and isospin-orderings, (see also Fig.1). Such aspect comes from that the condensed π^0 field generates a deep periodic potential $V_{\pi^0} \propto \tau_3 \sigma_z \nabla_z \varphi_{\pi^0}$ with spin-isospin dependence and hence nucleons arrange in z-direction with the layer spacing $d=\pi/k_0$ so as to feel efficiently this potential. Then, z-part of the single particle wave function is given by Wannier or Bloch functions due to the periodic localization, while its $\vec{r}_\perp \equiv \{x,y\}$-part remains as the two-dimensional (2D) plane wave. Because in the Bloch-orbital description, the band gaps appear in z-direction, the pairing correlation is operative only in the $\vec{r}$ -space where the Fermi gas nature holds, and the resulting superfluidity becomes of 2D character in contrast with the usual three-dimensional (3D) case. This feature provides us with an interesting problem, low-dimensional superfluid in nucleon matter.

Another case to be discussed here is the charged pion (π^c) condensation of running wave mode[11] $\varphi_\pi c \propto \exp(ik_c z - i\mu_\pi t)$ with μ_π being the pion chemical potential. In this case, there arises no localization like π^0-case, and 3D-nature of superfluid remains, but one important difference comes out compared with the case without pion condensate: Nucleon system turns out to be described by the quasiparticles composed of neutrons and protons, which is due to the effect of the one-body potential $V_\pi c \propto \{\tau_+ \sigma_z \nabla_z \varphi_\pi c + h.c.\}$ brought about by $\varphi_\pi c$, with the isospin-flip operator $\tau_\pm = (\tau_1 \pm i\tau_2)/2$. This means that the pairing correlation should be represented based on these quasiparticles, and hence provides us with another new problem of superfluidity.

From the viewpoint of nucleonic correlation, pion condensed phases are realized by the particle-hole correlation through the One-Pion-Exchange Potential (OPEP) originating from the π-N P-wave interaction. On the contrary, the pairing correlation responsible for superfluidity is the particular particle-particle (hole-hole) one. Therefore at the simplest level, we can introduce it to the nucleon system under pion condensation, in the usual manner by adopting the full two-nucleon potential including the OPEP. By noticing the characteristics of the new single-particle basis, the energy gap equation is natrually derived along the line in the generalized BCS-Bogoliubov theory.[2a)12)]

2. Superfluidity under π^0 condensation

We start with presenting the ALS model[8)] illustrated in Fig. 1 which gives us a powerful tool to describe the nucleon system under π^0 condensation. This model is constructed from the following orthogonal basis functions $\{\phi_\alpha\}$:

$$\phi_\alpha(\vec{\xi}) = \Omega_\perp^{-1/2} e^{i\vec{q}_\perp \vec{r}_\perp} \phi_j(z) \chi_{\sigma\tau} \text{ (spin, isospin)}, \qquad (1)$$

where $\vec{\xi} \equiv \{\vec{r}, \text{spin, isospin}\}$, $\vec{q}_\perp \equiv \{q_x, q_y\}$, $\vec{r}_\perp \equiv \{x, y\}$ and $\Omega_\perp$ is the 2D normalization volume. As already mentioned, $\phi_j(z)$ is given by the Wannier or Bloch functions due to the periodic localization. We adopt here the Bloch orbital basis constructed from the wave functions $\phi(z-d\ell)$ localized around the lattice cite ℓd:

$$\phi_j(z) \equiv \phi_{q_z}^{(\sigma\tau)}(z) = (N_z N(q_z)/2)^{-1/2} \sum_\ell e^{iq_z d\ell} \phi(z-d\ell), \qquad (2)$$

where $\phi(z-d\ell) = (a/\pi)^{1/4} \exp[-a(z-d\ell)^2/2]$, $N(q_z) = \Sigma_n^{all} \exp(2iq_z dn - ad^2n^2)$ and $N_z \equiv \Omega_z/d$ is the total number of the layers with Ω_z being the normalization volume (length) in the z-direction. The reduced momentum q_z runs over the region $|q_z| \leq \pi/2d \equiv q_{zF}$. Due to the spin- and isospin-orderings, it should be noted that in (2) the summation is taken as ℓ=even (odd) for $\sigma\tau=1(-1)$. Hence the dependence of $\phi_j(z)$ on σ and τ appears. The ground state of nucleon system $|\Phi_N\rangle \equiv |\Phi_{ALS}\rangle = |\Phi_{ALS}^{(n)}\rangle \otimes |\Phi_{ALS}^{(p)}\rangle$ is given by the Slater determinant of $\{\phi_\alpha\}$, where the 2D Fermi gas state is occupied up to $|\vec{q}_\perp| \leq q_{\perp F}^{(i)}$ with $q_{\perp F}^{(i)}$ being the 2D Fermi momentum. Here i=n(p) stands for neutron (proton). $q_{\perp F}^{(i)}$ is given by $q_{\perp F}^{(i)} = (4dq_F^{(i)}/3)^{1/2} q_F^{(i)}$ with the usual 3D Fermi momentum $q_F^{(i)} = (3\pi^2 \rho_i)^{1/3}$. Soon after the onset of π^0 condensation, nucleon system becomes well localized and the Fermi surface turns out to be of cylindrical shape with the single particle spectra $\varepsilon(q_\perp, |q_z|)$ being independent on q_z; $\varepsilon(q_\perp, |q_z|) = \varepsilon(q_\perp)$.[9] Due to the existence of the large band gap in q_z-direction, it is possible to simplify the problem in such a way that the excitation of Cooper pairs $(\vec{q}_\perp, q_z; -\vec{q}_\perp, -q_z)$ into the states with $|q_z| > q_{zF}$ should be neglected. For the proton mixing ratios ρ_p/ρ under π^0-condensate, we presupose from the calculations only with the π-N P-wave interaction that they are likely around several % as in usual neutron star matter. Therefore, as far as we are concerned with the pairing correlation, we can discuss separately the neutron system and proton one, because $q_{\perp F}^{(p)} \ll q_{\perp F}^{(n)}$ results from $\rho_p \ll \rho_n \simeq \rho$. Hence hereafter we suppress the isospin labels.

Fig.1.

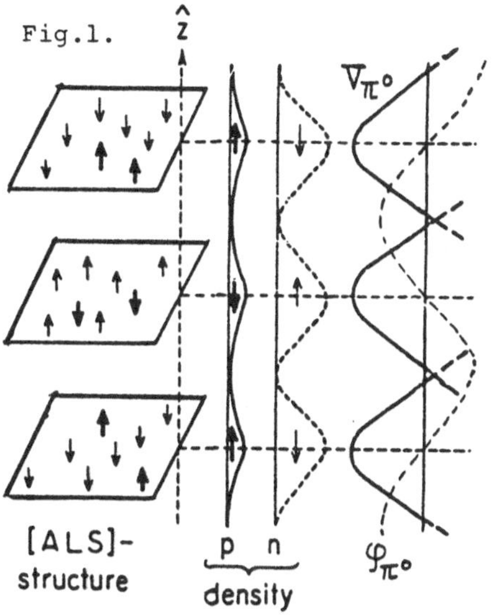

Neutrons

First we discuss the neutron superfluidity. We take the cylindrical coordinate representation suitable for the pairings of 2D-character.[1b)] Then we can express the wave function of the $(\vec{q}_\perp, q_z, \sigma_1; -\vec{q}_\perp, -q_z, \sigma_2)$-pair as

$$|\vec{q}\sigma_1; -\vec{q}\sigma_2\rangle = \phi_{q_z}^{(\sigma_1)}(z_1)\phi_{-q_z}^{(\sigma_2)}(z_2)\frac{1}{\Omega_\perp}\Sigma_S\Sigma_{m_S}(^1/_2\,^1/_2\;\sigma_1/_2\;\sigma_2/_2|Sm_S) \times \Sigma_{m_L} i^{m_L} J_{m_L}(q_\perp r_\perp) e^{-im_L\varphi_q} e^{im_L\varphi_r} \chi_{Sm_S}(1,2) \qquad (3)$$

after taking the partial wave expansion for exp $(i\,\vec{q}_\perp \vec{r}_\perp)$, where $\vec{q} \equiv \vec{q}_\perp + q_z\hat{z}$ and φ_q (φ_r) is the azimuthal angle of the vector $\vec{q}(\vec{r})$. In this case the pair state can be specified by a set of quantum numbers, $\tilde{\lambda} \equiv \{S, m_S, m_L\}$, instead of the specification to the 3D partial wave, $\lambda \equiv \{SLJm_J\}$, where S, L and J ($m_S$, m_L and m_J) denote the spin, orbital and total angular momenta (their z-components), respectively. Among possible pairings, $\tilde{\lambda} \equiv \{S=1, m_S=m_L=\pm 1\}$ is found to be most effective for the realization of neutron superfluidity at high densities ($\rho_n \gtrsim \rho_0$), because this pairing can utilize the strong attraction of the $^3P_2$-interaction at high momenta. The effect of $\tilde{\lambda}_1$-pairing contains the contributions from the usual 3D $\lambda \equiv \{^3P_2, {}^3F_2, \cdots\}$-pairs to satisfy $m_J=m_S+m_L=\pm 2$, whose main contributor is the $^3P_2$ one due to the centrifugal effect. So, this $\tilde{\lambda}_1$-pairing is the $^3P_2$-dominant one. By the use of (3), the pairing Hamiltonian $H^{\tilde{\lambda}_1}_{pair}$ for this particular pairing is written down. Finally our model Hamiltonian is given as follows:

$$H_{model} = H_0 + H_{pair}, \quad H_0 = \Sigma_{\vec{q}\sigma}\tilde{\varepsilon}(q_\perp)c^+_{\vec{q}\sigma}c_{\vec{q}\sigma}, \qquad (4),\ (5)$$

$$H^{\tilde{\lambda}_1}_{pair} = \frac{\pi}{\Omega_\perp}\Sigma_{\vec{q}'}\Sigma_{\vec{q}}\Sigma_\sigma \langle q_\perp'|V_{\tilde{\lambda}_1}(r_\perp;\, q_z',q_z)|q_\perp\rangle e^{im_L(\sigma)(\varphi_{q'}-\varphi_q)} \times c^+_{\vec{q}'\sigma}c^+_{-\vec{q}'\sigma}c_{-\vec{q}\sigma}c_{\vec{q}\sigma}, \qquad (6)$$

with $\langle q_\perp'|V_{\tilde{\lambda}_1}(r_\perp;q_z',q_z)|q_\perp\rangle \equiv \int_0^\infty r_\perp dr_\perp J_1(q_\perp' r_\perp)V_{\tilde{\lambda}_1}(r_\perp;q_z',q_z)J_1(q_\perp r_\perp)$ (6.a)

and $V_{\tilde{\lambda}_1}(r_\perp;q_z',q_z) \simeq \mathring{V}_{\tilde{\lambda}_1}(r_\perp)\times 2d/\Omega_z$

$$\equiv 2d/\Omega_z\left(\frac{a}{2\pi}\right)^{1/2}\int dz\,[e^{-im_L\varphi_r}\langle\chi_{Sm_S}|V(1,2)|\chi_{Sm_S}\rangle e^{im_L\varphi_r}]_{\tilde{\lambda}=\tilde{\lambda}_1}e^{-az^2/2}, \qquad (6.b)$$

where $c^+_{\vec{q}\sigma}$ denotes the creation operator for a neutron in the state ($\vec{q}$, σ), $\tilde{\varepsilon}(q_\perp) = \varepsilon(q_\perp)-\varepsilon(q_{\perp F})$, $m_L(\sigma)=1(-1)$ for $\sigma=\uparrow(\downarrow)$ and V(1,2) is the two nucleon potential. In the above expressions, the 2D effective pairing potential $V_{\tilde{\lambda}_1}$ is expressed by using the approximation to take into ac-count only the pairs within the same layer, because the $\tilde{\lambda}_1$-pairing is composed of two neutrons with equal spins and hence is dominant for the pairs in the same layer due to the localization. Under this approximation, $V_{\tilde{\lambda}_1}$ becomes independent on q_z and q_z'. As seen from (6.b), the spin-orbit part $V_{LS}(r)\vec{L}.\vec{S}$ in V(1,2), which is responsible for the strong $^3P_2$-attraction, appears in such a combination as $m_S m_L V_{LS}(r)$ with $V_{LS}(r) < 0$. This is the reason we have preferntially taken the same sign for m_L and m_S as in $\tilde{\lambda}_1$. If we introduce the suitable Bogóliubov transformation U_B as

$$U_B=e^{iS}, \quad iS = \frac{1}{2}\Sigma_{\vec{q}}\Sigma_\sigma\{\theta(\sigma\vec{q})c^+_{\vec{q}\sigma}c^+_{-\vec{q}\sigma}-h.c.\}, \qquad (7)$$

with $\theta(\sigma,\vec{q})=\beta(q_\perp, |q_z|)\exp(im_L(\sigma)\varphi_q)$ and β being real, we can formulate the problem by following the well known BCS-type theory. Finally, the energy gap equation for the $^3P_2$-dominant pairing is derived as

$$\Delta_1(q_\perp,|q_z|) = -\frac{\pi}{\Omega_\perp}\sum_{\vec{q}'} \langle q_\perp'|V_{\tilde{\lambda}_1}(r_\perp;q_z',q_z)|q_\perp\rangle \Delta_1(q_\perp',|q_z'|)/E_1(q_\perp',|q_z'|) \quad (8)$$

where $E_1(q_\perp,|q_z|) \equiv \sqrt{\tilde{\varepsilon}^2(q_\perp)+\Delta_1^2(q_\perp,|q_z|)}$ is the quasiparticle energy. Here we compare (8) with the usual 3D case with paying attention to $V_{\tilde{\lambda}}(r_\perp,q_z',q_z) = V(r_\perp)$ noted already. Then the following points are remarked: (i) The weight in the integrals leads to $q_\perp'(r_\perp)$ instead of $q'^2(r^2)$, (ii) the effective potential matrix element is taken for the Bessel function instead of the spherical one, and (iii) the energy gap function depends on $q_\perp$ instead of q, that is, $\Delta_1(q_\perp, |q_z|) = \Delta_1(q_\perp)$, reflecting faithfully the 2D character of the superfluidity. Numerical calculations are performed by using the effective mass (m_N^*) approximation as $\tilde{\varepsilon}(q_\perp) = \hbar^2(q_\perp^2-q_{\perp F}^2)/2m_N^*$ and by adopting the typical ALS-parameters[8] (d,a)=(in fm, in fm^{-2}) as (1.6, 1.95), (1.4, 4.08) and (1.4, 5.10) according to $\rho_n=\rho_0$, $2\rho_0$ and $3\rho_0$. As an example, results[1b] obtained for the G3RS ^3O-3 potential are shown in Fig.2 in terms of the critical temperature T_c; $\kappa_B T_c = 0.57\ \Delta_1(q_{\perp F})$. The results for G3RS ^3O -2 potential are also shown in comparison with the usual 3D $^3P_2$-superfluidity. As far as the effective mass parameter m^* ($\equiv m_N^*/m_N$)$\gtrsim$0.7, we can expect that neutrons in the densities $\rho_n=(1\text{-}3)\rho_0$ considered are still in the superfluid state of the $^3P_2$-dominance in the interior of neutron stars with the internal temperature $T_i \sim 10^8$ °k. In spite of the dimensional change from the usual Fermi gas, T_c are not so different. The results mentioned above can be confirmed by such an approach using Fourier expansion for localized waves.[1a)]

Protons[1d)]

We can formulate the proton superfluidity in the same manner as neutron case. But an important difference arises: Owing to the low density of protons ($\rho_p \ll \rho_n$) the proton superfluid, if possible, should be caused by the pairing correlation to utilize efficiently the $^1S_0$-interaction strongly attractive at low momenta. This feature is in contrast to the neutron case. Following the pairing scheme presented, the most provable candidate is the $\tilde{\lambda}_0 \equiv \{S=0,\ m_S=m_L=0\}$-pairing where the $^1S_0$-one dominates. In this case, the pairing correlation is operative on the antiparalell pair ($S=m_S=0$) composed of two protons in the neighbouring layers, especially in the nearest layers. The resulting energy gap function depends on q_z in addition to $q_\perp$ as $\Delta_0(q_\perp,|q_z|)=\Delta_0(q_\perp)\cos q_z d$, which comes from the q_z-dependence of $V_{\tilde{\lambda}_0}$ ($\propto \cos(q_z'-q_z)d$) under the nearest layer approximation. Critical temperature is related to the proton gap $\Delta_0(q_{\perp F})$ as $\kappa_B T_c \simeq 0.57\ \Delta_0(q_{\perp F})/\Gamma_\perp$ with $\ln \Gamma_\perp = \ln 2 - 0.5$. For the typical proton mixing = 0.05, we show in Fig.3, as an example, the results obtained by using the G3RS ^1E-2 potentials, where the results for $^1S_0$-superfluidity of protons realized from the ordinary Fermi gas are also shown in comparison. Calculations indicate $T_c \gtrsim 10^9$ °k unless the relevant m* is unexpectedly reduced by π^0-condensation. Thus, we can also expect that protons with small mixture in the neutron star interiors with π^0 condensation persist to be superfluid of the $^1S_0$-dominance. We note that the critical temperaturs are close to those obtained in the absence of π^0 condensate.

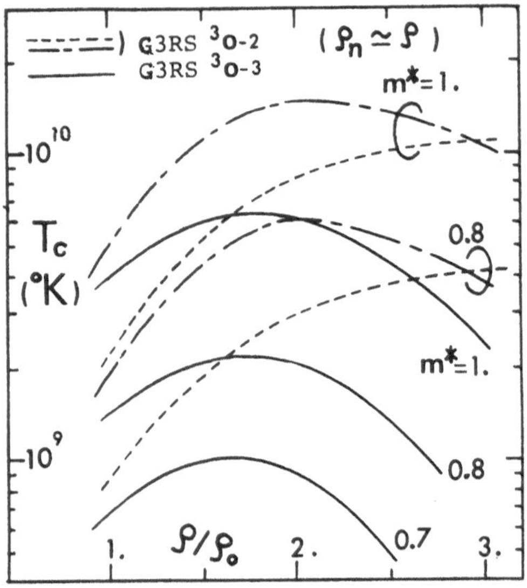

Fig.2. Critical temperature for neutron $^3P_2$-dominant superfluidity under π^0 condensation; ---- : without π^0 condensate.

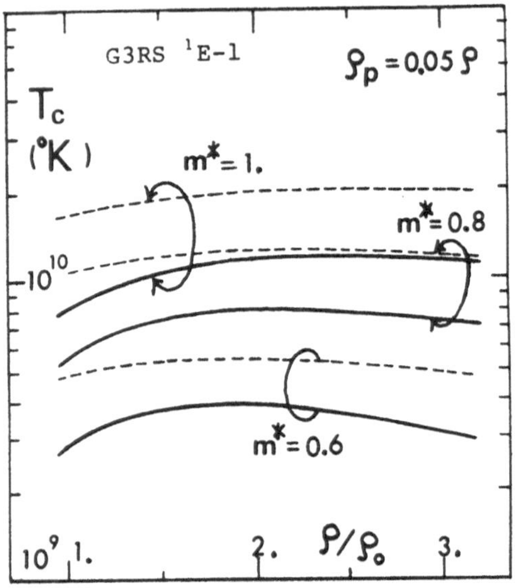

Fig.3. Critical temperature for proton $^1S_0$-dominant superfluidity under π^0 condensation; ---- : without π^0 condensate.

3. Superfluidity under π^c condensation[1c)]

Fore simplicity, we consider the case with π^c condensate realized from pure neutron matter. Due to the reason mentioned in 1, the nucleon normal mode is described by the quasiparticles (η,ζ) composed of neutrons and protons (n,p):

$$\tilde{\eta}_\beta = u_\beta \tilde{n}_\beta + ie(\sigma) v_\beta \tilde{p}_{\beta_-} , \quad \tilde{\zeta}_\beta = u_\beta \tilde{p}_{\beta_-} + ie(\sigma) v_\beta \tilde{n}_\beta \tag{9}$$

where $\tilde{\eta}_\beta$, $\tilde{\zeta}_\beta$, $\tilde{n}_\beta$ and $\tilde{p}_\beta$ stand for the annihilation operator for η, ζ, n and p respectively, with $\beta \equiv (\vec{q},\sigma)$, $\beta_- \equiv (\vec{q}-k\hat{z}, \sigma)$. Also $u_\beta^2 + v_\beta^2 = 1$ with u_β and v_β being real and $e(\sigma)=1(-1)$ for $\sigma=\uparrow(\downarrow)$. By solving selfconsistently the nucleon field equation together with the π^c field one under the charge neutrality condition, and also by the energy minimization, we get all the quantities necessary to represent the system. According to the growth of the condensate, there arises large band gaps between the single particle spectrum E_- for η-particle and the one E_+ for ζ-particle as $E_+ - E_- \simeq 140\ \rho/\rho_0$ MeV when $\rho \gtrsim \rho_0$. Thus it is energetically profitable to take as $|\Phi_N\rangle = \Pi_\beta^{occ} \eta_\beta^+ |0\rangle$ where only the lower band (E_-) is occupied. Because of the existence of this band gap, the excitation of a Cooper pair ($\vec{q}\sigma, -\vec{q}\sigma'$) from η-particle states into ζ-particle ones can be safely neglected. So, we can restrict ourselves to the η-particle (new neutron) space. Pairing intefaction Hamiltonian between η-particles, $H^{\eta\eta}_{pair}$, is to be deduced from that of the original two nucleon interactions, H^{NN}_{pair} :

$$H^{NN}_{pair} \rightarrow H^{\eta\eta}_{pair} = \frac{1}{2}\Sigma_{\beta_1'\beta_2'}\Sigma_{\beta_1\beta_2}\{\langle\beta_1'\beta_2'|V(1,2)|\beta_1\beta_2\rangle_{T=1}(1-u^2v^2)$$
$$\langle\beta_1'\beta_2'|V(1,2)|\beta_1\beta_2\rangle_{T=0}u^2v^2\}\eta^+_{\beta_1'}\eta^+_{\beta_2'}\eta_{\beta_2}\eta_{\beta_1} \qquad (10)$$

where $(\beta_1;\ \beta_2) \equiv (\vec{q}\sigma_1;\ -\vec{q}\sigma_2)$ and T denotes the total isospin of two nucleon state. In the derivation of (10), the relations; $\tilde{n}_\beta \rightarrow u_\beta \tilde{\eta}_\beta$, $\tilde{p}_{\beta_-} \rightarrow -ie(\sigma)v_\beta\tilde{\eta}_\beta$ coming from (9), and the β-independence of $(u_\beta,\ v_\beta)$-factors under the good approximation to neglect nucleon recoil effects, are used. Owing to high density ($\rho \gtrsim \rho_0$) of the η-particles, the most probable candidate for the pairing correlation to cause superfluidity should be the $^3P_2$ (T=1) one. Then the model Hamiltonian necessary for our purpose can be represented as $H_{model} = H_0^{(\eta)} + H^{\eta\eta}_{pair}\ (^3P_2)$, where

$$H^{\eta\eta}_{pair}(^3P_2) = \frac{1}{2}\Lambda\Sigma_{\vec{q}',\vec{q}}\Sigma_{\sigma_1'\sigma_2'}\Sigma_{\sigma_1\sigma_2}\langle\vec{q}'\sigma_1',\ -\vec{q}'\sigma_2'|V(1,2)|\vec{q}\sigma_1,\ -\vec{q}\sigma_2\rangle_{\lambda=\lambda_1}$$
$$\times\ \eta^+_{\vec{q}'\sigma_1'}\ \eta^+_{-\vec{q}'\sigma_2'}\ \eta_{-\vec{q}\sigma_2}\ \eta_{\vec{q}\sigma_1} \qquad (11)$$

with $\Lambda \equiv (1-u^2v^2)$ and $\langle\ \rangle_{\lambda=\lambda_1}$ denoting the matrix elements for $^3P_2$-pairs. $H_0^{(\eta)}$ corresponding to (5) is expressed as $\Sigma_\beta \tilde{\varepsilon}^\eta(q)\eta^+_\beta\eta_\beta$ with $\tilde{\varepsilon}^\eta(q) = \hbar^2(q^2-q_F^2)/2m_N^*$. In this way the problem becomes quite the same as the one previously studied for the case without pion condensation. The only difference, which characterizes the superfluidity under π^c-condensate, is the appearence of the attenuation factor Λ which originates from that η-particles are composed of neutrons and protons. Λ is estimated as $\Lambda = 1-(\tilde{\rho}-1)^2/4\tilde{\rho}^2$ with $\tilde{\rho} \equiv \rho/\rho_0$ and diminishes the $^3P_2$-interaction by about 20 % compared with the case without pion condensate. The results for the maximum m_J coupling ($|m_J|=2$) obtained by using the Mongan's I-potential are shown in Fig.4 where the results for the $^3P_2$-superfluidity realized from the ordinary Fermi gas are also shown in order to see the effects of Λ. Critical temperature is given by $\kappa_B T_c = 0.57\delta/\Gamma_1$ with δ denoting the $^3P_2$-gap and $\ln\Gamma_1 = 1.201$. The π^c-condensed phase begins to

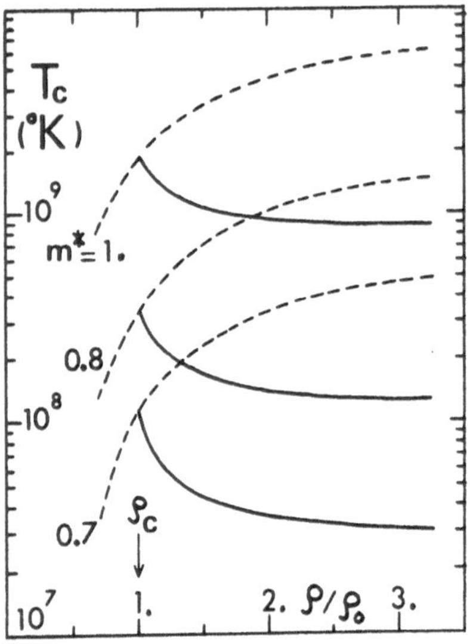

Fig.4. Critical temperature for $^3P_2$-superfluidity under π^c condensation; ---- : without π^c condensate.

appear at the density $\rho_c \simeq \rho_0$ when a simple model calculation is adopted. It is to be noted that the growth of π^c condensate reduces strongly the magnitudes of the energy gaps. T_c become smaller by an order of magnitude than those in the absence of pion condensation and the existence of nucleon superfluid is delicate depending sensitively on m*; it is expected for $m^* \gtrsim 0.8$ but not for $m^* \lesssim 0.7$. The nucleon superfluidity is known to be less realizable in the π^c-condensed phase than in the π^0-condensed one, since in the former phase the pairing correlation encounters the characteristic attenuation effct.

4. Concluding Remarks

Unless the realistic m* in nuclear medium with pion condensates result in smaller values than those for ordinary phase, we can expect that neutrons are in the $^3P_2$-dominant superfluid with $T_c \sim 10^{8-9}$ °K and also protons are in the $^1S_0$-dominant superfluid $T_c \sim 10^{9-10}$ °K, even under the situation that the π^0 condensation is at work in the interiors of neutron stars. While under π^c condensate the existence of superfluidity depends critically on m*: It is likely for $m^* \gtrsim 0.8$ ($T_c \gtrsim 10^8$ °K) but unlikely for $m^* \lesssim 0.7$ ($T_c < 10^8$°K). In this sense, the determination of m* in pion condensed phase becomes very important. Recently we have found the important fact that due to the 2D character of Fermi gas in the π^0-condensed phase, the m* becomes remarkably larger ($m^* \gtrsim 0.9$ at $\rho \simeq (1-3)\rho_0$) than that for the ordinary phase or the π^c-condensed one with the 3D character.[13] This comes mainly from that in the former phase, the momentum ($q_\perp$) dependence of the kinetic energy near the Fermi surface is more steep owing to $q_{\perp F} > q_F$, while that of the single particle potential becomes by far moderate, than the respective parts in the latter phase. Therefore we remark that the tendency mentioned above becomes more prominent.

So far we have focused our attention to the representative two types of pion condensate, to draw charactristics of the superfluidities under pion condensates. It has been recognized, however, that the simultaneous condensations of π^0 and π^c are most plausible with allowing π^c condensate in $\vec{r}_\perp$-space of the ALS model (π^0 condensate).[14] In the existence of superfluids possible under such situation? Because the π^c condensate turns out to occur in the space of 2D Fermi gas and hence the relevant m* is large, we may well expect that the existence is allowable, without changing T_c drastically from the values previously obtained regardless of pion condensation and with extending the existent region up to higher densities. To draw final conclusions, more careful investigations have to be done, by getting realistically m* and ALS parameters according to density variation and also by including the possible modifications of pairing interaction due to pion condensate.

The author would like to express his sincere thanks to Professor R. Tamagaki for his enlightening discussions and collaborations. He is indebted to Dr. T. Tatsumi, Mr. T. Kunihiro and Mr. H. Furukawa for their cooperative discussions.

References

1) a) T. Takatsuka and R. Tamagaki, Prog. Theor. Phys. 62 (1979) 1655.
b) R. Tamagaki, T. Takatsuka and H. Furukawa, Prog. Theor. Phys. 64 (1980) 2107.
c) T. Takatsuka and R. Tamagaki, Prog. Theor. Phys. 64 (1980) 2270.
d) T. Takatsuka and R. Tamagaki, submitted to Prog. Theor. Phys..
2) a) R. Tamagaki, Prog. Theor. Phys. 44 (1970) 905.
b) T. Takatsuka and R. Tamagaki, Prog. Theor. Phys. 46 (1971) 114.
c) T. Takatsuka, Prog. Theor. Phys. 48 (1972) 1517; 50 (1973) 1754, 1755.
3) M. Hoffberg, A.E. Glassgold, R.W. Richardson and M. Ruderman, Phys. Rev. Letters 24 (1970) 775.
4) N.C. Chao, J.W. Clark and C.H. Yang. Nucl. Phys. A179(1972), 320.
5) As review articles; A.B. Migdal, Rev. Mod. Phys. 50 (1978) 107; G.E. Brown and W. Weise, Phys. Reports. C27 (1976) 1; G. Baym and D.K. Campbell, "Mesons in Nuclei" Vol. 3 (1979), chap.3.
6) S. Tsuruta, Phys. Reports 56 (1979) 237.
7) D. Pines, Proceedings of the 16 th Solvay Congress on Physics (1974 Editions de l' Université de Bruxells), P. 174.
8) T. Takatsuka, K. Tamiya, T. Tatsumi and R. Tamagaki, Prog. Theor. Phys. 59 (1979) 1933. As review article, R. Tamagaki, Nucl. Phys. A 328 (1979) 352.
9) T. Matsui, K. Sakai and M. Yasuno, Prog. Theor. Phys. 60 (1978), 442; 61 (1979), 1093.
10) T. Tatsumi, Prog. Theor. Phys. 63 (1980), 1252.
11) R.F. Sawyer and D.J. Scalapino, Phys. Rev. D7 (1972) 953.
12) J.A. Sauls and J.W. Serene, Phys. Rev. D17 (1978), 1524.
13) T. Tamiya and R. Tamagaki (in preparation); T. Takatsuka, Y. Saito and J. Hiura (in preparation).
14) T. Tamiya and R. Tamagaki, Prog. Theor. Phys. 60 (1978) 1753; T. Tatsumi (in preparation)

SPIN-ISOSPIN ORDER: CRITICAL DENSITY IN NUCLEAR MATTER AND A POSSIBLE REALIZATION IN NUCLEI

F. Palumbo
INFN, Laboratori Nazionali di Frascati, Frascati, Italy.

1. INTRODUCTION

At the beginning of the past decade different authors[1,2] predicted an ordered phase of nuclear matter at sufficiently high density, originally called Nuclear binding by the OPEP or Pion condensate. I will refer to it here as Spin-isospin ordered phase (SIOP).

It is now widely accepted that SIOP can occur only at densities higher than normal density, which excludes SIOP in nuclei. Current research is therefore mostly devoted to predict and detect precursor phenomena of SIOP in nuclei[3], or to study the possibility of realizing SIOP in heavy-ions collisions.

The above conclusion is grounded on calculations on nuclear matter and qualitative arguments about nuclei. I want to discuss the soundness of the assumptions at the basis of these calculations and arguments explaining why in my view they should not be taken as conclusive. I will then present some recent developments related to nuclei.

2. NUCLEAR BINDING BY THE OPEP

The OPEP is strong and long range, but it does not contribute to the binding energy in Hartree approximation unless the nuclear system is in a SIOP. This is due to its spin-isospin dependence

$$V_{OPE} = \frac{1}{3} f^2 \mu \vec{\tau}_1 . \vec{\tau}_2 \left\{ -\frac{4\pi}{3} \partial(\vec{r}) \vec{\sigma}_1 . \vec{\sigma}_2 + \left[\vec{\sigma}_1 . \vec{\sigma}_2 + \left(1 + \frac{3}{\mu r} + \frac{3}{\mu^2 r^2}\right) S_T \right] \frac{e^{-\mu r}}{\mu r} \right\}. \tag{1}$$

In fact assuming the spins quantized along the z-axis

$$\langle V_{OPE} \rangle_{direct} = \frac{1}{2} \sum_{\tau_3(1)\sigma_3(1)} \sum_{\tau_3(2)\sigma_3(2)} \int d\vec{r}_1 \int d\vec{r}_2 \langle \tau_3(1)\sigma_3(1)\tau_3(2)\sigma_3(2) . \Big| V_{OPE} \Big| \tau_3(1)\,\sigma_3(1)\,\tau_3(2)\,\sigma_3(2) \rangle \, \varrho_{\tau_3(1)\,\sigma_3(1)}(\vec{r}_1) \, \varrho_{\tau_3(2)\,\sigma_3(2)}(\vec{r}_2) , \tag{2}$$

where $\varrho_{\tau_3 \sigma_3}$ is the one-body density matrix of nucleons of isospin τ_3 and spin σ_3. The spin-isospin matrix element is

$$\langle \tau_3(1)\sigma_3(1)\tau_3(2)\sigma_3(2) \Big| V_{OPE} \Big| \tau_3(1)\sigma_3(1)\tau_3(2)\sigma_3(2) \rangle = \frac{1}{3} f^2 \mu \left\{ -\frac{4\pi}{3} \partial(\vec{r}) + \left[1 + \left(1 + \frac{3}{\mu r} + \frac{3}{\mu^2 r^2}\right)\left(3 \frac{z^2}{r^2} - 1\right) \right] \frac{e^{-\mu r}}{\mu r} \right\} \tau_3(1)\sigma_3(1)\tau_3(2)\sigma_3(2) . \tag{3}$$

It is convenient to introduce the spin-isospin density operator

$$S_{ik} = \bar{\psi}\tau_i\,\sigma_k\psi. \qquad (4)$$

Eq. (2) can be rewritten in terms of the average value of S_{33}

$$\langle S_{33}\rangle = \sum_{\tau_3\sigma_3} \varrho_{\tau_3\sigma_3}\,\tau_3\sigma_3\,, \qquad (5)$$

$$\langle V_{OPE}\rangle_{direct} = \frac{1}{2}\int d\vec{r}_1 \int d\vec{r}_2\, \langle S_{33}(\vec{r}_1)\rangle \langle S_{33}(\vec{r}_2)\rangle\, \frac{1}{3} f^2\mu \left\{ -\frac{4\pi}{3}\,\delta(\vec{r}) + \left[1 + \left(1 + \frac{3}{\mu r} + \frac{3}{\mu^2 r^2}\right)\left(3\frac{z^2}{r^2} - 1\right)\right] \frac{e^{-\mu r}}{\mu r} \right\}. \qquad (6)$$

Eqs. (5) and (6) show that if $\varrho_{\tau_3\sigma_3}$ is independent of $\tau_3\sigma_3$, $\langle V_{OPE}\rangle_{direct}=0$. It is therefore natural to think that a SIOP should be favored at sufficiently high density. At high density in fact, the exchange potential energy per particle grows like the density ϱ, the kinetic energy per particle like $\varrho^{5/3}$ and the direct potential energy per particle like ϱ^2. This last term is therefore dominating, and if the OPEP were the whole N-N interaction, a SIOP would necessarily be established which would eventually lead to nuclear collapse.

Spin-isospins of the nucleons can be ordered only if the nucleons are to some extent localized. The factor $(\frac{3z^2}{r^2} - 1)$ in eq. (6), requiring maximum asymmetry between the z-axis and the x-y plane, favors localization along one direction only, the direction of spin quantization. It turns out, as the result of a variational calculation employing the OPEP (with a monopole regularization) as the only N-N interaction, that the localization should be complete, qualifying the phase transition as a first order one. This result was already contained in the original paper[1], but it has been fully exploited by Tamagaki and coworkers[4]. Nuclear matter in SIOP is crystal-like along the z-direction as shown in Fig. 1.

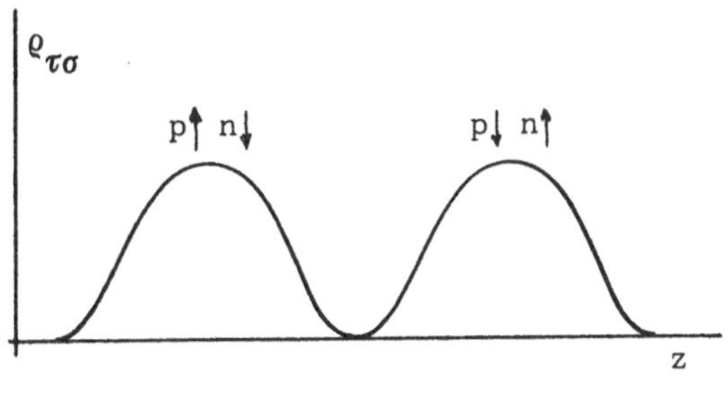

FIG. 1 - Spin-isospin density with the OPEP.

The critical density is determined by comparing the ground state energy in the normal state with the ground state energy in the ordered state, with the result $\varrho_c \sim 0.5\ \varrho_o$.

Of course use of the OPEP as the whole N-N potential is unrealistic, and one has to investigate the effect of the other components of the interaction. In view of the crystal-like character of the ordered phase particularly important is expected to be the role of the core of the N-N potential. It can be important in two ways. It can change the structure of the ordered phase and it can change the critical density.

The first possibility has been investigated(5) using the OBEP of the Ueda and Green school, and exploring different configurations. The structure which is energetically favored by these potentials is represented in Fig. 2.

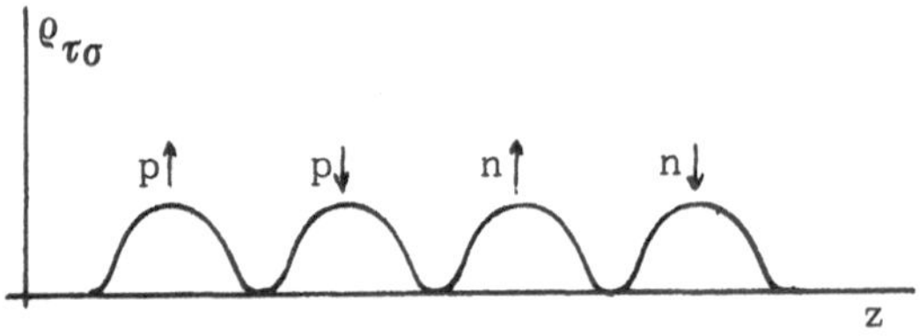

FIG. 2 - Spin-isospin density with the OBEP.

This result has been obtained by neglecting the short-range correlations in the w.f.. And this leads us to the second point: How will these correlations change the critical density?

If one assumes that the short-range correlations are the same in the normal as in the ordered phase, their effect is to increase the critical density above ϱ_0. This assumption, however, is far from justified. It is worse than to use the same short-range correlations for a fluid and crystal, because here short-range correlations must be spin-isospin dependent and anysotropic.

Use of this so far unjustified assumption can be avoided if one determines the critical density as the point of instability of the normal state. This is what has been done in the study of Pion condensation. It does not require the evaluation of any quantity related to the ordered state, but rests an another assumption to be discussed later.

3. PION CONDENSATION

The pion propagator in the nuclear medium is

$$D(\vec{k},\omega,\varrho) = \left[\omega^2 - k^2 - \mu^2 - \Pi(\vec{k},\omega,\mu)\right]^{-1} \tag{7}$$

where $\vec{k}$ is the momentum, ω the energy, μ the mass and Π the selfenergy which is a function of the nuclear density ϱ. The normal state becomes unstable when the pion propagator has a pole at energy $\omega = 0$. In this case it costs no energy to produce pions.

The equivalence between pion condensation and the phase described in the previous section is established by the identity of the ground state. If there is pion condensation, the average value of the pion field is different from zero

$$(\Delta + \mu^2)\langle\varphi_i\rangle = \frac{f}{\mu}\,\partial_k \langle S_{ik}\rangle \tag{8}$$

and therefore $< S_{ik} > \neq 0$. Condensation of charged pions is related to superconductivity in layers[4] and will not be discussed here.

In order to establish the equivalence completely we observe that both in pion condensation and in nuclear binding by the OPEP parity and isospin are broken. In the case of pion condensation this is obvious because the pion field is pseudoscalar and isovector. In the other case parity is broken as a consequence of breaking of translational invariance, while breaking of isospin follows from the fact that the operators S_{ik} are isovectors. In fact applying the Wigner-Eckart theorem we have

$$< T\, T_z \,|\, S_{3k} \,|\, T\, T_z > \propto \frac{T_z}{T(T+1)} \quad . \tag{9}$$

For symmetric nuclear matter $T_z = 0$, and $< S_{3k} >$ cannot be different from zero for a state of definite T.

This shows that isospin breaking must be a characteristic feature of (static, see below) SIOP also in nuclei, since the above argument does not depend on the system being finite or infinite. Parity breaking, on the contrary, is a consequence of $< \varphi_i > \neq 0$ only for an infinite system, where parity is defined w.r. to arbitrary points, but not for nuclei, where parity is defined only w.r. to the c.m..

Established the equivalence between Pion condensation and Nuclear binding by the OPEP, let us turn to the determination of the critical density. If the pion self-energy is evaluated in RPA neglecting short-range correlations between nucleons, the critical density is found to be lower than ϱ_0. In the present case, however, introduction of short range correlations is much easier, as already noted, because we need to deal only with the normal state. This has been done using the Landau parameter, and this approximation has been recently checked to be very good[6]. The effect of the short range correlations is to increase the critical density up to $\varrho_c \simeq 2\varrho_0$.

This procedure is correct, however, only if the phase transition is of second order. This is the assumption I was talking about.

In order to appreciate the difference between first order and second order phase transitions in the present context, let us refer to a well-known case, the vapor-liquid phase transition, whose phase diagram is reported in Fig. 3.

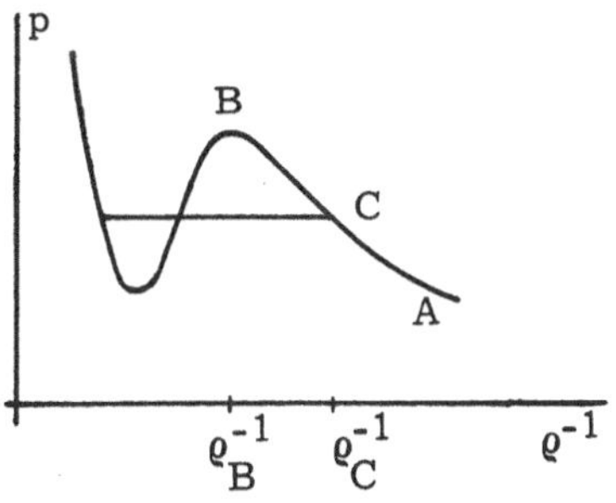

FIG. 3 - The vapor-liquid phase diagram.

This a first order phase transition, occuring at density ϱ_c. The critical density is determined by the crossing of two curves, giving the free energy of the vapor and liquid phase, respectively. Each curve is analytic, and the phase transition is due to the fact that the physical system is described by one analytic function before the transition and by a different one afterwards.

Actually the vapor can be made to follow the curve AB beyond the critical point C by means of an adiabatic compression. This phase of supersaturated vapor can be also described theoretically by means of the RPA. The point B will appear in this case as a second order phase transition occuring at $\varrho_B > \varrho_C$. At this point there is in fact a singularity in the second derivative of the Gibbs potential with respect to the pressure telling that the compressibility is infinite

$$k_T = -\frac{1}{V}\left.\frac{\partial V}{\partial P}\right|_T = -\frac{1}{V}\frac{\partial^2 G}{\partial p^2} \; . \tag{10}$$

A similar situation could occur in our case. A calculation of the type just described looking at the pole of the pion propagator can tell us very little concerning the possibility of a first order phase transition at lower density. This in fact is related to large quantum fluctuations around the mean field approximation of RPA. Dyugaev has studied this problem[7] with the conclusion that the phase transition is actually of first order, but at a critical density very close to the critical density it would have as a second order one. It is the estimate of the difference between these critical densities that seems to me very uncertain, due to the difficulty of properly taking into account the core effects, if they are large.

Note that the possibility I am considering is relevant also to experiment. Suppose in fact that the critical density is higher than the experimental density, that the phase transition is of first order and that one tries to reach it in heavy-ion collisions. If the experimental conditions correspond to an "adiabatic compression" the critical density for the first order phase transition is overcomed without effect until the second order phase transition is realized at higher density.

Therefore looking for the instability of the normal state is not an alternative procedure w.r. to the comparison of the energies of the normal and the ordered state, but rather a complementary one. Both should be used to determine the characther of the phase transition and the critical density.

A last remark is in order about precursor phenomena. If $\varrho_c > \varrho_o$, and the phase transition is of first order, such phenomena do not exist. We will see, however, that something very similar to them can exist in nuclei, due to their finite size.

There are many important points left which I do not have time to discuss, including the effects of isobars and the problem of the convergence of the sum of the bubble diagrams (bubbles into bubbles)[6].

4. SIOP IN NUCLEI

I mentioned at the beginning arguments against the existence of SIOP in nuclei[8]. Such arguments are i) that almost degenerate parity doublets should exist on account of parity breaking in nuclear matter and ii) that the levels with the quantum numbers of the pion should be lowered.

I will discuss these points after presenting a possible mechanism for SIOP in nuclei. According to my previous analysis I do not consider yet settled the value of the critical density, and therefore I will have in mind both the case that $\varrho_c > \varrho_o$ and $\varrho_c < \varrho_o$.

Let us separate the nucleus into two parts with spin-isospin order, for instance one part containing spin-up protons and spin-down neutrons, the other spin-down neutrons and spin-up protons. Let us denote by d the distance between their c.m. and by V(d) the separation energy (Fig. 4).

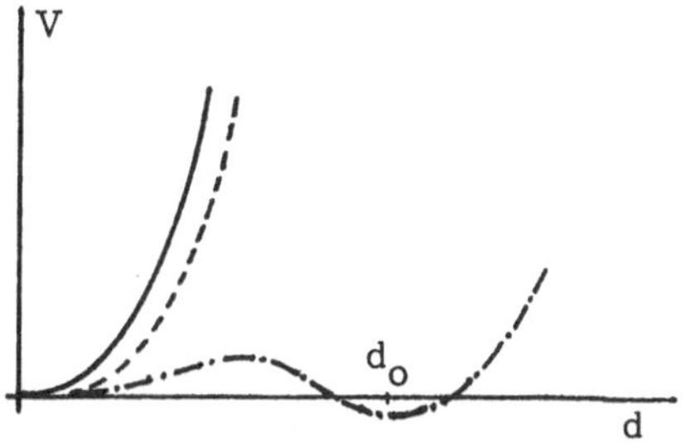

FIG. 4 - The potential separation energy: The solid line is for the disordered separation in the zero-point potion, the dashed line for the spin-isospin ordered separation in the zero-point motion, the dot-dashed line for the static separation.

This must be compared with the separation energy $V_o(d)$ of the nucleus into two parts each of which has no spin-isospin order. Such separation actually takes place in the zero-point motion, and what we investigate is whether the spin-isospin ordered separation is favored or not w.r. to the disordered separation.

If $V(d) > V_o(d)$ the disordered separation is preferred. If $V_o(d) > V(d) > 0$ the zero-point motion will take place between two spin-isospin ordered phases. We talk in this case of nonstatic order. The average value of the pion field is proportional to the average value of d, and therefore vanishes. The order parameter is $\langle \varphi^2 \rangle \propto \langle d^2 \rangle$.

The extreme possibility is that V(d) becomes negative with a minimum at $|d| = d_o$. In this case we have a static order with order parameter d_o and an average pion field $\langle \varphi \rangle \propto d_o$.

Only the case of nonstatic order has been investigated[9]. Due to the known difficulties with short-range correlations, only the OPEP has been taken into account. As a result it is not possible to predict reliably wheter the nonstatic order is actually realized. It is however possible to predict a number of characteristic features the nucleus should have if it were in the ordered phase.

It turns out that oscillations must be one-dimensional, and along the direction of spin quantization (oscillations in the perpendicular plane can also occur but not associated with spin-isospin). This direction must coincide with the symmetry axis for an oblate nucleus, and must be perpendicular to it for a prolate nucleus. In any case this kind of correlation is only possible for nuclei with $A < 60$.

The signature of this mode is the lowering of the excitation energy and the enhancement of the B(M2). Typical values are reported in the Table 1.

A	∂	$\hbar\omega$ (MeV)	B(M2; K=0 → K=0) W.u.	B(M2; K=0 → |K|=2) W.u.
12	-0.4	12	20	0
	0.4		5	16
28	-0.4	11	28	0
	0.4		8	23

Table 1 Excitation energy in MeV and B(M2) in W.u. for two nuclei at different values of the deformation parameter ∂. For $\partial < 0$ the nucleus is oblate and the oscillation takes place along the symmetry axis, while for $\partial > 0$ the nucleus is prolate and the oscillation takes place along a direction perpendicular to it.

The zero-point correlation just described can coexist with other spin-isospin correlations, for instance a breathing mode of spin-up protons and spin-down neutrons against spin-down neutrons and spin-up protons. This would renormalize M2 transitions and would presumably enhance M1 transitions.

It so appears that nonstatic spin-isospin order enhances the e.m. transition amplitudes and lowers the energy of unnatural parity levels, which is considered a precursor to Pion condensation. This point needs further investigation. We see, however, that no parity doublets are to be expected. (I already observed, moreover, that parity breaking is not to be expected even in the presence of static SIOP). In addition, lowering of the levels of unnatural parity is a signature of nonstatic SIOP only, while nothing has been proved, as far as I know, concerning static SIOP. I should also emphasize that all the mentioned effects are to be expected only in deformed nuclei. I cannot see how observations concenring spherical nuclei (the famous level[(8)] of ^{16}O at 12.78 MeV) can be relevant to SIOP in any of its possible realizations.

I will conclude this discussion of nonstatic order by mentioning that it is energetically more favored the higher the density. Also if it is not actually realized in nuclei, it could therefore be excited under compression.

The extreme possibility of static order according to the mechanism outlined has not yet been studied in detail. An entirely different possibility has been considered by G. Do Dang[(10)], who has studied a nucleus made only of spin-up protons and spin-down neutrons. The necessary density has been estimated[(6)] to be twice the experimental density.

I will conclude my talk by quoting an experiment which in my view can set an upper bound on the amplitude of spin-isospin density fluctuations in nuclei in an almost model independent way. If so the experiment, though very difficult, does not suffer from the ambiguities of many other tests proposed, which depend on the details of s.p. w.f., exchange currents, and so on.

The idea[11] is that nuclei in static SIOP should give rise to a coherent scattering of neutrinos at values of the momentum transfer where the coherent scattering by normal nuclei is negleageable. This is due to the axial current. This current in the Weimberg-Salam model is

$$J_\mu = \frac{1}{2} \overline{\psi} \gamma_\mu \gamma_5 \tau_3 \psi , \qquad (11)$$

and its spatial components in the nonrelativistic approximation become just the order parameter

$$J_k = -i S_{3k} . \qquad (12)$$

This effect could also have interesting astrophysical consequences.

REFERENCES

(1) F.Calogero, in "The Nuclear Many-Body Problem", Ed. by F. Calogero and C. Ciofi degli Atti (Roma, 1972), Vol. 2, p. 535; F. Calogero and F. Palumbo, Lett. Nuovo Cimento 6, 663 (1973).
(2) A.B. Migdal, ZEFT 63, 1993 (1972); Soviet Phys. JETP 36, 1052 (1973); R.F. Sawyer, Phys. Rev. Letters 29, 382 (1972); D.J. Scalapino, Phys. Rev. Letters 29, 386 (1972).
(3) S.A. Fayans, E.E Sapershtein and S.V. Tolokonnikov, J. Phys. G3, L51 (1977); M. Gyulassi and W. Greiner, Ann. Phys. (N.Y.) 109, 485 (1977); M. Ericson and J. Delorme, Phys. Letters 76B, 192 (1978).
(4) R. Tamagaki, Nuclear Phys. A328, 352 (1979).
(5) F. Calogero, F. Palumbo and O. Ragnisco, Nuovo Cimento 29A, 509 (1975).
(6) W.H. Dickhoff, A. Faessler, J. Meyer-ter-Vehn, H. Müther, to be published.
(7) A.M. Dyugaev, IEPT Lett. 22, 83 (1975).
(8) S. Barshay and G.E. Brown, Phys. Letters 47B, 107 (1973).
(9) N. Lo Iudice and F. Palumbo, Phys. Rev. Letters, submitted to.
(10) G. Do Dang, Phys. Rev. Letters 43, 1708 (1979).
(11) F. Palumbo, Frascati Preprint 80/47(P), to be published.

GROUP THEORY AND COLLECTIVE DEGREES OF FREEDOM IN NUCLEAR MANY BODY SYSTEMS

M. Moshinsky
Instituto de Física, UNAM
Apdo. Postal 20-364, México 20, D.F.

The author wishes to thank the organizers of the II International Conference on "Recent Progress in Many Body Theories" and, in particular, Dr. Manuel de Llano, for the opportunity of presenting some of the research work done recently in the field indicated in the title. As much of the material has already been published or is in press, we would like here only to summarize the main ideas and give a list of references where they are presented in detail.

It is well known that the pioneering work of Bohr and Mottelson (BM)[1] introduced collective degrees of freedom, associated with the quadrupole vibrations of the liquid drop, in the many body nuclear system, described at that time mainly by the nuclear shell model[2]. These collective degrees of freedom proved to be extremely useful in systematizing our knowledge of nuclei, and for a quarter of century they represented, together with the shell model, the standard procedure for attacking any nuclear structure problem. The importance of the BM approach led the author and his collaborators[3,4] to analyze the group theory underlying it and, with its help, they were able to find the explicit analytic expression for the eigenstates of the five dimensional BM oscillator Hamiltonian characterized by the irreducible representation (irreps) of the $U(5)\supset O(5)\supset O(3)$ chain groups.

About five years ago an independent approach to collective degrees of freedom in nuclei was developed by Arima and Iachello[5] through the Interacting Boson Approximation (IBA). In this procedure pairs of protons or neutrons outside closed shells were assumed to act as bosons with Hamiltonian involving one and two body interactions. As the bosons were of the s and d type, one had six states and a unitary group U(6) was present in the picture. Different chains of subgroups of this main group corresponded to the description of nuclei in the vibrational, rotational and triaxial limit.

Again the author and his collaborators became interested[6] in the deeper group theory underlying this model and were in fact able to obtain analytically the matrix elements of the Casimir operators of the subgroups U(3) and O(6) of U(6) in the basis characterized by the irreps of the chain of groups $U(6) \supset U(5) \supset O(5) \supset O(3)$.

In view of the fact that the BM model[1] and its extensions[7], and the IBA gave essentially equivalent predictions for nuclear structure problems, the question arose whether in some way they were not equivalent. The author proved that any IBA Hamiltonian involving one and two body interactions could be written in the BM language as a linear combination of 1, β^2, β^4 and $\beta^3 \cos 3\gamma$ with coefficients that depend on the BM oscillator Hamiltonian $\hat{n}$ and the square of the angular momentum L^2. Thus in a basis in which $\hat{n}$, L^2 are diagonal[3,4] the matrix elements of IBA Hamiltonians can be calculated straightforwardly by the procedures used in the BM model.

Other approaches for relating the IBA and BM model, which appeared, at first sight, to involve a very different viewpoint, were developed recently by Ginocchio and Kirson[9] and by Klein and Vallieres[10]. As a result of discussions at the present Conference with Klein, and in a preceeding one on Nuclear Physics with Dieperink, Ginocchio and Talmi, the present author and his collaborators[11] were able to show, under certain assumptions, that all of these approaches were equivalent.

All the previous discussion concerned what could be called macroscopic theories of collective motions as ad-hoc assumptions were made in relation with the collective degrees of freedom though, at least in the case IBA, an effort was made to derive them from more microscopic considerations[12].

To the author, a more basic microscopic approach was proposed from different viewpoints by Filippov[13] and his group and by Vanagas[14]. In both cases they start from a transformation of coordinates in the many body system, introduced by Dzublik et al.[15] and by Zickendraht[16], which automatically separates the three Euler angles θ_i, and three deformation parameters ρ_i, $i = 1,2,3$, from the rest. Given then any two body Hamiltonian one can project out its collective part as function of ρ_i, θ_i and their derivatives, either from the scalar representation[14] of the O(A-1) group associated with A-1 Jacobi vectors associated with the nucleons, or by projecting on the lowest representation of O(A-1) consistent with the Pauli principle[13].

The question arises whether these microscopic approaches can be correlated with the IBA and thus, from the discussion of the previous paragraphs, also with the BM model. The author with Chacón and Vanagas considered this question and showed that it is possible to find a representation in quantum mechanics of a canonical transformation that relates the microscopic approach associated with the scalar representation[14] of O(A-1) with the IBA model. Considering then a microscopic description in two dimensional space and also an IBA model in the same number of dimensions, that was called σ-δ (instead of s-d) IBA model, the author and Seligman[18)] obtained explicitly the canonical transformation when projecting the collective part from the three body problem i.e. A=3. Later the author and Chacón[19)] showed that the results continue to hold when one projects from the general A-body problem.

The problem stands at the stage outlined in the previous paragraphs. Many interesting possibilities are opening up in different directions, but we prefer to outline them in future publications as fully realized ideas, rather than state them now as interesting surmises.

REFERENCES

1. A. Bohr, Mat. Fys. Medd. Dan. Vid. Selsk. 26, 14 (1952) Rotational States in Nuclei, Thesis, Copenhagen, 1954; A. Bohr and B. Mottelson, Mat. Fys. Medd. Dan. Vid. Selsk. 27, 16 (1953).
2. M.G. Mayer and J.H.D. Jensen, "Elementary Theory of Nuclear Shell Structure", (John Wiley, New York, 1955).
3. E. Chacón, M. Moshinsky and R.T. Sharp, J. Math. Phys. 17, 668 (1976).
4. E. Chacón and M. Moshinsky, J. Math. Phys. 18, 870 (1977).
5. A. Arima and F. Iachello, Ann. Phys. (N.Y.) 99, 253 (1976); 111, 201 (1978); 123, 468 (1979).
6. O. Castaños, E. Chacón, A. Frank and M. Moshinsky, J. Math. Phys. 20, 35 (1979).
7. G. Gneuss, M. Seiwert, J. Maruhn and W. Greiner, Z. Physik 296, 147 (1980).
8. M. Moshinsky, Nucl. Phys. A338, 156 (1980).
9. J.N. Ginocchio and W. Kirson, Phys. Rev. Lett. 44, 1744 (1980), Nucl. Phys.
10. A. Klein and M. Vallieres, (Private Communication).
11. O. Castaños, A. Frank, P.O. Hess and M. Moshinsky, Phys. Rev. Lett. (Submitted for publication).
12. A. Arima, T. Ohtsuka, F. Iachello and I. Talmi, Phys. Lett. 66B, 205 (1977).
13. G.F. Filippov, Fiz. Elem. Castits, At. Yadra 4, 992 (1973), Sov. J. Part. Nuc. 4, 405 (1974).
14. V. Vanagas, "The Microscopic Nuclear Theory", Lecture Notes, Dept. of Physics, University of Toronto, 1977.
15. A.Ya. Dzublik, V.I. Ovcharenko, A.I. Steshenko and G.F. Filippov, Yad. Fiz. 15, 869 (1972); Sov. J. Nucl. Phys. 15, 487 (1972).
16. W. Zickendraht, J. Math. Phys. 12, 1663 (1971).
17. E. Chacón, M. Moshinsky and V. Vanagas, J. Math. Phys., February or March 1981.
18. M. Moshinsky and T.H. Seligman, J. Math. Phys. (Submitted for publication).
19. E. Chacón and M. Moshinsky, KINAM, (México) (Submitted for publication).

List of Papers not contained in the Proceedings

Perspectives in Many-Body Physics, K. A. Brueckner

Diagrammatic Alternative to HNC for Bosons, A. Lande

Spin-Polarized Quantum Systems, L. H. Nosanow

Aspects of Solid-like Structure in Pion Condensation, M. da C. E. Ruivo

Nuclear-Matter Approach to Two-Body Friction in Heavy-Ion Collisions, R. Sartor

Phase Transition to Quark Matter from Hot Nuclear Matter, P. J. Siemens

List of Participants

V.C. Aguilera-Navarro, Instituto de Fisica teorica, Rua Pamplona 145, CEP o14o5, Sao Paolo, Brasil

B.J. Alder, Theoretical Physics Division, Lawrence Livermore Laboratory, University of California, P.O.Box 8o8, Livermore, CA 9455o, U.S.A. *

M. Alexanian, Physics Department, Montana State University, Bozeman, Montana 59717, U.S.A.

L. Andrade, Facultad de Ciencias, Universidad Nacional Autonoma de Mexico, Ciudad Universitaria, Mexico 2o, D.F., Mexico

A. Bagchi, Xerox Corporation, Bldg. 147, 8oo Phillips Road, Webster, N.Y., U.S.A.

G.A. Baker, Los Alamos Scientific Lab, Mail Stop 457, Los Alamos, New Mexico 87545, U.S.A.

R. Baquero, Instituto de Fisica, U.A. de Puebla, Mexico

R. Barrera, Instituto de Fisica, Universidad Nacional Autonoma de Mexico, Apartado Postal 2o-364, Mexico 2o, D.F., Mexico

B.R. Barrett, Department of Physics, The University of Arizona, Tucson, Arizona 85721, U.S.A. *

M. Bauer, Instituto de Fisica, Universidad Nacional Autonoma de Mexico, Apartado Postal 2o-364, Mexico 2o, D.F., Mexico *

R.L. Becker, Physics Division Bldg 6oo3 X-1o, Oak Ridge National Laboratory, Oak Ridge, Tennessee 3783o, U.S.A.

M. Berrondo, Instituto de Fisica, Universidad Nacional Autonoma de Mexico, Apartado Postal 2o-364, Mexico 2o, D.F., Mexico

R.F. Bishop, UMIST, Department of Mathematics, P.O.Box 88, Manchester M6o 1QD, England

J.P. Blaizot, CEN-Saclay, Service de Physique Theorique, B.P. No 2, 9119o Gif-Sur-Yvette, France

B.H. Brandow, Group T-11, Los Alamos Scientific Lab, P.O.Box 1663, Los Alamos, New Mexico 87544, U.S.A.

K.A. Brueckner, Department of Physics Bo19, University of California, La Jolla, California 92o37, U.S.A.

J.R. Buchler, Department of Physics and Astronomy, University of Florida, Gainesville, Florida 32611, U.S.A.

C. Campbell, School of Physics and Astronomy, University of Minnesota, 116 Church St. S.E., Minneapolis, Minn. 55455, U.S.A.

J.P. Carbotte, Physics Department, McMaster University, Hamilton, Ontario, Canada

D. Ceperley, NRCC Bldg 5oD, Lawrence Berkeley Lab, University of California, Berkeley, California 9472o, U.S.A.

J.W. Clark, Department of Physics, Washington University, St. Louis, MO 6313o, U.S.A.

F. Coester, Physics Division Bldg 2o3, Argonne National Lab, 97oo South Cass Ave., Argonne, Illinois 6o439, U.S.A.

S.A. Coon, Physics Department, College of Liberal Arts, The University of Arizona, Tucson, Arizona 85721, U.S.A.

J. Dabrowski, Institute of Nuclear Research, Nuclear Theory Department, Hoza 69, Pl-oo-681, Warsaw, Poland

J.P. Daudey, Laboratoire de Physique Quantique, Université Paul Sabatier, 118 Route de Narbonne, 31o77 Toulouse, France

B. Day, Argonne National Lab, Bldg 2o3, Argonne, Illinois 6o439, U.S.A.

M. de Llano, Instituto de Fisica, Universidad Nacional Autonoma de Mexico, Apartado Postal 2o-364, Mexico 2o, D.F., Mexico

A.E.L. Dieperink, IKO, Oosterringdijk 18, Amsterdam, Netherlands

K. Emrich, Institut für Theoretische Physik, Ruhr-Universität Bochum, 463o Bochum, West-Germany

R.D. Etters, Department of Physics, Colorado State University, Fort Collins, Colorado 8o523, U.S.A.

S. Fantoni, Department of Physics, University of Illinois, Urbana, Illinois 618o1, U.S.A.

A.L. Fetter, Department of Physics, Stanford University, Stanford, California 943o5, U.S.A.

J. Flores, Instituto de Fisica, Universidad Nacional Autonoma de Mexico, Apartado Postal 2o-364, Mexico 2o, D.F., Mexico

M. Fortes, Instituto de Fisica, Universidad Nacional Autonoma de Mexico, Apartado Postal 2o-364, Mexico 2o, D.F., Mexico *

B. Friman, Department of Physics, University of Illinois, Urbana, Illinois 618o1, U.S.A.

J.L. Gammel, Department of Physics, Saint Louis University, 221 North Grand Blvd, St. Louis, MO 631o3, U.S.A.

A. Gersten, Department of Physics, Ben Gurion University, P.O.Box 653, Beer-Sheva 8412o, Israel

M.D. Girardeau, Department of Physics, University of Oregon, Eugene OR. 974o3, U.S.A.

P. Goldhammer, Department of Physics and Astronomy, University of Kansas, Lawrence, Kansas 66o45, U.S.A.

E.P. Gross, Department of Physics, Brandeis University, Waltham, Mass. o2154, U.S.A. *

R. Guardiola, Departamento de Fisica Nuclear, Facultad de Ciencias, Universidad de Granada, Spain

P. Haensel, Copernicus Astronomical Center, Polish Academy of Sciences, Ul. Bartycka 18, oo-716 Warsaw, Poland

S. Hernandez, Depto. de Fisica, Facultad de Ciencias Exactas, Universidad de Buenos Aires, 1428 Buenos Aires, Argentina

K. Holinde, Institut für Theoretische Kernphysik, Universität Bonn, Nußallee 14-16, D-53oo Bonn, West-Germany

L. Jacobs, Instituto de Fisica, Universidad Nacional Autonoma de Mexico, Apartado Postal 2o-364, Mexico 2o, D.F., Mexico

R. Jauregui, Instituto de Fisica, Universidad Nacional Autonoma de Mexico, Apartado Postal 2o-364, Mexico 2o, D.F., Mexico

A. Kallio, Department of Theoretical Physics, University of Oulu, SF 9o1o1 Oulu, Finnland

G. Kalman, Department of Physics, Boston College, Chestnut Hill, Mass. o2167, U.S.A.

M.H. Kalos, Courant Institute of Mathematical Sciences, New York University, 251 Mercer Street, New York, N.Y. 1oo12, U.S.A.

J. Keller, Fac. Quimica, Universidad Nacional Autonoma de Mexico, Mexico 2o, D.F., Mexico

A. Klein, Department of Physics, University of Pennsylvania, Philadelphia, PA. 19174, U.S.A.

D.J. Klein, Department of Physics, University of Texas, Austin, TX 78712, U.S.A.

S. Köhler, Physics Department, University of Arizona, Tucson, Arizona 85721, U.S.A. *

S.E. Koonin, Kellog Radiation Lab, Cal. Tech. Pasadena, California 91125, U.S.A.

E. Krotscheck, Department of Physics, State University of New York, Stony Brook, N.Y. 11794, U.S.A.

H.G. Kümmel, Institut für Theoretische Physik, Ruhr-Universität Bochum, Postfach 1o2148, D-463o Bochum 1, West-Germany

K.E. Kürten, School of Physics and Astronomy, University of Minnesota, 116 Church St. S.E., Minneapolis, Minn. 55455, U.S.A.

A. Lande, Institute for Theoretical Physics, University of Groningen, P.O.Box 8oo, W.S.N., Groningen, Netherlands

L. Lantto, Physics Department, State University of New York, Stony Brook, N.Y. 11794, U.S.A.

A. Lejeune, Institute de Physique Nucleaire Theorique, Sart-Tilman, B 4ooo Liège I. Belgique

J.S. Levinger, Department of Physics, RPI, Troy N.Y. 12181, U.S.A. *

Elliot Lieb, Department of Physics, Princeton University, P.O.Box 7o8, Princeton, N.J. o8554, U.S.A.

J. Lomnitz-Adler, Instituto de Fisica, Universidad Nacional Autonoma de Mexico, Apartado Postal 2o-364, Mexico 2o, D.F., Mexico

A. Lumbroso, Bldg 7o, Lawrence Radiation Lab., Berkeley, California 9472o, U.S.A.

C. Mahaux, Institut de Physique, Université de Liège, Sart-Tilman, B-4oo Liège I, Belgique

O. Monica, Instituto de Fisica, Universidad Nacional Autonoma de Mexico, Apartado Postal 2o-364, Mexico 2o, D.F., Mexico

A.B. Mondragon, Instituto de Fisica, Universidad Nacional Autonoma de Mexico, Apartado Postal 2o-364, Mexico 2o., D.F. Mexico *

H.J. Monkhorst, Quantum Theory Project, University of Florida, Gainesville, Florida 32611, U.S.A.

G. Monsivais, Instituto de Fisica, Universidad Nacional Autonoma de Mexico, Apartado Postal 2o-364, Mexico 2o, D.F., Mexico

M. Moshinsky, Instituto de Fisica, Universidad Nacional Autonoma de Mexico, Apartado Postal 2o-364, Mexico 2o, D.F., Mexico

S.A. Moszkowski, UCLA-Physics Department, Los Angeles, California 9oo24, U.S.A.

J.W. Negele, 6-3o2, M.I.T., Cambridge, Mass. o2139, U.S.A.

L.H. Nosanow, Division of Materials Research, National Science Foundation, Washington, D.C. 2o55o, U.S.A.

M. Olvera, Instituto de Fisica, Universidad Nacional Autonoma de Mexico, Apartado Postal 2o-364, Mexico 2o, D.F., Mexico

J.C. Owen, Department of Theoretical Physics, University of Manchester, Manchester M13 9PL, England

F. Palumbo, Centro Studi Nucleari Della Cassaccia, Cas. Post 24oo, ooloo Roma, Italia,

V.R. Pandharipande, Department of Physics, University of Illinois, Urbana, Illinois 618o1, U.S.A.

D. Pines, Department of Physics, University of Illinois, Urbana, Illinois 618o1, U.S.A.

A. Plastino, Departamento de Fisica, Facultad de Ciencias Exactas, Universidad Nacional, C.C. 67, 19oo La Plata, Argentina

L. Reatto, Instituto di Fisica, Universita di Milano, Via Celoria 16, Milano, Italia

J. Recamier, Instituto de Fisica, Universidad Nacional Autonoma de Mexico, Apartado Postal 2o-364, Mexico 2o, D.F., Mexico

M.L. Ristig, Institut für Theoretische Physik, Universität Köln, Zülpicher Str. 77, 5 Köln, West-Germany

O. Rojo, UPIICSA, Instituto Politecnico Nacional, Mexico 9, D.F., Mexico *

J. Ros-Pallares, Departamento de Fisica Nuclear, Universidad de Granada, Granada, Spain

S. Rosati, Instituto di Fisica, Universita di Pisa, 561oo Pisa, Italia

M. Da C.E. Ruivo, Departamento de Fisica, Universidade da Coimbra, Coimbra, Portugal

K.A. Sage, Department of Physics, University of Arizona, Tucson, Arizona 85721, U.S.A.

D. Sandler, Kellog Radiation Lab 1o6-38, California Institute of Technology, Pasadena, CA 91125, U.S.A.

L.E. Sansores, Instituto de Investigacion de Materiales, UNAM, Mexico 2o, D.F., Mexico

R. Sartor, Institut de Physique, Sart-Tilman, Universitê de Liêge, 4ooo Liêge I, Belgique

W.A. Seitz, Dept. of Marine Science, Moody College, Texas A&M University System, Galveston, TX 77553, U.S.A.

R. Seki, Physics Dept., California State University, Northridge, CA 9133o, U.S.A.

T.H. Seligman, Instituto de Fisica, Universidad Nacional Autonoma de Mexico, Apartado Postal 2o-364, Mexico 2o, D.F., Mexico *

P.J. Siemens, Physics Department, Texas A&M University, College Station, Texas 77843, U.S.A.

R.A. Smith, Physics Department, Texas A&M University, College Station, Texas 77843, U.S.A.

T. Takatsuka, College of Humanities and Social Sciences, Iwate University, Morioka o2o, Japan

R. Tamagaki, Department of Physics, Kyoto University, Kyoto, Japan

J.P. Vary, Physics Department, Ames Laboratory, Iowa State University, Ames, Iowa 5oo11, U.S.A.

B. VerWest, Department of Physics, Texas A&M University, College Station, TX 77843, U.S.A.

A. Wanda, Instituto de Fisica, Universidad Nacional Autonoma de Mexico, Apartado Postal 2o-364, Mexico 2o, D.F., Mexico

C.W. Woo, Provost, Revelle College, University of California, La Jolla, California 92o93, U.S.A.

J.G. Zabolitzky, Institut für Theoretische Physik, Ruhr-Universität Bochum, D-463o Bochum 1, West Germany *

* Session Chairman

Lecture Notes in Physics

Vol. 114: Stellar Turbulence. Proceedings, 1979. Edited by D. F. Gray and J. L. Linsky. IX, 308 pages. 1980.

Vol. 115: Modern Trends in the Theory of Condensed Matter. Proceedings, 1979. Edited by A. Pekalski and J. A. Przystawa. IX, 597 pages. 1980.

Vol. 116: Mathematical Problems in Theoretical Physics. Proceedings, 1979. Edited by K. Osterwalder. VIII, 412 pages. 1980.

Vol. 117: Deep-Inelastic and Fusion Reactions with Heavy Ions. Proceedings, 1979. Edited by W. von Oertzen. XIII, 394 pages. 1980.

Vol. 118: Quantum Chromodynamics. Proceedings, 1979. Edited by J. L. Alonso and R. Tarrach. IX, 424 pages. 1980.

Vol. 119: Nuclear Spectroscopy. Proceedings, 1979. Edited by G. F. Bertsch and D. Kurath. VII, 250 pages. 1980.

Vol. 120: Nonlinear Evolution Equations and Dynamical Systems. Proceedings, 1979. Edited by M. Boiti, F. Pempinelli and G. Soliani. VI, 368 pages. 1980.

Vol. 121: F. W. Wiegel, Fluid Flow Through Porous Macromolecular Systems. V, 102 pages. 1980.

Vol. 122: New Developments in Semiconductor Physics. Proceedings, 1979. Edited by F. Beleznay et al. V, 276 pages. 1980.

Vol. 123: D. H. Mayer, The Ruelle-Araki Transfer Operator in Classical Statistical Mechanics. VIII, 154 pages. 1980.

Vol. 124: Gravitational Radiation, Collapsed Objects and Exact Solutions. Proceedings, 1979. Edited by C. Edwards. VI, 487 pages. 1980.

Vol. 125: Nonradial and Nonlinear Stellar Pulsation. Proceedings, 1980. Edited by H. A. Hill and W. A. Dziembowski. VIII, 497 pages. 1980.

Vol. 126: Complex Analysis, Microlocal Calculus and Relativistic Quantum Theory. Proceedings, 1979. Edited by D. Iagolnitzer. VIII, 502 pages. 1980.

Vol. 127: E. Sanchez-Palencia, Non-Homogeneous Media and Vibration Theory. IX, 398 pages. 1980.

Vol. 128: Neutron Spin Echo. Proceedings, 1979. Edited by F. Mezei. VI, 253 pages. 1980.

Vol. 129: Geometrical and Topological Methods in Gauge Theories. Proceedings, 1979. Edited by J. Harnad and S. Shnider. VIII, 155 pages. 1980.

Vol. 130: Mathematical Methods and Applications of Scattering Theory. Proceedings, 1979. Edited by J. A. DeSanto, A. W. Sáenz and W. W. Zachary. XIII, 331 pages. 1980.

Vol. 131: H. C. Fogedby, Theoretical Aspects of Mainly Low Dimensional Magnetic Systems. XI, 163 pages. 1980.

Vol. 132: Systems Far from Equilibrium. Proceedings, 1980. Edited by L. Garrido. XV, 403 pages. 1980.

Vol. 133: Narrow Gap Semiconductors Physics and Applications. Proceedings, 1979. Edited by W. Zawadzki. X, 572 pages. 1980.

Vol. 134: γγ Collisions. Proceedings, 1980. Edited by G. Cochard and P. Kessler. XIII, 400 pages. 1980.

Vol. 135: Group Theoretical Methods in Physics. Proceedings, 1980. Edited by K. B. Wolf. XXVI, 629 pages. 1980.

Vol. 136: The Role of Coherent Structures in Modelling Turbulence and Mixing. Proceedings 1980. Edited by J. Jimenez. XIII, 393 pages. 1981.

Vol. 137: From Collective States to Quarks in Nuclei. Edited by H. Arenhövel and A. M. Saruis. VII, 414 pages. 1981.

Vol. 138: The Many-Body Problem. Proceedings 1980. Edited by R. Guardiola and J. Ros. V, 374 pages. 1981.

Vol. 139: H. D. Doebner, Differential Geometric Methods in Mathematical Physics. Proceedings 1981. VII, 329 pages. 1981.

Vol. 140: P. Kramer, M. Saraceno, Geometry of the Time-Dependent Variational Principle in Quantum Mechanics. IV, 98 pages. 1981.

Vol. 141: Seventh International Conference on Numerical Methods in Fluid Dynamics. Proceedings. Edited by W. C. Reynolds and R. W. MacCormack. VIII, 485 pages. 1981.

Vol. 142: Recent Progress in Many-Body Theories. Proceedings. Edited by J. G. Zabolitzky, M. de Llano, M. Fortes and J. W. Clark. VIII, 479 pages. 1981.

Selected Issues from

Lecture Notes in Mathematics

Vol. 728: Non-Commutative Harmonic Analysis. Proceedings, 1978. Edited by J. Carmona and M. Vergne. V, 244 pages. 1979.

Vol. 729: Ergodic Theory. Proceedings 1978. Edited by M. Denker and K. Jacobs. XII, 209 pages. 1979.

Vol. 730: Functional Differential Equations and Approximation of Fixed Points. Proceedings, 1978. Edited by H.-O. Peitgen and H.-O. Walther. XV, 503 pages. 1979.

Vol. 731: Y. Nakagami and M. Takesaki, Duality for Crossed Products of von Neumann Algebras. IX, 139 pages. 1979.

Vol. 733: F. Bloom, Modern Differential Geometric Techniques in the Theory of Continuous Distributions of Dislocations. XII, 206 pages. 1979.

Vol. 735: B. Aupetit, Propriétés Spectrales des Algèbres de Banach. XII, 192 pages. 1979.

Vol. 738: P. E. Conner, Differentiable Periodic Maps. 2nd edition, IV, 181 pages. 1979.

Vol. 742: K. Clancey, Seminormal Operators. VII, 125 pages. 1979.

Vol. 755: Global Analysis. Proceedings, 1978. Edited by M. Grmela and J. E. Marsden. VII, 377 pages. 1979.

Vol. 756: H. O. Cordes, Elliptic Pseudo-Differential Operators – An Abstract Theory. IX, 331 pages. 1979.

Vol. 760: H.-O. Georgii, Canonical Gibbs Measures. VIII, 190 pages. 1979.

Vol. 762: D. H. Sattinger, Group Theoretic Methods in Bifurcation Theory. V, 241 pages. 1979.

Vol. 765: Padé Approximation and its Applications. Proceedings, 1979. Edited by L. Wuytack. VI, 392 pages. 1979.

Vol. 766: T. tom Dieck, Transformation Groups and Representation Theory. VIII, 309 pages. 1979.

Vol. 771: Approximation Methods for Navier-Stokes Problems. Proceedings, 1979. Edited by R. Rautmann. XVI, 581 pages. 1980.

Vol. 773: Numerical Analysis. Proceedings, 1979. Edited by G. A. Watson. X, 184 pages. 1980.

Vol. 775: Geometric Methods in Mathematical Physics. Proceedings, 1979. Edited by G. Kaiser and J. E. Marsden. VII, 257 pages. 1980.

Vol. 779: Euclidean Harmonic Analysis. Proceedings, 1979. Edited by J. J. Benedetto. III, 177 pages. 1980.

Vol. 780: L. Schwartz, Semi-Martingales sur des Variétés, et Martingales Conformes sur des Variétés Analytiques Complexes. XV, 132 pages. 1980.

Vol. 782: Bifurcation and Nonlinear Eigenvalue Problems. Proceedings, 1978. Edited by C. Bardos, J. M. Lasry and M. Schatzman. VIII, 296 pages. 1980.

Vol. 783: A. Dinghas, Wertverteilung meromorpher Funktionen in ein- und mehrfach zusammenhängenden Gebieten. Edited by R. Nevanlinna and C. Andreian Cazacu. XIII, 145 pages. 1980.

Vol. 786: I. J. Maddox, Infinite Matrices of Operators. V, 122 pages. 1980.

Vol. 787: Potential Theory, Copenhagen 1979. Proceedings, 1979. Edited by C. Berg, G. Forst and B. Fuglede. VIII, 319 pages. 1980.

Vol. 791: K. W. Bauer and S. Ruscheweyh, Differential Operators for Partial Differential Equations and Function Theoretic Applications. V, 258 pages. 1980.

Vol. 792: Geometry and Differential Geometry. Proceedings, 1979. Edited by R. Artzy and I. Vaisman. VI, 443 pages. 1980.

Vol. 793: J. Renault, A Groupoid Approach to C*-Algebras. III, 160 pages. 1980.

Vol. 798: Analytic Functions, Kozubnik 1979. Proceedings. Edited by J. Ławrynowicz. X, 476 pages. 1980.

Vol. 799: Functional Differential Equations and Bifurcation. Proceedings 1979. Edited by A. F. Izé. XXII, 409 pages. 1980.

Vol. 801: K. Floret, Weakly Compact Sets. VII, 123 pages. 1980.

Vol. 802: J. Bair, R. Fourneau, Etude Géometrique des Espaces Vectoriels II. VII, 283 pages. 1980.

Vol. 804: M. Matsuda, First Order Algebraic Differential Equations. VII, 111 pages. 1980.

Vol. 805: O. Kowalski, Generalized Symmetric Spaces. XII, 187 pages. 1980.

Vol. 807: Fonctions de Plusieurs Variables Complexes IV. Proceedings, 1979. Edited by F. Norguet. IX, 198 pages. 1980.

Vol. 809: I. Gumowski and Ch. Mira, Recurrences and Discrete Dynamic Systems. VI, 272 pages. 1980.

Vol. 810: Geometrical Approaches to Differential Equations. Proceedings 1979. Edited by R. Martini. VII, 339 pages. 1980.

Vol. 816: L. Stoica, Local Operators and Markov Processes. VIII, 104 pages. 1980.

Vol. 819: Global Theory of Dynamical Systems. Proceedings, 1979. Edited by Z. Nitecki and C. Robinson. IX, 499 pages. 1980.

Vol. 821: Statistique non Paramétrique Asymptotique. Proceedings, 1979. Edited by J.-P. Raoult. VII, 175 pages. 1980.

Vol. 826: Ph. G. Ciarlet et P. Rabier, Les Equations de von Kármán. VI, 181 pages. 1980.

Vol. 827: Ordinary and Partial Differential Equations. Proceedings, 1978. Edited by W. N. Everitt. XVI, 271 pages. 1980.

Vol. 831: Representation Theory I. Proceedings, 1979. Edited by V. Dlab and P. Gabriel. XIV, 373 pages. 1980.

Vol. 832: Representation Theory II. Proceedings, 1979. Edited by V. Dlab and P. Gabriel. XIV, 673 pages. 1980.

Vol. 836: Differential Geometrical Methods in Mathematical Physics. Proceedings, 1979. Edited by P. L. García, A. Pérez-Rendón, and J. M. Souriau. XII, 538 pages. 1980.

Vol. 837: J. Meixner, F. W. Schäfke and G. Wolf, Mathieu Functions and Spheroidal Functions and their Mathematical Foundations Further Studies. VII, 126 pages. 1980.

Vol. 838: Global Differential Geometry and Global Analysis. Proceedings 1979. Edited by D. Ferus et al. XI, 299 pages. 1981.

Vol. 840: D. Henry, Geometric Theory of Semilinear Parabolic Equations. IV, 348 pages. 1981.

Vol. 841: A. Haraux, Nonlinear Evolution Equations- Global Behaviour of Solutions. XII, 313 pages. 1981.

Vol. 842: Séminaire Bourbaki vol. 1979/80. Exposés 543–560. IV, 317 pages. 1981.

Vol. 843: Functional Analysis, Holomorphy, and Approximation Theory. Proceedings. Edited by S. Machado. VI, 636 pages. 1981.

Vol. 845: A. Tannenbaum, Invariance and System Theory: Algebraic and Geometric Aspects. X, 161 pages. 1981.

Vol. 849: P. Major, Multiple Wiener-Itô Integrals. VII, 127 pages. 1981.